AF580864

Handbook of Applied Photometry

To Our Readers

This book incorporates some unique features which facilitate access to photometry information on the World Wide Web. The book has a Web Companion, a news group for ongoing discussion about the book and related issues in photometry. The authors and editor will periodically post new information related to this book on the Web and respond to the most frequently asked questions sent in by readers. We also intend to post a complete list of typographical errors and errata submitted for this volume, and to keep this handbook current for as long as possible by posting new developments and some original material on the Web. In addition, both the book and its Web Companion contain a list of related Web sites which offer useful photometry information.

To access this additional information or to submit your comments on the book, point your Web browser to the home page of the Optical Society of America (OSA), at URL http://www.osa.org.org/HOMES/DISCUSS/PHOTOMET/index.htm. You can also access this information by going to the OSA home page, selecting the Technical Homes, selecting Photonics, selecting Discussion Groups, and then selecting the Handbook of Applied Photometry. From the OSA Home Page (http://www.osa.org) follow the menu prompts to Photonics Technical Home—Discussion Groups—Handbook of Applied Photometry.

We also welcome your comments on this book. Please send your e-mail to photometry@osa.org to have your comments posted to the discussion group. To our knowledge, this is the first technical handbook to incorporate direct Web access; it should be an interesting experiment, and I look forward to further dialog about this process with you, the reader.

Casimer DeCusatis

Handbook of Applied Photometry

Sponsored by the
Optical Society of America

Editor
Casimer DeCusatis
IBM Corporation
Poughkeepsie, New York

American Institute of Physics

Springer

Library of Congress Cataloging-in-Publication Data
Handbook of Applied Photometry / editor, Casimer DeCusatis.
p. cm.
Includes bibliographical references and index.
ISBN 1-56396-416-3
1. Photometry—Industrial applications. 2. Radiation—Measurement. 3. Optical measurements. I. DeCusatis, Casimer.
TA152.H36 1997 96-35231
535' .22' 0287—dc20 CIP

Printed on acid free paper.

Printed and bound by Maple-Vail Book Manufacturing Group, York, PA.
Printed in the United States of America.

9 8 7 6 5 4 3

ISBN 1-56396-416-3 Springer-Verlag New York Berlin Heidelberg SPIN 10640098

Contents

PART 2: METHODOLOGY

Contributors

Randall P. Bergin *Independent Testing Labs (ITL), 3386 Long Horn Road, Boulder, Colorado 80302*

Theodore W. Cannon *National Renewable Energy Laboratory, 1617 Cole Boulevard, Golden, Colorado 80401*

Carolyn J. Sher DeCusatis *formerly Lighting Research Center, Rensselaer Polytechnic Institute, Troy, New York 12181*

Casimer DeCusatis *IBM Corporation, Department 48KA/MS P343, 522 South Road, Poughkeepsie, New York 12603*

David C. Gross *Osram Sylvania Inc., 71 Cherry Hill Drive, Beverly, Massachusetts 01915*

Rainer Köhler *Bureau International des Poids et Mesures, Pavillon de Breteull, 92312 Sevres Cedex, France*

Yoshi Ohno *NIST, A320/220, Gaithersburg, Maryland 20899*

Justin J. Rennilson *Advanced Retro Technology (a part of Gamma Scientific Inc.), 2733 Via Orange Way, Suite 104, Spring Valley, California 91978*

János D. Schanda *CIE Central Bureau, Kegelgasse 27, A-1030 Vienna, Austria*

William E. Schneider *Optronic Laboratories, Inc., 4470 35th Street, Orlando, Florida 32811*

Richard Young *Optronic Laboratories, Inc., 4470 35th Street, Orlando, Florida 32811*

Preface

Great discoveries and achievements invariably involve the cooperation of many minds.

—ALEXANDER GRAHAM BELL

Photometry is the branch of science that deals with the measurement of light. It is usually viewed as a subset of the broader field of radiometry, which also includes the measurement of radiation outside the visible spectrum. Much of our information about the world comes to us through light, and the observation or detection of light is perhaps one of the most fundamental processes in nature. Fundamental questions whose answers require photometric measurements cut across many different disciplines. What is the best way to design floodlights for an outdoor stadium, highway signs for a dark night, or exit signs that can be seen in a smoke-filled room? How is it possible to derive a standard for measuring light, or to build an accurate reference lamp whose light will serve as a standard for scientists around the world? How much energy and money can be saved by adding dimmer switches in a large office building? Why does a full moon appear to be much brighter than any other phase of the moon? Why does it appear to a movie camera that Elizabeth Taylor has violet eyes? These are only some of the questions which we hope to answer in this handbook.

I first became interested in creating a handbook of photometry while attending a meeting at the National Institute of Science and Technology (NIST) of the Council of Optical Radiation Measurement (CORM) in 1993. I had spoken with many other scientists and engineers working on photometry applications, and found a common need for a reference book that would address both the fundamentals of optical measurements and some of the more advanced, practical topics in this field. While most everyone agreed that certain ''classic'' references in this field were valuable, notably John Walsh's *Photometry* (Dover Publications, New York, 1958), it's also true that the most recent edition of this book was published in 1958. There have been many accomplishments worth noting in recent years; however, few people have either the time or credentials to compile all this material alone. In addition, the technical community has become fragmented by the range of problems mentioned earlier; it would be a challenge to write a book that balanced the different perspectives of the lighting designer, the optical physicist, the government standards laboratories, and the applications engineer, among others. Since I have had good experience with collaborative efforts in the past, I decided to locate experts in this field who might be interested in compiling a Handbook of Photometry. Because of time and space limitations, some more advanced topics such as astronomy applications have been omitted. Since each chapter is structured to read as a complete unit, there is some duplication of fundamental concepts such as the photometric curve; this should also help to re-emphasize important material. I have tried to allow each author to keep his or her own unique style and voice, with the result that some chapters will be more rigorous and some less formal and more conversational. This is a deliberate attempt to reflect the wide diversity of this field; I feel that the interplay of different perspectives is a unique strength of this book, and that the result will appeal to a broader range of readers since everyone will find some part of the book that will reflect their personal preferences. As editor, I have tried to make the many voices combine in unison, like a choir, while not sacrificing their individuality. As the opening quote from Bell suggests, I hope that we have created an enduring classic reference for others in this growing field.

But this handbook was not intended only for the established photometry specialist. There has been a need in the optics community for a practical, user-friendly handbook introducing the field of photometry. As the work force becomes more mobile and corporations downsize, many optics professionals are asked to assume responsibility for areas in which they have not been specifically trained. Too often, engineers or scientists with adequate background in mathematics, physics, and technology will be faced with a photometry problem outside their area of expertise. Without some guidance they have difficulty locating the relevant literature, for although much has been written it is often not at an introductory level. There has been a great deal of work documenting photometric measurement standards (far too much, in fact, to make it practical to reprint it all here) but it is underutilized because only a few people in the field know where to find it or are prepared to read it. These works are not necessarily readily available, and new research is often scattered in many different journals. Many of the existing works are intimidating, fail to explain terminology or basic principles, or focus on the problems of the advanced researcher rather than introducing a basic understanding of the field. Lighting designers, engineers in related fields, and others all face these common problems.

Given this situation, it was my intention to write a practical guide that states common design formulas, rules of thumb, and the assumptions that are part of photometric measurements. Most texts properly treat photometry as a subdiscipline of radiometry; I wanted to focus instead on the problems particular to optical measurements. This handbook is divided into three parts, which may be studied as separate units if desired. The first four chapters contain introductory material and define terminology and units used in photometry. The next four chapters review the fundamentals of the field, including practical examples for those readers interested in making accurate photometric measurements. Finally, the last three chapters contain advanced material suitable for researchers or practitioners investigating state-of-the-art problems in this field.

In preparing this material, I have attempted to create a book that is both easy to use and yet valuable to the seasoned optical scientist. The introductory chapter contains some simple worked-out examples to give a feel for the magnitude and importance of quantities involved and to illustrate practical applications that the user may encounter. Later chapters contain case studies taken from practical experiences in a photometric laboratory. We discuss experimental techniques, stressing an understanding of the instrument being used and noting common problems encountered (shadow falling on a detector, glare, calibration, geometry problems, etc.). Many good illustrations are used to demonstrate common problem areas. The handbook contains a discussion of photometry standard documents and a guide to interpreting them once they have been obtained from the standard bodies (the sheer volume of material makes it impractical to reprint all the standards; nor should this be necessary, once the reader knows how to locate and interpret them). It also contains a thorough bibliography including references to other books, current literature, Illuminating Engineering Society (IES) handbook, standardized documents, and a brief historical review. The intended goal is to give a reader the necessary tools and background to delve further into the literature and understand it. This book also covers areas that have until now been neglected in many formal books, such as retroreflection and a discussion of how to implement the various photometric standard documents available in the literature.

Finally, this handbook incorporates a new feature: reader feedback via the Internet and the World Wide Web. This book has a Web Companion, a newsgroup for ongoing discussion about the book and related issues in photometry. The authors and editor will periodically post new information related to this book on the Web and respond to the most frequently asked questions sent in by readers. We also intend to post a complete list of typographical errors and errata submitted for this volume, and to keep this handbook current for as long as possible by posting new developments and some original material on the Web. To access this additional information or to submit your comments on the book, point your Web browser to the home page of the Optical Society of America, at the URL http://www.osa.org/HOMES/DISCUSS/PHOTOMET/index.htm or to post comments to this forum send e-mail to photometry@osa.org. From the OSA home page follow the menu prompts to Phototonics Technical Home—Discussion Groups—Handbook of Applied Photometry. To our knowledge, this is the first technical handbook to incorporate direct Web access; it should be an interesting experiment, and I look forward to further dialog about this process with you, the reader. For further information about photometry on the Web, try the pages listed in Table 1 or consult your Web searcher or index of choice. Although the list in Table 1 is by no means a complete list, and does not represent any sort of endorsement by the authors, editor, or publisher, it does have some useful sites.

I must acknowledge the efforts of all the chapter authors who have worked together to create this book. Without their tireless efforts over the course of almost a year, this handbook would not have been possible for any one individual to produce. I am grateful for the many different perspectives they present, from diverse background including the CORM Photometry Subcommittee within NIST, the National Renewable Energy Laboratory, the International Bureau of Weights and Measures, the Illuminating Engineers Society, and the Commission Internationale de l'Eclairge (CIE). Large and small corporations, standards bodies, and

TABLE 1. *Some web sites for photometry information.*

Organization	URL
The Optical Society of America (OSA) OpticsNet	HTTP://WWW.OSA.ORG
The American Institute of Physics (AIP)	HTTP://WWW.AIP.ORG
The Society of Photo-optic Instrumentation Engineers (SPIE)	HTTP://WWW.SPIE.ORG
The Institute of Electrical and Electronics Engineers (IEEE)	HTTP://WWW.IEEE.ORG
The Commission International De l'Eclairage (International Commission on Illumination, or CIE)	HTTP://WWW.HIKE.TE.CHIBA-U.AC.JP/IKEDA/CIE/HOME.HTML
or the U.S. National Committee of the CIE	HTTP://CHELSEA.IOS.COM/KDTJEK/CIEUSA.HTML
The International Electrotechnical Commission (IEO)	HTTP://WWW.HIKE.TE.CHIBA-U.AC.JP/IKEDA/IEC/HOME.HTML
The Society of Illumination Engineering (IES)	HTTP://WWW.AECNET.COM/IES
National Institute of Standards & Technology (NIST)	HTTP://WWW.NIST.GOV
International Society of Weights & Measures (Bureau International des Poids et Measures, BIPM)	HTTP://WWW.BIPM.FR
The American National Standards Institute (ANSI)	HTTP://WWW.ANSI.ORG
The American Physical Society (APS)	HTTP://WWW.APS.ORG
The National Conference of Standards Laboratories (NCSL)	HTTP://TS.NIST.GOV/TS/HTDOCS/NCSL/NCSLPAGE.HTML
The International Standards Organization (ISO)	HTTP://WWW.ISO.CH/WELCOME.HTML
International Association of Lighting Designers	HTTP://WWW.AECNET.COM
U.S. Department of Commerce Lighting Industry Reports	GOPHER://UNA.HH.LIB.UMICB.EDU
The Pennsylvania State University, Dept. of Architectural Engineering, Lighting Programs	HITP://WWW.ENGR.PSU.EDU/WWW/DEPT/ARC/SERVER/AETOP.HTML
The Lighting Research Center at Rensselaer Polytechnic Institute[a]	HTTP://WWW.RPI.EDU/DEPT/LRC/LRC.HTML
The Lighting Reseource	HTTP://WWW.WEBCOM. COM/-LIGHTARC
General Electric	HTTP://WWW.GE.COM
Lawrence Berkley Labs, Energy and the Environment	HTTP://EONDE.LBL.GOV

[a]One of the few university programs in the country to offer degree programs in lighting, as well as a resource for the National Lighting Program, specifier reports, and good links to many other lighting sites.

academia are all represented. In the spirit of professional cooperation, both the Optical Society of America and the American Institute of Physics have combined to sponsor and publish this volume; I am grateful for their support and patience. Finally, as always, I dedicate this book to the people who give meaning to my life and taught me to look for wonder in the world; my wife and 2-year-old daughter Anne, my parents, my godmother Isabel, and her mother Mrs. Crease.

Casimer DeCusatis
May 1996

1

Light and Radiation

Theodore W. Cannon

National Renewable Energy Laboratory, 1617 Cole Blvd., Golden, Colorado 80401

Dedicated to my ophthalmologist, Dr. Robert Stofac, M.D. for professional dedication to the eyesight of his patients.

1.1 INTRODUCTION

This book is about *photometry*—the science of measuring light. Before describing the measurement process in subsequent chapters, it is appropriate to discuss light and radiation. In the discussion of these topics, we will introduce some elementary concepts that will be covered in depth in later chapters. The reader who is familiar with these topics may wish to skip this chapter; the reader who desires to learn about or review them will find the chapter of benefit. For additional terminology, see Refs. [1–3].

1.2 LIGHT TO THE BEHOLDER

To those who are sighted, light provides a large part of the information they receive about their universe. To specialists working in a light-related area, it has a more specific nature, as follows.

- Medical eye specialists (ophthalmologists, optometrists, opticians) deal with the manipulation of light entering the eye in order to produce clear vision. They use light as a diagnostic tool and laser light to correct certain problems of the eye surgically.
- Other medical specialists use specialized lights and optical fibers to study other areas of the body that are of interest. Carefully controlled lasers provide a means of removing and cauterizing tissues.
- To communications engineers, light is the means of getting information from one point to another, generally through optical fibers transmitting information at much higher rates than wires or cables.
- Solar engineers deal with conversion of sunlight into electrical, thermal, biological, or wind energy.
- Material scientists use light to probe certain properties of the materials they are investigating.
- Entertainers use light effects to direct attention to their performances and to dazzle the audience with laser light shows.
- Architects use light to illuminate the interior and exterior of structures of their design and to provide vision to those inside.
- Illumination engineers specialize in optimizing illumination in areas where it is essential to be able so see clearly and comfortably.
- Astronomers use light to look back toward the beginning of time which provides, along with certain types of invisible radiation, knowledge of creation and information about the mysteries of the universe.
- Physicists may consider light to encompass radiation beyond the detection limits of the eye: ultraviolet and infrared radiation, as well as the visible. Optical spectra provide information about atomic and molecular processes too small to be seen.
- To theologians or philosophers, light may be the very essence of being. The Bible refers to light 232 times, starting with the creation of light in Genesis 1 and ending with the abolition of light near the end of the book of Revelation.

1.3 WHY PHOTOMETRY?

The detection of light is a fundamental process, yet to measure it requires great finesse. The eye, working together with the visual cortex of the brain, constitutes a very powerful sensory organ. We describe visual sensations by their color, brightness, shape, and contrast, but these sensations are not very quantitative [4]. The eye, with its built-in adaptations, can only be used to describe light in a very qualitative way.

An instrument for measuring the amount of light is called a *photometer*. A simple bench photometer can be constructed for visually comparing the luminous intensities of two sources of light. The sources are mounted near the extremities of an optical bench with a movable reflective surface between. If equal flux densities from both sources are observed at a point between the two sources, the ratio of their luminous intensities equals the square of the ratio of their distances from the surfaces; see discussion of the inverse square law in Sec. 1.7.3. Thus, to use the eye as even a simple photometric sensor requires supplementary equipment, and it is not very precise at that. Colors may also be matched visually, but direct use of vision to quantify light and color is quite imprecise.

Use of certain optical detectors, together with special filters that, in combination, have a response close to that of the average human eye, greatly improve the accuracy, repeatability, and ease of use of photometers.

If the end result of an application involving light is simply the human perception of light, why is it necessary to have photometry at all? One could, for example, purchase lamps that would give "enough" light of the right "color" and "intensity" for a given situation or computer monitors that would produce "enough" light of the "right" color and "contrast" to be useful. The science of photometry has been developed to quantify light and its properties accurately. Some reasons for employing the science of photometry are as follows.

- Perceptions of light by the eye and brain are subjective and vary from individual to individual. Generally, it is difficult for two or more observers to reach a consensus on brightness, contrast, color, etc., from the same observation. Properties of light observed without instruments are very imprecise.
- Photometry allows us to compare light sources and other light-related products. For example, light sources are rated according to their average expected lifetime, light output, and color and spatial distributions. Common household incandescent light bulbs are rated in units of total light output, called *luminous flux*, measured in lumens. A typical 100-watt (W) (electrical power) light is labeled 1710 average lumens (lm), lifetime 750 hours (h). In selecting light fixtures for a given industrial application, the illumination engineer can refer to manufacturer's information that contains photometric data, including the angular distribution of light for a variety of fixtures.
- Quantification is necessary to analyze and predict the results of changes during the development of light-related products. For example, as the composition of phosphors in a color cathode ray tube screen is modified, the resultant changes

in color output can be numerically compared to determine if the predicted results have been achieved or to predict what changes are necessary to achieve the desired colors.

- Specifications must often be written before purchasing or even deciding on light-related products for a specific application. For example, when specifying road signs, the reflective and color properties of the paint must be specified so that they can be seen under a variety of lighting and road conditions. The manufacturer of the paint must meet measurable specifications that can be tested by photometric laboratories or the purchaser.
- Before doing business in national and international markets, it is often necessary for product specifications to meet certain test standards. A number of standardizing organizations are set up to certify laboratories in the photometry area and photometric practices must be rigidly adhered to by the certified laboratories.
- Reference documents have been issued by a variety of organizations such as the American National Standards Institute (ANSI), American Society for Testing and Materials (ASTM), and Commission Internationale de L'Eclairage (CIE). These documents provide specifications, methods, practices, etc., for implementing photometry. Measured properties of light described by these documents include color, luminous intensity, spatial and spectral distribution of light sources, reflection and transmission properties of optical materials, color and reflective properties of displays (i.e., television monitors, flat panel displays), standard light sources, etc. Applications include light sources, lighting environments, displays, reflective materials, light meters and exposure controls for cameras, sensitivity of photographic films and electronic imagers, etc.

The measurement of transmittance and reflectance of surfaces and media as a function of wavelength is referred to as *spectrophotometry*. Specialized instruments (e.g., spectrophotometers, goniospectrometers, flame spectrophotometers, etc.) used in this area will be discussed in later chapters [5–7].

The science of measuring radiant electromagnetic energy is referred to as *radiometry*. Radiometry includes the measurement of energy at shorter wavelengths than the visible (e.g., ultraviolet), longer than the visible (infrared, microwave region), as well as the visible. Radiometry is, therefore, broader in scope than photometry and often requires different hardware and techniques than photometry [8].

1.4 LIGHT AS RADIATION

Radiation, in the broadest physical sense, is the dissemination of energy from a source. For example, it can be used to describe heat, acoustic waves, nuclear particles, and radio waves as well as light. In this book, we use a more restrictive definition, the dissemination of visible energy, i.e., energy that can be detected by the human eye. The interaction between oscillating electric and magnetic fields

TABLE 1.1. *Units most commonly used for the wavelength of light.*

Name	Symbol	Value (meters)	Range
Micrometer	μm	10^{-6}	0.380–0.780
Nanometer	nm	10^{-9}	380–780
Ångstrom	Å	10^{-10}	3800–7800

allows energy to propagate as radiation from a source (in the case of light, a luminous body) through transparent media to a receiver (in the case of light, the eye or a manmade detector). The rapidity with which these interactions occurs is described as the *frequency* of the radiation. The overall scope of the frequency spectrum is extensive, from a few hertz[1] (Hz) through cosmic rays with frequencies in the zettahertz region; visible light occupies only a minute portion of this domain from about 384 to 789 THz.

1.5 LIGHT AS A WAVE

Light is generally described in terms of *wavelength* rather than frequency. Wavelength is the distance between any two successive points of a periodic wave for which the oscillation has the same phase. The most common units for wavelength are shown in Table 1.1.

The light practitioner generally views *light* as radiation over a range of wavelengths from about 380 to 780 nanometers (nm), the range normally detected by the human eye and the definition used in this book. Visually, there is some variation in these limits. The Illumination Engineering Society (IES) has defined light as *visually evaluated radiant energy* [9].

The relationship between wavelength λ and frequency ν is given by

$$\nu = \frac{C}{\lambda}, \tag{1.1}$$

where c is the speed of light [299 792 458 meters/second (m/s) in a vacuum, computational value 3.00×10^8 m/s].

Example 1

Calculate the frequency of light near the wavelength of greatest sensitivity to the human eye ($\lambda=550$ nm).

$$\nu = \frac{c}{\lambda} = \frac{3.00\times10^8 \text{ m/s}}{550\times10^{-9} \text{ m}} = 5.45\times10^{14} \text{ cycles/s} = 545 \text{ THz}. \tag{1.2}$$

[1]Frequency, in cycles per second, is generally expressed in hertz, abbreviated Hz, where 1 hertz=1 cycle per second. One kilohertz (kHz)=1000 Hz, one terahertz (THz)=10^{12} Hz, and one zettahertz (ZHz)=10^{21} Hz.

TABLE 1.2. *The electromagnetic spectrum.*

Region	Wavelength
Electric power	∞–5×10^6 m
Radio or television	1×10^4 m–1 cm
Infrared	1000–0.78 μm
Visible (light)	0.78–0.39 μm
red light	6470–7000 Å
orange light	5850–6740 Å
yellow light	5750–5850 Å
max. visibility	5560–5750 Å
green light	4912–5560 Å
blue light	4240–4912 Å
violet light	4000–4240 Å
Ultraviolet	0.39–0.032 μm
X ray	0.032–0.000 01 μm
γ radiation	0.000 01–0.000 000 6 μm
Cosmic rays	0.0005 Å

Units of Frequency

1 Hz=1 cycle per second

1 kHz=1000 Hz

1 mHz=1 000 000 Hz=1000 kHz

Units of Wavelength

1 Å=1 Ångstrom=10^{-10} m

1 μm=1 micrometer=1×10^{-6} m=10^4 Å

The sciences of electricity and magnetism developed separately until the nineteenth century when Hans Christian Oersted (1777–1851), Michael Faraday (1791–1867), and James Clerk Maxwell (1831–1879) understood the interaction between electric and magnetic fields well enough to develop the science of electromagnetism. These relationships between electricity and magnetism are very elegantly and succinctly expressed in basic equations developed by Maxwell [10] from experimental observations; they provide the foundation for all of electromagnetism. The most important regions of the frequency and wavelength domain covered by this *electromagnetic spectrum* are shown in Table 1.2.

The regions adjacent to the visible portion of the electromagnetic spectrum are the ultraviolet (UV, meaning ''beyond the violet'') and infrared (IR, meaning ''below the red'') regions. These regions can be subdivided as shown in Table 1.3.

1.6 LIGHT AS A STREAM OF PARTICLES

Experiments on the photoelectric effect by Millikan and others showed results that could not be explained by the wave theory of light. Radiation of sufficiently short wavelength, i.e., sufficiently high frequency, impinging on certain substances

TABLE 1.3. *Regions adjacent to the visible portion of the electromagnetic spectrum.*[a]

Name	Wavelength range
UV-C	100–280 nm
UV-B	280–315 nm
UV-A	315–400 nm
VIS	380–780 nm[b]
IR-A, near IR (NIR)	780–1400 nm
IR-B	1.4–3 μm
IR-C, far IR	3 μm–1 mm

[a]The wavelength boundaries of all of these regions are not consistent in all references. The boundaries used here are from the CIE.
[b]Varies with the individual. The overall limits extend from 360 to 800 nm.

causes bound electrons to be given off with a maximum velocity proportional to the frequency of the radiation. According to classical wave theory, the threshold at which the bound electrons are given off would be related to the intensity of the radiation. Einstein was able to show that the photoelectric effect could be explained by making the assumption that the energy in a light beam travels through space in concentrated bundles called *photons*. The energy of each photon, e_λ, in joules (J), is given by

$$e_\lambda = \frac{1.988\times 10^{-16}}{\lambda} \tag{1.3}$$

when λ is in nanometers. The threshold at which photoelectrons were given off was found to be related to the wavelength of the radiation rather than its intensity.

Example 2

Calculate the energy of a photon at 550 nm wavelength.[2]

$$E = \frac{1.988\times 10^{-16}}{550} = 3.61\times 10^{-19}\ \text{J} = 361\ \text{zJ}. \tag{1.4}$$

Example 3

Calculate the number of photons per second corresponding to 10 watts (W) of optical radiation at 550 nm wavelength. We will use the energy per photon from Eq. (1.4). Ten watts is equivalent to 10 J/s. The equivalent radiation in photons/second is the optical power divided by the energy per photon,

[2]One zeptojoule (zJ) $= 10^{-21}$ J.

$$\phi_p = \frac{10 \text{ J/s}}{3.61 \times 10^{-19} \text{ J/photon}} = 2.77 \times 10^{19} \text{ phontons/s.} \tag{1.5}$$

Whether light is treated as a wave or particle depends on the application. Generally, in photometry, the wave interpretation is used, although the particle interpretation is often useful when dealing with the response of detectors. The *quantum efficiency* of a detector is defined as the ratio of generated electrical charge to the number of incident photons, often measured in electrons per photon.

Example 4

A silicon detector has a quantum efficiency of 0.85 at a wavelength of 750 nm. For an incident radiation of 0.2 W at this wavelength, what electrical output current can be expected? First, the energy in joules per photon is calculated using Eq. (1.3) as

$$e_\lambda = \frac{1.988 \times 10^{-16}}{\lambda} = \frac{1.988 \times 10^{-16}}{750} = 2.65 \times 10^{-19} \text{ J/photon.} \tag{1.6}$$

Now 0.2 W is equivalent to 0.2 J/s, so the number of electrons per second is

$$\phi_p = \frac{(0.85 \text{ electron/photon}) \times (0.2 \text{ J/s})}{2.65 \times 10^{-19} \text{ J/photon}} = 6.42 \times 10^{17} \text{ electrons/s.} \tag{1.7}$$

The electrical charge on an electron is 1.6×10^{-19} coulomb (C) and an electrical current of 1 ampere (A) corresponds to a flow of charge of 1 C/s. The electrical current corresponding to 6.42×10^{17} electrons/s is then

$$I = (6.42 \times 10^{17} \text{ electrons/s}) \times (1.6 \times 10^{-19} \text{ C/electron}) = 0.109 \text{ C/s} = 0.103 \text{ A.} \tag{1.8}$$

1.7 QUANTIFYING LIGHT

The science of photometry deals with the quantitative measurement of the properties of light. Three basic quantities, viz., spectral content, spatial distribution, and polarization, must be known to completely describe light, as shown in Table 1.4. Additional quantification is necessary when dealing with modulated light, i.e., modulation frequency, amplitude, and type of modulation (e.g., pulse, sine wave).

1.7.1 Spectral Content

Every light sensor, including the eye, has a wavelength-dependent response called its *spectral response*. For electronic detectors, this response is calculated from the ratio of the output signal to the incident optical power of monochromatic light at selected wavelengths.

TABLE 1.4. *The three basic quantities that describe light.*

Property	Description	Instrument
Spectral content	Optical power as a function of wavelength	Spectroradiometer, spectroscope, or multifilter radiometer
Spatial distribution	Physical distribution of sources of light	Imaging or geometrically scanning radiometer
Polarization	Nonrandom orientation of electric and magnetic fields of the light wave	Polarimeter

Once the spectral response of a device has been determined, if the amount of optical power incident on the detector's surface at each wavelength is known, one can determine the detector's output current. Incident optical power per unit area per unit wavelength is known as *spectral irradiance.* A number of characteristics of light can be determined from a measurement of its spectral irradiance. Several examples, measured in W/(m^2 nm), will illustrate this.

Spectral irradiance can be measured using narrow-bandpass filter–detector combinations for each desired wavelength. A more flexible and common approach is to use either a *spectroradiometer* or a *spectrograph* [11,12]. Both types of instruments use either a prism or, more commonly, a diffraction grating to disperse the radiation into its wavelength-dependent (spectral) components. The former measures the light over a narrow wavelength band centered on the detector one wavelength at a time[3] as the prism or grating is rotated. The modern spectrograph employs a linear array of detectors, each one measuring a portion of the dispersed radiation, so that the entire spectrum is measured simultaneously without having to rotate the dispersive element. Older spectrographs used photographic film for detection.

Spectroradiometers and spectrographs are generally calibrated for their response to the incoming radiation by use of *spectral lamp standards*, incandescent lamps that have been carefully seasoned and selected for good stability and then calibrated by a standardizing laboratory such as the U.S. National Institute of Standards and Technology (NIST) [13–16]. The spectrum of a 1000-W, FEL-type tungsten-halogen spectral lamp standard, is shown as connected dots in Fig. 1.1. For comparison, a 3220 degrees Kelvin (K) blackbody fit of the data (see boxed text titled Blackbody Radiation) is shown. The blackbody equation may be used to extend the spectral values beyond the measured values using the coefficients derived from the measured peak intensity and wavelength at the peak intensity.

[3]Some spectroradiometers use two gratings or prisms.

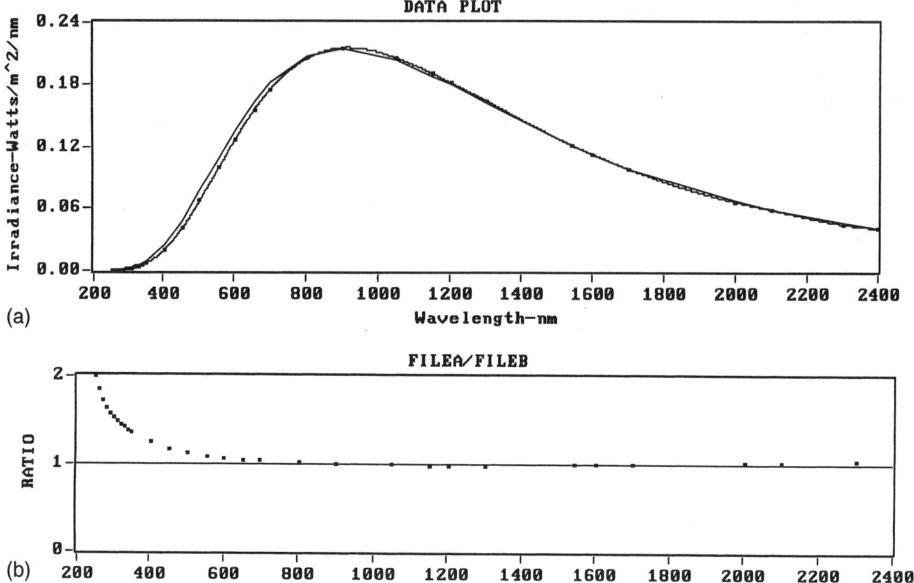

FIGURE 1.1. (a) Spectral irradiance of a NIST FEL type, 1000-W, spectral lamp standard. The 30 data points provided on the NIST Calibration Certificate are indicated by squares. The line through the squares is a cubic spline 2-nm interpolation. The other line is a 3220-K, peak-matched (at 900 nm) blackbody-curve fit to the data. (b) Ratio of the blackbody fit to the Calibration Certificate values of panel (a). The blackbody parameters can be used to approximate the spectral irradiance values beyond the measured wavelengths.

BLACKBODY RADIATION

Blackbody radiation plays such an important part in radiometry and photometry that a more detailed discussion is warranted. As we have seen, the spectral distribution of light sources can often be approximated using a blackbody fit to the measured data. Since blackbody spectra can be readily calculated based on physical principles, it was used to calibrate spectral radiation sources before the advent of the high-accuracy cryogenic radiometer, precision-stable lasers, and ~100% quantum efficiency solid-state detectors.

An explanation of the shape of blackbody radiation led to one of the greatest concepts in physics, the quantization of energy, including, of course, quantization of optical radiation. Blackbody radiation is used to calibrate spectral light standards as its radiation can be calculated from a precise measurement of temperature and application of Planck's radiation law. Blackbody radiation is also important to the practitioner of photometry as sources of light generally approximate blackbody radiation in their spectral characteristics and are often described in terms of their equivalent blackbody temperature. A fit of measured lamp data over a limited wavelength region to Planck's radiation law can be used to extend the spectral curve to longer or shorter wavelengths.

BLACKBODY RADIATION AND PLANCK'S QUANTUM HYPOTHESIS

Any inanimate body that completely absorbs all radiation falling upon it is called a *blackbody*. A blackbody maintained at constant temperature radiates the same amount of energy it receives. An approximation to a perfect blackbody can be realized in practice by constructing a hollow, insulated enclosure that contains a small hole in one wall. The hole serves as a port for incident energy to enter the cavity, in which case there is little chance of it being reflected out again, so it acts as a perfect absorber. If the cavity is heated, the blackbody acts as a source of radiation emitted through the hole.

Blackbody radiation curves are shown in Fig. a, which shows the spectral exitance (power per unit area per unit wavelength) versus wavelength for several blackbody temperatures. Note that the peak of each curve occurs at shorter and shorter wavelengths as the temperature, and resulting radiation, are increased.

The blackbody plays a very important part in radiation physics [26]. The *Stefan–Boltzmann law*

$$M_e = \sigma T^4 \tag{1}$$

relates the output power through the opening to the interior temperature T (in units of K), where M_e is the total radiant flux density (radiated power per unit area) and the Stefan–Boltzmann constant $\sigma = (5.6697 \pm 0.0029) \times 10^{-8}$ W/(m^2 K^4). This law was developed from the experimental observations of Josef Stefan (1835–1893) and explained using electromagnetic and thermodynamic arguments by Ludwig Boltzmann (1844–1906).

In 1893, Wien (1864–1928) was able to show, using classical theory, a relationship between the absolute temperature and the wavelength at which the peak

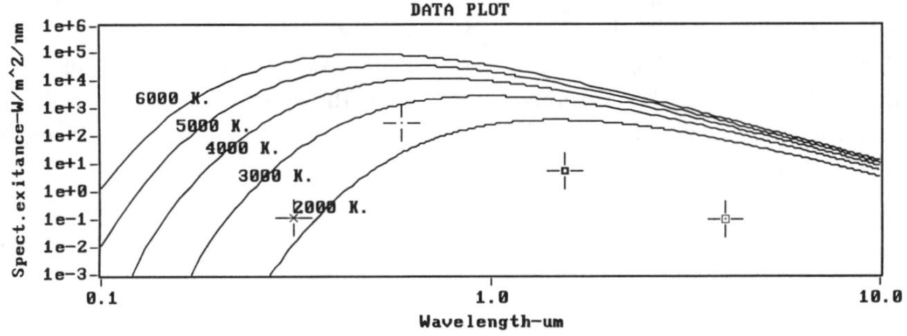

FIGURE a. Blackbody curves for 2000, 3000, 4000, 5000, and 6000 K.

of the radiation versus wavelength occurs, viz.,

$$\lambda_{\max}T = 2.8978 \times 10^6 \text{ nm K}. \tag{2}$$

This expression is known as Wien's displacement law and describes the hyperbola passing through the peak points of the curves shown in Fig. a.

When people began to calculate the blackbody curves of Fig. a using classical theory, they only met with limited success. Wien produced a formula that fitted observed data fairly well in the short-wavelength region but failed substantially at long wavelengths. Rayleigh (1824–1919) and Jeans (1877–1946) developed a description in terms of standing-wave modes of the field within the enclosure. This *Rayleigh–Jeans formula* only matched the measured curves in the very-long-wavelength region.

It remained for Max Planck (1858–1947) to explain the blackbody radiation curves, and in so doing he initiated quantum theory. Planck first matched the observed data with an empirical expression. He then worked to explain the form of the equation within the framework of thermodynamics. He assumed the walls of the blackbody to be in thermal equilibrium with the enclosed radiation field, the atoms of the walls behaving as electrical oscillators absorbing and emitting radiant energy. He assumed that all oscillator frequencies ν were possible and present in the emitted spectrum. He then made the *ad hoc* assumption that an oscillator could absorb or emit only discrete amounts of energy that were proportional to its frequency. Moreover, each such energy value had to be an integral multiple of an "energy element" $h\nu$, such that all possible oscillator energies E_m are given by

$$E_m = mh\nu, \tag{3}$$

where m is a positive integer, v is the frequency, and h is a constant determined by fitting the actual data. Using this reasoning and some statistical arguments, Planck derived the following formula for the spectral exitance:

$$M_{e\lambda} = \frac{2\pi hc^2}{\lambda^5}\left(\frac{1}{e^{hc/\lambda kT}-1}\right), \tag{4}$$

where k is Boltzmann's constant, 1.3807×10^{-23} J/K, and h is Planck's constant $(6.6256 \pm 0.0005) \times 10^{-34}$ J s. This equation shows extremely good agreement with experimental results. In the process of fitting the radiation curve, Planck had uncovered a finding of enormous significance, the quantum of energy. Since $c = \nu\lambda$, the energy of a photon of light in Joules, $h\nu$, can be calculated from the wavelength λ (nm) as

$$e = \frac{hc}{\lambda} = \frac{1.988 \times 10^{-16} \text{ J nm}}{\lambda}. \tag{5}$$

PRACTICAL BLACKBODIES

One brand of commercial blackbodies is available covering a temperature range from about −20 to 3000°C (253 to 3273 K), corresponding to peak wavelengths from about 685 nm in the visible to 11 500 nm in the IR region. These units are used for calibration of infrared thermometers, radiometers, heat-flux meters, thermal imaging systems, and spectrographic analyzers. The temperature of a blackbody can be accurately determined in a freezing point blackbody calibraion source.

The freezing temperatures of certain metals are known quite precisely, e.g., the freezing point of gold is 1064.18±0.04 °C (1337.36 K), corresponding to a peak wavelength of about 2200 nm. The variable-temperature gold blackbody will remain at this precise temperature during the freezing process. Radiation from the gold-point blackbody serves as a spectral radiance source of known distribution, calculated from Eq. (4) from which spectral irradiance can be derived geometrically for use in calibrating radiometric standards. Prior to the use of intensity-stabilized lasers and the cryogenic radiometer, a methodology based on the variable-temperature blackbody was used at the National Institute of Standards and Technology (NIST) for calibrating radiometric and photometric standards.

It is generally not practical to construct a blackbody out of a hollow sphere, unless the design can entrap the radiation within uniformly heated walls and have a low aperture size in relation to the overall surface area. The design of these blackbody simulators is discussed in the literature.

Wavelength calibrations of these instruments are made against emission lines from certain gases. Wavelength calibration sources, including Ar and Kr lamps, can be obtained from commercial sources.

Some representative special data, both natural (solar) and from several lamps are shown. Special software described in the boxed text titled Software for Photometry/ Radiometry was used to prepare Figs. 1.1–1.9

SOFTWARE FOR RADIOMETRY AND PHOTOMETRY

The acquisition, processing, analysis, and storage of radiometric and photometric data are facilitated using computers. Since a single spectral measurement (scan) can generate a large amount of data (e.g., 401 pairs of numbers for a 380–780-nm scan in 1-nm increments, 2700 or more for a measurement over the solar region), processing spectral data has traditionally been done using spreadsheets. Spreadsheet analysis can become cumbersome if spectra of many data formats are processed or if a number of analysis procedures are used on each one.

A computer program called *Quiklook for Radiometry* has been developed at the National Renewable Energy Laboratory to facilitate these operations. The user can easily manipulate the various functions by use of the user-friendly graphical user interface shown in Fig. b. Many of the manipulations are facilitated using a mouse, with minimal keyboard entry required. Features of this program include the following.

Automatic callup and plotting of data of various file formats by mouse-clicking the desired file names selected from a Quiklist file. File format details are transparent to the user.

The Quiklist file can be readily edited to add or change file designations, directory paths, axis units, scaling factors, and read formats.

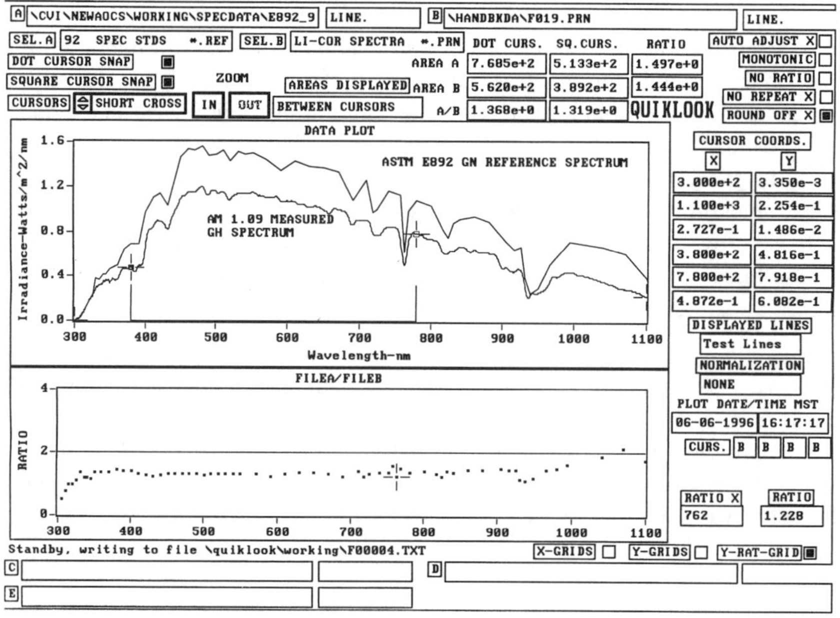

FIGURE b. Graphic user interface for the Quiklook spectral analysis program. Data for the ASTM 892 global reference spectrum and a global horizontal measured spectrum are plotted in the upper graph. The lower plot is the ratio of the measured to the reference spectrum. One pair of cursors (dot cursors) is set at the 300- and 1100-nm limits of the measured spectrum and the other pair (square cursors) at the 380- and 780-nm limits of the photopic region, also indicated by the vertical lines at the bottom of the top graph. The ratio cursor is set at 762 nm, the bottom of the oxygen absorption band. Total integrated areas (W/m^2) and their ratios for the areas defined by the two pairs of cursors are indicated for both plots in the upper right-hand top, the cursors' x and y coordinates in the upper right-hand column. See the text for a full description.

The ratio of any cursor-selected pair of curves is automatically plotted.

The total area under each cursor-selected curve is displayed.

Areas under the curves between wavelength values selected by cursors that are locked to the data plots are automatically displayed along with area ratios.

Histograms are automatically plotted of spectra along with their ratios for equal-size or selected wavelength bins over selected wavelength ranges.

Data can be displayed in radiometric, photometric, detector signal (i.e., volts or amperes), or other units.

Coordinates of cursors that are locked to selected plots are displayed.

Cursors can be selected and stepped along data plots using the keyboard arrow keys.

Reference spectral line wavelengths can be automatically displayed or superimposed on the plots and selected for detailed coordinate display.

The reference spectral lines can be used to identify visually the source of radiation or absorption by matching spectral or absorption lines with those of the measured spectra.

Spectral data containing either absorption or emission lines can be automatically adjusted to correct for wavelength errors in the data.

Data can be normalized to total area, selected areas, or peak values.

Numerical data values can be viewed in tabular form and edited if desired.

Edited data can be saved in various file formats.

Entire sections of cursor-selected data can be deleted.

Plots can be labeled.

Sections of cursor-selected data can be merged to form a composite file; the composite file can then be saved to a disk file.

Data can be shifted on the x axis (e.g., wavelength) by any amount in either the positive or negative direction.

Data can be interpolated to any given increment. Interpolated data can be saved to a file if desired.

The y axis can be linear or logarithmic.

File mathematics, i.e., adding, subtracting, multiplying, dividing, and averaging, are provided.

File statistics are displayed including average, median, maximum, minimum, standard deviation, and percent standard deviation for each plotted file.

Linear, polynomial, exponential, and blackbody fits to each plotted file can be plotted along with the fit parameters. The fitted data can then be processed in the same way as the original data.

The data can be smoothed to remove statistical noise.

Photopic quantities such as luminance and chromaticity values can be calculated from spectral data.

Processed data can be saved to files in selected formats.

Spectral mismatch factors for common photovoltaic materials referenced to ASTM standard outdoor spectra, and reference cells of known spectral response can be calculated and displayed for an input spectra.

Biological and other dose values based on selected erythemal or other action spectra can be displayed for an input spectra.

Hard copies of the computer screen, panels, data plots, ratio plots, histograms, ratio histograms, etc., can be printed directly from the program or saved to file for future printing.

Figure 1.2 shows the spectral irradiance of solar radiation outside of the earth's atmosphere (extraterrestrial radiation, the upper curve) and of sunlight incident on a vertical surface facing south in the northern hemisphere for a typical terrestrial, outdoor, cloud-free case. Data for both cases were generated using an atmospheric transmission model (see the boxed text titled Atmospheric Transmission Models) which shows very good agreement with measured data for cloudless cases. Note that the solar wavelength range is from the ultraviolet (UVB) at about 300 nm to the near infrared (IR-B) at 3000 nm. Ultraviolet energy below about 300 nm is so strongly absorbed by the earth's ozone layer that essentially none of it reaches the earth, and extraterrestrial radiation above 3000 nm contains an almost insignificant fraction of the total. Notice that there are a number of dips in the extraterrestrial and terrestrial plots due to absorption of energy within the sun's outer layers (the Fraunhofer lines) and addiitional dips in the terrestrial plot due to the presence of various atmospheric constituents (primarily oxygen, water, ozone, and carbon dioxide) in the atmosphere. These dips are called *absorption bands*. The water-vapor absorption bands at the longer wavelengths are so strong that they allow essentially no radiation to reach the ground near their minima. Superimposed over the entire region is an overall effect of atmospheric molecular scattering and absorption, which has the greatest effect at shorter wavelengths. The shape of the outdoor solar spectral irradiance curve is changing continuously due to changes in the length of the optical path with solar position and atmospheric composition.

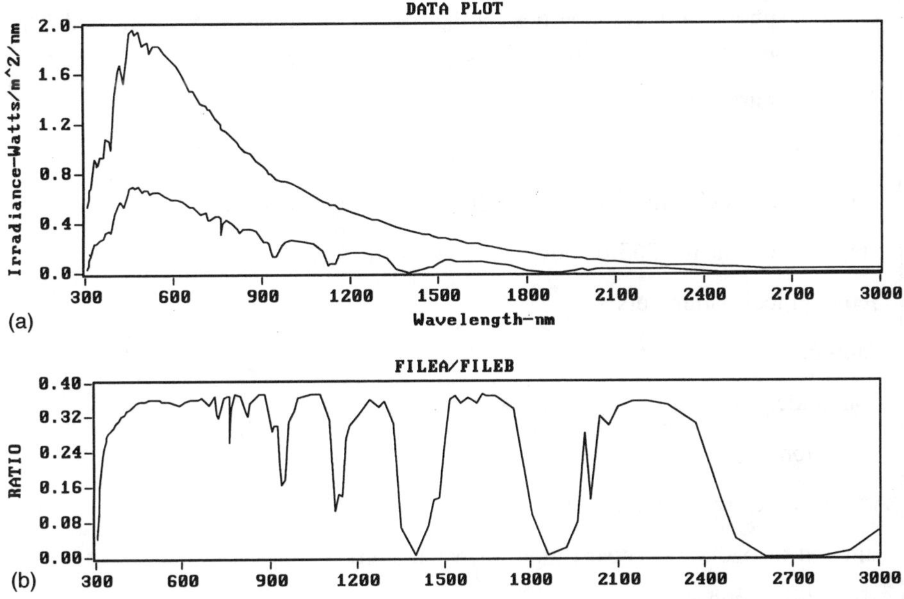

FIGURE 1.2. (a) Extraterrestrial solar spectral irradiance (upper curve) and modeled spectral irradiance on a south-facing, vertical surface at solar noon (air mass 1.04), Golden, Colorado, at the summer solstice. (b) The ratio of the two curves in panel (a) represents the atmospheric transmission as a function of wavelength. Note the seven water-vapor absorption bands centered near 724, 824, 937, 1120, 1395, 1860, and (with CO_2) 2700 nm. The rapid rolloff at the short-wavelength end is due to atmospheric ozone, the sharp absorption at 762.5 nm due to oxygen, and the absorption between 1900 and 2100 nm due to mixed gases.

ATMOSPHERIC TRANSMISSION MODELS

The cloud-free spectrum of solar energy reaching a surface of given orientation at a given location on the earth's surface at any time of the day can be calculated in units of irradiance [W/(m^2 nm)] using an atmospheric transmission model. Such a model is extremely useful for estimating the amount of illumination on a given surface using the V_λ function. Models vary in complexity from the so-called simple models, which treat the entire atmosphere as a single-layer filter, to very sophisticated multilayered models [27–29]. A simple model, SPCTRL2, developed at the National Renewable Energy Laboratory (NREL) for use on a PC-type computer, generates results that compare very favorably with measured spectra. This model has been written into a user-friendly form, Quikspec, that is easily controlled and vividly displays and plots single or multiple spectra using a Windows™ interface. The SPECTRL2–Quikspec model requires a number of input parameters to define the solar and receiver geometry and atmospheric optical properties for the specific situation being calculated. The input parameters are as follows.

Temporal parameters:

ime zone;

solar time, clock time, or air mass (morning or afternoon);

day of the year (0–365 or 366).

Geographical parameters:

latitude;

longitude;

elevation.

Atmospheric parameters:

amount of ozone (keyboard input or calculated);

turbidity at 500 nm;

amount of water vapor.

Surface albedo at different wavelengths.

Wavelengths:

minimum;

maximum;

increment.

Receiving surface orientation:

aspect angle;

tilt angle.

Receiving mode:

global horizontal;

global normal;

direct normal;

global tilt.

Solar tracking:
on or off for global or direct normal modes.

Outputs include spectral plots and a table of all inputs and calculated outputs (total irradiance, illuminance, ozone, etc.) for each run. Solar trajectories can be plotted on a polar plot for each case.

All modeled data can be saved to files for analysis using Quiklook, spead-sheets, etc.

The model basically calculates the amount of optical attenuation and scattering for the optical path determined by the solar position relative to the receiver. The attenuation coefficients are then applied to the extraterrestrial spectrum, determined from the measurements of Neckel and Labs [30], to determine the amount of energy received at each wavelength.

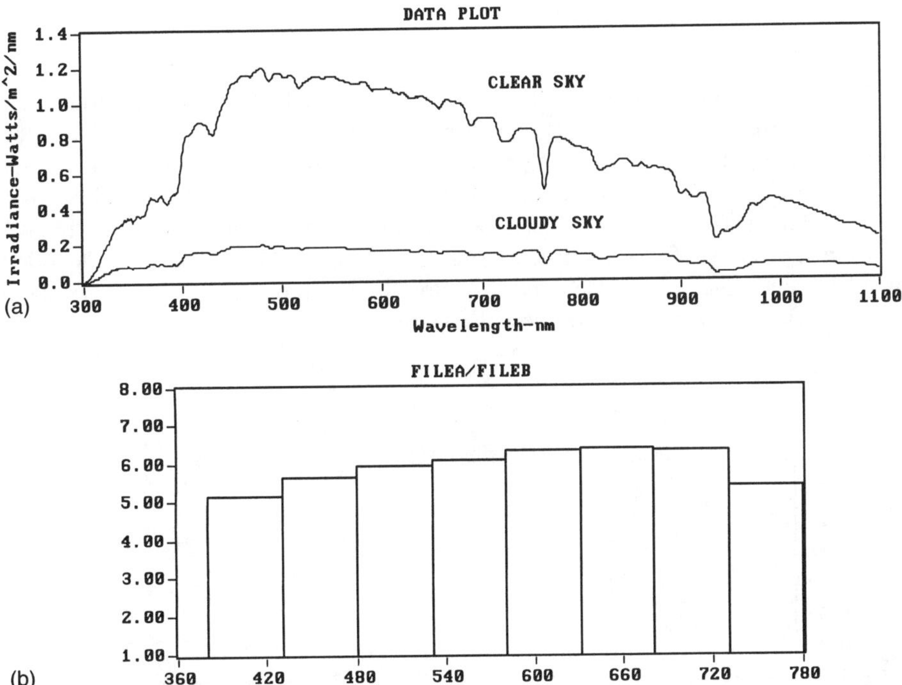

FIGURE 1.3. (a) Measured global solar spectral irradiance on a horizontal plane for clear-sky (upper curve) and cloudy-sky cases at 16:15 MST, air mass=1.1, Golden, Colorado, on 31 May and 2 June 1996, respectively. For the unclouded case, total irradiance (300–1100 nm) is 562 W/m^2, visible irradiance (380–780 nm) is 389 W/m^2, and illuminance is 81.2 klx. For the cloudy case, total irradiance a (300–1100 nm) is 99.0 W/m^2, visible irradiance (380–780 nm) is 65.5 W/m^2, and illuminance is 13.2 klx. (b) Ratio of clear-sky to cloudy-sky cases of panel (a). Note the greater attenuation of longer wavelengths by the cloud tends to "blue shift" the irradiance under cloudy skys.

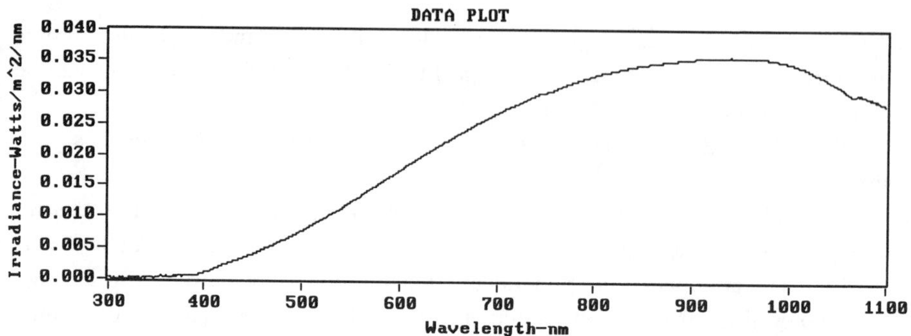

FIGURE 1.4. Measured spectral irradiance for a 60-W common incandescent task light with white reflector. The lamp is 15 in. above the spectroradiometer. Total irradiance (300–1100 nm) is 17.2 W/m^2, visible irradiance (380–780 nm) is 6.34 W/m^2, and illuminance is 1.00 klx.

The effect of cloud cover is shown in Fig. 1.3, which shows the spectral irradiance measured in a horizontal plane for both clear and cloudy days. Note that although the general features are present, the overall data are shifted to give relatively more energy at the short-wavelength end of the spectrum for the cloudy-sky case, resulting in a "gloomy" visual perspective, giving some people the "blues."

Data are shown for six artificial sources in Figs. 1.4–1.9: a common 60-W incandescent light bulb, a FHS-type projector lamp, a fluorescent light fixture, metal halideworking bay lamps, and two solar simulators.

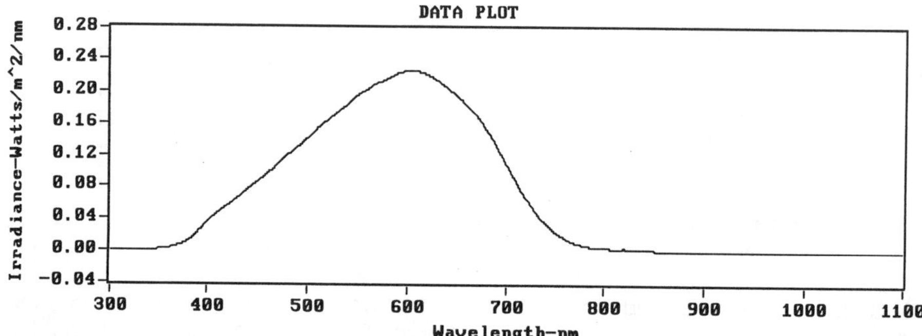

FIGURE 1.5. Measured spectral irradiance for a FHS-type projector bulb (no lens) at about 24 in. Note the rolloff of the long-wavelength irradiance at wavelengths longer than about 600 nm due to the dichroic reflector used in this lamp, thereby concentrating the energy in the visible region while protecting the projected film from excessive heat. Total irradiance (300–1100 nm) is 51.3 W/m^2, visible irradiance (380–780 nm) is 50.7 W/m^2, and illuminance is 14.2 klx.

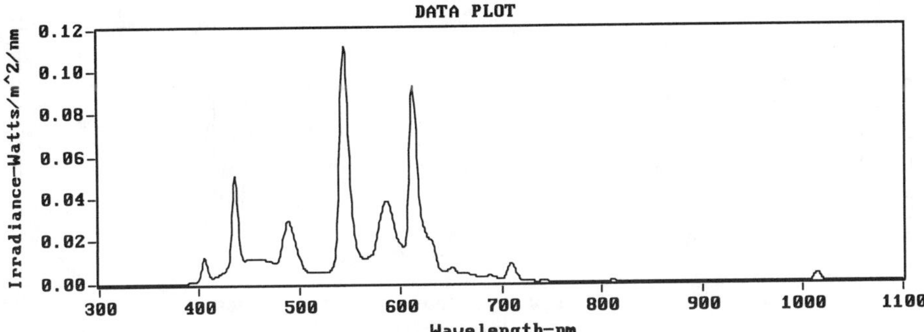

FIGURE 1.6. Measured spectral irradiance from a four-tube fluorescent lamp fixture. The measurement was made about 55 in. directly below the fixture. Total irradiance (300–1100 nm) is 5.67 W/m^2, visible irradiance (380–780 nm) is 5.60 W/m^2, and illuminance is 1.98 klx. The spectrum is composed of Hg lines superimposed on a continuum provided by the phosphor.

The measured spectral irradiance of a FHS-type projection lamp is shown in Fig. 1.5. A dichroic reflector behind the bulb is designed to reflect light over the visible region and transmit the longer wavelengths to reduce heating of the projected film. Note the falloff in spectral irradiance at wavelengths $\gtrsim$600 nm.

The fluorescent source of Fig. 1.6 is a fluorescent continuum with Hg excitation lines. Different phosphors are used to change the color of these lamps. Fluorescent lamps provide a ubiquitous, convenient reference for checking the wavelength calibration of spectroradiometers and spectrographs against the Hg lines.

The metal halide and Xe sources shown in Figs. 1.7–1.9 are rich in spectral lines. Xe and metal halide lamps with appropriate wavelength-dependent optical filters are used as solar simulators.

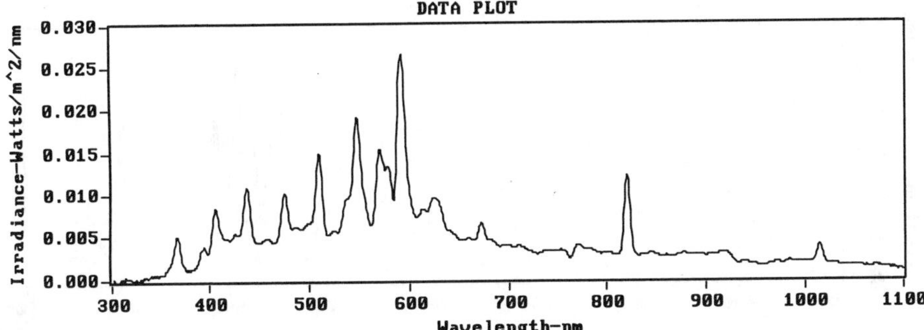

FIGURE 1.7. Measured spectral irradiance in a large working bay illuminated by metal halide lamps. The total irradiance (300–1100 nm) is 3.62 W/m^2, visible irradiance (380–780 nm) is 2.73 W/m^2, and illuminance is 0.771 klx.

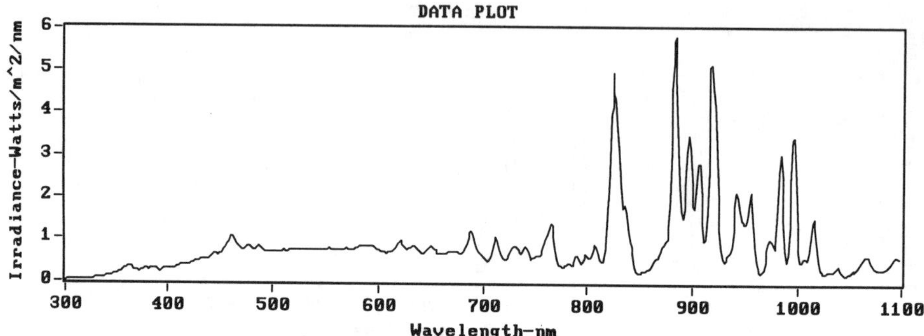

FIGURE 1.8. Measured spectral irradiance for a bank of filtered Xe arc lamps in a solar simulator. Total irradiance (300–1100 nm) is 5.67 W/m^2, visible irradiance (380–780 nm), is 5.60 W/m^2, and illuminance is 1.98 klx.

The atmospheric absorption lines shown in Figs. 1.2 and 1.3 and the spectral lines shown in Figs. 1.4–1.9 are useful for checking the wavelength calibration of spectroradiometers and spectrographs. The strongest lines over the silicon detector range 300–1100 nm for some of these cases are shown in Table 1.5, rounded off to the nearest nanometer.

The total amount of radiation over a range of wavelengths can be determined by adding the contributions at each wavelength. The total is known as *irradiance* or *total irradiance*, E_T, where

$$E_T = \int_{\lambda_{min}}^{\lambda_{max}} E_\lambda \, d\lambda. \tag{1.9}$$

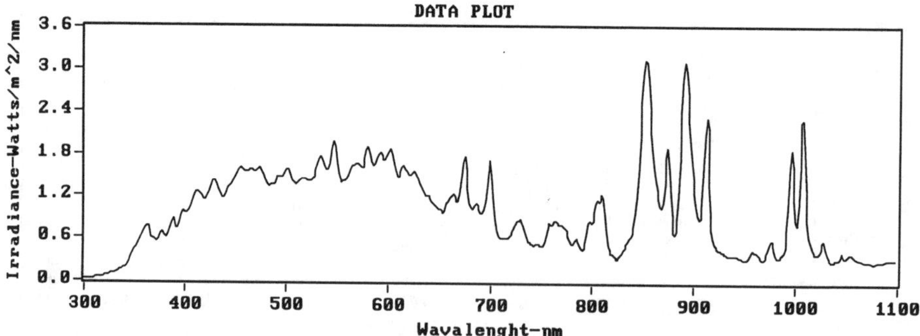

FIGURE 1.9. Measured spectral irradiance for a metal halide lamp solar simulator. Total irradiance (300–1100 nm) is 791 W/m^2, visible irradiance (380–780 nm) is 503 W/m^2, and illuminance is 120 klx.

TABLE 1.5. *Atmospheric absorption lines and emission lines for Hg and Xe, 300–1100 nm useful for wavelength calibration of the spectroradiometer. The most pronounced lines are shown in bold. Wavelengths are rounded off to the nearest nanometer.*

Atmospheric or solar absorption lines	Mercury emission lines	Xenon emission lines
340 solar	313	302
390 solar	365	312
430 solar	405	324
490 solar	436	337
520 solar	546	358
570 ozone	577	362
690 water vapor	579	378
724 water vapor	1014	408
762 oxygen		411
824 water vapor		**467**
936 water vapor		469
1120 water vapor		470
		484
		529
		542
		544
		547
		553
		567
		618
		619
		620
		621
		687
		708
		712
		730
		764
		789
		797
		823
		828
		835
		841
		882
		895
		916
		938
		951
		969
		980
		992

TABLE 1.5. (*Continued.*)

Atmospheric or solar absorption lines	Mercury emission lines	Xenon emission lines
		1011
		1052
		1053
		1055
		1071
		1084
		1090

E_T is measured in units of power per unit area, typically W/m^2. Since spectroradiometers measure at discrete wavelengths, the integral is generally evaluated as a sum of discrete values, generally using the trapezoidal rule for numerical integration.

1.7.2 Response of the Human Eye

The color response of the human eye depends on the individual and on the intensity of the light. At daylight intensities, the cones of the eye are the primary receptors and the response is called *photopic vision*, V_λ. At night, the rods become the primary receptors, and the eye's response changes to *scotopic vision*, V'_λ. The photopic and scotopic responses have been standardized by the Commission Internationale de l'Eclairage (CIE) and are tabulated in Table 1.6, and plotted in Fig. 1.10.

The peak value for V_λ occurs at 555 nm, and for V'_λ is shifted about 45 nm toward the short-wavelength end of the spectrum, to about 510 nm. Special optical filters are used by instrument manufacturers to give their photometers nearly the same response as the average eye.

Of course, the eye's response does not suddenly go from photopic to scotopic at sundown! The region of response between the two regions is called the region of *mesopic vision*. Twilight is a critical transition region as the eye becomes adapted to the darkness.

1.7.3 Spatial Distribution of Light

The manner in which an optical material transmits light or an optical detection system responds to light is generally a function of the direction of the radiation entering the system [17]. For example, the transmission of typical glass (index of refraction of 1.5) varies from about 98% for a plane wave of light entering at normal incidence to about 60% at an incidence angle of 80° to the normal to 0° at

TABLE 1.6. *Photopic V_λ and scotopic V'_λ spectral luminous efficiency values adjusted to unity at their peaks.*

Wavelength (nm)	Photopic response V_λ	Scotopic response V'_λ
380	0.000 04	0.000 589 2
390	0.000 12	0.002 209
400	0.000 4	0.009 29
410	0.001 2	0.034 84
420	0.004 0	0.096 6
430	0.011 6	0.199 8
440	0.023	0.328 1
450	0.038	0.455
460	0.060	0.567
470	0.091	0.676
480	0.139	0.793
490	0.208	0.904
500	0.323	0.982
510	0.503	0.997
520	0.710	0.935
530	0.862	0.811
540	0.954	0.650
550	0.995	0.481
560	0.995	0.328 8
570	0.952	0.207 6
580	0.870	0.121 2
590	0.757	0.065 5
600	0.631	0.033 15
610	0.503	0.015 93
620	0.381	0.007 37
630	0.265	0.003 335
640	0.175	0.001 497
650	0.107	0.000 677
660	0.061	0.000 312 9
670	0.032	0.000 148 0
680	0.017	0.000 071 5
690	0.008 2	0.000 035 33
700	0.004 1	0.000 017 80
710	0.002 1	0.000 009 14
720	0.001 05	0.000 004 78
730	0.000 52	0.000 002 546
740	0.000 25	0.000 001 379
750	0.000 12	0.000 000 076 0
760	0.000 06	0.000 000 042 5
770	0.000 03	0.000 000 241 3
780	0.000 015	0.000 000 139 0

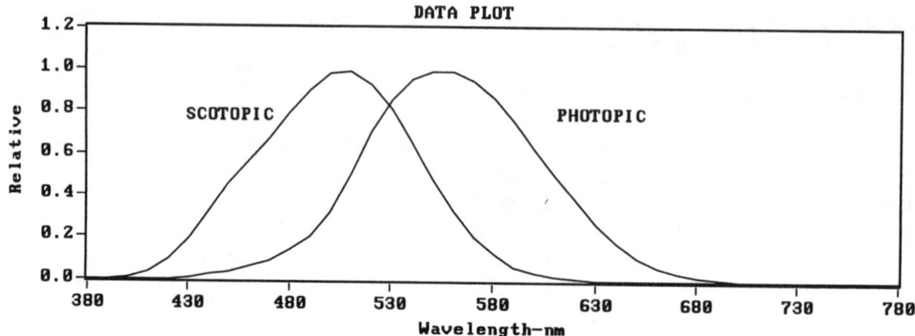

FIGURE 1.10. Photopic V_λ and scotopic V'_λ spectral luminous efficiency values adjusted to unity at the peak wavelengths.

grazing (90° to normal) incidence. The eye responds differently to perimeter vision than it does to directly viewed objects.

Detector systems such as radiometers and photometers are generally designed to have a good cosine response, defined such that

$$e = E \cos \theta, \tag{1.10}$$

where e is the output, E the amount of entering optical power, and θ the angle of incidence of the entering optical radiation as shown in Fig. 1.11. This relationship is a function of wavelength for certain types of cosine-response detectors. Excellent cosine response can be obtained by use of either translucent diffusers or

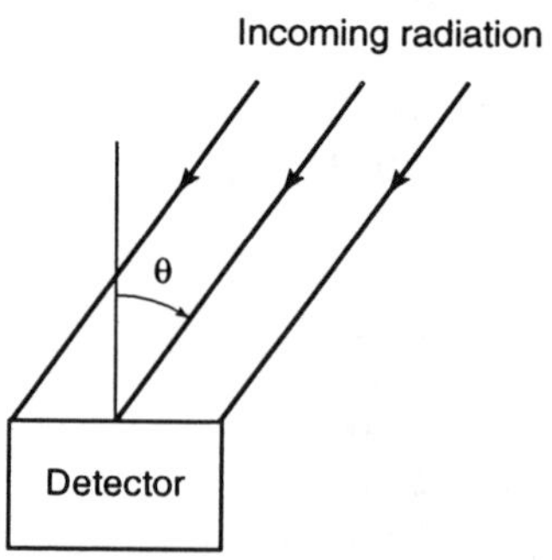

FIGURE 1.11. Geometrical relationships that determine the ideal cosine response of a radiation detection system.

integrating spheres[4] with a penalty of significant loss of signal, especially when integrating spheres are used .

The spatial distribution of sources of optical radiation can be measured using either an imaging radiometer or a mechanically scanning radiometer. Figure 1.12 shows the spatial distribution of photopically detected radiation from a cloudy sky recorded by an all-sky imaging radiometer developed at the Solar Energy Research Institute (now called the National Renewable Energy Laboratory, NREL) [18–20]. The imaging radiometer is calibrated in units of *luminance* (foot-Lamberts, fL). Additional radiation mapping instrumentation is described in the literature [21–23]. This type of data is useful for determining the distribution of natural light for various window designs (fenestration), global solar collectors, etc.

Light from a point source obeys the inverse square law,

$$E_T = \frac{I}{R^2}, \tag{1.11}$$

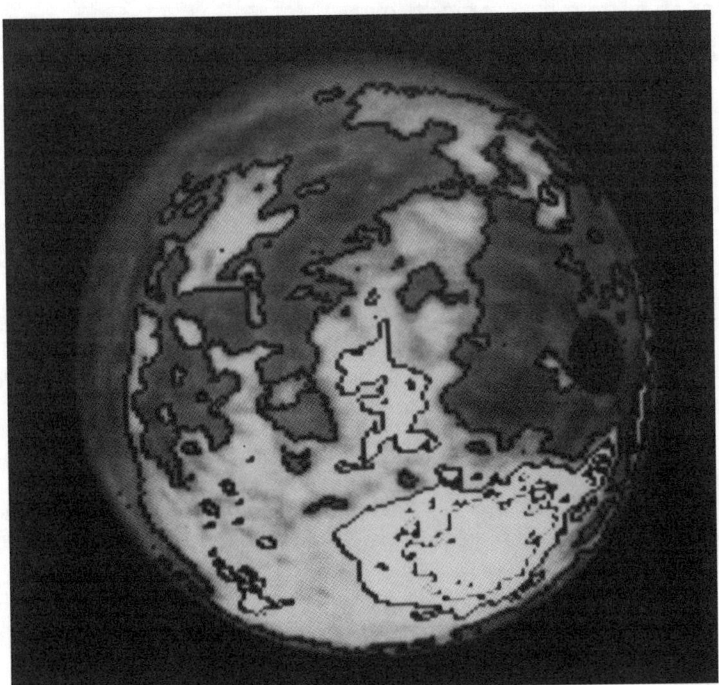

FIGURE 1.12. The spatial distribution of photopically filtered radiation (luminance) of a cloudy sky recorded by the NREL-developed imaging radiometer (a television-based all-sky flux mapper). Isocontour lines are at 500, 1500, 2500, and 3500 ft L.

[4]An integrating sphere for use in the visible region is a hollow sphere, generally having its interior coated with a highly reflective, scattering material such as barium sulfate or polytetrafluoroethylene (PFTE). Light enters through one port and exits through another after undergoing multiple reflections off the interior surface.

E being the irradiance (or illuminance) at distance R from a source of intensity (or luminous intensity) I on a surface perpendicular to the line between the point source and the surface where E is measured.

A common problem encountered in the laboratory is to calculate the irradiance or illuminance from a reflected beam or from an extended source (not a point source) as a function of distance when the irradiance or illuminance can be measured at two or more distances from some reference point. At a sufficient distance from the mirror or extended source, the light appears to come from a virtual point source at some distance Δ behind the mirror or extended source. The accuracy of this method depends on the distances from the mirror or source in relationship to the size of the extended source.

Example 5

The illuminance measured on a plane perpendicular surface 5 m from an extended source is 75 kilolux (klx) and at 10 m is 20 klx. Where is the virtual point source in relation to the extended source and approximately what will be the illuminance at 15 m from the extended source?
Since I is the same at all distances,

$$(75\ \text{klx})\times(5+\Delta)^2=(20\ \text{klx})\times(10+\Delta)^2,$$

$$18.75+7.5\Delta+0.75\Delta^2=20+4\Delta+0.2\Delta^2,$$

$$0=1.25-3.5\Delta-0.55\Delta^2,$$

$$\Delta=0.34\ \text{m}.$$

Thus, the virtual point source is 0.34 m behind the extended source. To determine the illuminance at 15 m from the extended source, from Eq. (1.11),

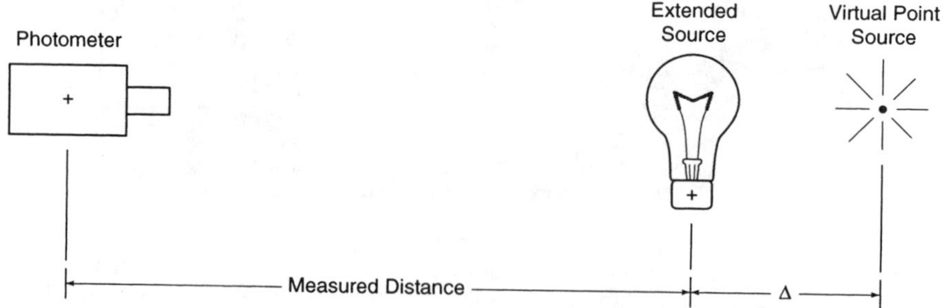

FIGURE 1.13. Geometry for determining the location of a virtual source using an extended source.

$$E=(75\ \text{klx})\times\left(\frac{5+0.34}{15+0.34}\right)^2=9.09\ \text{klx}. \qquad (1.12)$$

1.7.4 Polarization

A property of light that must be considered in certain photometric applications is *polarization*. For example, light from the sky is polarized to a degree depending on the position of the sun relative to the viewing direction. Light reflected by and transmitted through transparent surfaces (e.g., water or glass) is generally polarized.

Polarization occurs where there is nonrandom orientation of the electric field in a light wave. If the electric field (vector) is oriented in only one plane normal to the direction of propagation of the light, the light is said to be *plane-polarized*. Polarization is the only one of the three properties of light considered here that cannot be detected by the eye without the use of special material called a *polarizer*. A polarizer will (ideally) only transmit light having its electric vector parallel to its transmitting axis; see Fig. 1.14. Thus, it can convert unpolarized light to polarized light, or it can be used as an *analyzer* to determine if incident light is polarized. A very common analyzer material is used in Polaroid™ sunglasses. By looking at a blue sky through a pair of these glasses while rotating them around their visual axis, one can observe polarization as a variation in the brightness of the light from the sky; it is most pronounced when the observer is looking thought the glasses while facing at a right angle to the sun. Filters are made for cameras using this effect, which greatly enhances the contrast of scattered clouds against a dark blue-sky background.

Polarization is difficult to quantify [24]. Special equipment is generally required to measure the degree and type of polarization, and a special set of four parameters,

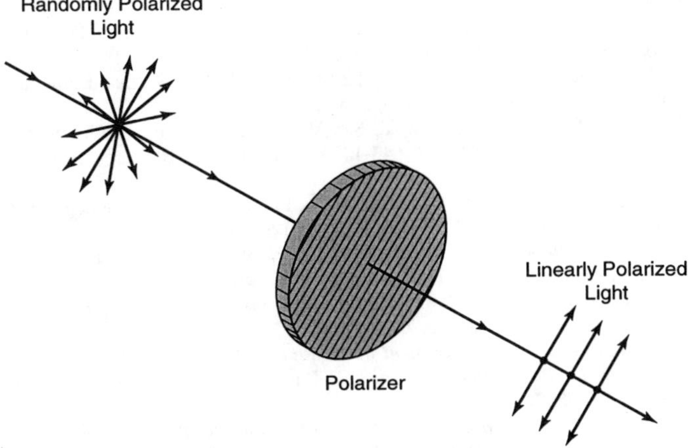

FIGURE 1.14. Randomly polarized light becomes linearly polarized after passing through the polarizer.

the *Stokes parameters*, are used to quantify the polarization. For additional information on these topics see subsequent chapters and *Introduction to Radiometry and Photometry* by McCluney [25].

REFERENCES

1. American National Standard Nomenclature and Definitions for Illuminating Engineering, ANSI/IES Report No. RP-16-1986, American National Standards Institute/Illuminating Engineering (1986).
2. CIE, International Lighting Vocabulary, CIE Pub. No. 17.4, IEC Pub. 50(845) (1987).
3. Parker, Sybil P., ed., *Optics Source Book* (McGraw-Hill, New York, 1988).
4. Guiness, A. E., ed., *ABC's of the Human Mind* (Reader's Digest Association, Inc., Pleasantville, NY, 1990).
5. Burgess, C., and Mielenz, K. D., eds., *Advances in Standards and Methodology in Spectrophotometry*, Analytical Spectroscopy Library—Volume 2 (Elsevier, New York, 1987).
6. Burgess, C., and Jones, D., eds., *Spectrophotometry, Luminescence and Colour, Science and Compliance* (Elsevier Science, B.V. Amsterdam, 1995).
7. Weidner, V. R., and Hsia, J. J., *Spectral Reflectance*, NBS Special Publication No. 250-8 (1987).
8. Nemhauser, R. I., Alexander, G., and Duda, R., ''Radiometry and Photometry: Once over Lightly,'' Opt. Spectra, April (1976).
9. Kaufman, J. E., ed. *IES Handbook*, 5th ed. (Illuminating Engineering Society of North America, New York, 1984).
10. Maxwell, J. C., *A Treatice on Electricity and Magnetism* (Dover, New York, 1960).
11. Schneider, W. E., ''Automated Spectroradiometric Systems: Components and Applications,'' Test and Measurements World, **6**, 159–169 (1985).
12. Nicodemus, F., ed., *Self-Study Manual on Optical Radiation Measurements*, Part I—*Concepts*, NBS Technical Note No. 910-1 (1976), Chaps. 1–3.
13. Walker, J. H., Saunders, R. D., Jackson, J. K., and McSparron, D. A., *NBS Measurements Services: Spectral Irradiance Calibrations*, NBS Publication No. SP250-20 (1987).
14. Schneider, W. E., and Goebel, D. G., ''Standards for Calibration of Optical Radiation Measurement Systems,'' Laser Focus/Electro-Optics, **20** (9), 82–96 (1984).
15. Early, E. A., and Thompson, A., ''Irradiance of Horizontal Quartz-Halogen Standard Lamps,'' J. Res. Natl. Inst. Stand. Technol. **101** 141–153 (1996).
16. Walker, J. H., Saunders, R. D., and Hattenburg, A. T., *Spectral Radiance Calibrations*, NBS Special Publication No. 250-1 (1987).
17. Nicodemus, F., ed., *Self-Study Manual on Optical Radiation Measurements*, Part I—*Concepts*, NBS Technical Note No. 910-2 (1978), Chaps. 4 and 5.
18. Robbins, C. L., Hunter, K. C., and Cannon, T., Mapping Sky and Surface Luminance Distribution Using a Flux Mapper, Energy and Buildings, **6**, 247–252 (1984).
19. Cannon, T. W., and Dwyer, L. D., ''An All-Sky Video Based Luminance Mapper for Daylighting Research,'' Technical Report SERI/TP-643-1324 (Solar Energy Research Institute, Golden, CO, 1981).
20. Weaver, N. L., Balcomb, J. D., and Spitzglas, M. S., Photometric Measurements in Atria Using the SERI Luminance Mapper Model 2, SERI/TR-274-3306 (1988).
21. Cannon, T. W., *Development of a Second-Generation, Video-Based Luminance Mapper for Daylighting Research*, SERI BLG1063/10-14-86 TP-3049 (1986).
22. Valko, P., *Angular Distribution of Sky Radiance and Diffuse Irradiance on Inclined Surfaces*, status report (Swiss Meteorological Institute, Zurich, 1987).
23. Nakamura, H., and Oki, M., *Measurement of Luminance Distribution Under Various Sky Conditions by Orthographic Projection Camera*, CIE Publication No. 36 (1976), Comp. Rendu 18a, Session London, 493–502.
24. Nicodemus, F., ed., *Self-Study Manual on Optical Radiation Measurements*, Part I—*Concepts*, NBS Technical Note No. 910-3 (1977), Chap. 6.
25. McCluney, R., *Introduction to Radiometry and Photometry* (Artech, Boston, 1994), p. 402.
26. Nicodemus, F., ed., *Self-Study Manual on Optical Radiation Measurements*, Part I—*Concepts*, NBS Technical Note No. 910-8 (1985), Chap. 12.

27. Bird, R. E., and Riordan, C., ''Simple Solar Spectral Model for Direct and Diffuse Irradiance on Horizontal and Tilted Planes at the Earth's Surface for Cloudless Atmospheres,'' J. Climate Appl. Meteorol. **25**, 87–97 (1986).
28. Gueymard, C., ''SMARTS2, A Simple Model of the Atmospheric Radiative Transfer of Sunshine: Algorithms and Performance Assessment,'' Profession Paper No. FSEC-PF-270-95 (1995).
29. Gueymard, C., ''Critical Analysis and Performance Assessment of Clear Sky Solar Irradiance Models using Theoretical and Measured Data,'' Solar Energy, **51**, 121–138 (1993).
30. Neckel, H., and Labs, D., Solar Physics **74**, 231–249 (1981).

2

Photometric and Radiometric Quantities

Rainer Köhler

Bureau International des Poids et Mesures, Pavillon de Breteuil, 92312 Sèvres Cedex, France

2.1 INTRODUCTION

Radiometry and photometry describe the propagation of energy by radiation through space. Radiometry treats this problem in a purely physical way, in terms of energy or power and the geometry within which the propagation takes place. In photometry the same problem is described, but the analysis is based not simply on the power propagated but on the visual effect on a human observer that this power would produce.

The understanding of photometry is often considered difficult because of the "strange" quantities and units in this field of optics. It is the aim of this chapter to show that the photometric quantities are not strange but merely the outcome of geometric and physiological considerations.

Although this book is concerned with photometry, the quantities used are derived in parallel with those in radiometry, with the thought that this is an essential underpinning to the understanding of the field.

2.2 NOMENCLATURE

Radiometry and photometry deal necessarily with a source of radiation, a receiver, and the space in between them. In photometry the receiver is the human eye or a detector approximating the human eye. For the rest of this chapter we adopt a nomenclature in which symbols describing properties of the source have the subscript S and all those concerned with the detector use the subscript D. The quantity dA_S thus describes an infinitesimal surface element of the radiation source and Ω_D the solid angle "seen" by the detector. By convention, radiometric quantities carry the subscript e (energetic) and photometric quantities the subscript v (visual). They may be omitted where there is no possibility of confusion.

2.3 PHOTOMETRY VERSUS RADIOMETRY

As does every other physical detector of radiation, the human eye reacts to electromagnetic radiation only in a certain part of the spectrum, that is, to a limited range of wavelengths or frequencies. Such a radiation of sufficient power within a wavelength range of about 380–830 nm can stimulate the eye. This part of the spectrum is described as the visible region, more simply it is usually called *light*. Radiation outside this region should not be referred to as light. There is no such thing as "ultraviolet light" or "infrared light." Moreover, complicating things further, the sensitivity of the human eye to radiation is not the same for each of the wavelengths (colors), the intensity of the light, or the field of view. Last but not least, even under identical conditions the perception is subject to the observer. This subjective nature of the visual system sets photometric quantities apart from purely physical quantities.

The aim in photometry is thus to measure light in such a way that the results correlate as closely as possible with the visual sensation that would be experienced by a standard human observer exposed to the same radiation [1]. The definitions and conventions derived in this chapter aim at enabling photometric measurements to be made by purely physical methods but also to yield results that correlate adequately with visual experience [2].

Three different types of human vision are distinguished:

1. Photopic vision, when the eye is adapted to high luminance[1] levels.
2. Scotopic vision, when the eye is adapted to low luminance levels.
3. Mesopic vision, when the eye is adapted to intermediate levels of luminance.

In order to fulfill the above-mentioned aim of photometry, the CIE [3] introduced two special functions, $V(\lambda)$ in 1924 and $V'(\lambda)$ in 1931, which describe the relative spectral sensitivity of the average human eye for photopic, $V(\lambda)$, and scotopic, $V'(\lambda)$ vision.

Relative spectral sensitivity here means the ratio of the perceived optical stimulus to the incident radiant power as a function of wavelength normalized to unity at the maximum of the function. No special function for the mesopic region has yet been defined. A more detailed discussion of the two functions is given in Ref. [1]. Figure 2.1 shows the two functions. It can be seen that the sensitivity of the eye for light-adapted vision is highest at a wavelength of 555 nm, that is, in the green

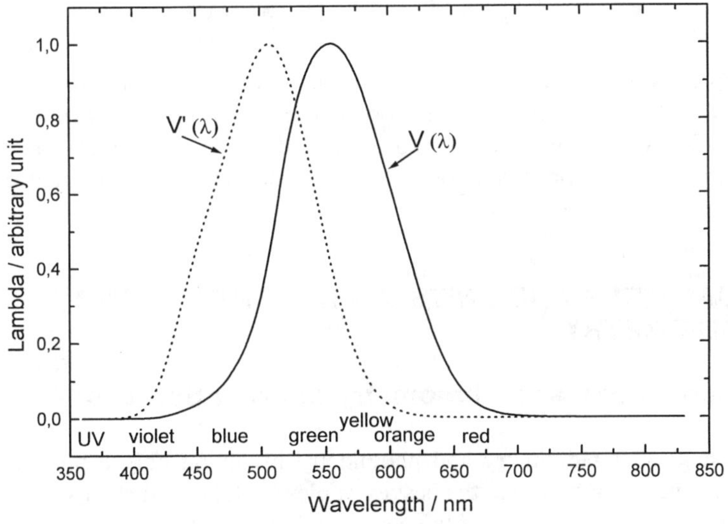

FIGURE 2.1. The $V(\lambda)$ and the $V'(\lambda)$ functions.

[1] As the term *luminance* has not yet been defined, we accept for the moment, without proof, that luminance corresponds closely to the subjective impression of brightness of a light source.

region. In the case of scotopic vision the maximum is situated at 507 nm.

The two $V(\lambda)$ functions allow us to derive a photometric quantity from the definition of the corresponding radiometric quantity. We start by the definition of spectral density of a radiometric quantity. We adopt the usual convention to denote energetic quantities with the subscript e and photometric quantities with the subscript v. We define the spectral density of a radiometric quantity which has the symbol X as

$$X_{e,\lambda}=\frac{dX_e}{d\lambda}, \tag{2.1}$$

where X can be any of the following quantities (which will be derived subsequently in this chapter): flux, energy, irradiance, radiance, or intensity. The effect on a human observer of light from a source with an arbitrary spectral distribution is the sum of the spectral density of the source quantity [Eq. (2.1)] multiplied with the sensitivity of the eye at the same wavelength as defined by the $V(\lambda)$ function. This means that the photometric quantity corresponding to a radiometric quantity is obtained from

$$X_v=K_m\int_{380\ \mathrm{nm}}^{830\ \mathrm{nm}} X_{e,\lambda}V(\lambda)\ d\lambda \tag{2.2}$$

for photopic vision and

$$X_v'=K_m'\int_{380\ \mathrm{nm}}^{830\ \mathrm{nm}} X_{e,\lambda}V'(\lambda)\ d\lambda \tag{2.3}$$

for scotopic vision. K_m and K_m' are proportionality constants that have been defined together with the respective $V(\lambda)$ functions. They are referred to as the maximum spectral efficacy and will be discussed in more detail in the chapter about flux.

We will now start the derivation of the different quantities and give the corresponding units for them.

2.4 QUANTITIES AND UNITS IN PHOTOMETRY AND RADIOMETRY

2.4.1 Radiometry and Photometry of Point Sources

We first look at radiometry and photometry involving only a point source. A point source need not be small, the best examples for this are distant stars. Although they are not tiny objects, some of them having diameters larger than the orbit of the earth, when seen from a large distance (as from the earth), they can certainly be treated as point sources! The essential criterion for a source to be treated as a point source is that the product of the lateral dimensions is small compared to the square of the distance to the source (Fig. 2.2):

$$d_xd_y\ll r^2. \tag{2.4}$$

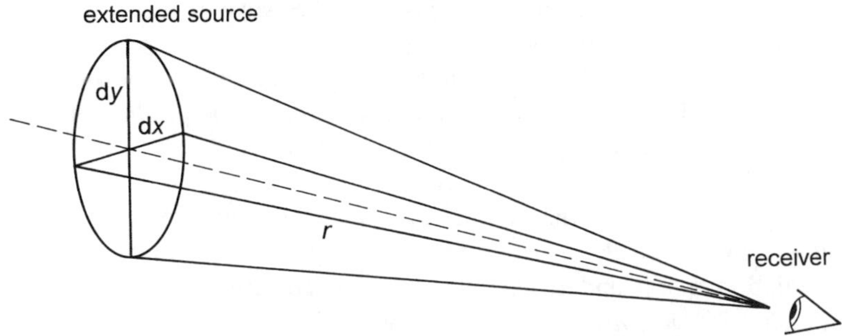

FIGURE 2.2. Geometry for the definition of a point source.

An ideal point source is one that emits isotropically in the sense that the flux emitted is the same in all directions.

We start with the definition of the different quantities in photometry and their respective units, with the photometric equivalent of the power corresponding closely to the physical quantity of power.

2.4.2 Radiant Flux (Radiant Power) and Luminous Flux

The radiant flux is the energy radiated by a source per unit of time: if Q denotes energy, then

$$\Phi = \frac{dQ}{dt}. \tag{2.5}$$

The unit of radiant flux Φ is the watt (1 W=1 J/s). Luminous flux has been given a special unit, the *lumen* (lm).

We see now how Eq. (2.2) transforms the purely physical quantity radiant flux Φ_e (in units of watts) into the corresponding photometric quantity luminous flux (in units of lumens). Explicitly for this case,

$$\Phi_v = K_m \int_{380\ \mathrm{nm}}^{830\ \mathrm{nm}} \Phi_{e,\lambda} V(\lambda) d\lambda. \tag{2.6}$$

It now becomes clear that K_m must have the unit lumen/watt (lm/W). For historical reasons its value was set so as to link the power perceived to the physical power in such a way that a monochromatic source of wavelength equal to that at the maximum of the $V(\lambda)$ curve (555 nm) and having a radiant flux of 1 W has a luminous flux of 683 lm. In other words, the constant K_m=683 lm/W fixes a relationship among the physical, radiometric quantity, and the physiological (human eye) related quantity, for a light-adopted eye under the particular field of view for which the $V(\lambda)$ curve holds. K'_m has the value of 1700 lm/W at a wavelength of 507 nm. The human eye is almost 2.5 times as sensitive at the wavelength of maximum sensitivity when it is dark adapted. In this case the wavelength of the maximum sensitivity shifts towards the blue.

The constants K_m and K'_m are referred to as the *maximum spectral efficacy* of radiation for photopic and scotopic visions, respectively. The product of the respective $V(\lambda)$ function and its K_m value is the *spectral luminous efficacy* function, shown in Fig. 2.3.

Equation (2.6) means that the radiometric, physical "power" is translated into a photometric, physiological "eye power" or "light power,"[2] that is, the effect of incident radiation on a (standard) human observer. The unit of the (nonexisting) eye power is called the *lumen* (a sort of "light watt"[3]).

The total *radiant* flux is the power a source transmits into all the surrounding space by radiation. The total *luminous* flux is the part of this power perceived as light. To measure the total radiant or luminous flux all the radiation emitted by the source must be detected. This can be done by placing the source inside an integrating sphere or by scanning and integrating the radiant field, using a goniophotometer.

The relation between photometric and radiometric quantity for flux is

$$\Phi_v = K_m \int_{380\ \text{nm}}^{830\ \text{nm}} \Phi_{e,\lambda} V(\lambda) d\lambda, \tag{2.7}$$

where Φ_e is measured in W and Φ_v in lm.

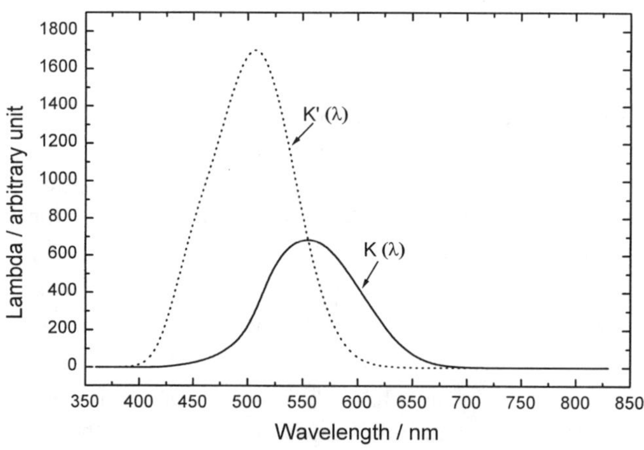

FIGURE 2.3. The spectral luminous efficacy functions.

[2]The term "eye power" does not exist in the literature and should be forgotten after having grasped its meaning.

[3]The same remark as for the preceding footnote applies here.

2.4.3 Radiant Energy and Luminous Energy

One of the most fundamental quantities of a system is that which describes the total energy emerging from a source, or a part of this source, integrated over time. The radiant energy Q_e describes the energy coming from a source in the form of electromagnetic radiation and is measured in joules (J). The luminous energy Q_v describes the part of the same radiant energy that is perceived as light. The luminous energy is sometimes also called the *quantity of light*. It has no unique unit, its unit being derived from the unit of luminous flux, the lumen (lm) to be lm s. The relation between the photometric and radiometric quantity for energy is

$$Q_v = K_m \int_{380\ \mathrm{nm}}^{830\ \mathrm{nm}} Q_{e,\lambda} V(\lambda) d\lambda. \tag{2.8}$$

The concept of energy is used in applications where the flow of energy is not constant, as in a pulsed laser. In some applications, e.g., in biology, it may be necessary to evaluate the total absorbed energy over a specified length of time. Such a quantity is called a *dose* of a radiation.

2.4.4 Radiant Intensity and Luminous Intensity

The radiant intensity is defined as the energy flux per unit solid angle[4] in a given direction from the source. The photometric unit is the *candela*.[5] Intensity is the most important photometric quantity, as its unit is a base unit of the *Système International* (SI):

$$I = \frac{d\Phi}{d\Omega}. \tag{2.9}$$

A filled detector with a given active area at a distance r_1 from a point source recording the flux $d\Phi$ as shown in Fig. 2.4 measures the intensity of the source. The detector area and the distance from the source define the solid angle. A detector of different area placed at a distance r_2 such that it intercepts the same cone of light (Fig. 2.5) does, by definition [Eq. 2.39)], cover the same solid angle as seen from the surface. The flux received by each of the detectors must be the same, hence they measure the same intensity.

---

[4]The plane and solid angle are derived in Appendix 2.3 of this chapter.

[5]See Appendix 2.4 of this chapter for the definition of the candela.

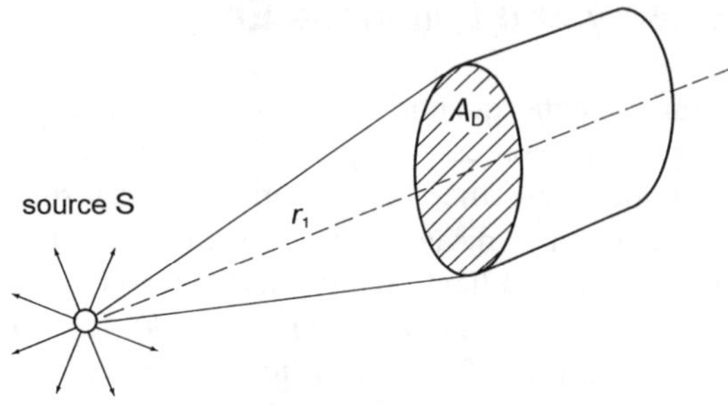

FIGURE 2.4. Measuring the intensity of a source (S) with a detector of active area A_D.

For historical reasons the term *intensity* presents a special problem. The proper definition of the SI base unit of intensity is that given in Eq. (2.9) and Appendix 2.4. As Palmer pointed out [4], the term intensity is used in many other contexts in physics. One definition that is often used is the one of *optical intensity*, meaning the square of the field amplitude. The prefix *optical* is often dropped, even when the subject is visual radiation, thus light. This can lead to confusion and should be avoided.

The relation between the photometric and radiometric quantity for intensity is

$$I_v = K_m \int_{380 \text{ nm}}^{830 \text{ nm}} I_{e,\lambda} V(\lambda) d\lambda \quad \text{candela,} \tag{2.10}$$

which defines the luminous intensity. The unit is the candela. I_e is measured in W/sr and I_v in cd=(lm/sr).

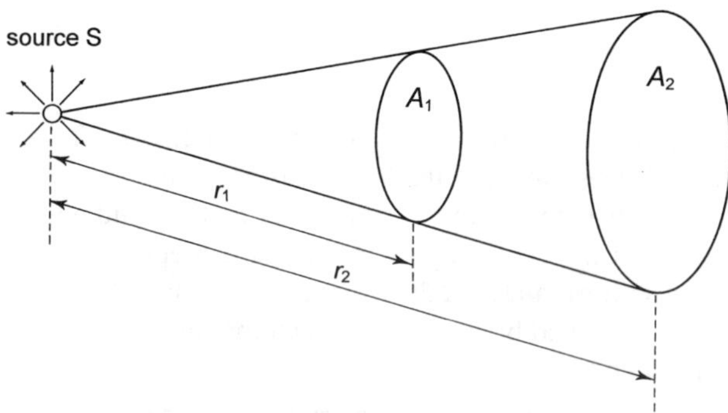

FIGURE 2.5. Constant intensity defined by the solid angle.

2.4.5 Irradiance and Illuminance

Irradiance is defined as the radiant flux per unit detector area. The luminous flux per detector area has been given the name illuminance. The unit of irradiance is W/m^2, and illuminance has been given a special unit, the *lux* ($=lm/m^2$). Irradiance is given as

$$E=\frac{d\Phi_S}{dA_D}. \tag{2.11}$$

An ideal point source emitting uniformly into space has a given irradiance (illuminance) at a surface at the distance r_1, given by $E_1=\Phi/A_1$ (see Fig. 2.6). Another surface having the same area and intercepting the same source of flux at a different distance r_2 consequently has an irradiance of $E_2=\Phi/A_2$.

Hence, from Eq. (2.9), the flux received per unit surface of each detector is $\Phi=I\Omega$. As the solid angle is the same, the respective distances are $A_i=\Omega r_i^2$ [from Eq. (2.39)]. Combining these expressions we obtain:

$$E=\frac{\Phi}{A_i}=\frac{I\Omega}{r_i^2\Omega}=\frac{I}{r_i^2}. \tag{2.12}$$

This result is known as the *inverse square law of irradiance*. In deriving this law we have taken no account of the fact that the irradiance is constant only over spherical surfaces, according to the definition of the solid angle in Fig. 2.16 and Eq. (2.39). Most actual radiometric and photometric problems involve plane receiving surfaces. The illuminance of a plane surface irradiated by a point source can be derived by first considering the case of an infinitesimal surface element where the line between the element and the source is not normal to the element itself (the element is not perpendicular to the source) (see Fig. 2.7).

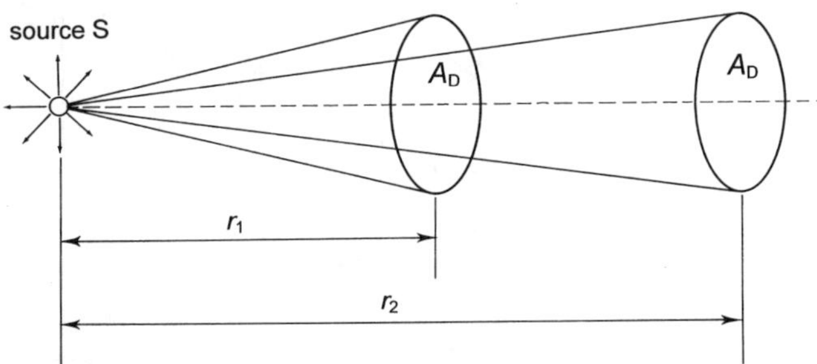

FIGURE 2.6. The inverse square law for irradiance. .

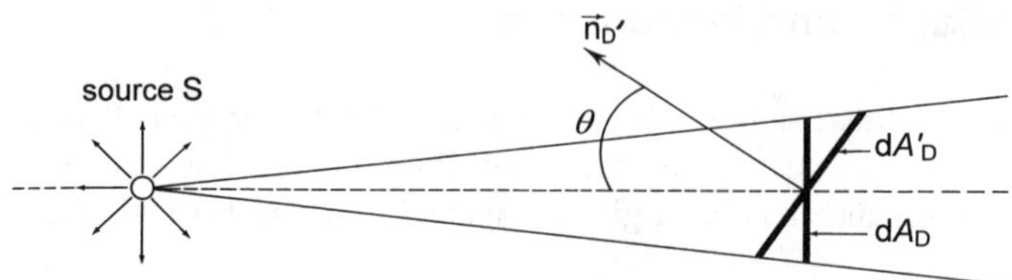

FIGURE 2.7. Irradiance of a plane that is not perpendicular to the source.

The area of element dA'_D is larger than that of element dA_D although, seen from the source, they cover the same solid angle

$$dA'_D = \frac{dA_D}{\cos\theta} \tag{2.13}$$

so the illuminance decreases by the factor cos θ:

$$E' = \frac{\Phi}{A'_D} = \frac{\Phi}{A_D/\cos\theta} = E\cos\theta. \tag{2.14}$$

From this result we can derive a general expression for the irradiance of a surface element by a point source (see Fig. 2.8).

Observe that the distance from the source to the element dA' is greater than the distance to the element dA perpendicular to the source by the factor

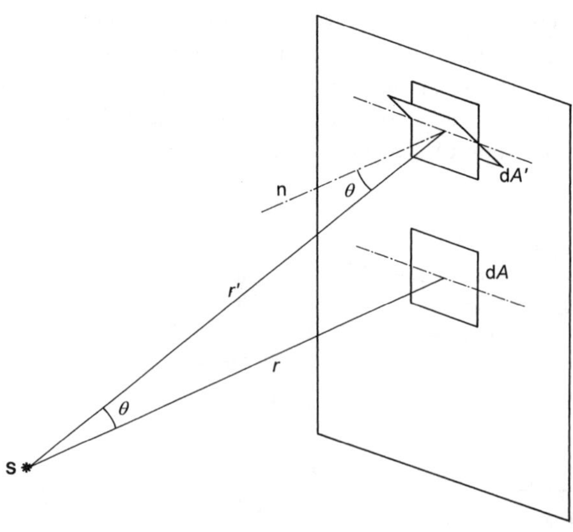

FIGURE 2.8. Irradiance of a plane by a point source.

$$r' = \frac{r}{\cos\theta}. \tag{2.15}$$

Inserting Eqs. (2.15) and (2.14) into Eq. (2.12) we obtain

$$E' = \frac{I}{r^2}\cos^3\theta = E\cos^3\theta. \tag{2.16}$$

This result is known as the $\cos^3$ law in radiometry and photometry since it requires that the irradiance of a plane by a point source falls off across the plane as $\cos^3\theta$ from the point at which the illumination is normal to the plane.

The relation between the photometric and radiometric quantity for irradiance is

$$E_v = K_m \int_{380\ \text{nm}}^{830\ \text{nm}} E_{e,\lambda} V(\lambda)\, d\lambda, \tag{2.17}$$

where E_v is measured in lux and E_e in W/m^2.

2.4.6 Radiant Exitance and Luminous Exitance

Radiant and luminous exitances have the same relationship for an emitter as irradiance to a detector. Exitance is a source quantity:

$$M = \frac{d\Phi}{dA_S}. \tag{2.18}$$

The reciprocity theorem in optics requires that the same considerations as noted above for irradiance apply for exitance. Perhaps the most well-known result in physics that can be expressed in terms of exitance is Planck's law for the spectral radiation of a blackbody:

$$M(T,\lambda) = \frac{c_1}{\lambda^5\left[\exp\left(\frac{c_2}{\lambda T}\right) - 1\right]}, \tag{2.19}$$

where $c_1 = 2\pi hc^2$ and $c_2 = hc/k$. Here $h = 6.6262\times10^{-34}$ J s (Planck's constant), $c = 2.9979\times10^8$ m/s (the speed of light), and $k = 1.3801\times10^{-23}$ J/K (Boltzmann's constant).

The relation between the photometric and radiometric quantity for exitance is

$$M_v = K_m \int_{380\ \text{nm}}^{830\ \text{nm}} M_{e,\lambda} V(\lambda)\, d\lambda, \tag{2.20}$$

where M_v is measured in lm/m^2 and M_e in W/m^2.

2.4.7 Radiance and Luminance

Consider the radiation from a point source emitting into a solid angle $d\Omega_S$ as in Fig. 2.9. The radiance (luminance) at a given point in a given direction is the flux per unit solid angle and per unit projected area perpendicular to the specified direction.

$$L=\frac{d^2\Phi}{d\Omega_S dA_S \cos\theta}, \tag{2.21}$$

where dA_S is the elemental area containing the given point, $d\Omega_S$ is the elemental solid angle containing the given direction, and θ is the angle between the normal to the elemental area and the given direction.

The importance of radiance and luminance is twofold: (a) the conservation of radiance and (b) luminance corresponds closely to the visual sensation of brightness. To derive the first property it is useful to define first the concept of the geometric extent or *étendue géométrique*.

Geometric Extent

Consider the geometry in Fig. 2.10. A source S and a detector D are placed in a homogeneous medium at a distance of many wavelengths. The laws of geometrical optics can be applied in this case. The only radiation that a surface element of the detector D can "see" is specified by rays between D and the source S. The element of solid angle subtended at the source by the receiver area depends on the distance and the inclination of the receiver element with respect to the shortest line between S and D and is given as

$$d\Omega_S = dA_D \cos\theta_D / r^2. \tag{2.22}$$

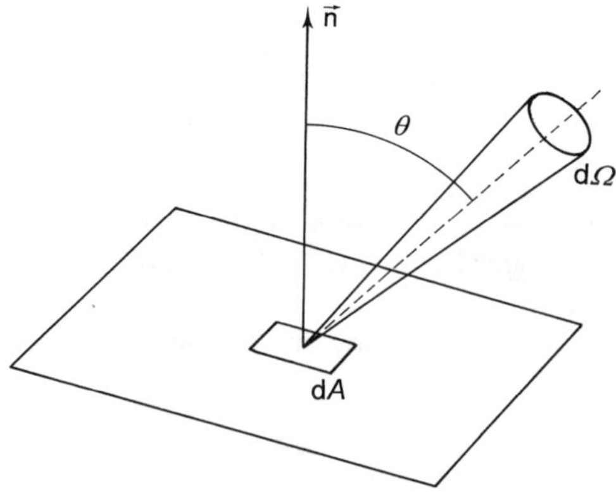

FIGURE 2.9. The definition of radiance (luminance).

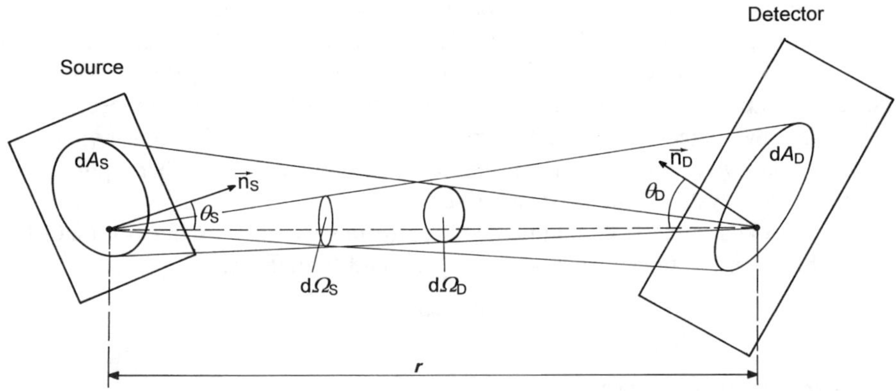

FIGURE 2.10. Definition of the geometric extent.

The same is true when observed from the receiver:

$$d\Omega_D = dA_S \cos\,\theta_S / r^2. \tag{2.23}$$

The quantity

$$d^2G = dA_S \cos\,\theta_S d\Omega_S = dA_S \cos\,\theta_S dA_D \cos\,\theta_D / r^2 = dA_D \cos\,\theta_D d\Omega_D \tag{2.24}$$

is called the geometrical extent of the rays defined by dA_S and dA_R. Considering the totality of the source and the receiver the extent is obtained by integrating over all space:

$$G = \int\int d^2G = \int\int \frac{\cos\,\theta_S \cos\,\theta_D}{r^2}\, dA_S dA_D. \tag{2.25}$$

Conservation of Radiance and Luminance

Inserting the first part of Eq. (2.24) into the definition of luminance [Eq. (2.21)], we obtain

$$L_S = \frac{d^2\Phi_S}{d^2G}. \tag{2.26}$$

In a similar way considerations of the receiving side show that the luminance can be expressed as

$$L_D = \frac{d^2\Phi_D}{d^2G}. \tag{2.27}$$

We see from Eqs. (2.26) and (2.27) that the geometric extent seen from the source and seen from the detector is the same, $L_D = L_S$. If radiation is neither gained nor lost in the medium in which the propagation of energy takes place between the source and the detector, then it must be that $\Phi_S = \Phi_D$. That means that luminance is a quantity that is conserved in the system; it is the same at the receiver as it was at the source. We can even say that luminance is not a source quantity nor is it a detector quantity, it is purely a geometric quantity of the beam connecting the source and the detector. It can be shown that this law of the conservation of luminance is even true in case of the presence of lenses or other optics. The luminance of a system cannot be increased.

In the (rare) case where the luminance is uniform we obtain, by integration of Eq. (2.26), the basic relationship between the flux and luminance:

$$\Phi = LG. \tag{2.28}$$

Luminance Is Brightness

As luminance cannot be increased by an optical system, the purpose of such a system must be to redirect the flux. If you look at a page of this book and consider that this page has a certain luminance (Fig. 2.11), then it follows from the definition of luminance [Eq. (2.21)] and the definition of irradiance [Eq. (2.11)] that

$$E_R = \frac{d\Phi}{dA_D} = L d\Omega \cos\theta, \tag{2.29}$$

which just means that the eye converts the luminance to an illuminance on the retina. The retina, as do almost all other detectors of radiation, reacts to the flux density of the radiant field, that is, to the illuminance. The principal physiological sensation of brightness is linked to the luminance of the source we are looking at. More succinctly, we can say that the eye compares luminances [5].

The relation between photometric and radiometric quantity for luminance is

$$L_v = K_m \int_{380\ \text{nm}}^{830\ \text{nm}} L_{e,\lambda} V(\lambda) d\lambda, \tag{2.30}$$

where L_v is measured in cd/m^2 and L_e in W/m^2 sr.

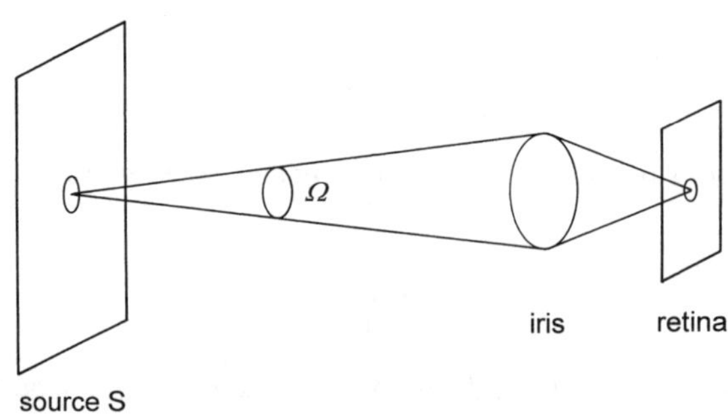

FIGURE 2.11. The eye looking at a source luminance.

2.5 RADIOMETRY AND PHOTOMETRY OF EXTENDED SOURCES

Real-world radiometric and photometric problems rarely deal with point sources. O'Shea [6] proposes that an extended source can be treated as a large collection of identical, uniformly distributed, point sources (Fig. 2.12).

Suppose that each small unit area dA of an extended source radiates with the same, constant intensity $I_S = \Phi_S/\Omega_S$ for all angles. The radiance of the source in the direction of the normal to the surface (θ=0) is then $L(0) = \Phi_S/(\Omega_S dA_S)$ [Eq. (2.21)]. If one observes the source at some angle θ other than zero, the individual sources will appear to be denser by a factor of cos θ giving a brightness

$$L_S(\theta) = \frac{\Phi}{dA\Omega \cos\theta} = \frac{L(0)}{\cos\theta}, \tag{2.31}$$

that is, the source will appear brightest when observed at a grazing angle. This is certainly not true for the sources we come across every day. Most sources have constant brightness over a large range of observation angles and the brightness may even decrease at very large angles.

We can approximate real sources by assuming that the brightness is independent of the observation angle. This means that the intensity radiated by each individual point source has to fall off as cos θ:

$$I(\theta) = \frac{\Phi}{\Omega} \cos\theta. \tag{2.32}$$

Sources obeying this law are called *Lambertian sources*. Most sources are not truly Lambertian, but can be approximated by a Lambertian model, at least for a considerable range of the observation angle.

If a plane is placed at some distance r from an extended source (Fig. 2.13), then the irradiance of the plane from a point on the source is $E = I_S/r^2$ [Eq. (2.12)]. If the intensity of this source point is $I = L_S dA$ [Eq. (2.21)], it follows that $E = L_S dA/r^2$. Summing over all sources we obtain the whole source area so that we finally find

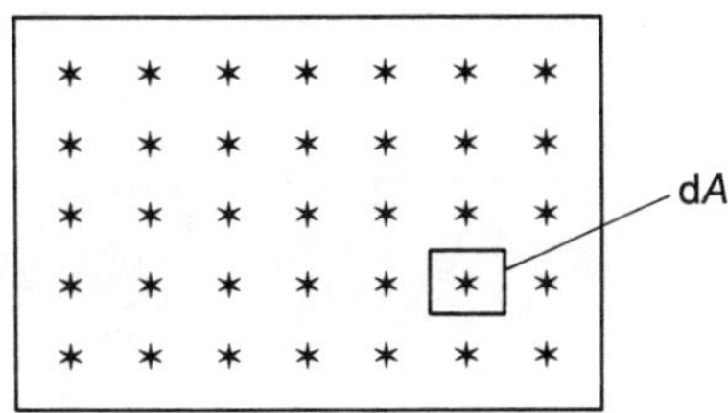

FIGURE 2.12. An extended source constructed from many point sources.

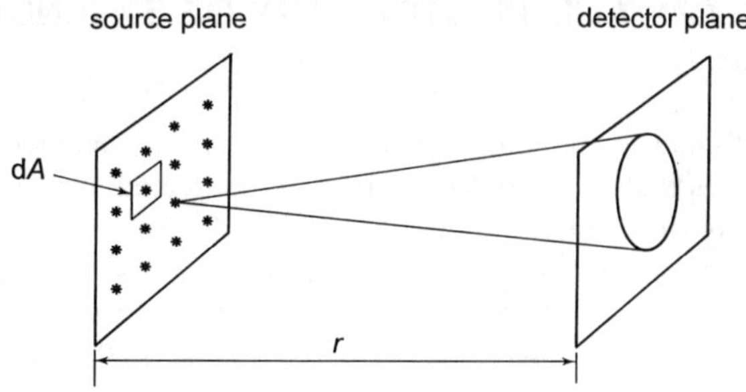

FIGURE 2.13. Irradiance of a plane by an extended source.

$$E=L_S\frac{A_S}{r^2}=L_S\Omega_D. \tag{2.33}$$

Finally we derive the total excitance of a Lambertian source (Fig. 2.14). We rewrite the definition of radiance [Eq. (2.21)] in the form

$$d^2\Phi=L_S dA_S\cos\theta\,d\Omega=\frac{L_S dA_S\cos\theta\,dA_D}{r^2} \tag{2.34}$$

$$=L_S dA_S\cos\theta\sin\theta\,d\theta d\psi, \tag{2.35}$$

which can be integrated to give

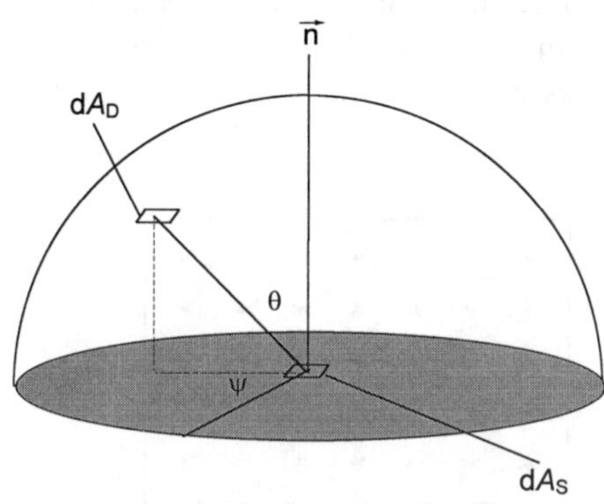

FIGURE 2.14. Hemispherical irradiance by a Lambertian source.

$$M=\int \frac{d\Phi}{dA_S}=L_S\int_0^{\pi/2}\int_0^{2\pi} \cos\theta \sin\theta \, d\theta d\psi \tag{2.36}$$

to find the fundamental result:

$$M=\pi L_S. \tag{2.37}$$

2.5.1 Geometric Extent of an Extended Source

In deriving the geometric extent we considered the source and the receiver as integrals of small areas. It is easy to see, using the conservation of radiance (luminance) that the concept of the geometric extent can be used to describe sources illuminating a diaphragm and sources behind an aperture. This is important as in many practical cases no source surface element can be defined, e.g., when considering the radiance of the sky. The luminance in this case can be calculated in terms of the received luminance, e.g., of the eye. This is a useful consequence of the conservation of luminance.

2.6 TYPICAL FIGURES AND VALUES

In this short section a number of typical values will be presented to give the reader an idea.

Typical Radiometric Quantities

Total flux from a 100-W incandescent light bulb:	82 W
Typical He–Ne laser:	5 mW
Radiant flux from a 40-W fluorescent lamp:	23.2 W
Solar irradiance (in earth orbit):	1357 W/m^2
Terrestrial solar irradiance:	600–900 W/m^2
Radiance of sun at its surface:	2.3×10^7 W/(m^2 sr)

Typical Photometric Quantities

Total luminous flux from a 100-W light bulb:	1740 lm
Luminous flux from a 5-mW He–Ne laser:	0.818 Im
Luminance of a blackbody at 6500 K:	3×10^9 cd/m^2

APPENDIX 2.1: RADIOMETRIC AND PHOTOMETRIC QUANTITIES

Radiometric Quantities

Quantity	Symbol	Definition	Unit
Radiant energy	Q_e		J
Radiant flux, radiant power	Φ_e	$\Phi=dQ/dt$	W
Irradiance	E_e	$E=d\Phi_R/dA_R$	W/m^2
Exitance	M_e	$M=d\Phi_S/dA_S$	W/m^2
Radiant intensity	I_e	$I=d\Phi/d\Omega$	W/sr
Radiance	L_e	$L=d^2\Phi/(d\Omega\ dA\cos\theta)$	W/(sr m^2)

Photometric Quantities

Quantity	Symbol	Definition	Unit
Quantity of light, luminous energy	Q_v	$Q_v=K_m\int V(l)Q_l dl$	lm s
Luminous flux	Φ_v	$\Phi_v=dQ_v/dt$	lm (lumen)
Illuminance	E_v	$E_v=d\Phi_{v,R}/dA_R$	lx (lux)
Luminous excitance	M_v	$M_v=d\Phi_{v,S}/dA_S$	lm/m^2
Luminous intensity	I_v	$I_v=d\Phi_v/d\Omega$	cd (candela)
Luminance	L_v	$L_v=d^2\Phi_v/(d\Omega\ dA\cos\theta)$	cd/m^2

APPENDIX 2.2: LIST OF SYNBOLS

Subscripts

S	Source quantity
D	Detector quantity
e	Energetic (radiometric quantity)
v	Visual (photometic quantity)

Quantities

dA	Infinitesimal area element
Q	Energy (luminous or radiant)
Φ	Flux (luminous or radiant)
E_e	Irradiance

E_v	Illuminance
M	Excitance (luminous or radiant)
I	Intensity (luminous or radiant)
L_e	Radiance
L_v	Luminance
r	Distance
Ω	Solid angle

APPENDIX 2.3: GEOMETRICAL QUANTITIES

Radiometry and photometry deal with the propagation of light in space so a definition of the relevant quantities must begin with a definition of the geometric quantities that describe the space of propagation.

Plane Angle

Generally, light sources emit in all directions in the space around them (one exception is the beam of a laser). To describe this the concept of the solid angle is used. The solid angle is derived similarly to the plane angle in Fig. 2.15.

The plane angle, Fig. 2.15, is defined as the ratio of the length of the arc l of a part of a circle to the length of the radius r of the same circle, centered at S:

$$\theta=\frac{l}{r}. \tag{2.38}$$

As the circumference of a circle is $2\pi r$, this definition implies that a complete circle covers a plane angle of 2π. The plane angle as defined in Eq. (2.38) is dimensionless, but it has been given a unit, named *radian* (rad), so that the full circle has 2π radians. More generally, the plane angle for lines that are not part of a circle is defined by the projection of this line onto the circle (curve C in Fig. 2.15).

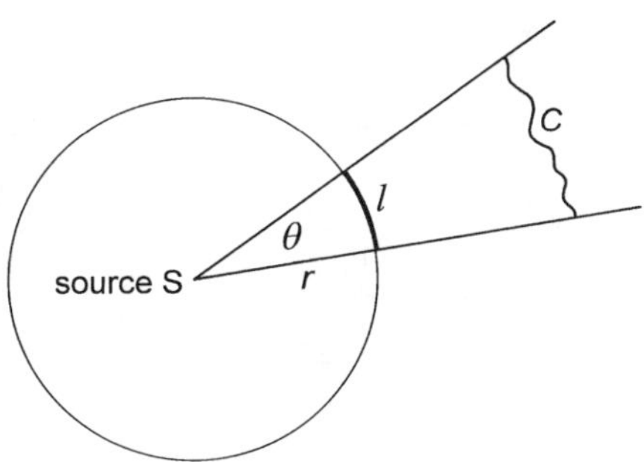

FIGURE 2.15. The plane angle.

Solid Angle

Passing into three-dimensional space, the solid angle is defined as the ratio of the area of a part of a sphere to the square of the radius of the sphere. Note that this definition ensures that the square of the radius keeps the solid angle dimensionless like the plane angle (Fig. 2.16):

$$\Omega = \frac{A}{r^2}. \tag{2.39}$$

The quantity in Eq. (2.39) has also been given a unit, the *steradian* (sr). It is easy to see that a full sphere has exactly 4π steradians. Again, if the surface C is not part of the sphere, it is its projection onto the sphere which is to be used.

APPENDIX 2.4: DEFINITION OF THE CANDELA

The Comité Consultatif de Photométrie et Radiométrie (CCPR) recommended in 1977 during its ninth session [7] the use of the value 683 lm/W for the luminous efficacy of radiation of a frequency of 540.014×10^{12} Hz, corresponding to a wavelength of 555 nm in standard air [8]. This recommendation was adopted in the same year by the Comité International des Poids et Mesures (CIPM). The Conférence Générale des Poids et Mesures (CGPM) redefined the candela with only a small change compared to the recommendation as follows [9]:

> The candela is the luminous intensity, in a given direction, of a source that emits monochromatic radiation of frequency 540×10^{12} Hz and that has a radiant intensity in that direction of 1/683 W/sr.

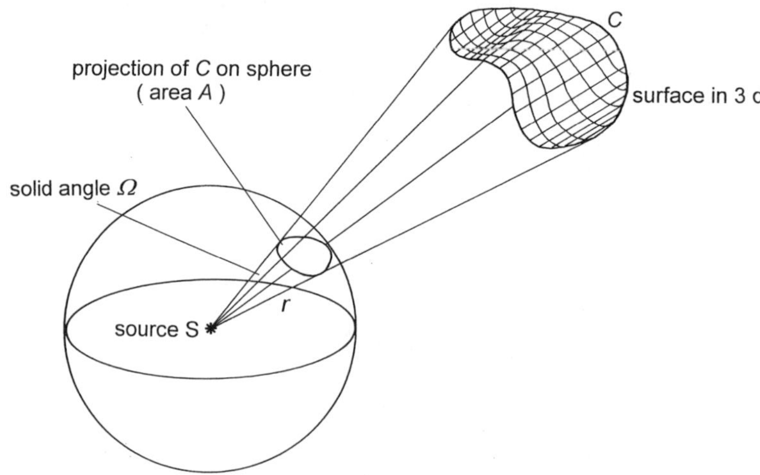

FIGURE 2.16. The solid angle.

This definition of the candela applies for photopic, scotopic, and mesopic vision. The frequency 540×10^{12} Hz corresponds to a wavelength of 555.016 nm in standard air.

In the beginning, photometric standards were light sources, the earliest ones being candles, and the name *candela* has been retained as the name of the photometric base unit. From 1948 until 1979, the thermal radiation from a black body at the temperature of freezing platinum ($\approx$2041 K was used for the definition. Today the definition of the candela is in terms of monochromatic radiation. The value (1/683) watt per steradian in the present definition was chosen so that there was continuity in the photometric units as maintained in laboratories in 1979, when the present definition was adopted. The 1979 definition gives no indication as to how the candela should be realized. This was deliberate to allow new techniques to be applied without having to change the definition of the base unit. The candela is realized today in national standards laboratories using radiometric methods based upon absolute detectors. However, standard lamps are still used to maintain the photometric units; they provide either a known luminous intensity in a given direction, or known luminous flux.

REFERENCES

1. *The Basis of Physical Photometry*, CIE Publication No. 18.2 (1983).
2. *Principles Governing Photometry*, Bureau International des Poids et Measures (BIPM) Monograph (1983).
3. CIE is the Commission Internationale de l'Éclairage (International Commission on Illumination).
4. Palmer, J. M., ''Getting Intense on Intensity,'' Metrologia **30**, 371–372 (1993).
5. Terrien, J., ''Que sais-je, photométrie''.
6. O'Shea, D. C., *Elements of Modern Optical Design* (Wiley, New York, 1985).
7. BIPM, Comité Consultaty de Photometrie et Radiométrie (CCPR), 9th Session, 1997.
8. Edlén, B., ''The refractive index of air,'' Metrologia **2**, 71–80 (1966).
9. BIPM, Comptes rendus des séances de la 16e Conférence Générale des Poids et Mesures, Paris, October, 1979.

3

Photometric Standards

Yoshi Ohno

NIST, A320/220, Gaithersburg, Maryland 20899

3.1 HISTORY OF PHOTOMETRIC STANDARDS

The history of the standards for light dates back to the early nineteenth century, when the flame of a candle was used as a unit of luminous intensity that was called *the candle*. The candle power, the old name for the luminous intensity, originated from the use of candles. As early visual photometers were improved, it was determined that candles were not reproducible to the accuracy of measurement even when the composition, form, and rate of burning were carefully specified. Numerous efforts were made to use controlled flame lamps. In the mid-nineteenth century, the standard candles were gradually superseded by various other flame standards such as the carcel lamp, the pentane lamp, and the Hefner lamp [1].

Despite careful specifications of manufacturing details and numerous determinations of the correction factors, none of the flame standards proved adequate for accurate photometry. In the late nineteenth century, suggestions were made to construct some form of standard depending on the radiation given by a specified area of surface at a given temperature, such as the melting platinum standard known as the Violle standard. This standard utilizing molten platinum, however, was found unsatisfactory because of variations in the surface emissivity and the freezing point caused by contamination. About the same time, the use of an incandescent filament lamp as a standard was proposed. But it was found to be impractical because it was not possible to specify and manufacture such a lamp to the extreme accuracy required for an absolute standard.

In early twentieth century, to improve the Violle standard, investigations on platinum point blackbodies began at some national laboratories. The blackbody consisted of a cylindrical radiator made of pure fused thoria (about 45 mm long), which was immersed in pure molten platinum maintained at the temperature of solidification (2042 K). The entire blackbody was heated in a high-frequency induction furnace with 7 kW power to bring it to the melting point. An agreement was first established in 1909 among the national laboratories of France, Great Britain, and the United States to use this method. The unit was recognized as *the international candle*. This standard was adopted by the Commission Internationale de l'Eclairage (CIE) in 1921 [2]. After a successful realization of the candle in 1931 [3], this method became universally recognized. In 1948, it was adopted by the Conférence Générale des Poids et Mesures (CGPM) [4] with a new Latin name *candela*. In 1967, CGPM adopted a more precise definition of the candela [5]:

> The candela is the luminous intensity, in the perpendicular direction, of a surface of 1/600 000 square meter of a blackbody (full radiator) at the temperature of freezing platinum under a pressure of 101 325 newtons per square meter.

The candela also became one of the base SI units (Systéme International d'Unités) when the SI was established in 1960 [6].

Although this definition served to establish the uniformity of photometric measurements in the world, difficulties in fabrication of the blackbody and in improving accuracy were addressed. Since the mid-1950s, suggestions were made to define the candela in relation to the optical watt so that complicated source standards would not be needed. There were many efforts to determine the constant that would provide a numerical relationship between the photometric quantities and the radiometric quantities [7,8].

In 1979, the new definition of the candela was adopted by CGPM [9], defining the candela in relation to the radiant power (watt) by introducing the constant K_m as described in the later sections of this chapter. The 1979 redefinition of the candela has allowed the use of appropriate techniques to derive the photometric units from the radiometric scales.

After the new definition, most national laboratories have realized the candela based on the absolute responsivity of detectors rather than blackbody radiation. Before the international intercomparison of photometric units held by the Consultatif Comité de Photométrie et Radiométrie (CCPR) in 1985 [10], many national laboratories realized the candela by using room-temperature electrical substitution radiometers (ESRs). This intercomparison showed a $\pm 1\%$ variation of the national units of candela, which was slightly better than previous intercomparisons, but the improvement was less than expected. In the early 1980s, the silicon photodiode self-calibration technique [11,12] was developed and used extensively for realization of photometric units. Absolute cryogenic radiometers are now used in national laboratories to provide radiometric scales with uncertainties on the order of 0.01%. The candela is now realized based on cryogenic radiometers at several national laboratories. With these recent improvements in technology, a smaller variation of national units is expected, and will be the subject of another CCPR international intercomparison of photometric units planned for 1998.

3.2 PHOTOMETRY, PHYSICAL PHOTOMETRY, AND RADIOMETRY

The primary aim of photometry is to measure visible radiation or light, in such a way that the results correlate as closely as possible with what the visual sensation would be of a normal human observer exposed to that radiation. Until about 1940, visual comparison techniques of measurements were predominant in photometry, where typically an observer was required to match the brightness of two visual fields viewed either simultaneously or sequentially.

In modern photometric practice, almost all measurements are made with photodetectors and is referred to as physical photometry. In order to achieve the aim of photometry, one must take into account the characteristics of the human vision. The relative spectral responsivity of the human eye is similar for most observers but can vary depending on individuals and on the viewing conditions. A relative spectral responsivity of the human eye was first adopted by the CIE in 1924 [13] and subsequently redefined as a part of the colorimetric standard observers in 1931 [14]. This human spectral responsivity function is called *the spectral luminous efficiency function*, or the $V(\lambda)$ function. This function is defined as an average of many human observers for the 2° field of view in the fovea centralis, under relatively high luminance levels. This human vision model is called *the CIE Standard Photometric*

Observer for photopic vision. The $V(\lambda)$ function gained wide acceptance, and was republished by CIE in 1983 [15] and published by the Comité International des Poids et Mesures (CIPM) in 1982 [16] to supplement the 1979 definition of candela. Thus a photodetector, the spectral responsivity of which is approximated to the $V(\lambda)$ function, replaced the role of the human eye in photometry. Since 1924, all the measurements of physical photometry have been based on the $V(\lambda)$ function as a standard for human observers. The $V(\lambda)$ function is defined in the range 360–830 nm and has a peak value normalized to 1.0 at 555 nm (Fig. 3.1). The tabulated data of the function at every 1 nm are published in Refs. [15, 16, and 17 (disk)]. In most cases, the region 380–780 nm is used for calculation with negligible errors because the $V(\lambda)$ function falls to values below 10^{-4} outside this region.

The $V(\lambda)$ function represents the spectral responsivity of human vision in a 2° field of view at relatively high luminance levels (higher than several cd/m^2). The human vision in this level is called *photopic vision.* The spectral responsivity of human eyes deviates significantly at very low luminance levels (less than $\sim 10^{-3}$ cd/m^2) [17b] when the rods in the eyes are the dominant receptors. This type of vision is called *scotopic vision.* Its spectral responsivity, peaking at 507 nm, is designated as the $V'(\lambda)$ function, and was defined by CIE in 1951 [18], recognized by CIPM in 1976 [19], and republished by CIPM in 1982 [16]. The perception between photopic vision and scotopic vision is called *mesopic vision.* Mesopic vision has been extensively studied [20], but has not been officially defined. In current practice, almost all the photometric quantities are still measured in the units of photopic vision even at such low luminance levels except for special measurements for research purposes. All the photometric quantities discussed in this chapter are for photopic response.

Photometry is now based on radiometry, a science of measuring optical radiation in quantities such as radiant flux or radiant power (unit: W). As defined in the definition of candela by CGPM in 1979 and CIPM in 1982, a photometric quantity X_v is defined in relation to the corresponding radiometric quantity $X_{e,\lambda}$ by the equation

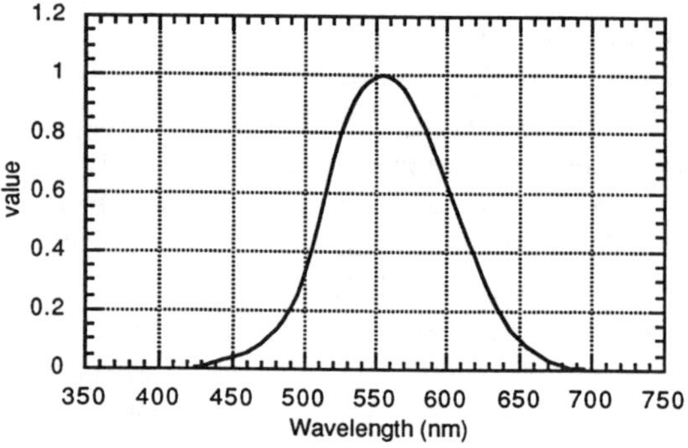

FIGURE 3.1. CIE $V(\lambda)$ function.

$$X_v = K_m \int_{360\ \text{nm}}^{830\ \text{nm}} X_{e,\lambda} V(\lambda) d\lambda. \tag{3.1}$$

The constant K_m relates photometric quantities and radiometric quantities, and is called *the maximum spectral luminous efficacy (of radiation) for photopic vision.* The value of K_m is defined to be 683 lm/W as described in the next section.

3.3 SI UNITS AND THE INTERNATIONAL LEGAL METROLOGY SYSTEM

The base of all the photometric quantities is the candela. The candela was first defined by the CGPM in 1948 based on the radiation emitted by a blackbody source at the temperature of the freezing point of platinum. The definition of the candela adopted by the CGPM in 1979 is

> The candela is the luminous intensity, in a given direction, of a source that emits monochromatic radiation of frequency 540×10^{12} Hz and that has a radiant intensity in that direction of 1/683 W/sr.

540×10^{12} Hz corresponds to the wavelength 555.016 nm in standard air. Frequency is used in this definition in order to make the definition independent of the refraction index of media. It should be noted that the $V(\lambda)$ function is not implicitly stated in the SI definition of candela. However, the photometric quantities are defined by CIPM [16] using the $V(\lambda)$ function for practical realization of photometric units. According to this SI definition of candela, the value of K_m in Eq. (3.1) is calculated as $683 \times V(555.000\ \text{nm})/V(555.016\ \text{nm}) = 683.002$ lm/W [15]. K_m is normally rounded to 683 [lm/W] without affecting accuracy of real measurements.

Other photometric units such as the lumen (luminous flux) and lux (illuminance) are derived from the candela. The definitions of all the photometric quantities and units are given in Refs. [16] and [21], and described in Chap. 2. Although English units as shown in Table 3.1 are still widely used, use of the SI units in all photometric measurements is recommended. The definitions of the English units are described below for conversion purposes only.

The definition of foot lambert is such that the luminance of a perfect diffuser is 1 fL when illuminated at 1 fc. In the SI unit, the luminance of a perfect diffuser will be $1/\pi$ cd/m^2 when illuminated at 1 lx. For convenience of changing from English units to the SI units (metric system), the conversion factors are listed in Table 3.2. For example, 1000 lx is the same illuminance as 92.9 fc, and 1000 cd/m^2 is the same

TABLE 3.1. *English photometric units and definition.*

Unit	Quantity	Definition
Foot candle (fc)	Illuminance	Lumen per square foot (lm/ft^2)
Foot lambert (fL)	Luminance	$1/\pi$ candela per square foot ($1/\pi$ cd/ft^2)

luminance as 291.9 fL. Conversion factors to and from many other units are given in Refs. [22, 23]. Further information of the SI units is found in Refs. [23, 24].

To better understand the international metrology system, it is useful to know the relationship between such organizations as CGPM, CIPM, CCPR, CIE, and Bureau International des Poids et Measures (BIPM). These acronyms arise from the French versions of the organization names. In English, their names would be CGPM, General Conference of Weights and Measures; CIPM, International Committee for Weights and Measures; CCPR, Consultative Committee of Photometry and Radiometry; BIPM, International Bureau of Weights and Measures; and CIE, International Commission on Illumination. All the SI units are officially defined by CGPM, which is the decision-making body for the Treaty of the Meter (Convention du Mètre), signed in 1875. The decision of CGPM legally governs the metrology system in the world to those countries signatory to the Treaty of the Meter or agreeing to its usage. CIPM is a committee under CGPM, charged with the management of the international system of units and related fundamental units, and consists of subcommittees for each technical field. CCPR is a subcommittee under CIPM that recommends the definitions of units in photometry and radiometry and holds international intercomparisons of photometric units and radiometric scales among national laboratories. CCPR consists of representatives of interested national standardizing laboratories. BIPM is a metrology laboratory under the supervision of CIPM, with staff and facilities in Paris. CIE, originally organized to promote uniformity and quality of optical measurements, is an academic society in the field of lighting science. Many definitions developed by CIE, such as the $V(\lambda)$ function, the color-matching functions, and the standard illuminants, were adopted by CGPM and/or by the International Organization for Standardization (ISO) as international standards. CIE has recently been officially recognized by ISO and the International Electrotechnical Commission (IEC) as a standards-creating body in the field of optical radiation.

3.4 LUMINOUS-INTENSITY STANDARDS

During the previous definition of the candela from 1948 to 1979, a platinum-point blackbody was used to realize the candela. Now the candela is most often realized

TABLE 3.2. *Conversion between English units and SI units.*

To obtain the value of	Multiply the value of	By
lx from fc	fc	10.764
fc from lx	lx	0.092 90
cd/m^2 from fL	fL	3.426 3
fL from cd/m^2	cd/m^2	0.291 86
m (meter) from feet	ft	0.304 80
mm (millimeter) from inch	in.	25.400

based on the absolute responsivity of detectors as provided in the 1979 redefinition of the candela. In this method, referred to as the detector-based candela, calibrated detectors provide the illuminance unit and the candela is deduced from the illuminance and the distance from the source to the photometer. On the other hand, the candela can still be realized based on the radiation from a blackbody of a known temperature (the source-based method). In this section, the theories and procedures for establishing the unit of candela are described, followed by characteristics of standard lamps and standard photometers used as transfer standards.

3.4.1 Detector-Based Candela Realization

Most national laboratories currently realize the candela based on the absolute responsivity of detectors. Some laboratories use room-temperature ESRs [25,26], others use the silicon photodiode self-calibration technique, 100% quantum efficient silicon photodiode trap detectors [27–29], or absolute cryogenic radiometers [30,31]. Cryogenic radiometers are now considered the most accurate means for establishing radiometric scales [32,33]. The cryogenic radiometer is cooled by liquid helium to 5 K and works on the principle of electrical substitution. As an example, the construction of a cryogenic radiometer used at National Institute of Standards and Technology (NIST) is shown in Fig. 3.2. The details of the realization of the candela and other photometric measurements at NIST are described elsewhere [34]. Cryogenic radiometers are used to realize the candela with stated

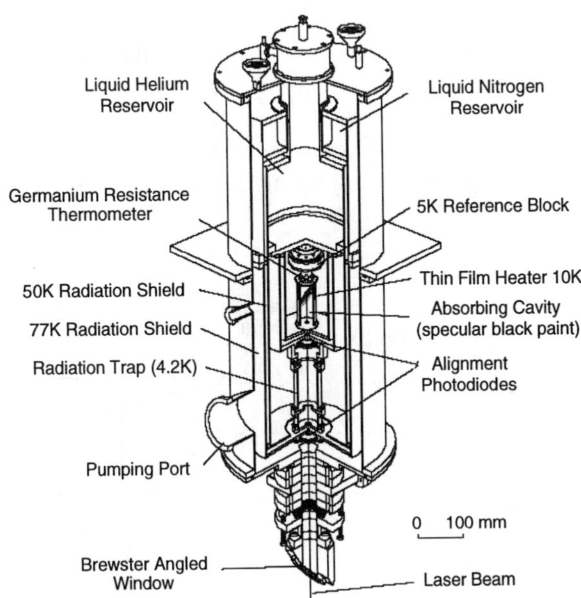

FIGURE 3.2. Construction of the NIST high accuracy cryogenic radiometer.

uncertainties of 0.2–0.4% (expanded uncertainty with $k=2$) [30,31].

The principles of the detector-based realization of the candela are described below. A standard photometer, consisting of a silicon photodiode, a $V(\lambda)$-correction filter, and a precision aperture, is shown in Fig. 3.3.

First, the absolute spectral responsivity $s(\lambda)$ (in A/W) of the photometer is determined based on the absolute spectral responsivity scale. The area of the aperture A is measured by using a dimension measuring instrument. The illuminance responsivity s_v (in A/lx) of the photometer is then obtained by

$$s_v = \frac{A \int_\lambda S(\lambda) s(\lambda) d\lambda}{K_m \int_\lambda S(\lambda) V(\lambda) d\lambda}, \tag{3.2}$$

where $S(\lambda)$ is the spectral power distribution of the light to be measured, $V(\lambda)$ is the spectral luminous efficiency function, and K_m is the maximum spectral luminous efficacy (683 lm/W). Planckian radiation at 2856 K (CIE Illuminant A [17,35]) is normally used for $S(\lambda)$.

The calibrated photometer provides the unit of illuminance. When the photometer is used to measure a light source, the luminous intensity I_v (in cd) of the source is given by

$$I_v = \frac{d^2}{\Omega_0} \frac{y}{s_v}, \tag{3.3}$$

where d is the distance (in m) from the light source to the reference plane (aperture surface) of the photometer, Ω_0 is the unit solid angle (in sr), and y is the output current (in A) of the photometer.

The procedure for the detector-based realization of the candela used at NIST is shown in Fig. 3.4 as an example. A cryogenic radiometer acts as the absolute radiometric base at the top of the chain. The spectral responsivity scale is established on silicon photodiode trap detectors based on laser beam measurements using the cryogenic radiometer at several wavelengths. Then the absolute spectral respon-

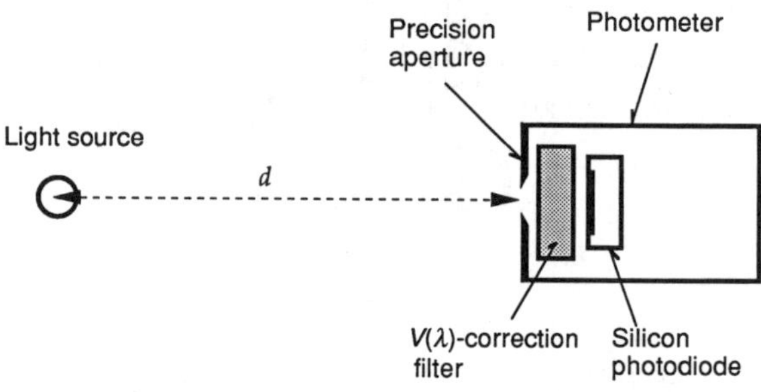

FIGURE 3.3. Geometry for the detector-based candela realization.

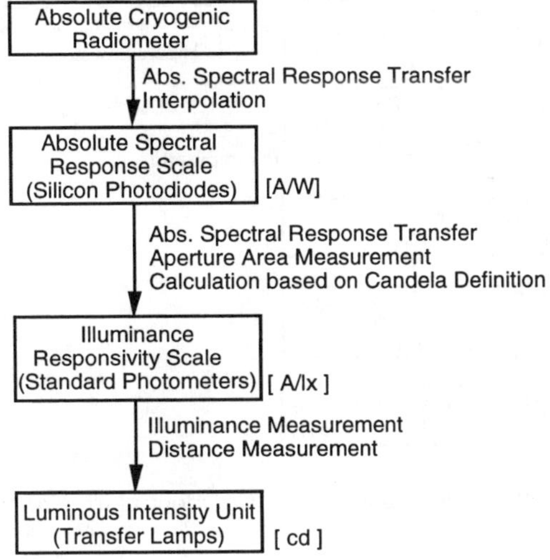

FIGURE 3.4. The detector-based candela realization procedure at NIST.

sivities $s(\lambda)$ of the standard photometers are measured using a monochromatic beam underfilling the precision entrance aperture of the photometer. Corrections are made for the spatial nonuniformity of spectral response over the apertured area. The illuminance responsivity (in A/lx) of each photometer is then calculated using Eq. (3.2).

3.4.2 Source-Based Candela Realization

Even though the candela is commonly realized based on absolute detectors, blackbodies can still be used to realize the candela [36]. In this case, the candela is derived from the spectral irradiance scale based on a blackbody of a known temperature. Based on Planck's Law, the spectral radiance of a blackbody at a temperature T is given by

$$L_e(\lambda,T)=c_1 n^{-2}\pi^{-1}\lambda^{-5}[\exp(c_2/n\lambda T)-1]^{-1} \tag{3.4}$$

where $c_1=2\pi hc^2=3.741\,7749\times10^{-16}$ W m^2, $c_2=hc/k=1.438\,769\times10^{-2}$ m K (from Ref. [21]), h is Planck's constant, c is the speed of light in vacuum, k is the Boltzmann constant, $n(=1.000\,28)$ is the refractive index of standard air [15,37], and λ is the wavelength.

As an example, Fig. 3.5 shows the procedure to realize the spectral irradiance scale at NIST [38] using a gold-point blackbody operating at its freezing point temperature 1337.33 K. The spectral radiance of the blackbody at a certain wavelength is then transferred to ribbon filament lamps (the gold point secondary standards), which are used to determine the temperature of a variable temperature

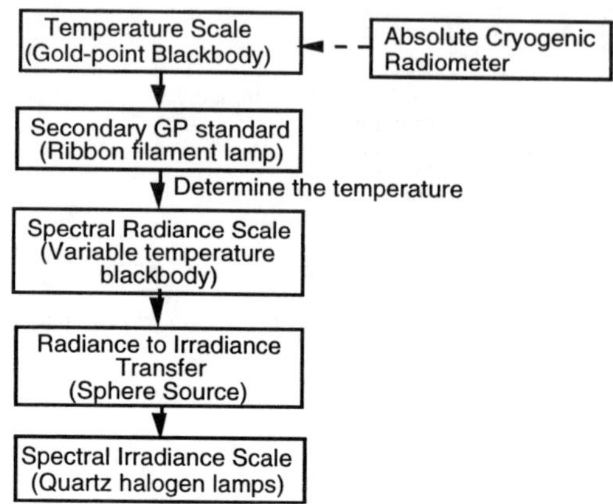

FIGURE 3.5. Procedures for realization of the spectral irradiance scale at NIST.

blackbody operated at temperatures up to 2500 K. A small integrating sphere source is then used to transfer from spectral radiance to spectral irradiance. The sphere source is equipped with an exit aperture of known area A placed at distance d from the monochromator entrance aperture. The spectral irradiance $E_e(\lambda)$ of the sphere source is obtained from the spectral radiance $L_e(\lambda)$ by

$$E_e(\lambda)=A\ L_e(\lambda)/d^2. \tag{3.5}$$

The spectral irradiance of the sphere source is transferred to a group of working standard lamps. High-power quartz halogen lamps, known for their stability and higher power in the UV, operating at 3000 to 3200 K are commonly used as spectral irradiance standards.

Once the spectral irradiance scale is established, the luminous intensity I_v of the lamp is calculated from the spectral irradiance $E_e(\lambda)$ by

$$I_v=d^2K_m\int_\lambda E_e(\lambda)\mathrm{d}\lambda, \tag{3.6}$$

where K_m is the maximum spectral luminous efficacy (683 lm/W), and $V(\lambda)$ is the spectral luminous efficiency function. The luminous intensities of the spectral irradiance lamps are transferred to luminous intensity standard lamps operating at 2856 K that serve as luminous intensity primary standards. This method was used at NIST until 1991.

It should be noted that, unless the blackbody temperature is determined radiometrically, the source-based scale is dependent on the temperature scale. In 1990, the International Practical Temperature Scale (IPTS68) was revised to International Temperature Scale (ITS90) [39], and the gold-point temperature changed from 1337.58 to 1337.33 K. This change caused a shift of source-based photometric scales by 0.35%. The temperature scale is now determined most accurately using radiometry. The gold-point temperature was determined using absolutely calibrated detectors in 1989 [40], and this value was adopted in the ITS90. Therefore, in ITS90, there should not be a significant difference between the source-based photometric units and the detector-based photometric units. However, there is no guarantee that the international temperature scale will not change again in the future. A future direction is that the temperature of a high-temperature blackbody will be determined radiometrically rather than depending on a fixed-point blackbody [41,42].

3.4.3 Luminous-Intensity Transfer Standard Lamps

Transfer standard lamps are used to transfer the unit of luminous intensity from one laboratory to another (for example, from a national laboratory to an industrial laboratory). Luminous-intensity standard lamps should have reproducible and stable output for repeated and long-time use, exhibit uniform angular intensity distribution, and be equipped with a special base or some other means to allow precise alignment of the lamp. To satisfy these requirements, specially designed lamps are manufactured as standard lamps, or certain types of general production lamps are carefully selected for use as standard lamps.

Lamp Types

Luminous-intensity standard lamps are incandescent lamps or quartz halogen lamps. Many varieties of standard lamps manufactured in the past are no longer available, or are prohibitedly expensive. A shortage of high-quality, reasonably priced standard lamps is a worldwide problem facing photometry. Figure 3.6 shows some of the standard lamps widely used in the United States and Europe.

Lamp (a) is a GE Airway Beacon type,[1] 120-V gas-filled incandescent lamp, which has been widely used in the United States. This lamp is equipped with a single-coil filament having a monoplane structure and a bipost base that allows precise alignment of the lamp using an alignment device (a mirror mounted on the bipost) and a laser. The lamp comes in different wattages from 100 to 1000 W with a clear or inside-frosted bulb. Lamp (b) is an Osram Sylvania 1000-W modified FEL-type quartz halogen lamp potted on a bipost base. This lamp has a coiled-coil filament mechanically clamped at both ends with no middle support. The lamp bulb is either clear or frosted. This lamp is designed for operation at ~110 V, 8.1 A for a distribution temperature of ~3100 K and can be operated at ~85 V, 7.2 A for 2856 K. Lamp (c) is an Osram Wi41/G type, 30 V, 175 W, gas-filled incandescent lamp having a reverse-conical shape bulb, and equipped with an apertured black mask coated on the bulb so that only the filament is seen on the optical axis,

[1]Specific firms and trade names are identified in this paper to specify the experimental procedure adequately. Such identification does not imply recommendation or endorsement by the National Institute of Standards and Technology, nor does it imply that the materials or equipment identified are necessarily the best available for the purpose.

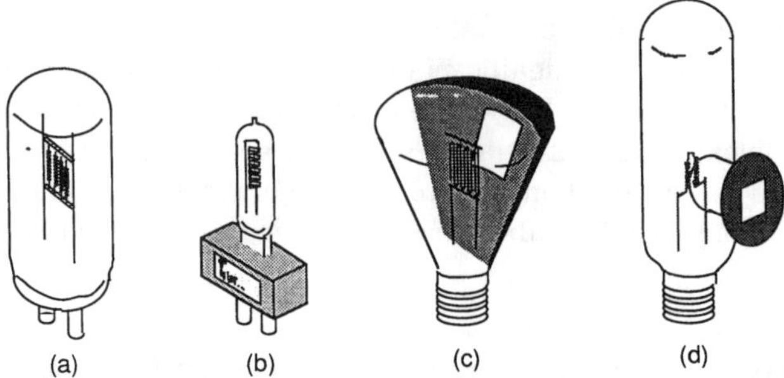

FIGURE 3.6. Various types of luminous intensity standard lamps.

shielding out all the internal reflections from the bulb and other structures. This lamp has a straight-wire filament in the monoplane structure, which provides robustness and smooth angular intensity distributions and allows for precise alignment using a telescope. This lamp is designed for operation at ~2750 K. Lamp (d) is a Polaron LIS type, 12.7 V, 320 W, gas-filled incandescent lamp having a flat window on a cylindrical port positioned away from the filament, which reduces the effect of blackening and also allows precise alignment using a laser beam. An aperture mask is attached on the window to shield light other than that from the filament. The lamp has a self-standing, low-voltage, thick filament that makes the lamp robust and reproducible against shocks. This lamp is designed for operation at 2856 K.

Conventional gas-filled standard lamps tend to have a large bulb in order to reduce blackening and provide better aging characteristics. Quartz halogen lamps are often used as spectral irradiance standards for their higher color temperature[2] (higher output in the UV). They are now used as photometric standards (operated at 2856 K) due to their low aging rate, compact size, and low-cost availability. Some quartz halogen lamps are stable within a color-temperature range of 2000–3200 K [43].

Lamp Seasoning

New incandescent lamps decrease in luminous intensity by 10% or more in the first few percent of the lamp life, depending on type of lamps. When standard lamps are purchased, the lamps must be seasoned or run at the rated current for at least 5% of the rated life of the lamps. The required seasoning time depends on the type of lamp and conditions. A general guideline is 50 h for normal gas-filled incandescent lamps for use at 2856 K. For quartz halogen lamps to be used at 2856 K, the lamps should first be seasoned at the rated current (usually ~3200 K) for about 24 h, then

[2]Throughout this chapter, *color temperature* is used to represent correlated color temperature and does not necessarily follow the CIE definition of the term [44]. See Sec. 3.7 for details.

seasoned again at 2856 K for an additional 48 h or so. After seasoning, the aging characteristics of the lamp (see the next section) should be tested to see if the seasoning has been sufficient to achieve the desired stability.

Standard lamps are usually operated on DC power in order to have much better stability of power supplies and better accuracy of electrical measurements than on AC power and also to avoid problems of flicker under AC operation. During the seasoning process, the tungsten filament of the lamp is partially recrystallized according to the electrical polarity applied. If the polarity is changed, the lamp may undergo an unstable state again. Therefore, the polarity should be kept the same when the lamps are seasoned and when they are used. The polarity should be marked on the lamp base or it should be clearly defined in a document.

Lamp Characteristics and Screening

Each individual standard lamp should be characterized and evaluated to account for manufacturing variations, as some lamps (individually or in an entire batch) are not acceptable for use as standard lamps. As an example, Table 3.3 lists the criteria for lamp screening used at NIST. New types of lamps or unknown lamps should be tested for all of these characteristics. Depending on the type of lamp and its history, some of the characteristics need not be tested for each individual lamp.

Before seasoning, the lamps are visually inspected to see if the filament is mounted straight and leads are welded or tightly contacted, the base is rigid, and the bulb has no visible spots. After seasoning, the aging characteristic of the lamp (drift of luminous intensity as a function of operating time) is tested, and the aging rate (change/h) is calculated. The aging of a lamp occurs due to blackening of the bulb, thinning of the filament, and other factors. The lamps are tested under continuous operation for 24–48 h or longer, depending on the measurement accuracy of the system. Luminous intensity and other lamp parameters are continuously measured during the stability test. A computer-feedback control of the lamp current can provide better stability than that of a power supply itself for a long period of time. The photometer temperature should be monitored during the test and the photometer signal should be corrected for temperature change.

An example of the aging characteristic of a selected quartz halogen lamp is shown in Fig. 3.7. It should be noted that, in this case, the lamp current is kept

TABLE 3.3. *Criteria for lamp screening used at NIST.*

Criteria	Typical requirement for standard lamps
Visual appearance	Straight filament, clean bulb, etc.
Aging rate	<0.02%/h in luminous intensity
Angular uniformity	<0.3% in ±1° (compared to center)
Reproducibility	<0.3% (after realignment)
Storage stability	<0.3% in one month

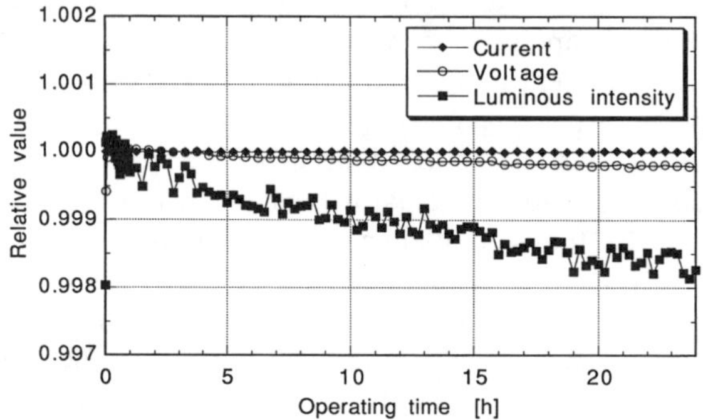

FIGURE 3.7. Aging characteristics of an FEL-type lamp operated at 2856 K.

constant (to within ±0.002%). If the lamp voltage is kept constant, the luminous intensity would change in the opposite direction. The lamp voltage and luminous intensity do not always change as shown in the figure. An FEL-type quartz halogen lamp operated at 3100 K increases its lamp voltage and light output as the length of operating time continues (Fig. 3.8). The physical cause of aging is a complex subject involving the evaporation of the filament, the filament interaction with the fill gas, and optical properties of the glass envelope. These issues are outside the scope of this chapter.

The aging rate of gas-filled lamps is normally larger than for quartz halogen lamps. Figure 3.9 shows an example of an Airway Beacon type lamp. The beginning of the aging curve is usually not linear since it includes the lamp stabilization process. Some lamps exhibit a sharp decrease, e.g., 0.5% in the first 30 min. Such lamps require a long stabilization time and are not suited for standard use, whereas lamps with a linear aging curve tend to be reliable. Because of the aging characteristics as mentioned above, standard lamps need to be recalibrated periodically

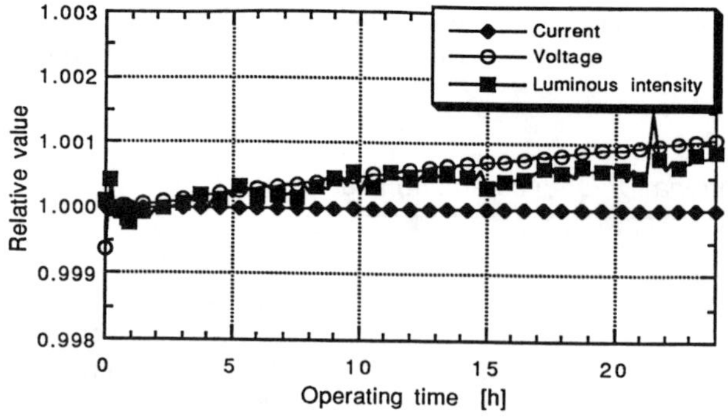

FIGURE 3.8. Aging characteristics of an FEL-type lamp operated at 3100 K.

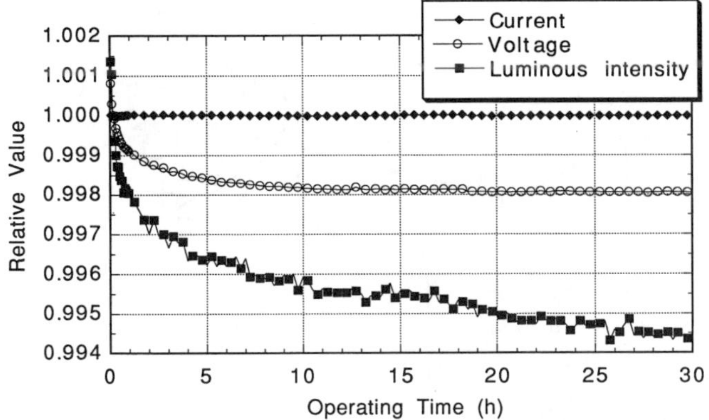

FIGURE 3.9. Typical aging characteristics of an Airway Beacon type lamp.

(typically every 30 to 50 h of total operating time), depending on the type of lamps and uncertainty required.

Another important characteristic is the angular intensity distribution. Sharp changes of the angular intensity distributions make the lamp alignment very critical. This is caused by shadowing of filament coils in a clear bulb. Lamps with frosted bulbs, or those with straight-wire filaments have less problems and usually need not be tested. The angular intensity distributions can be tested by horizontal rotation and vertical swing of the lamp, or by measuring the illuminance distribution on a plane along the optical axis, at considerable distance from the lamp. In this case, as known as cosine-cubed law [1], the illuminance distribution $E_v(x,y)$ as shown in Fig. 3.10 can be converted to the angular luminous intensity distribution $I_v(\theta_x, \phi_y)$ by

$$I_v(\theta_x, \theta_y) = E_v(x,y)d^2/(\cos\,\theta_x \cos\,\theta_y)^3,$$

$$\theta_x = \arctan(x/d), \tag{3.7}$$

$$\theta_y = \arctan(y/d).$$

Figure 3.11 shows the angular intensity distribution of a typical selected 1000-W FEL lamp with a clear bulb. Lamps meeting the stability and uniformity criteria are calibrated for luminous intensity. Measurements are repeated three times with remounting and reburning the lamp each time to check if the luminous intensity reproduces within acceptable range. Lamps exhibiting poor reproducibility in the remeasurements with a normal course of handling are discarded.

Another problematic characteristic of incandescent lamps sometimes observed is storage stability. While most lamps tend to exhibit good reproducibility when operated repeatedly in a short period of time, some lamps exhibit a change of characteristics after a long period of storage. In such a case, the lamps tend to show poor reproducibility in luminous intensity and lamp voltage in the first few burnings after storage and may not reproduce the previous values even after sufficient stabi-

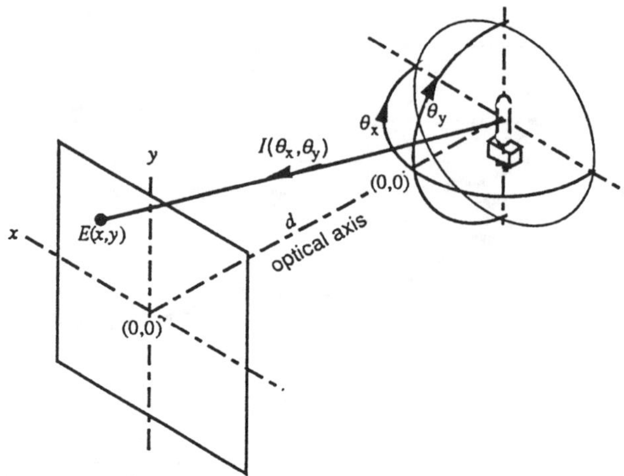

FIGURE 3.10. Conversion from the illuminance distribution $E_v(x,y)$ to luminous-intensity distribution $I(\theta_x, \phi_y)$.

lization. Such lamps should be disposed of from the calibration source inventory. Storage stability of lamps should be tested at least one month after their last use.

Operation and Handling of Standard Lamps

Standard lamps should be handled carefully to avoid mechanical shocks to the filament. Before operation, the bulb of the lamp is cleaned with a soft, lint-free cloth to remove dust accumulated from the packing material. The lamp bulb should not be touched with bare hands. Gloves should be used to avoid fingerprints. Special attention should be paid to quartz halogen lamps since water droplets or oily

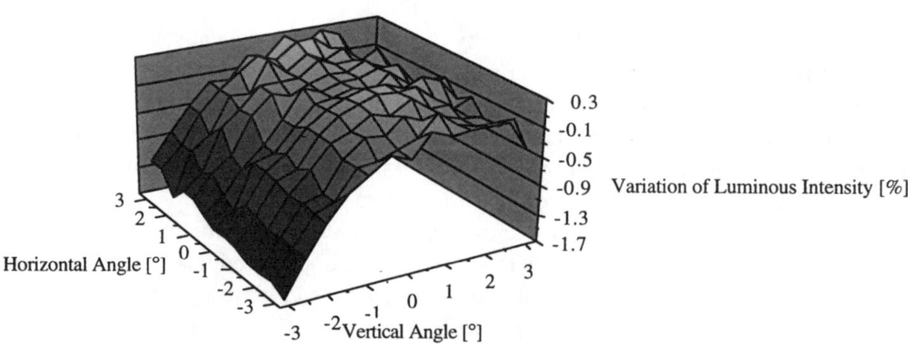

FIGURE 3.11. Spatial nonuniformity of a typical FEL-type lamp.

deposits on the bulb can cause permanent white spots on the quartz envelope after burning the lamp. Ethyl alcohol is used only when oily deposits such as fingerprints are to be removed. Lamps should be kept in a container when not in use.

The lamps are operated on DC power with a specified electrical polarity. The lamps are calibrated at a specified current. The lamp voltage is not used since it is difficult to reproduce due to the different structures and conditions of the sockets among users. However, it may be useful to monitor the lamp voltage (measured on the same socket) in order to detect changes in the lamps.

3.4.4 Illuminance Transfer Standard Photometers

The quality of commercial $V(\lambda)$-corrected detectors (photometers) has been improved significantly with the availability of high-quality silicon photodiodes. As a result, some types of commercially available photometers can be used as photometric transfer standards instead of traditional luminous-intensity standard lamps. Standard lamps are sensitive to mechanical shocks, change with burning time, and drift during stabilization period. Well-maintained photometers are less subject to such problems and can provide a dynamic range of several orders of magnitude. The short-term stability of photometers is usually superior to lamps, and although the long-term stability has not been tested for many different types of photometers, some types of photometers exhibit satisfactory stability (~0.1% per year). It should be noted, however, that some other types of photometers have shown changes by more than 1% in a year, making their use difficult for standards work. In general, for luminous intensity and illuminance measurements, use of standard photometers are recommended, but the photometers should be calibrated frequently (at least once a year) until the long-term stability data are accumulated. It should also be noted that photometers do not transfer the luminous intensity unit itself. Photometers only provide the illuminance unit. One can use photometers for luminous intensity standards only if distance can be measured accurately.

Requirements for Standard Photometers

A standard photometer consists of a detector (generally, a silicon photodiode), a $V(\lambda)$-correction filter, an aperture, and in some cases, a diffuser. The $V(\lambda)$-correction filter matches the total spectral responsivity of the photometer (photodiode +filter+diffuser) to the $V(\lambda)$ function. The photometer head does not necessarily need cosine correction because the photometer is normally used with an incandescent standard lamp placed on the optical axis of the photometer at a sufficient distance to provide normal incident light with small divergence angle. An important requirement of a standard photometer is that its reference plane is accurately and clearly defined. If the reference plane of the photometer is erroneously defined, the inverse square law does not accurately predict the signal as a function of distance. In order to establish an accurate reference plane, a standard photometer should have an aperture in its front as shown in Fig. 3.12(a), referred to as nondiffuser-type

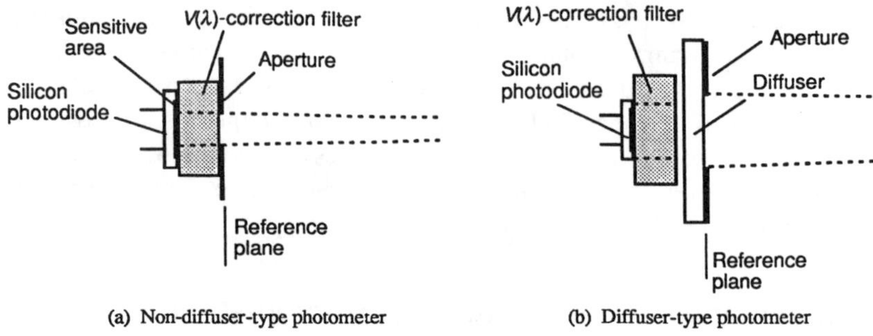

FIGURE 3.12. Construction of standard photometers.

photometer. A diffuser can be added between the aperture and the $V(\lambda)$-correction filter as shown in Fig. 3.12(b), referred to as a diffuser-type photometer. The reference plane of the photometer head is the plane that includes the sharp edges of the aperture.

Photometers equipped with neither an aperture nor a diffuser are not recommended for use as standard photometers. When such a photometer must be calibrated, its reference plane should be determined by a photometric method (see "Operation and Handling of Standard Photometers" in Sec. 3.4.4). If the photodiode of the photometer is overfilled with radiation, the photodiode surface will be close to but not exactly a correct reference plane since the optical length is shortened by the $V(\lambda)$-correction filter.

Nondiffuser-Type Photometers

Standard photometers with a limiting aperture as shown in Fig. 3.12(a) are often used by national laboratories to realize and maintain the illuminance unit. This type of photometer allows spectral responsivity measurement using a collimated monochromator output beam (as described in Sec. 3.4.1). This type of a photometer exhibits a narrow acceptance angle as shown in Fig. 3.13, which is advantageous in blocking stray light from the ambient but disadvantageous for use with a large-size lamp at shorter distances due to departure from cosine response. In this type of photometer, both the $V(\lambda)$-correction filter and the photodiode must be much larger than the aperture so that the photodiode is underfilled.

Diffuser-Type Photometers

Standard photometers equipped with a diffuser as shown in Fig. 3.12(b) are also commonly used. Illuminance meters equipped with a diffuser for cosine correction

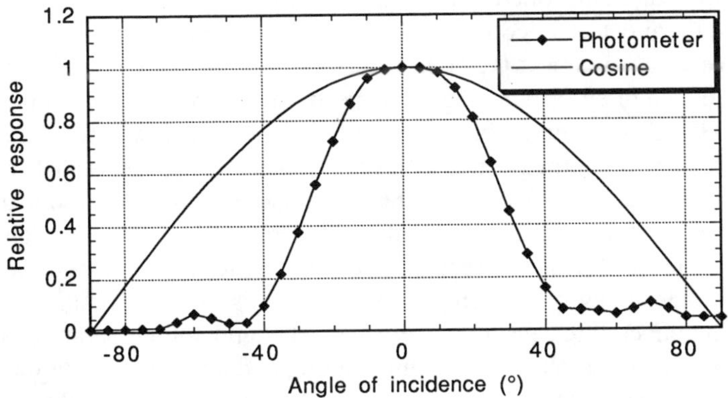

FIGURE 3.13. Angular responsivity of a nondiffuser-type photometer.

can also be used as standard photometers if they have a flat diffuser that provides the reference plane precisely. Illuminance meters having a dome-shaped diffuser are not adequate for standard photometers. The material of the diffuser should be chemically stable and not subject to UV degradation. Opal glass is generally preferred in terms of long-term stability. Diffuser-type photometers are more subject to stray light due to a large acceptance angle, but less subject to errors for a large-size lamp at shorter distances. The spectral responsivity should be measured with the detector surface overfilled by uniform irradiation, since the spectral responsivity over the diffuser area tends to be nonuniform. This type of photometer can employ a partial filter[3] placed at some distance from the diffuser. A diffuser is generally required for partial filters due to their spatial non-uniformity of spectral transmittance.

Temperature Consideration

The responsivity of a photometer is a function of temperature. The transmittances of colored glass filters tend to change significantly with temperature. Typical $V(\lambda)$-correction filters can have a temperature coefficient as high as 0.1%/°C. It is recommended that standard photometers be equipped with either a temperature sensor to make corrections or a temperature-controlling device to keep the photometer temperature constant [45]. In this respect, standard photometers are characterized as

- Temperature-controlled type.
- Temperature-monitored type.
- No-sensor type.

[3] A partial filter is a filter made of multiple layers of different filters, on top of which a number of small size filter chips of different colors are placed and partially cover the active area of the detector, also called a mosaic filter.

The temperature-controlled type usually incorporates a temperature sensor and a heater or a thermoelectric cooler to maintain the photometer temperature within a small range (e.g., within ±0.2 °C). This type is highly recommended, but most expensive of the three. When a heater is used, the reference temperature is usually set to 30–35 °C, and the ambient temperature must be lower than the reference temperature. The photometer should have an indicator that shows that the temperature controller is properly working. The temperature-monitored type usually incorporates a temperature sensor connected to the detector-filter package. The temperature reading allows for the correction of errors due to the temperature differences. The absolute accuracy of the temperature sensors is not important. The no-sensor type has no temperature sensor or controller. The ambient temperature is measured and assumed to be the photometer temperature, and an approximate correction for temperature is made. The correction may not be as accurate as the temperature-monitored type. It takes a few hours for a photometer to reach equilibrium with the ambient temperature.

Characterization of Standard Photometers

Relative Spectral Responsivity

No photometer can be matched perfectly to the $V(\lambda)$ function, and an error occurs when a photometer measures a light source having a spectral distribution different from the calibration source (normally the CIE Illuminant A). The degree of the spectral mismatch with the $V(\lambda)$ function is evaluated by the term f_1' given in the CIE Publication No. 69 [46] (see "Relative Spectral Responsivity" in Sec. 5.2.4). It is recommended that a standard photometer has a f_1' value of less than 3%. The f_1' is an evaluation index and cannot be used for correction purposes. In order to make spectral mismatch corrections, standard photometers must be characterized for the relative spectral responsivity to obtain the spectral mismatch correction factor ccf*. The procedure for the spectral mismatch correction is described in "Spectral Mismatch of Photometers" in Sec. 5.1.3.

Temperature Dependence

The responsivities of photometers change depending on the temperature of their optical components. Measurement errors may occur if a photometer is used at an ambient temperature different from when it was calibrated. Unless it is a temperature-controlled photometer, or unless the ambient temperature is precisely controlled, the temperature dependence of the photometer should be evaluated and corrections should be made. The procedure for making corrections for temperature variation of photometers is given in "Photometer Temperature Variation" in Sec. 5.1.3.

Linearity

High-quality silicon photodiodes recently available have linear response over several orders of magnitude, and the linearity of a standard photometer is usually not a problem at illuminance levels less than 10^3 lx. However, standard photometers should be evaluated for their linearity over the entire illuminance range in which the photometers are to be used. Refer to ''Photometer Temperature Variation'' in Sec. 5.1.3 for the procedures for linearity measurement.

Long-Term Stability

The responsivity of high-quality standard photometers are very stable over a relatively short period of time. It should be noted, however, that the responsivity of photometers can change over a long period of time. While specific types of silicon photodiodes are known to be very stable over time, the transmittance of $V(\lambda)$-correction filters tend to change over time. In some cases, the surface of the filter forms some kind of cloudy deposit.

The long-term drift of photometers can only be measured by periodically calibrating the photometer against other reliable standards (e.g., reproducible standard lamps with their burning time strictly limited, or an absolute radiometer). An example of the long-term stability of actual photometers is shown in Fig. 3.14. In this case, photometers A, B, and C exhibit significant drift. The responsivity of these photometers can often be restored by cleaning the filter surfaces.

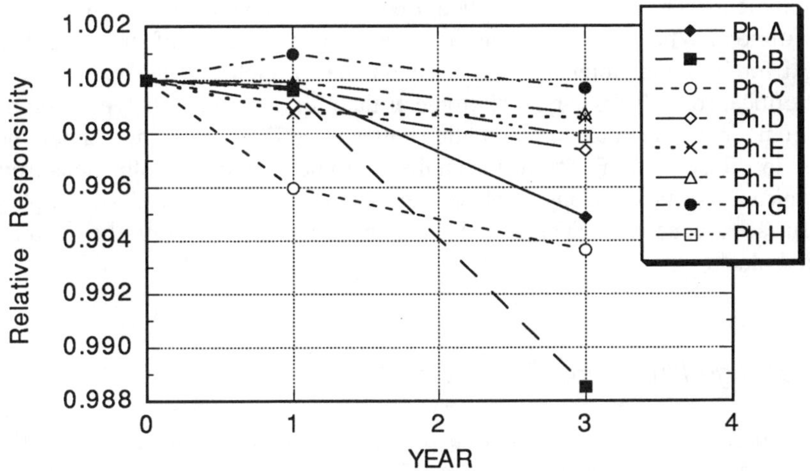

FIGURE 3.14. Long-term drift of various photometers.

Operation and Handling of Standard Photometers

Calibration

The photometric responsivity of a standard photometer in V/lx or A/lx is normally calibrated against reference standard photometers using the CIE Illuminant A. The ambient temperature (typically 25 °C), the photometric distance, the illuminance level, and the reference plane of the photometer are recorded.

Photometer Temperature

When a no-sensor-type photometer is used, the photometer should be set up in its measurement location with its power turned on for at least 1 h before measurement. It is recommended to use the photometer at a temperature within ±1 °C from the temperature at which the photometer had been calibrated. When the ambient temperature is different by more than 2 °C, a correction should be applied using the temperature coefficient of the photometer. One should not hold a photometer with bare hands before or during measurements since that would cause the photometer to heat up. The ambient temperature should always be stated in the test results.

Reduction of Stray Light

Care should be taken to minimize stray light as in any other photometric measurements. The measurement should be made in a darkroom, or in a light-tight compartment. When standard photometers are used, reduction of stray light is more critical for luminous-intensity measurement than for photometer calibration, and more critical for a diffuser-type photometer than for a nondiffuser-type photometer. Any components such as an automatic shutter or additional aperture should not be placed too close (within ~20 cm) from the photometer since the reflections from the photometer's surfaces strongly reflect back from a short distance. All the photometer surfaces, except for aperture edges or diffuser surface, should be anodized or painted black.

Use at High Illuminance Levels

Even though some photometers have a linear response up to levels of 10^5 lx or higher, one should be careful about the effect of the heat from incandescent sources. Even if the photometer is a temperature-controlled type or a temperature-monitored type, the heating up of the $V(\lambda)$-correction filter by incoming radiation will not be eliminated or corrected if the radiation is too high. Nondiffuser-type photometers, with the $V(\lambda)$-correction filter exposed to radiation, tend to be more sensitive than the diffuser-type photometers. The effect of heat can be evaluated by measuring the change of the photometer signal after the photometer is exposed to a high illuminance field from a stabilized source. If there is a heat effect, the photometer signal will gradually change and stabilize in about 30 min. When used at illuminance levels higher than $\sim 10^3$ lx, the photometer should be exposed to the radiation only long enough to take readings.

Maintenance

Standard photometers with apertures tend to catch dust particles on the filter surface, which can cause a non-negligible error, especially when the aperture size is small. Before using the photometer, the filter surface should be inspected for cleanliness. If dust is observed, it should be removed with an air spray. Before using a diffuser-type photometer, the surface of the diffuser can be cleaned using a lens tissue in order to remove possible dust particles on the surface. Commercial instruments usually have instruction for cleaning and maintenance. The user should adhere to the instructions to avoid damage to optical surfaces by contact or inappropriate cleansing agents. Standard photometers are usually stored in a dessicator when not used. It is preferable for temperature-controlled photometers to be powered during storage to avoid repeated thermal cycling.

It is recommended that a group of more than three standard photometers be used to maintain the photometric units at each laboratory. The photometers can be periodically cross-checked with one another to detect unexpected changes of responsivity. A great advantage of the photometers is that, unlike standard lamps, their operating hours are not limited. However, standard photometers should be calibrated at least once a year until the long-term stability data are established that indicate an appropriate calibration cycle.

Determination of the Reference Plane

The reference plane of a photometer can be determined by the photometric method as described below. The method requires a standard photometer with its reference plane accurately defined and a photometric bench with a distance measurement capability.

First, the signal of the standard photometer V_{s1} and the signal of the test photometer V_{t1} are taken when both photometers are placed at a distance y from the lamp as shown in Fig. 3.15, y need not to be known. Both photometers are aligned so that their front surface is on the same plane. x_1 in Fig. 3.15 is the position of the reference plane of the standard photometer measured from its front surface, and x is the position of the reference plane of the standard photometer measured from its front surface.

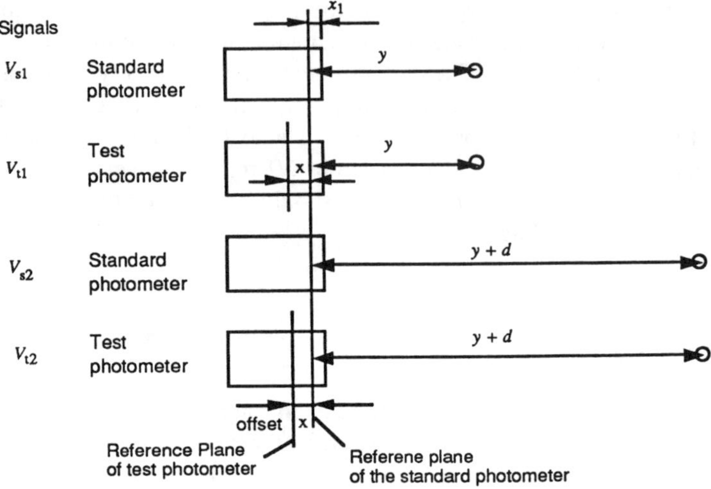

FIGURE 3.15. Determination of the photometer reference plane.

Then, the signal of the standard photometer V_{s2} and the signal of the test photometer V_{t2} are taken when both photometers are moved to a distance $y+d$ from the lamp. The distance shift d must be accurately measured. d should be larger than y to keep the measurement uncertainty reasonably small. Then the following equations are formed according to the inverse square law:

$$\frac{V_{s1}}{V_{s2}}=\left(\frac{y+d}{y}\right)^2, \tag{3.8}$$

$$\frac{V_{t1}}{V_{t2}}=\left(\frac{y+d+x}{y+x}\right)^2. \tag{3.9}$$

By solving Eqs. (3.8) and (3.9), the offset x of the reference plane of the test photometer is given by

$$x=d\left(\frac{1}{\sqrt{V_{t1}/V_{t2}}-1}-\frac{1}{\sqrt{V_{s1}/V_{s2}}-1}\right). \tag{3.10}$$

With the offset of the standard photometer given as x_1, the reference plane of the test photometer from its front surface is given by $x+x_1$.

3.5 LUMINOUS-FLUX STANDARDS

3.5.1 Goniophotometric Method

The total luminous flux of a light source is obtained by angular integration of luminous intensity over the 4π solid angle or by spatial integration of illuminance over a closed surface around the source as given by

$$\Phi_v = \int_{\Omega} I_v d\Omega \tag{3.11}$$

or

$$\Phi_v = \int_{A} E_v dA. \tag{3.12}$$

Goniophotometers are traditionally used to establish the luminous-flux unit, the lumen, as shown in Fig. 3.16. The goniophotometer measures either the luminous-intensity distribution $I(\theta,\phi)$ (in cd) or the illuminance distribution $E_v(\theta,\phi)$ (in lx) of a source, and the luminous flux Φ_v (in lm) of the source is given by

$$\Phi_v = \int_{\phi=0}^{2\pi} \int_{\theta=0}^{\pi} I_v(\theta,\phi) \sin\theta \, d\theta \, d\phi \tag{3.13}$$

or

$$\Phi_v = r^2 \int_{\phi=0}^{2\pi} \int_{\theta=0}^{\pi} E_v(\theta,\phi) \sin\theta \, d\theta \, d\phi, \tag{3.14}$$

where r (in m) is the radius of the spherical surface.

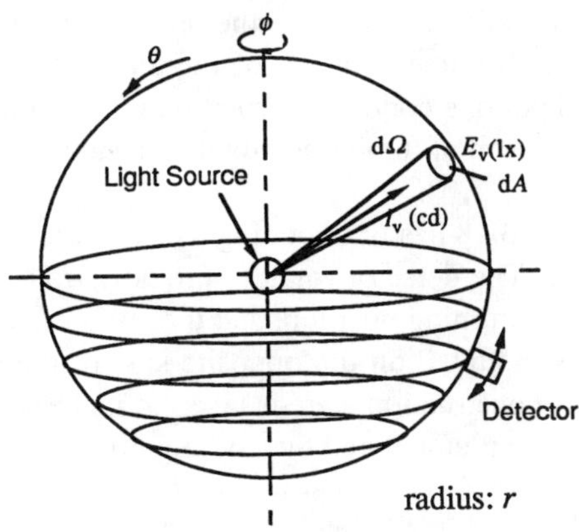

FIGURE 3.16. Goniophotometric method for total luminous-flux measurement.

The detector of the goniophotometer can be calibrated either on the goniophotometer or on the photometric bench. On the goniophotometer, the detector is calibrated against a luminous intensity standard lamp that is positioned precisely to the center of the detector rotation, and the orientation of the lamp relative to the detector (placed to a horizontal direction) is accurately aligned. The detector is calibrated for response to known luminous intensity. An advantage of this method is that the radius of the detector rotation need not be measured. When the detector is calibrated on the photometric bench, the detector is detached from the goniophotometer, and calibrated for illuminance reponsivity on the bench. As shown in Eq. (3.14), the radius of the detector rotation needs to be accurately measured, but in turn, precise alignment of the lamp is not necessary.

In order to realize the luminous-flux unit, a low uncertainty in the total flux measurement must be achieved and requires a special goniophotometer designed for this purpose. Goniophotometers are widely used for measurement of luminous-intensity distribution of luminaires, but many of these instruments are not suited for the lumen realization. For example, the dead angle[4] of the detector tends to be too large in these instruments. Mirror-type goniophotometers should not be used due to the sensitivity to polarization. The fixed-detector-type goniophotometer with the lamp rotating should not be used because the lamp output can alter with movement of the lamp burning position. Figures 3.17 and 3.18 show two common types of goniophotometers used at national laboratories for realization of the luminous-flux unit. For precise measurement of luminous flux, it is essential to keep the burning position of the lamp constant and to make the dead angle of the instrument negligibly small.

The three-axis goniophotometer shown in Fig. 3.17 is an ideal type recommended by CIE [47]. The burning position of the lamp can be preset to any angle, and once the lamp position is set, the lamp does not move or even rotate during measurement. The axes can be controlled in several different ways, but usually, the intermediate frame (ϕ frame) rotates faster with slower movement of the innermost frame (θ frame). The detector moves to a horizontal direction in which the variation of the luminous intensity is usually small, thus the measurement can be made faster reducing the problem of the time constant of the amplifier. The detector signal is continuously integrated per one continuous rotation of ϕ frame, which allows faster and more accurate integration of luminous flux than a stop-and-go type scan used in other arrangements.

The type shown in Fig. 3.18 has two rotating axes, is less expensive to build, and is more commonly used. The detector rotates vertically (to scan the θ angle) while the lamp is held at a fixed burning position, and the lamp is slowly rotated to set the ϕ angle. A light trap is mounted on the opposite side of the detector to minimize stray light errors. In this construction, care is taken to make sure that the lamp is not affected by vibration to the filament and the cooling effect of air. This two-axis type

[4]*Dead angle* is the angular region (angle from the light source) where the detector cannot be positioned, or measurement cannot be made accurately, due to shadows of a mechanism such as a lamp holder.

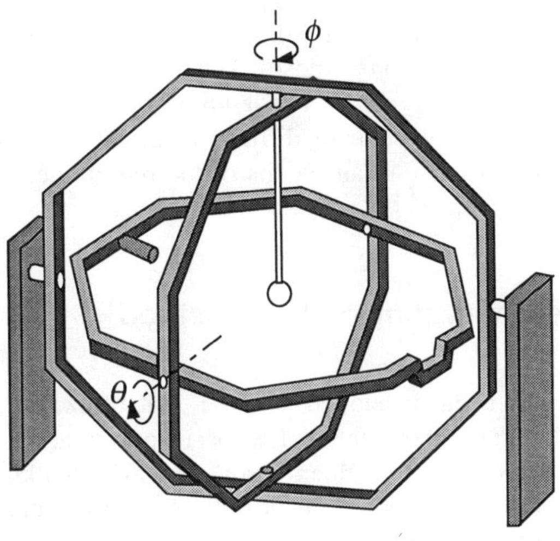

FIGURE 3.17. Three-axis goniophotometer.

is not used for fluorescent lamps since the lamp temperature will not be kept stable due to the rotation of the lamp. Also, continuous rotation of the detector is more difficult because the detector moves to the direction in which the luminous intensity tends to vary sharply. The two-axis type is usually operated on the stop-and-go mode, and requires a longer time for measurements.

With any type of goniophotometer, care must be taken to ensure high accuracy. To minimize stray-light errors, the instrument is normally installed in a darkroom, the detector is equipped with aperture screens in its front to limit the field of view

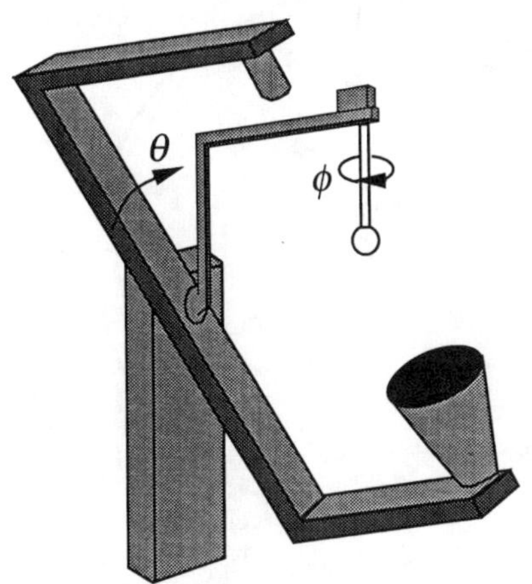

FIGURE 3.18. Two-axis goniophotometer.

to the minimum required, and a light trap or black velvet plate is placed on the opposite side of the detector. Even with this equipment, stray-light errors are evaluated experimentally for correction purposes. The dead angle of the goniophotometer is evaluated and corrections are made. The lamp socket and holder are made to have as high reflectance (either specular or diffuse) as possible while all other mechanisms are painted flat black. Further details of the requirements for goniophotometers are described in Ref. [47].

3.5.2 Absolute Integrating-Sphere Method

A new method has been developed at NIST that utilizes a special integrating sphere instead of a goniophotometer. The basic principle of this method is to calibrate the total flux of a lamp inside the sphere against the known amount of the flux introduced from a light source outside the sphere through an opening. This method was first proposed through a theoretical analysis using a computer simulation technique [48], then experimentally verified [49], and was actually applied to the realization of the luminous flux unit in 1995 [50].

Figure 3.19 shows a setup for the absolute integrating-sphere method. The flux from the external source is introduced through a calibrated aperture placed in front of the opening. The internal source, a lamp to be calibrated, is mounted in the center of the sphere. Two baffles are used to shield the detector and the opening from direct illumination by the internal source. The detector is exposed to the "hot spot" (the first reflection of the introduced flux from the external source) in order to equalize the sphere responsivity for the internal source and that for the external source. Baffle 2 is aligned so that neither surface is viewed by the detector.

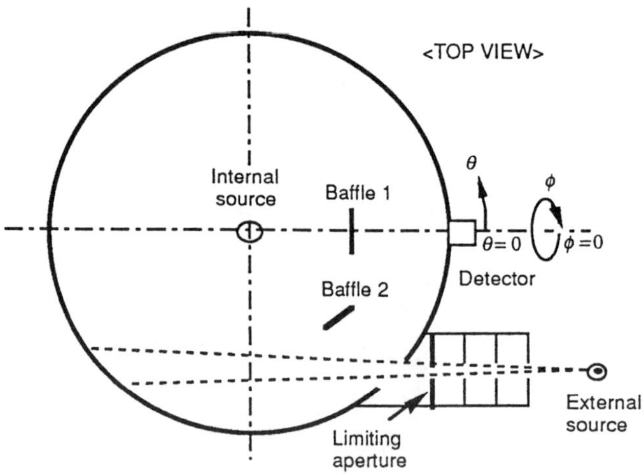

FIGURE 3.19. Setup for the absolute integrating-sphere method.

In this method, the external source and the internal source are operated alternately, and the total luminous flux Φ_i of the internal source is obtained by comparison to the luminous flux introduced from the external source as given by

$$\Phi_i = cE_a A y_i / y_e, \tag{3.15}$$

where E_a is the average illuminance (in lx) from the external source over the limiting aperture of known area, A, y_i is the detector signal for the internal source, and y_e is the detector signal for the external source. c is a correction factor for various nonideal behaviors of the integrating sphere. The determination of this correction factor is an important part of this method.

The response of the integrating sphere is not uniform over the sphere wall due to baffles and other structures inside the sphere, and also due to nonuniform reflectance of the sphere wall due to contamination. The light from the external source is incident at 45° while the light from the internal source is normal. When the incident angle is different, the diffuse reflectance of the sphere coating changes [51], which affects the sphere responsivity. When the spectral power distribution of the internal source is different from that of the external source, a spectral mismatch error occurs. All these corrections are made to determine the correction factor c. A self-absorption correction is not necessary if the internal source to be calibrated stays in the sphere when the external source is measured.

The correction for the spatial nonuniformity is essential to reduce the uncertainty of this method to an acceptable level. The spatial responsivity distribution function (SRDF), $K(\theta,\phi)$ of the sphere, is defined as the sphere response for the same amount of flux incident on a point (θ,ϕ) of the sphere wall or on a baffle surface, relative to the value at the origin, $K(0,0)$. $K(\theta,\phi)$ can be obtained by measuring the detector signals while rotating a narrow beam inside the sphere. The rotating lamp must be insensitive to burning position. $K(\theta,\phi)$ is further normalized for the sphere response to an ideal point source. The normalized SRDF, $K^*(\theta,\phi)$, is defined as

$$K^*(\theta,\phi) = 4\pi K(\theta,\phi) \Big/ \int_{\phi=0}^{2\pi} \int_{\theta=0}^{\pi} K(\theta,\phi) \sin\theta \, d\theta \, d\phi. \tag{3.16}$$

Figure 3.20 shows an example of the SRDF, $K^*(\theta,\phi)$, of the NIST 2-m integrating sphere having coating reflectance of ~96%. Using $K^*(\theta,\phi)$, the spatial correction factor scf_e for the external source with respect to an isotropic point source is given by

$$\mathrm{scf}_e = 1/K^*(\theta_e, \phi_e), \tag{3.17}$$

where (θ_e, ϕ_e) is the point on which the center of the illuminated area by the external source is located. The spatial correction factor scf_i for the internal source with respect to a point source is given by

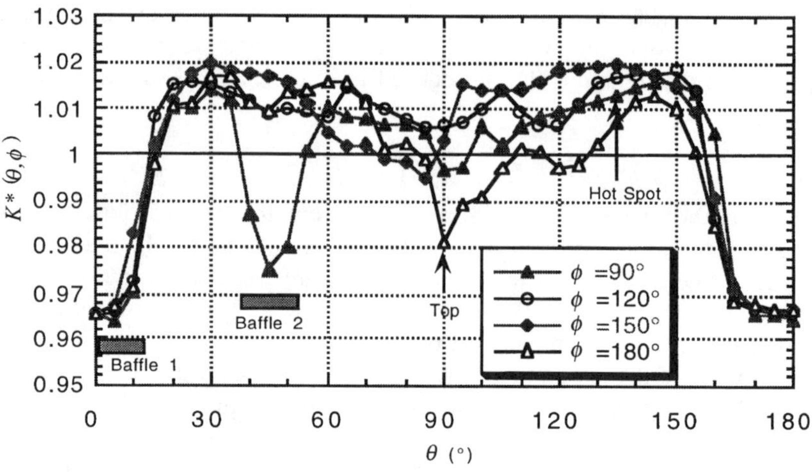

FIGURE 3.20. Measured SRDF of a NIST 2-m integrating sphere as shown in Fig. 3.19.

$$\mathrm{scf}_i = 1 \Bigg/ \int_{\phi=0}^{2\pi} \int_{\theta=0}^{\pi} I^*(\theta,\phi) K^*(\theta,\phi) \sin\theta \, d\theta \, d\phi, \tag{3.18}$$

where $I^*(\theta,\phi)$ is the normalized luminous intensity distribution of the internal source given by

$$I^*(\theta,\phi) = I_{\mathrm{rel}}(\theta,\phi) \Bigg/ \int_{\phi=0}^{2\pi} \int_{\theta=0}^{\pi} I_{\mathrm{rel}}(\theta,\phi) \sin\theta \, d\theta \, d\phi, \tag{3.19}$$

where $I_{\mathrm{rel}}(\theta,\phi)$ is the relative luminous intensity distribution of the internal source. $I^*(\theta,\phi)$ is normalized so that its total luminous flux is equal to 1 lm.

A goniophotometer is not necessarily essential in this method. Most of the luminous-flux standard lamps have fairly uniform angular intensity distributions and scf_i can be assumed unity (when the sphere reflectance is higher than 95%). Even when Eq. (3.18) is applied, only relative luminous-intensity distribution is necessary, and its accuracy is not critical. For example, the data for a group of lamps of the same type can be represented by one lamp. Once the distribution data are taken, they are used for the lifetime of the lamps.

In addition to the spatial nonuniformity correction described above, a correction is made for the sphere responsivity variation for different angles of incidence for the internal source (0°) and for the external source (45°). If the color temperatures of the internal source and external source are different, a spectral mismatch correction is applied using the method given in ''Spectral Mismatch Correction'' in Sec. 5.3.4.

In this case, the relative spectral throughput of the sphere is measured and combined with the spectral responsivity of the detector.

3.5.3 Luminous-Flux Transfer Standards

Requirements for Standard Lamps

The unit of luminous flux can be transferred only by lamps. Specially designed gas-filled, incandescent lamps of various power levels are used as luminous-flux standard lamps. Most of these lamps have either a medium screw base (E27) or a mogul screw base (E40). Unlike luminous-intensity standards, the precise alignment is not necessary for luminous-flux lamps. However, most of the general production lamps are not suitable for standard lamps due to their fragile filament and insufficient, unstable filament supports. Selected luminous-flux standard lamps reproduce to better than $\pm 0.1\%$.

Figure 3.21 shows three examples of luminous-flux standard lamps. Lamp type (a) has a coiled filament mounted in a circular and zigzag shape, which makes the angular intensity distribution fairly uniform. The filament is supported by many isolated supports that are welded to the filament so that they will not cause unstable contacts to the filament. Lamp (b) has a circular arrangement of filament. Lamp type (c) has a similar filament arrangement with an opal bulb which probably provides the most uniform angular intensity distributions. A low-voltage lamp design makes the filament robust and rigid and reproducible against mechanical shocks. Lamps with a frosted (or opal) bulb are preferred for goniophotometric measurements for their smooth angular intensity distribution curves, but they do not allow visible inspection of the filament. Figure 3.22 shows the angular intensity distributions of a type-(c) luminous-flux lamp.

Transfer to different types of lamps in an integrating sphere is more difficult than

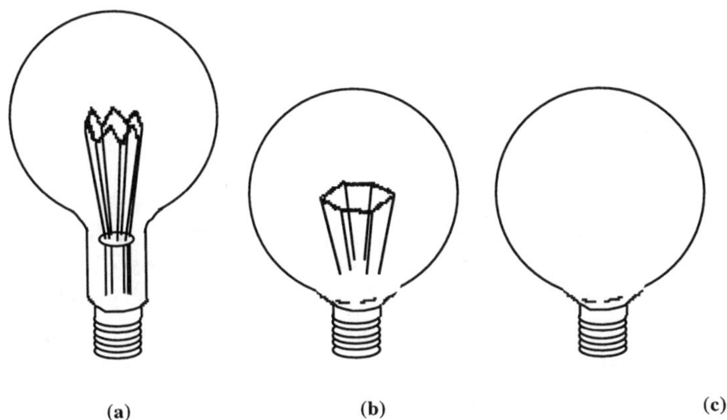

FIGURE 3.21. Luminous-flux standard lamps.

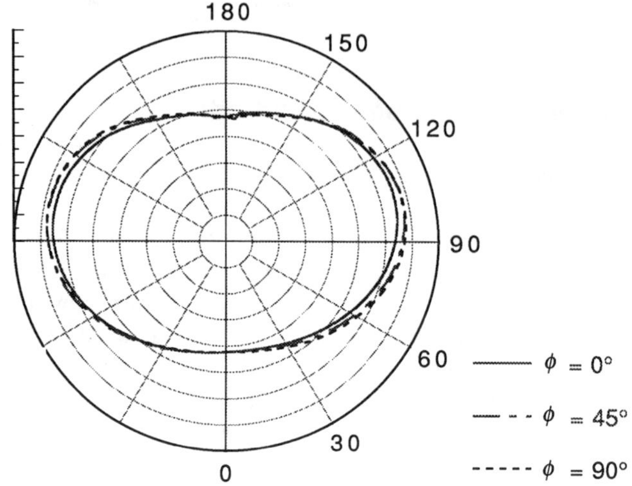

FIGURE 3.22. Angular luminous-intensity distribution of a flux standard lamp.

luminous intensity. Several corrections or uncertainty considerations are required for the integrating sphere such as self-absorption, spectral mismatch, angular intensity distributions, length of the lamp, etc. For these reasons, various types of standard lamps including discharge lamps are required by industrial laboratories. Like-to-like comparison is safer and preferred. Although the types of standard lamps provided by national laboratories are limited, linear fluorescent lamps are often provided as standards. Selected fluorescent lamps from general production reproduce to $\pm 1\%$. Compact fluorescent lamps are difficult to use as transfer standards since they tend to lose reproducibility after transportation. High-intensity discharge (HID) lamps are normally not used as transfer standards except for some special cases due to their insufficient reproducibility. Since corrections for integrating spheres are often difficult in industrial laboratories, insufficient accuracy of the measurement for these discharge lamps is a problem.

Seasoning and Screening

Incandescent standard lamps for luminous flux are seasoned in a similar manner as luminous intensity standard lamps (see ''Lamp Seasoning'' in Sec. 3.4.3). Luminous-flux standard lamps are normally operated in the base-up position, and seasoning should be conducted in the same burning position. Standard lamps are screened for aging rate, reproducibility, and storage stability, in a similar manner as luminous-intensity standard lamps (see ''Lamp Characteristics and Screening'' in Sec. 3.4.3). Figure 3.23 shows an example of the aging characteristic of a luminous-flux standard lamp. Aging characteristics vary largely depending on the type of lamp and operating color temperature. Traditionally, luminous-flux lamps have

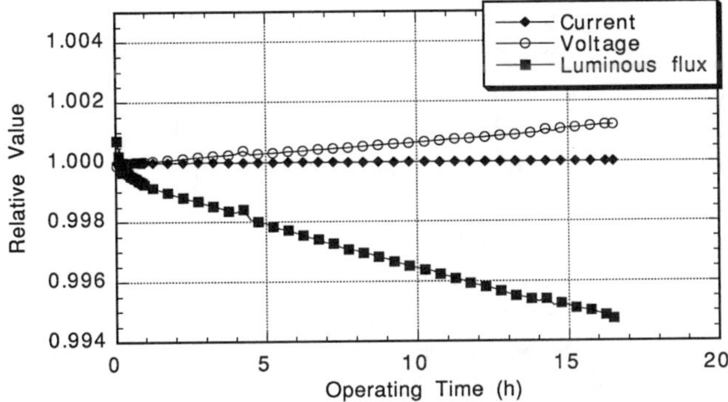

FIGURE 3.23. An example of the aging characteristic of a luminous-flux standard lamp.

been designed and used at color temperatures of 2700–2800 K in order to reduce the aging rate. Linear fluorescent lamps are normally seasoned for 750 h and screened for reproducibility.

Operation and Handling of Standard Lamps

Basically the same precautions for luminous intensity standard lamps (described in ''Operation and Handling of Standard Lamps'' in Sec. 3.4.3) apply to the luminous-flux standard lamps. The lamps should be handled carefully to avoid mechanical shocks to the filament. The bulb of the lamps should not be touched with bare hands. Before use, the bulb of the lamps is cleaned with a soft, lint-free cloth to remove dust from packing material.

Incandescent standard lamps should be operated using DC power with specified polarity and current. The lamp current should be ramped up and down slowly. Photometric measurements should be made after the lamp has stabilized, typically 10 min after turning on.

Luminous-flux standard lamps are normally operated in the base-up position. The standard lamps must always be operated in the same burning position, and should not be tilted since some lamp operations are tilt sensitive.

Fluorescent standard lamps are operated using AC power at a specified current, using a reference ballast of a specified impedance [52,53]. Current is normally used to determine the operating conditions since it is the most accurately reproducible parameter, even though the lamp power has the strongest correlation with the total luminous flux.

For rapid-start lamps, calibration is performed with the cathode heat on or off depending on the desired measurement conditions. Fluorescent lamps are very sensitive to the ambient temperature. The ambient temperature (measured behind a baffle at the same height of the lamp) is controlled to within 25±1 °C. The lamps are normally stabilized for 15 min before starting measurements. The recommended procedures for the operation of the fluorescent lamps are given in Ref. [54].

For miniature lamps, the size of sockets tends to be much larger relative to the size of the lamps. When a miniature lamp is mounted in a socket, the total flux may decrease significantly due to the absorption by the socket surfaces. Sometimes it makes sense to calibrate a miniature lamp together with a socket and always use the lamp with that particular socket. In such case, the combination of the lamp and the socket is considered as a standard source.

All the standard lamps, including fluorescent lamps, are operated at specified current rather than specified voltage because lamp voltage, in general, does not reproduce well due to different sockets used among users. However, lamp voltages reproduce fairly well on the same socket, and the lamp voltage is a useful indication to check if lamps have changed.

3.6 LUMINANCE STANDARDS

3.6.1 Detector-Based Realization of the Luminance Unit

A luminance standard can be established by using an illuminance standard photometer and an integrating-sphere source, with less uncertainty and difficulty than the traditional method using a diffuse reflectance or transmittance standard (see Sec. 3.6.2). As an example, Fig. 3.24 shows the geometry and the principles of the realization of a luminance unit used at NIST [55]. A limiting aperture with known area A (in m^2) is mounted in front of the opening of the integrating-sphere source.

The illuminance standard photometer measures the illuminance E_v (in lx) at distance d (in m) from the aperture reference plane. The average luminance L_v (in cd/m^2) over the aperture plane is given by

$$L_v = k\, E_v\, d^2/A, \tag{3.20}$$

where k is a geometrical correction factor determined by the radius r_a of the aperture, the radius r_d of the detector sensitive area, and the distance d, as given by

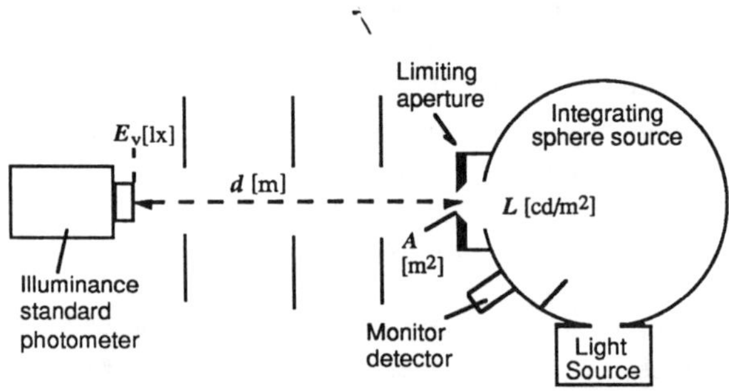

FIGURE 3.24. Configuration for the luminance unit realization at NIST.

$$k \simeq 1+\left(\frac{r_a}{d}\right)^2+\left(\frac{r_d}{d}\right)^2, \quad r_a, r_d<\frac{d}{10}. \tag{3.21}$$

The correction, $1-k$, is negligible (less than 0.02%) if $r_a/d<0.01$ and $r_b/d<0.01$. The aperture should be placed close to the sphere opening in order to reduce the diffraction loss to a negligible level caused by the aperture [56]. Geometrical factors in various geometries are found elsewhere [57].

The sphere source is normally operated at a distribution temperature of 2856 K. If the source is operated at a different color temperature, a spectral mismatch correction (see ''Characterization of Standard Photometers'' in Sec. 3.4.4) should be applied to the illuminance standard photometer.

3.6.2 Method Using a Diffuse Reflectance or Transmittance Standard

A luminance standard is traditionally established using a diffuse reflectance (transmittance) standard and a luminous-intensity standard lamp [46]. Pressed barium sulfate, pressed polytetrafluoroethylene (PTFE), and opal glass are commonly used as reference standards.

These reference diffusers are calibrated for a luminance coefficient q or a luminance factor β. The luminance coefficient is the ratio of luminance to illuminance in a given angle of incidence and at a given angle of viewing. The luminance factor is the ratio of the luminance of a material at a given geometry to that of a perfect diffuser and is equal to π times the value of luminance coefficient.

Diffusers can be calibrated either photometrically or spectrophotometrically. The geometry of the photometric method is shown in Figs. 3.25(a) and 3.25(b). The diffuser is uniformly illuminated by a reference source at a normal incidence, and the illuminance E_0 on the diffuser surface is first measured using a reference photometer. Then the same photometer is used to measure the illuminance E_1 from the diffuser surface within the limiting aperture of known area A. The photometer is placed at normal viewing of the transmitting diffuser and at a specified angle (normally 45°) for a reflecting diffuser. The luminance coefficient q (in sr^{-1}) is obtained by

$$q=\frac{L_v}{E_0}=\frac{kE_1d_p^2}{AE_0\Omega_0}, \tag{3.22}$$

where k is the geometrical correction factor as given in Eq. (3.21), and Ω_0 is the unit solid angle. The luminance factor β is given by

$$\beta=\pi q. \tag{3.23}$$

Although the principle of this method is simple, the actual measurement is not easy. The photometer should be sufficiently sensitive and linear over a large range since the level of E_1 is very low (0.3 lx with $d_p=1$ m, $r_a=0.01$ m, $E_0=1000$ lx)

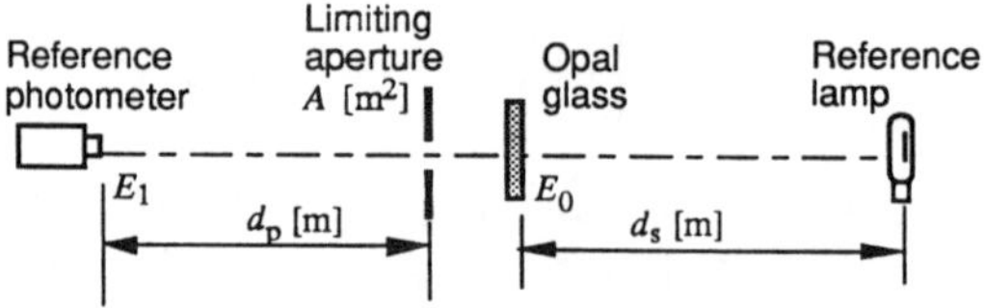

(a) calibration of diffuse transmittance

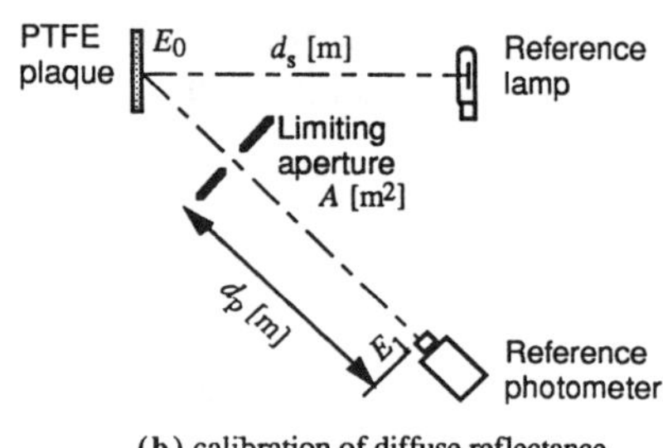

(b) calibration of diffuse reflectance

FIGURE 3.25. Photometric method of material calibration.

compared to the level of E_0. Extreme care should be taken to avoid errors of stray light when E_1 is measured, and to minimize interreflections between the aperture surface and the diffuser to a negligible level. The absolute accuracy of the reference photometer is not relevant, but spectral mismatch corrections must be applied if the diffuser alters the spectral power distribution of the light.

The reference lamp is normally operated at 2856 K, but it should be noted that the color temperature of the reflected (transmitted) light is shifted due to the spectral reflectance (transmittance) of the diffuser. The color temperature typically decreases by only 3–5 K with PTFE, but 100–200 K with opal glass. The reference lamp is sometimes operated so the color temperature of the reflected (transmitted) light becomes 2856 K, since this color temperature is recommended for luminance meter calibration. The values q and β should be reported with the color temperature of the reference source.

In the spectrophotometric method, the spectral bidirectional reflectance distribution function (BRDF) or spectral bidirectional transmittance distribution function (BTDF) at a normal incidence and at a given angle of viewing (normally 45°) is measured over the visible region. BRDF (BTDF) $r(\lambda)$ (in sr^{-1}) is the ratio of radiance to irradiance at a given wavelength and in a given geometry, and is a radiometric term corresponding to luminance coefficient. The luminance coefficient q is calculated by

$$q=\int_{\lambda} r(\lambda)S(\lambda)V(\lambda)d\lambda \Big/ \int_{\lambda} S(\lambda)V(\lambda)d\lambda, \tag{3.24}$$

where $V(\lambda)$ is the spectral luminous efficiency function, $S(\lambda)$ is the relative spectral power distribution of a reference source, and usually the CIE Illuminant A is used.

When a reference diffuser is illuminated at E (in lx) by the source having the spectral power distribution $S(\lambda)$, the luminance L (in cd/m^2) on the diffuser surface in the specified direction is given by

$$L=qE \tag{3.25}$$

or

$$L=\beta E/\pi. \tag{3.26}$$

3.6.3 Luminance Transfer Standards

Opal Glass

Opal glass has been the most widely used luminance standard because of its low cost, long-term stability, and ease of handling. The opal glass, illuminated by a known level of illuminance, provides a luminance scale. Luminous-intensity standard lamps are often used to provide a known illuminance, but an illuminance standard photometer can also be used with a reference lamp of known color temperature.

Despite the simplicity of the material, accurate measurement with an opal glass is not as easy as generally believed. An opal glass is sensitive to stray light from both sides of the glass. Extreme care should be taken to minimize ambient reflection from the front and the back sides. An opal glass should be uniformly illuminated over its entire surface area, since light incident on one part of the glass affects the luminance on other parts by volume diffusion. For example, a holder for the glass should not block the light falling on the edges of the glass. No labels should be affixed on the opal glass surface after calibration. Reference opal glasses are often equipped with an aperture to aviod these problems. Calibration will not be valid if the aperture is removed. Also, one should pay attention to the position of the reference plane of the opal glass. The illuminance should be measured on the same plane as was calibrated.

Opal glasses are normally calibrated for the CIE Illuminant A. It should be noted that the color temperature of the transmitted light is shifted due to the spectral transmittance of the opal glass and typically lowered by 100 to 200 K.

Integrating-Sphere Sources

Integrating-sphere sources are commonly used as luminance standards as well as spectral radiance standards. For this purpose, integrating-sphere sources are often equipped with an interchangeable aperture or a variable aperture at the entrance port for the light source, so that the luminance can be varied without significantly changing the color temperature. It would not be adequate to change luminance levels by changing the lamp current since the color temperature changes significantly. The sphere sources for luminance standards are normally operated at a color temperature of 2856 K.

It is recommended that the sphere sources for luminance standards be equipped with a temperature-stabilized monitor detector. The monitor-detector output is calibrated for a unit of luminance and is used to maintain the unit, usually with better stability and reproducibility than when relying on the internal lamp, since the monitor detector eliminates such factors as contamination and aging of the sphere coating as well as the aging of the lamp.

If the sphere has no monitor detector, the same cautions as for the luminous intensity standard lamps should be exercised in terms of lamp burning time (see ''Lamp Characteristics and Screening'' in Sec. 3.4.3). In that case, the sources need to be recalibrated every 50 to 100 h of operation depending on the desired accuracy. When a monitor detector is used, the sphere source needs to be calibrated periodically (once every year is recommended) since the detector can experience long-term drift even when not used as described in ''Operation and Handling of Standard Photometers'' in Sec. 3.4.4.

When the monitor detector is not temperature-stabilized, one should be aware of its responsivity drift as the sphere source heats up. The stability of the monitor detector can be tested by continuously monitoring its output by comparing it with an external luminance meter as the sphere source is turned on and then warms up. One can determine the required stabilization time for the detector when the ratio of the two signals has become stable.

Integrating-sphere sources are often equipped with an interchangeable aperture to accommodate different sizes of the exit port. It should be noted that, if the aperture is removed or changed, the luminance will change due to the effect of interreflections between the aperture surface and the sphere. This change is automatically corrected if a monitor detector is used. If there is no monitor detector, this change should be measured and corrected by using an external luminance meter. One should also be aware of the spatial nonuniformity of luminance over the exit port, which can be affected by different apertures. The exit port of the sphere source should always be capped when not used to avoid contamination.

Luminance Meters

Luminance meters having high-quality optics and electronics can also be used as transfer standards for luminance. Such high-quality luminance meters are as stable and reproducible as a sphere source with a monitor detector and are often more convenient to transport between laboratories. There are several important aspects of luminance meters to consider when used as transfer standards.

Among several potential problems, the out-of-measurement-field response can probably be the most troublesome. If the rejection of light outside the measurement field is not sufficient, the luminance-meter responsivity varies depending on the angular size of the target, which makes it difficult to consistently reproduce measurements. This characteristic can be tested using the method given in Ref. [46]. If this characteristic is not satisfactory, the geometry between the luminance meter and the target source should be precisely recorded in order to reproduce measurements.

The long-term stability of the luminance meter's responsivity is an important consideration. This characteristic can be tested by periodic calibration of the instruments against high-accuracy standards. The relative spectral responsivity of the luminance meter should be well matched to the $V(\lambda)$ function. It is recommended that the f_1' value (see "Characterization of Standard Photometers" in Sec. 3.4.4) for a reference luminance meter be less than 3%. Another consideration is that luminance meters used for standards need to have either a display of four digits or more or an analog output in order to read the signal with sufficient resolution. The gain factors in different ranges should be precisely calibrated.

3.7 COLOR-TEMPERATURE STANDARDS

3.7.1 Definitions of Terms

Color temperature is a sophisticated concept to express color of a light source by using just one number. According to CIE [44], color temperature is defined as "the temperature of a Planckian radiator whose radiation has *the same* chromaticity as that of a given stimulus." This means that the test source and the blackbody have the same appearance to a human observer. Note that the spectral power distributions of the source are not necessarily identical or even similar to that of a blackbody. Strictly speaking, however, the chromaticity coordinate of most of the light sources including incandescent lamps do not fall exactly on the Planckian locus. The CIE definition does not say how close the chromaticity should be to be considered as "the same." Therefore, the official definition of color temperature is somewhat vague, and this term is considered as a general term to introduce the concept.

In common practice of photometry, either *distribution temperature* or *correlated color temperature* is used to report measurement results. Another similar concept is a radiance temperature, which is defined as "the temperature of the Planckian radiator for which the radiance at the specified wavelength has the same spectral concentration as for the thermal radiator considered" [14]. Radiance temperature is used only for blackbodies and related standard lamps.

Distribution temperature is defined as "the temperature of the Planckian radiator whose relative spectral distribution $S_t(\lambda)$ is the same or nearly the same as that of the radiation considered in the spectral range of interest" [44]. Practically speaking, distribution temperature is a concept to represent the relative spectral power distribution of a quasi-Planckian source, such as an incandescent lamp, using one number. CIE [58] provides a more explicit definition as below. The distribution temperature T_d of a source is the temperature T of the Planckian radiator for which the following integral is minimized by adjustment of a and T:

$$\int_{\lambda_1}^{\lambda_2}[1-S_t(\lambda)/aS_b(\lambda,T)]^2 d\lambda, \tag{3.27}$$

where $S_t(\lambda)$ is the relative spectral distribution of the radiation being considered, and $S_b(\lambda,T)$ is the relative spectral distribution of the Planckian radiator at temperature T as given by

$$S_b(\lambda,T)=\lambda^{-5}[\exp(c_2/\lambda T)-1]^{-1}. \tag{3.28}$$

It is also specified that the wavelength region λ_1 to λ_2 shall be 380–780 nm, and the wavelength interval for calculation shall be less than 10 nm. The document [58] also specifies that the difference of the relative spectral power distribution of the radiation considered and that of a Planckian radiation should be less than 10% in order to use distribution temperature.

Correlated color temperature (CCT) is used for sources whose spectral power distribution is significantly different from that of Planckian radiation, such as discharge lamps. CCT is defined as ''the temperature of the Planckian radiator whose perceived color most closely resembles that of a given stimulus at the same brightness and under specified viewing conditions.'' Practically, CCT is obtained on the CIE 1960 (u,v) diagram by finding the temperature of a blackbody radiation whose chromaticity coordinate is closest to that of the light source in question [59].

If the relative spectral power distribution of the given radiation is identical to that of the Planckian radiator, the values of color temperature (in official definition), CCT, distribution temperature, and radiance temperature would be all the same. The term *color temperature* is often used informally to represent correlated color temperature. It may be confusing that, for incandescent lamps, both distribution temperature and CCT are used. However, this is usually not a problem since the differences between distribution temperature and CCT of typical incandescent lamps are very small (less than 2 or 3 K). However, in scale comparisons or other rigorous analyses, one should take into account the small differences between the two. An important difference to note is that, by definition, CCT gives only the color of a light source, whereas distribution temperature gives the spectrum of a light source within a defined uncertainty.

3.7.2 Realization of the Color Temperature Scale

The color temperature scale is normally derived from the spectral irradiance scale which is based on the International Temperature Scale or the radiance temperature scale [60]. An example of the scale realization procedure is shown in Fig. 3.26. At the top is the International Temperature Scale (ITS90). The radiance of a gold-point blackbody at its solidification temperature, 1337.33 K, at a certain wavelength is calculated using Planck's equation. The radiance is transferred to the secondary gold-point (GP) standard lamps. Then the radiance of the variable-temperature blackbody (VTBB) operating at a much higher temperature (e.g., 2500 K) is measured at the same wavelength against the secondary GP standard lamp, and

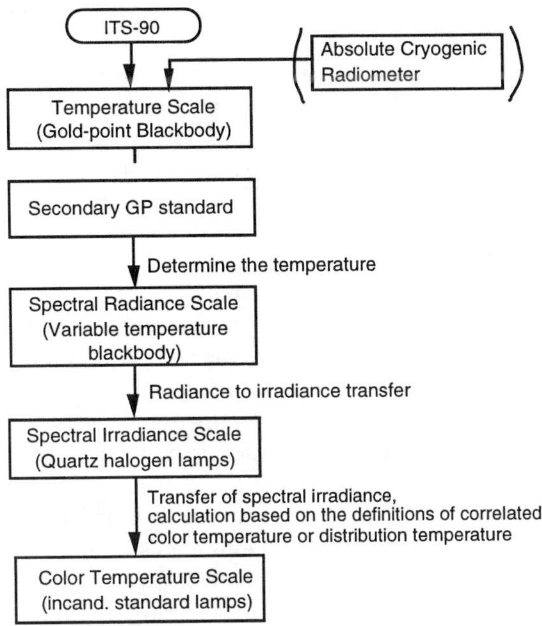

FIGURE 3.26. An example of the procedure for realization of the spectral irradiance scale and distribution temperature scale.

the temperature of the VTBB is determined based on the Planck's equation. The radiances at other wavelengths are computed using the Planck's equation, and the spectral radiance scale is established. The spectral radiance on the VTBB is then transferred to spectral irradiance on quartz halogen standard lamps via an integrating-sphere source. Finally, a group of color-temperature standard lamps are calibrated for spectral irradiance, and the correlated color temperature or the distribution temperature is computed.

Instead of relying on ITS90, some national laboratories determine the gold-point temperature radiometrically using absolutely calibrated detectors [40]. The radiance temperature scale is then realized independent of the International Temperature Scale, which may be subject to change.

The spectral irradiance of the color-temperature standard lamps are calibrated at several operating currents for the color temperature range 2000–3200 K, and used to calibrate color-temperature-measuring instruments such as a tristimulus colorimeter, a two-channel colorimeter, and a diode-array spectroradiometer.

3.7.3 Color-Temperature Transfer Standards

Luminous-intensity transfer standard lamps as described in ''Lamp Types'' in Sec. 3.4.3 are normally calibrated for color temperature and provide the color temperature scale also. These types of lamps are sometimes used only as color-temperature transfer standards. All the cautions in handling luminous-intensity standard lamps (''Operation and Handling of Standard Lamps'' in Sec. 3.4.3) apply, the only difference being that the alignment of color-temperature standard lamps are less critical and usually do not require a special alignment device. Gas-filled incandescent lamps are normally used for the color-temperature range 2000–2900 K, and quartz halogen lamps are normally used from 2800 to 3200 K. Some FEL-type quartz halogen lamps are also suitable for use in the range 2000–3200 K [43].

The aging of lamps affects their color temperature. Examples of the aging characteristics of FEL-type quartz halogen lamps and 300-W gas-filled incandescent standard lamps are shown in Figs. 3.27(a) and 27(b).

Another important characteristic for color-temperature standard lamps is the relative spectral power distribution. Deviation of the relative spectral power distribution from the Planckian curve can cause a serious measurement error when the red-to-blue ratio substitution method (see Sec. 5.4.2) is used to transfer from one type of lamp to another. Figure 3.28 shows the ratio of the relative spectral power distribution of a typical FEL-type lamp and two types of gas-filled lamps to that of the Planckian radiation of the same color temperature (normalized to unity at the peak wavelength). In this case, if the gas-filled lamps are calibrated against the FEL-type lamp with a blue-response detector peaking at 450 nm and a red detector peaking at 600 nm for example, an error of more than 10 K can be introduced. Care must be taken when the color temperature is transferred to a different type of lamp. To minimize this error, experience shows that the peak wavelengths of the blue-response detector and the red-response detector should be selected as 460 and 680 nm, respectively.

Unless a double monochromator with sufficient stray-light rejection is used, the transfer of color-temperature standards should be made in comparison to the standard lamps of the same or similar color temperature. For example, a diode-array

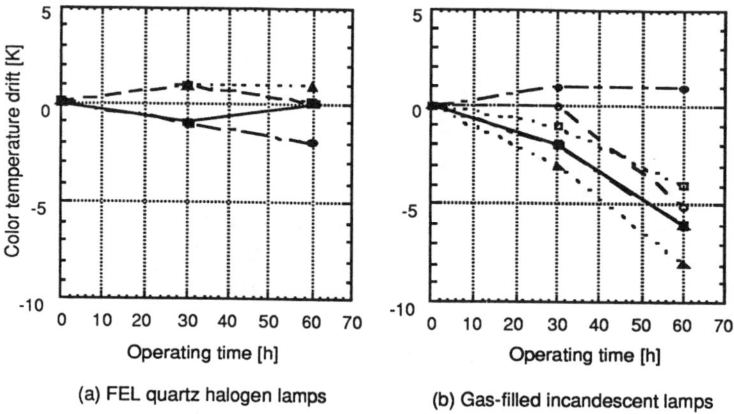

FIGURE 3.27. Aging characteristics of incandescent lamps in terms of color temperature.

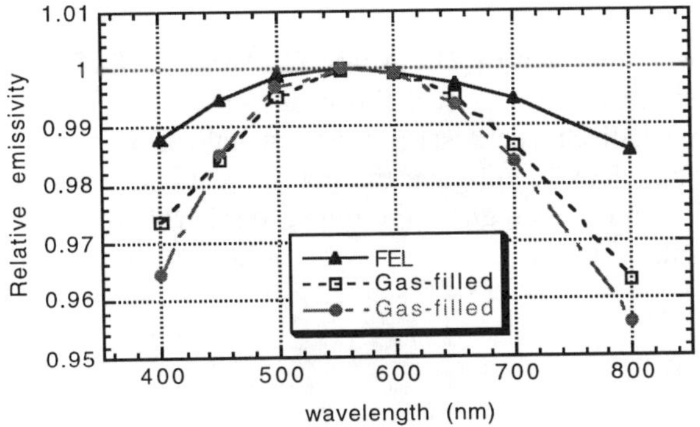

FIGURE 3.28. Ratio of the relative spectral power distribution of incandescent lamps to that of the Planckian radiation at the same color temperature (normalized to unity at the peak wavelength).

type spectroradiometer is often used for fast measurements. If such an instrument is calibrated using a 2856-K source and measures much lower color temperatures, errors can occur due to the internal stray light and other behaviors of the system. Very low energy in the blue region tends to be overestimated by the stray light. Figure 3.29 shows an example of such errors in a commercial diode-array system. For this reason, color-temperature standard lamps calibrated at several different color temperatures are often maintained.

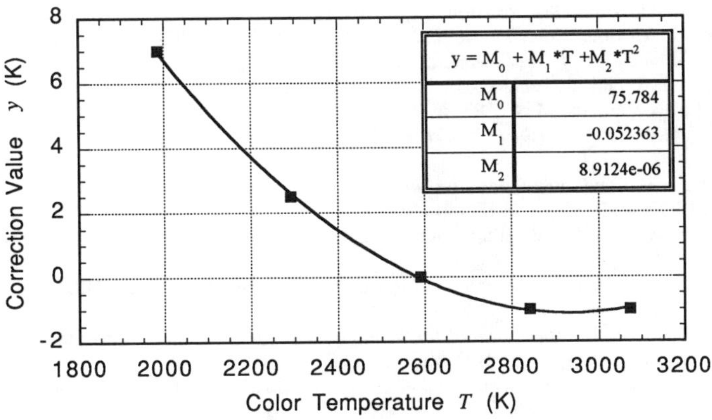

FIGURE 3.29. An example of color-temperature measurement errors for a diode-array spectroradiometer.

ACKNOWLEDGMENTS

This chapter was reviewed by Dr. Albert C. Parr, Chief of Optical Technology Division of NIST, Ms. Sally Bruce, Physical Scientist at NIST, and Dr. Georg Sauter of the Physikalisch-Technische Bundesanstalt, Germany. They spent many days reviewing this chapter and gave me many precious comments and extensive discussions, which contributed greatly to improve this chapter to the final form. The author is very grateful to them.

REFERENCES

1. Walsh, J. W. T., *Photometry* (Constable, London, 1953).
2. CIE Proc. **5**, 41 (1921).
3. Wensel, H. T., Roeser, W. F., Barbrow, L. E., and Caldwell, R. R., Bur. Stand. (U.S.) J. Res. **6**, 1103 (1931).
4. 9e CGPM Compte Rendu, 54 (1948).
5. 13e CGPM Compte Rendu, 104 (1967).
6. *The International System of Units*, 6th ed. (BIPM, Sévres, France, 1991).
7. Blevin, W. R., and Steiner, B., Metrologia **11**, 97 (1975).
8. Blevin, W. R., ''The Candela and the Watt,'' CIE Proc. P-79-02 (1979).
9. CGPM, Comptes Rendus des Séances de la 16e Conférence Générale des Poids et Mesures, Paris, 1979 (BIPM, Sèvres, France, 1979).
10. CIPM, Comité Consultatit de Photométrie et Radrométrie IIe Session—1986 (Bureau International des Poids et Mesures, Pavillon de Bretevil, F-92310, Sevres, France, 1986), pp. 140–164.
11. Geist, J., ''Quantum Efficiency of the *p-n* Junction in Silicon as an Absolute Radiometric Standard,'' Appl. Opt. **18**, 760–762 (1979).
12. Zalewski, E. F., and Geist, J., ''Silicon Photodiode Absolute Spectral Response Self-Calibration,'' Appl. Opt. **19**, 1214–1216 (1980).
13. CIE Compte Rendu, 67 (1924).
14. CIE Compte Rendu, Table II, 25–26 (1931).
15. *The Basis of Physical Photometry*, CIE Publication No. 18.2 (1983).
16. CIPM, Comité Consultatif de Photométrie et Radiométrie 10e Session—1982 (Bureau International Des Poids et Mesures, Pavillon de Breteuil, F-92310, Sèvres, France, 1982). The same content is published in BIPM Monographie—*Principles Governing Photometry* (Bureau International Des Poids et Mesures, Pavillon de Breteuil, Sèvres, France, 1983).
17. (a) *Photometric and Colorimetric Tables*, CIE Disk D001 (1988). (b) *Light as a True Visual Quantity: Principles of Measurement*, CIE Publication No. 41 (1978).
18. CIE Compte Rendu, **3**, Table II, 37–39 (1951).
19. CIPM Procés-Verbaux **44**, 4 (1976).
20. *Mesopic Photometry: History, Special Problems and Practical Solutions*, CIE Publication No. 81 (1989).
21. *ISO Standards Handbook, Quantities and Units*, 3rd ed. (1993); this book contains ISO 31-0:1992 through ISO 31-13:1992, and ISO 1000:1992.
22. *Lighting Handbook*, 8th ed. (Illuminating Engineering Society of North America, New York, 1993), pp. 946–949.
23. Taylor, B. N., *Guide for the Use of the International System of Units (SI)*, NIST Special Publication No. 811 (1995).
24. Taylor, B. N., ed., *Interpretation of the SI for the United States and Metric Conversion Policy for Federal Agencies*, NIST Special Publication No. 814 (1991).
25. Boivin, L. P., Gaertner, A. A., and Gignac D. S., ''Realization of the New Candela (1979) at NRC,'' Metrologia **24**, 139–152 (1987).
26. Carreras, C., and Corrons, A., ''Absolute Spectroradiometric and Photometric Scales Based on an Electrically Calibrated Pyroelectric Radiometer,'' Appl. Opt. **20**, 1174–1177 (1981).
27. Gardner, J. L., ''Radiometric and Photometric Standards in Australia,'' in CIE Proceedings of the 22nd Session, Melbourne, 1991, Vol. 1, Div. 2, pp. 5–8.

28. Cromer, C. L., Eppeldauer, G., Hardis, J. E., Larason, T. C., and Parr, A. C., ''National Institute of Standards and Technology Detector-Based Photometric Scale,'' Appl. Opt. **32**, 2936–2948 (1993).
29. Ohno, Y., ''Silicon Photodiode Self-Calibration Using White Light for Photometric Standards—Theoretical Analysis,'' Appl. Opt. **31**, 466–470 (1992).
30. Goodman, T. M., and Key, P. J., ''The NPL Radiometric Realization of the Candela,'' Metrologia **25**, 29–40 (1988).
31. Cromer, C. L., Eppeldauer, G., Hardis, J. E., Larason, T. C., Ohno, Y., and Parr, A. C., ''The NIST Detector-Based Luminous Intensity Scale,'' J. Res. NIST **101**, 109–131 (1996).
32. Martin, J. E., Fox, N. P., and Key, P. J., Metrologia **21**, 147–155 (1985).
33. Gentile, T. R., Houston, J. M., Hardis, J. E., Cromer, C. L., and Parr, A. C., ''The NIST High-Accuracy Cryogenic Radiometer,'' Appl. Opt. **35**, 1056–1068 (1996).
34. Ohno, Y., *Photometric Calibrations,* NIST Special Publication No. 250-37 (1997)
35. *CIE Standard Colorimetric Illuminants* CIE/ISO 10526 (1991).
36. Sapritsky, V., ''A New Standard for the Candela in the USSR,'' Metrologia **24**, 53–59 (1987).
37. Blevin, W. R., ''Corrections in Optical Pyrometry and Photometry for the Refractive Index of Air,'' Metrologia **8**, 146 (1972).
38. Walker, J. H., Saunders, R. D., Jackson, J. K., and McSparron, D. A., *Spectral Irraduance Calibrations*, NBS Special Publication No. 250-20 (1987).
39. Mielenz, K. D., Saunders, R. D., Parr, A. C., and Hsia, J. J., ''The New International Temperature Scale of 1990 and Its Effect on Radiometric, Photometric, and Colorimetric Measurements and Standards,'' in CIE Proceedings of the 22nd Session, Melbourne, 1991, Div. 2, (1991), Vol. 1, Div. 2, pp. 65–68.
40. Mielenz, K. D., Saunders, R. D., and Shumaker, J. B., ''Spectroradiometric Determination of the Freezing Temperature of Gold,'' J. Res. NIST, **95**, 49–67 (1990).
41. Goodman, T. M., *et al.*, ''Establishment of a New Absolute Spectral Irradiance Scale at NPL, Using a Radiometrically Calibrated Blackbody,'' in CIE Proceedings, 23rd Session, 1995, pp. 83–86.
42. Sapritsky, V. I., *et al.*, ''Candela and Lumen Realization on the Basis of the Blackbody Achievements,'' in CIE Proceedings, 23rd Session, New Delhi, 1995, pp. 83–86.
43. Ohno, Y., and Jackson, J. K., ''Characterization of Modified FEL Quartz-Halogen Lamps for Photometric Standards,'' Metrologia **32**, 693–696 (1996).
44. *CIE International Lighting Vocabulary*, CIE Publication No. 17.4 (1987).
45. Eppeldauer, G., ''Temperature Monitored/Controlled Silicon Photodiodes for Standardization,'' in *Surveillance Technologies*, SPIE Proc. **1479**, 71–77 (1991).
46. *Methods of Characterizing Illuminance Meters and Luminance Meters*, CIE Publication No. 69 (1987).
47. *Measurements of Luminous Flux*, CIE Publication No. 84 (1989).
48. Ohno, Y., ''Integrating Sphere Simulation—Application to Total Flux Scale Realization,'' Appl. Opt. **33**, 2637–2647 (1994).
49. Ohno, Y., ''New Method for Realizing a Total Luminous Flux Scale using an Integrating Sphere with an External Source,'' J. IES **23**, 88–98 (1994).
50. Ohno, Y., ''Realization of NIST Luminous Flux Scale Using an Integrating Sphere with an External Source,'' in CIE Proceedings, 23rd Session, New Delhi, 1995, pp. 87–90.
51. Venable, W. H., Hsia, J. J., and Weidner, V. R., ''Establishing a Scale of Directional-Hemispherical Reflectance Factor I: The Van den Akker Method,'' J. Res. NBS **82**, 29–55 (1977).
52. *Fluorescent Lamp Ballasts—Methods of Measurement*, ANSI C82.2-1984.
53. *Rapid Start Types Dimensional and Electrical Characteristics*, ANSI C78.1-1991.
54. *IES Approved Method for the Electrical and Photometric Measurements of Fluorescent Lamps*, IES LM-9-1988.
55. Ohno, Y., Cromer, C. L., Hardis, J. E., and Eppeldauer, G., ''The Detector-Based Candela Scale and Related Photometric Calibration Procedures at NIST,'' J. IES **23**, 88–98 (1994).
56. Blevin, W. R., ''Diffraction Losses in Radiometry and Photometry,'' Metrologia **7**, 39–44 (1970).
57. *Lighting Handbook*, 8th ed. (Illuminating Engineering Society of North America, City, 1993), Chap. 9, Lighting Calculations, p. 423.
58. CIE Collection in Photometry and Radiometry Publication No. 114 (1994).
59. *Colorimetry*, 2nd ed. CIE Publication No. 15.2 (1986).
60. Mielenz, K. D., Saunders, R. D., Parr, A. C., and Hsia, J. J., ''The 1990 NIST Scale of Thermal Radiometry,'' J. Res. NIST **95**, 621 (1990).

4

Fundamentals of Detectors

Carolyn J. Sher DeCusatis
Lighting Research Center, Rensselaer Polytechnic Institute, Troy, New York 12180

We are extremely fortunate to work at a time where many commercial detectors are available for photometric applications. Choosing the right detector for your application can sometimes be confusing because of the many options which are available. In this chapter, we will provide a general overview of detectors for photometric applications; this information should enable the reader to study the literature in more depth if necessary. The chapter takes a practical approach, beginning with some general guidelines on how to choose the right detector. This is followed by a description of detector terminology and how to read a detector specification. Several common detection elements are reviewed in detail, including photodiodes, photomultipler tubes, charge coupled devices, and thermal detectors. Accessories such as filters, diffusers, imaging optics, and electronics are briefly overviewed. Finally, we conclude with a discussion of noise and calibration that applies to all types of detector systems.

4.1 CHOOSING THE RIGHT DETECTOR

There is a broad and often bewildering variety of detector options available. Here are some general factors to consider when choosing a detector.

4.1.1 What Am I Trying To Detect?

A photometric standard document will state the type of detector required—for example, illuminance meter, luminance meter, spectroradiometer, etc. However, when conducting research where the photometric aspect is not central to the experiment, such as measuring human response to color or the visibility of road signs, it is important to properly select a photometric target measurement and plan the detection appropriately. We will briefly overview several types of detection systems and their uses.

Illuminance meters measure how much flux hits a surface. This type of meter is often used to determine the ambient light level of a room. Appropriate placement of the detector, particularly with respect to shadowing, can be very important. These meters are available in a large range of price and quality. Some do not have a separatable detector head; the detection element is on the same box as the readout. This is a rather durable design, and useful for field experiments. However, the experimental setup may inadvertantly block the sensor when taking a reading. Also, this type of detector often does not have a memory of measurements, or computer compatibility. Detectors with a separatable head are often more delicate than those with recessed heads. For many applications they are strong enough, but if there may be stress on the wiring connecting the head to the device, it can be a source of failures. Some detectors of this kind are placed on wands (''lollipop detectors'') for ease in temporary remote measurements. These detectors are easy to step on or otherwise break off in the field, which will place extreme stress on the wiring connected to the housing, not to mention making a sickening crunch sound. The quality of the filters on this type of detector varies widely. This is especially important if measurements will be in the edges of the photopic curve—very blue or very red. Assumptions may be made by the manufacturer on the ambient light source when developing the specification.

Luminance meters measure the ''photometric brightness'' of a surface. This type of detector is often used to determine how a target appears to a standard subject. They are available in convient ''point and shoot'' packages, like automatic cameras.

Despite this, it is important to read their manuals, especially if measurements will be made under unusual lighting conditions, such as at night in a parking lot lit by mercury arc lights.

Color meters measure the Commission Internationale de l'Eclairage (CIE) color coordinates using filters. They are available in both illuminance and luminance meter design.

Spectroradiometers measure the spectra of light using gratings. The typical detection element is a charge coupled device (CCD) chip, compared to the above-mentioned detectors, which all tend to run on p-i-n diodes, although other detection elements may be available. Spectroradiometers tend to require an AC power source for the motor that moves their gratings, as well as being connected to personal computers (not necessarily laptops!), which can make them awkward and bulky in the field (as compared to the laboratory).

4.1.2 Which Detector Types Fit My Wavelength and Signal Requirements?

We show here in Fig. 4.1 a chart of the sensitivity and wavelength range of several detection elements [1]. For descriptions of the detection elements, see "The Detection Element" in Sec. 4.3. This discussion is placed first because it is important to know which detector elements might suit your application before discussing how all the detectors work. The characteristics of the signal, such as signal-to-noise ratio, amplitude, time response, and frequency bandwidth, will all influence the choice of detector. The detector system's linear range, field of view, noise equivalent power (NEP), and other factors will affect the type of output obtained. We will discuss these issues in detail in the following sections.

4.1.3 What "Quality" of Detector Do I Need?

There is no official quality standard for photometric equipment. In other words, it is not possible to state that a detector is "laboratory quality" and know that it will be sufficient for a certain class of experiment. Therefore, as an experimenter it is necessary to read the specification of each detector and see if it meets the experimental requirements. In the next section, we will define quantities needed to compare different detectors. When determining the experimental requirements, it is important to consider the propogation of error from all parts of the system, not only the detector. It is often possible to get information about a detector that is not published in the advertising literature by talking directly to the company representatives. In particular, sales literature may only state a number, while the representatives may have access to a graph determining how the number was calculated. This can be valuable when discussing the photopic accuracy of filter-type detectors or for evaluating more sensitive types of detectors which may be required to perform near the limits of their published specifications.

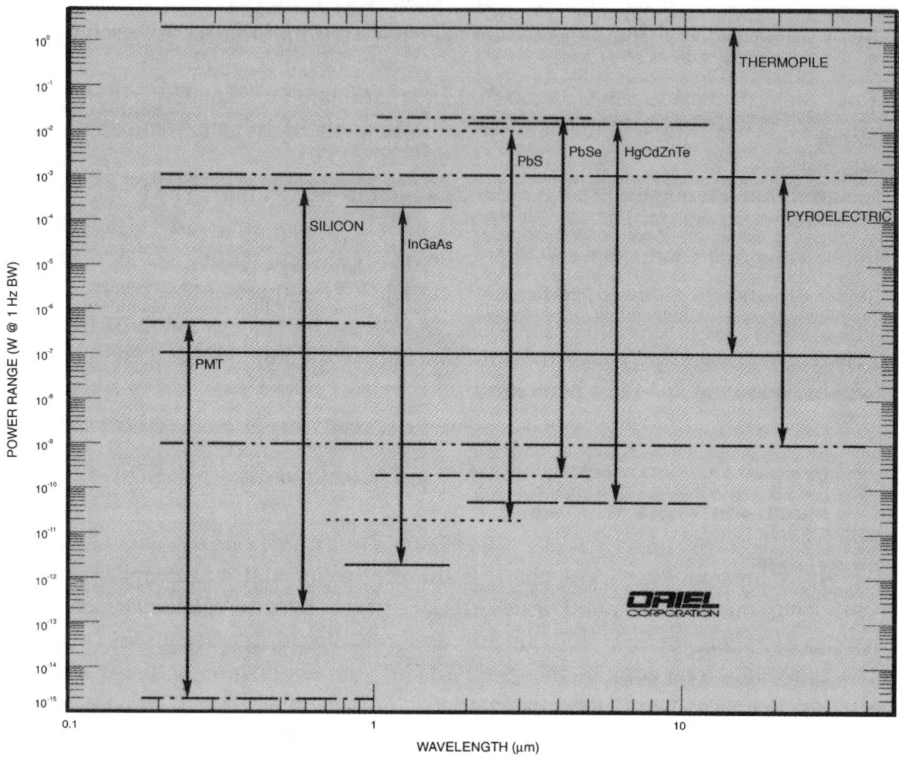

FIGURE 4.1. Sensitivity and wavelength ranges of various detectors. [Reprinted by permission from *ORIEL Instruments Catalog,* volume II (ORIEL Instruments, Stratford, CT, 1994).]

4.1.4 Where Do I Find Manufacturers?

Manufacturer listings exist in such books as *The Photonics Buyer's Guide* [2], or the *Lasers and Optronics Technology and Industry Reference* issue [3]. At conferences, such as the Optical Society of America or the American Physical Society's annual meetings, there are exhibit halls where the manufacturers demonstrate their products. Information can also be acquired by talking to other people doing similar experiments. These people can be located through societies, such as The Optical Society of America (OSA) [4], the American Physical Society (APS) [5], the Council for Optical Radiation Measurements (CORM) [6], and the International Commission on Illumination (CIE) [7] and through their journal publications. While manufacturers will send you sales literature on their products, it will take you time to acquire these data. Most researchers form their own libraries of this information for future use. Finding a researcher with a full set of files can save you a lot of time, even if the files are out of date.

4.1.5 Will the Electronics Be Compatible with My Other Equipment?

It is possible to be delayed months waiting for adaptors between different types of electrical connectors (sometimes cutting the wiring invalidates manufacturer calibrations). When ordering equipment, be certain that it is compatible with existing systems or can be made compatible by minor adjustments. This includes being aware if your detector has a computer compatible output; keep your future automation needs in mind.

4.2 DETECTOR SPECIFICATIONS

In this section we will define the terminology used in a typical detector specification. Every detector specification should include a picture and/or physical description of the part, including dimensions and construction (i.e., plastic housing). I have tried to be inclusive in my list of terms, which means not all of these quantities will apply to every detector specification. Since specifications are not standardized, it is impossible to include all possible terms used; however, most detectors are described by certain standard figures of merit which will be discussed in this section. It is important to consider the manufacturer's context for all values; a detector designed for a specific application may not be appropriate for a different application even though the specification seems appropriate. If you are trying to use a detector in a novel way, discuss it with the company representative and technical staff that designed the detector; a detector can be a major investment and should be treated appropriately.

4.2.1 Terminology and Characteristics

Active area and *effective sensing area* are just what they sound like: the size of the detecting surface of the detection element. The uniformity of response refers to the percentage change of the sensitivity across the active area.

Dark current is the current that flows in a working photodetector when there is no optical radiation incident on the detector. It is usually simply measured and then subtracted from the flux, like background. However, the dark current is temperature dependent, so one measurement at the beginning of the experiment is usually not sufficient. It is not a good idea for the anticipated signal to be a small fraction of the dark current; rms noise in the dark current may mask your signal. This is less of an issue if you are using signal chopping as part of your amplification electronics.

Operating temperature is the temperature range over which a detector is accurate and will not be damaged by being powered. However, there may be changes in sensitivity and dark current which must be taken into account: read the manual. Storage temperature will have a considerably larger range; basically, it describes the temperature range under which the detector will not melt, freeze, or otherwise be damaged or lose its calibration.

4.2.2 Figures of Merit for Detector Comparison

There are several figures of merit used to characterize the performance of different detectors. *Responsivity*, or *response*, is the sensitivity of the detector to input flux. It is given by

$$R(\lambda)=D(\lambda)/\phi(\lambda), \tag{4.1}$$

where D is the detector output signal (in amps) and flux is the incident light signal on the detector (in watts). Thus, the units of responsivity are amps per watt. *Dark current* is usually subtracted out from $D(\lambda)$. *Responsivity* is defined at a specific wavelength; the term *spectral responsivity* is used to describe the variation at different wavelengths. Responsivity versus wavelength is often included in a specification as a graph, as well as placed in a performance chart at a specified wavelength.

Photopic response is the total response of a detector that is filtered with the appropriate photometric filter. It can be calculated as follows:

$$R=\frac{\int_{\lambda}E(\lambda)R(\lambda)V(\lambda)d\lambda}{\int_{\lambda}E(\lambda)d\lambda}, \tag{4.2}$$

where $E(\lambda)$ is a standard irradiance distribution, $V(\lambda)$ is the transmitance of the photopic filter, and $R(\lambda)$ is the response of the unfiltered detector. Naturally, the photopic response of a detector is more likely to be measured using a standard light source, than calculated using the spectral response. The typical units of photopic response are microamps/foot candle.

NEP stands for *noise equivalent power*. It is the amount of flux that would create a signal of the same strength as the root mean square (rms) detector noise. In other words, it is a measure of the minimum detectable signal; for this reason, it is the most commonly used version of noise equivalent detector input:

$$\mathrm{NEP}(\lambda)=\sigma(\lambda)/R(\lambda), \tag{4.3}$$

where σ is the rms noise current and R is the responsivity, define above. The ratio S/N is the signal-to-noise ratio. We will discuss noise further in a later section.

Sometimes people find it easier to work with *detectivity*, which is the reciprocal of NEP. The higher the detectivity, the smaller the signal a detector can measure; this is a convenient way to characterize more sensitive detectors. Detectivity and NEP vary with the inverse of the square of active area of the detector, as well as with temperature, wavelength, modulation frequency, signal voltage, and bandwidth. *Normalized detectivity* is detectivity multiplied by the square root of the product of active area and bandwidth; this product is usually constant, and allows comparison of different detector types independent of size and bandwidth limits. This is because most detector noise is white noise, and white-noise power is proportional to the bandwidth of the detector electronics. Thus the noise signal is proportional to the square root of bandwidth. Also, note that electrical noise power is usually proportional to detector area and the voltage, which provides a measure of that noise is proportional to the square root of power. *Normalized detectivity* is given by:

$$D_N = D\sqrt{A \times \mathrm{BW}} = \frac{\sqrt{A \times \mathrm{BW}}}{\mathrm{NEP}}, \tag{4.4}$$

where A is the active area in cm^2, and BW is the bandwidth. The units are $(\mathrm{cm\ Hz})^{1/2}\ \mathrm{W}^{-1}$. Normalized detectivity is a function of wavelength and spectral responsivity; it is often quoted as *normalized spectral responsivity.*

Bandwidth (BW) is the range of frequencies over which a particular instrument is designed to function within specified limits. Bandwidth is often adjusted to limit noise; it is frequently chosen as 1 Hz, so NEP is quoted in W/Hz.

Quantum efficiency (QE) is the ratio of the number of signals produced by the detection element to the number of incident photons. It will depend on the material from which the detector is made. It affects your measurement indirectly, through the responsivity, and can be characterized by

$$\mathrm{QE} = \frac{R(\lambda)hc}{\lambda q} = 1.2398\,\frac{R(\lambda)}{\lambda}, \tag{4.5}$$

where R is the responsivity as a function, of wavelength, h is Planck's constant (6.626×10^{-34} W/sec^2 give value), c is the speed of light (3×10^8 m/s), λ is the wavelength, and q is the charge of an electron. If wavelength is in microns and R is responsivity flux, then the units of quantum efficiency are amperes per watt.

Response time is the time it takes for a detector's output to rise after a sudden constant irradiance. The *rise time* is the time it takes for a detector's output to become $1/e$ when suddenly subjected to a constant irradiance; the *fall time* is the time it takes for detector output to fall to $1/e$ of the initial value when the signal is turned off.

Linearity range is the range of incident radiant flux over which the signal output is a linear function of the input. The lower limit of linearity is NEP, and the upper limit is *saturation.* Saturation is when the detector begins to form less signal output the same increase of input flux. When a detector begins to saturate, it has reached the end of its linear range. *Dynamic range* can be used to describe nonlinear detectors, like the human eye. For example, neutral-density filters (see Sec. 4.3.2) can be used to increase the dynamic range of a detector system by creating islands of linearity, whose actual flux is determined by dividing output signals of the detector by the transmission of the filter. Without filtering, the dynamic range would be limited to the linear range of the detector, which would be less because the detector would saturate without the filter to limit the incident flux. The units of linear range are incident radiant flux or power (watts or irradiance).

4.2.3 Other Important Terms

Measuring the response of a detector to flux is known as *calibration*. Some detectors can be self-calibrated, others require manufacturer calibration. Calibration certificates are supplied by most manufacturers; they are dated, and have certain time limits.

The *gain*, also known as the *amplification*, is the ratio of electron hole pairs generated per incident photon. Often detector electronics allows you to adjust the gain.

Wiring and pin output diagrams tell you how to operate the equipment, by schematically showing how to connect the input and output leads.

4.2.4 Typical Specification Examples

The following section provides a quick reference to some of the terminology that may be found in a detector specification. If you are unfamiliar with the different types of detectors, skip to the following section for a description of how they work.

Photodiodes

A sample specification for a photodiode is given in Fig. 4.2 [8]. *Bias voltage* (V_B) is the voltage applied to a silicon photodiode to change the potential photoelectrons must scale to become part of the signal. It is discussed more fully in "*p-i-n* Diodes" in Sec. 4.3.1. Bias voltage is basically a set operating characteristic in a packaged detector, in this case −24 V. *Shunt resistance* is the resistance of a silicon photodiode when not biased. *Junction capacitance* is the capacitance of a silicon photodiode when not biased. *Breakdown voltage* is the voltage applied as a bias which is large enough to create signal on its own. When this happens, the contribution of photoelectrons is minimal, so the detector cannot function. However, once the incorrect bias is removed the detector should return to normal.

Photomultiplier Tubes (PMTs)

A sample specification for a photomultiplier tube given in Fig. 4.3 [9]. The transmission of the window material affects the spectrum of the light that reaches the *photocathode*, the light-sensitive element. The *photocathode material* is the

Part Number	Total Area	Active Area	Storage & Operating Temp.	Responsivity	Shunt Resistance	Dark Current	Breakdown Voltage[1]	Capacitance[2]	NEP	Response Time[5]
				(min)	(min)	(max)	(min)	(typ)	(typ)	
	(mm^2)	(in)	(°C)	(A/W)	(M-Ohm)	(nA)	(V)	(pF)	(W/√Hz)	(nsec)
SD 100-33-22-223	5.1	0.100 (dia.)	-40 to 110	0.2 @ 550nm	300	6.4 @ 5V	50	87 @ 0V	4.2×10^{-12} (3)	15 @ 5V
SD 100-34-23-123	5.1	0.100 (dia.)	-40 to 110	0.01 @ 254nm	650	6.0 @ 5V	5	100 @ 0V	6.0×10^{-12} (3)	45 @ 0V
SD 290-32-31-243	42.6	0.300 x 0.220	-20 to 75	0.5 @ 900nm	N/A	320 @ 24V	75	32 @ 24	7.0×10^{-13} (4)	30 @ 24V
SD 170-31-22-223	14.8	0.152 x 0.152	-40 to 110	0.2 @ 550nm	500	18 @ 10V	50	250 @ 0V	3.5×10^{-13} (3)	90 @ 0V

1. Typical values listed. Minimum value shall be 50% of typical.
2. Typical values are listed in the table. Maximum value is 20% higher than the typical value.
3. Test conditions are peak wavelength and V_B=10mV.
4. Test conditions are V_B=-24V and 900nm.
5. Response times were measured at 950nm with a 50Ω load.

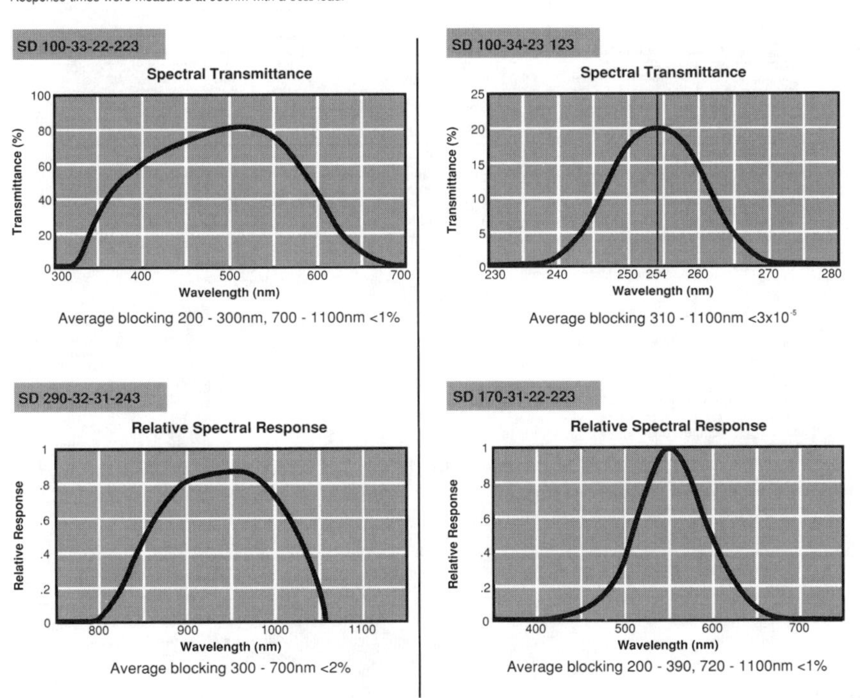

FIGURE 4.2. Specification for a filter-detector-type photodiode. [Reprinted by permission from the API Catalog (Advanced Photonix, Camarillo, CA, Inc., 1996).]

Type No. (A)	Remarks	Spectral Response: Curve Code (B)	Spectral Response: Range (nm)	Spectral Response: Peak Wavelength (nm)	Photocathode Material (C)	Window Material (D)	Outline No. (E)	Dynode Structure / No. of Stages (F)	Socket (G)	Maximum (H): Anode to Cathode Voltage (Vdc)	Maximum (H): Average Anode Current (mA) (J)	Cathode Sensitivity, Luminous: Min. (μA/lm)	Cathode Sensitivity, Luminous: Typ. (μA/lm)
1-1/8 inch (28 mm) Dia. Types													
R1459	For VUV detection, MgF_2 window, Cs-I Photocathode	100M	115 ~ 200	140	Cs-I	MF	1	B/11	E678-14C	2500	0.01	—	—
R1460	Variant of R1459 with Cs-Te photocathode	200M	115 ~ 320	210	Cs-Te	MF	1	B/11	E678-14C	1500	0.01	—	—
R431S	Solar blind response, Cs-Te photocathode, short length	200S	160 ~ 320	210	Cs-Te	Q	2	B/11	E678-14C	1500	0.01	—	—
R268	Bialkali photocathode, visible response, 11-stage	400K	300 ~ 650	420	Bi	K	1	B/11	E678-14C	1500	0.01	60	95
R434	For scintillation counting, 9-stage				Bi	K	4	B/9	E678-14C	1250	0.1	50	95
R1759	Variant of R434 with 7-stage dynodes				Bi	K	4	B/7	E678-14C	1250	0.1	50	95
*R3082	For high energy physics, high speed response, low TTS			420	Bi	K	3	L/10	E678-14C	1500	0.2	60	95
R269	Variant of R268 with UV glass window	400U	185 ~ 650	420	Bi	U	1	B/11	E678-14C	1500	0.1	40	95
R292	Variant of R268 with synthetic silica window	400S	160 ~ 650	420	Bi	Q	1	B/11	E678-14C	1500	0.1	40	95
*R3172	For high energy physics, high speed response, low TTS				Bi	Q	3	L/10	E678-14C	1500	0.2	60	95
R1282	High temp. bialkali photocathode, ruggedlized type	401K	300 ~ 650	375	H Bi	K	2	B/11	E678-14H	2500	0.1	—	40
R374	Wide spectral response, multialkali photocathode	500U	185 ~ 850	420	M	U	1	B/11	E678-14C	1500	0.1	80	150
R1104	High gain variant of R374				M	U	1	B/11	E678-14C	1500	0.1	80	140
R453	General purpose type of R374 with relaxed dark spec				M	U	1	B/11	E678-14C	1500	0.1	80	120
R376	Variant of R374 with synthetic silica window	500S	160 ~ 850	420	M	Q	1	B/11	E678-14C	1500	0.1	80	150
*R2228	Extended red multialkali photocathode	501K	300 ~ 900	650	M	K	1	B/11	E678-14C	1500	0.1	100	200
R316	For red to IR detection, S-1 response	700K (S-1)	400 ~ 1200	800	Ag-O-Cs	K	1	B/11	E678-14C	1500	0.01	10	20
R316-02	High red-sensitivity variant of R316, QE 0.08% at 1.06μm				Ag-O-Cs	K	1	B/11	E678-14C	1500	0.01	10	20

(A) *: Newly listed in this catalog.
(B) Typical spectral response characteristics are shown on pages 76 and 77.
(C) Photocathode materials
- Bi : Bialkali
- H Bi : High temperature bialkali
- M : Multialkali

(D) Window materials
- MF : MgF_2
- Q : Synthetic silica
- K : Borosilicate glass
- U : UV glass

(E) Basing diagram symbols are explained on page 17.
(F) Dynode Structure
- B : Box-and-grid
- L : Linear focused

(G) A socket will be supplied with a tube.
(H) The maximum ambient temperature range is −80 to +50°C except high temperature bialkali photocathode types which withstand up to +175°C. When using tubes with glass base at −30°C or below, see precautions on page 74.
(J) Averaged over any interval of 30 seconds maximum.
(K) At the wavelength of peak response.
(L) Voltage distribution ratios used to measure characteristics are shown on page 62.
(M) Anode characteristics are measured with the supply voltage and the voltage distribution ratio specified by Note (L).

a : At 122 nm
b : At 254 nm
c : At 1000 A/lm
d : Measured using a red filter Toshiba IR-D80A.
e : At 4 A/lm

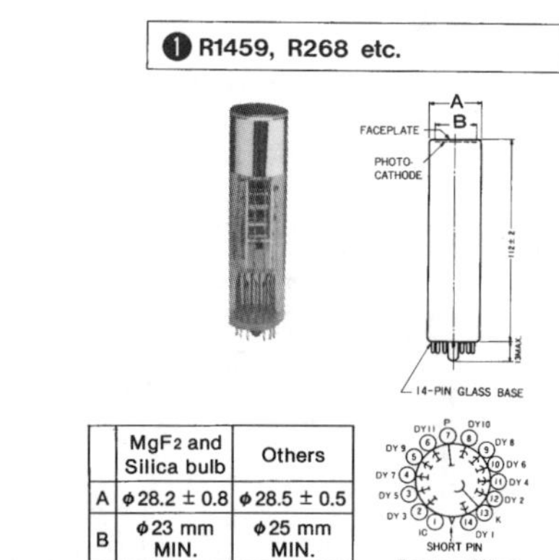

	MgF_2 and Silica bulb	Others
A	ϕ28.2 ± 0.8	ϕ28.5 ± 0.5
B	ϕ23 mm MIN.	ϕ25 mm MIN.

R2228 has a plano-concave faceplate.

FIGURE 4.3. Specification for a typical photomultiplier tube. [Reprinted by permission from *Photomultiplier Tubes Catalog* (Hamamatsu Photonic Systems, Bridgewater, NJ, 1990).]

initial element that converts light to electrons; therefore the window material andthe photocathode material determine the spectral response of the detector. The *dynode structure* and the *number of stages* determine the multiplication of the initial photoelectrons into an amplified signal.

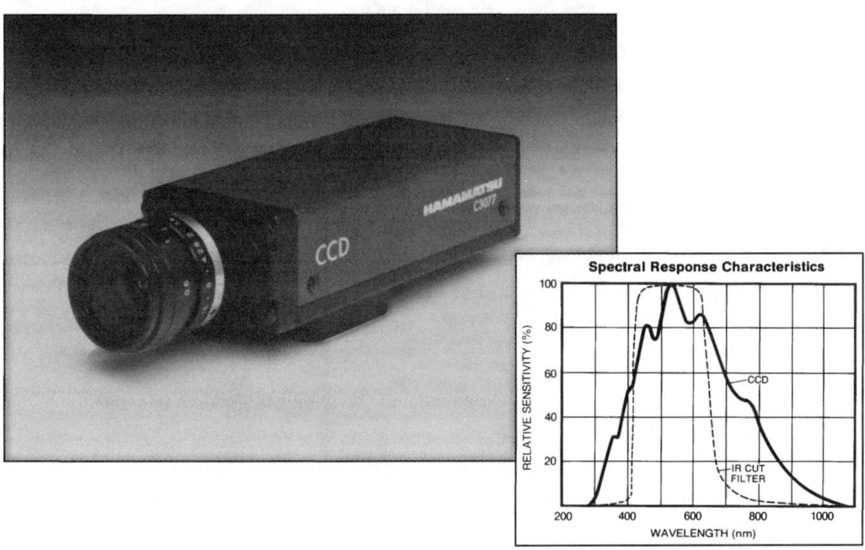

Small, Lightweight CCD Video Camera, High Resolution, High Sensitivity, & High S/N Ratio

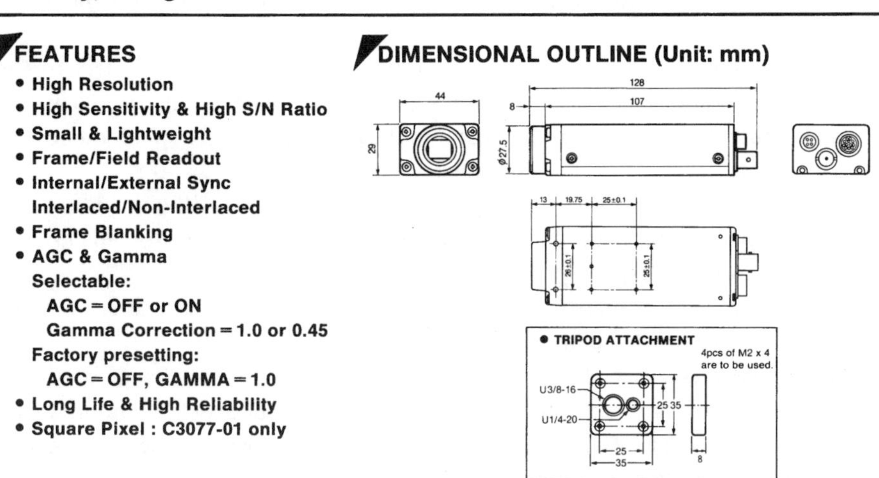

FIGURE 4.4. Specification for a typical charge coupled device (CCD) detector. (Reprinted by permission from Product Spec. Hamamatsu C3077 CCD Camera System. Hamamatsu Photonic Systems, Bridgewater, NJ.)

Charge-Coupled Devices (CCDs)

A sample specification for a charge-coupled detector is given in Fig. 4.4 [10]. *Horizontal resolution* is the number of individual pixels that can be distinguished in a horizontal scanning line. *Lag* is the number of frames an image remains in a camera tube after its initial formation; it is particularly a problem when trying to view moving objects under low illumination. The focusing system in a CCD camera may cause *geometric distortion*, which refers to the fact that the image is not a true to scale reproduction of the object. An example of this is barrel distortion, which has less image to pixel in the center of the picture than in the edges. Altering the lens system of a camera, for example, by adding close-up lenses, may increase the distortion.

CCD cameras often save luminance information logarithmically. *Gamma* is the base of the logarithm. *Automatic gain control* (AGC) is how a camera with minimal logic can set its own gain, and change it as required: this is good if you are a photographer, but it must be disabled for luminance information to have meaning.

4.3 INSIDE THE DETECTOR

4.3.1 Detection Elements

In the next three sections I will discuss the parts of a detector. The detection element, where the light is first sensed, is the natural place to start. We will begin with an overview of the different types of detection elements before going into more detail about each one.

The most common detection element for illuminance and luminance detectors is the p-i-n photodiode. The usual p-i-n material for detection in the visible is silicon, possibly doped for enhanced signal in the blue. Germanium photodiodes are used for signals near and in the ultraviolet.

Occasionally photomultiplier tubes are used for measuring illuminance, especially in laboratory equipment. Photomultiplier tubes are more sensitive than p-i-n diodes, but also more fragile.

Spectroradiometers primarily use CCD chips as their sensors. There are some CCD-based luminance detectors in experimental use.

Thermal detectors measure light by measuring the temperature change produced by the absorption of radiation. They are sensitive in the visible and the infrared, and are most typically used for measurements in the infrared, for example, radiometers and spaceborne horizon sensors, as well as burglar alarms and fire detection systems. Examples of thermal detectors are bolometers, pyroelectric detectors, and thermopiles. Another detector used primarily in the IR for applications similar to thermal detectors are photoconductive detectors.

p-i-n Diodes

The p-i-n diode is a solid-state device; to understand its function, we must first describe a bit of semiconductor physics. Several excellent introductory references to solid-state physics, semiconductors, and condensed matter are available. In a solid-state device, the electron potential can be described in terms of *conduction bands* and *valence bands*, rather than individual potential wells (see Fig. 4.5) [11]. The highest energy level containing electrons is called the *Fermi level.* If a material is a conductor, the conduction and valence bands overlap and charge carriers (electrons or holes) flow freely; the material carries an electrical current. An insulator is a material for which there is a large enough gap between the conduction and valence bands to prohibit the flow of carriers; the Fermi level lies in the middle of the forbidden region between bands, called the band gap. A *semiconductor* is a material for which the band gap is small enough that carriers can be excited into the conduction band with some stimulus; the Fermi level lies at the edge of the valence band (if the majority of carriers are holes) or the edge of the conduction band (if the majority carriers are electrons). The first case is called a *p-type semiconductor*, the second is called *n type*. These materials are useful for optical detection because incident light can excite electrons across the band gap and generate a photocurrent.

The p-i-n diode is a photoconductive device formed from a sandwich of three layers of crystal, each layer with different band structures caused by adding impurities (doping) to the base material, usually silicon or germanium. The layers are doped in this arrangement: p-type (or proton) on top, intrinsic, meaning undoped, in a thin middle layer, and n-type (or negative) type on the bottom. For a silicon crystal a typical p-type impurity would be boron, and indium would be a p-type impurity for germanium [12]. The change in potential at the interface has the effect

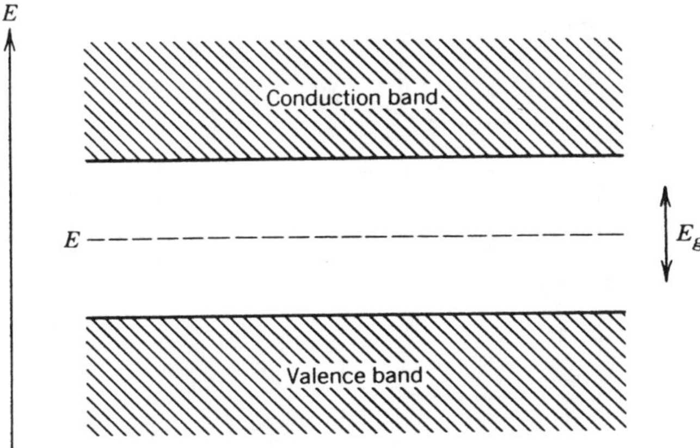

FIGURE 4.5. Allowed electron energies in an insulator (an example of the band structure of a solid). (From E. L. Dereniak and D. G. Crowe, *Optical Radiation Detectors.* © 1984 John Wiley & Sons. Reprinted by permission of John Wiley & Sons, Inc.)

of influencing the direction of current flow, creating a diode.

The structure of a typical p-i-n photodiode is shown in Fig. 4.6 [13]. The p-type and n-type silicon form a potential at the intrinsic region; this potential gradient depletes the junction region of charge carriers, both electrons and holes, and results in the conduction band bending. The junction drives holes into the p-type material and electrons into the n-type material. The difference in potential of the two materials determines the energy an electron must have to flow through the junction. When photons fall on the active area of the device, they generate carriers near the junction, resulting in a voltage difference between the p-type and n-type regions. If the diode is connected to external circuitry, a current will flow that is proportional to the illumination.

Photodiodes can be operated either with or without a bias voltage. Unbiased operation is called the photovoltaic mode; the $1/f$ noise is lower and the NEP is better at low frequencies. Signal-to-noise ratio is superior to the biased mode of operation for frequencies below about 100 kHz [14]. Biasing (connecting a voltage potential to the two sides of the junction) will sweep carriers out of the junction region faster and change the energy requirement for carrier generation to a limited extent. Biased operation (photoconductive mode) can be either forward or reverse bias. Reverse bias of the junction (positive potential connected to the n side and negative connected to the p side) reduces junction capacitance and improves response time; for this reason it is the preferred operation mode for pulsed detectors. A p-i-n diode used for photodetection may also be forward biased (the positive potential connected to the p side and the negative to the n side of the junction), to

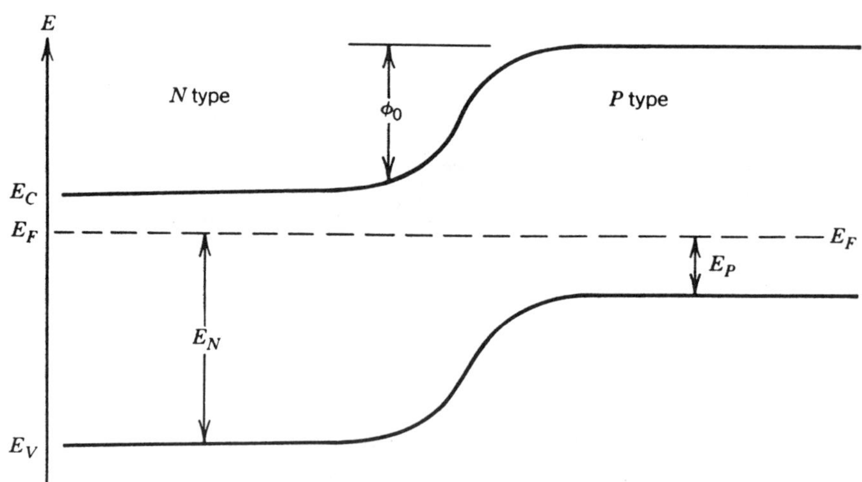

FIGURE 4.6. *p-n* junction energy diagram. (From E. L. Dereniak and D. G. Crowe, *Optical Radiation Detectors.* © 1984 John Wiley & Sons. Reprinted by permission of John Wiley & Sons, Inc.)

make the potential scaled for current to flow less, or in other words to increase the sensitivity of the detector (Fig. 4.7) [15].

p-i-n diodes can be used to detect light because when photon flux irradiates the junction, the light creates electron hole pairs with their energy determined by the wavelength of the light. Current will flow if the energy is sufficient to scale the potential created by the *p-i-n* junction. This is known as the photovoltaic effect [16]. It should be mentioned that the material from which the top layers of a *p-i-n* diode is constructed must be clear (and clean!) to allow the passage of light to the junction. If the bias voltage is increased significantly, the photogenerated carriers have enough energy to start an avalanche process, knocking more electrons free from the lattice which contribute to amplification of the signal. This is known as an avalanche photodiode (APD); it provides higher responsivity, especially in the near infrared, but also produces higher noise due to the electron avalanche process.

The sensitivity of a *p-i-n* diode can vary widely by quality of manufacture. A typical *p-i-n* diode size ranges from 5 mm×5 mm to 25 mm×25 mm. Ideally, the detection surface will be uniformly sensitive. At NIST there is a detector profiler that, by using extremely well-focused light sources, can determine the sensitivity of a detector's surface. For most applications, such as illuminance measurements, it is required that the detector is uniformly illuminated.

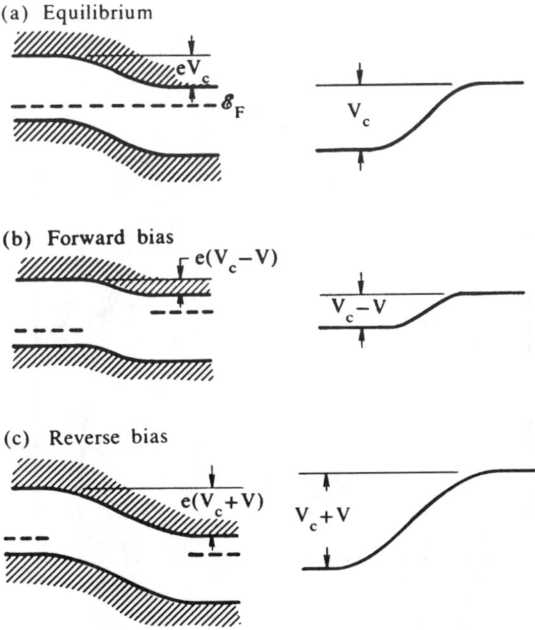

FIGURE 4.7. (a) The spatial dependencies of the electrostatic potential and the electron energy levels. (b) The same as (a) but for a forward biased *p-n* junction. (c) The same as (a) but for a reverse biased *p-n* junction. [Reprinted by permission from G. Burns, *Solid State Physics* (Academic Press, San Diego, CA, 1985), p. 330.]

The spectral responsivity of an uncorrected silicon photodiode is shown in Fig. 4.8 [17]. The typical QE curve is also shown for comparison. An ideal silicon detector would have zero responsivity and QE for photons whose energies are less than the band gap, or wavelengths much longer than about 1.1 μm. Just below the long-wavelength limit, this ideal diode would have 100% QE and responsivity close to 1 A/W; responsivity versus wavelength would be expected to follow the intrinsic spectral response of the material. In practice, this does not happen; these detectors are less sensitive in the blue region, which can sometimes be enhanced by clever doping, but not more than an order of magnitude. This lack of sensitivity is because there are fewer short wavelength photons per watt, so responsivity in terms of power drops off, and because more energetic blue photons may not be absorbed in the junction region. Filters are used so detectors will respond photometrically, or to the CIE color coordinates; however, the lack of overall sensitivity in the blue region can potentially create noise problems when measuring a low-intensity blue signal. In the deep ultraviolet, photons are often absorbed before they reach the sensitive region by detector windows or surface coatings on the semiconductor. The departure from 100% QE in real devices is typically due to Fresnel reflections from the detector surface. The long-wavelength cutoff is more gradual than expected for an ideal device because the absorption coefficient decreases at long wavelengths, so more photons pass through the photosensitive layers and do not contribute to the QE. As a result, QE tends to roll off gradually near the band-gap limit. This response is typical for silicon devices. Other common materials are InGaAs, which

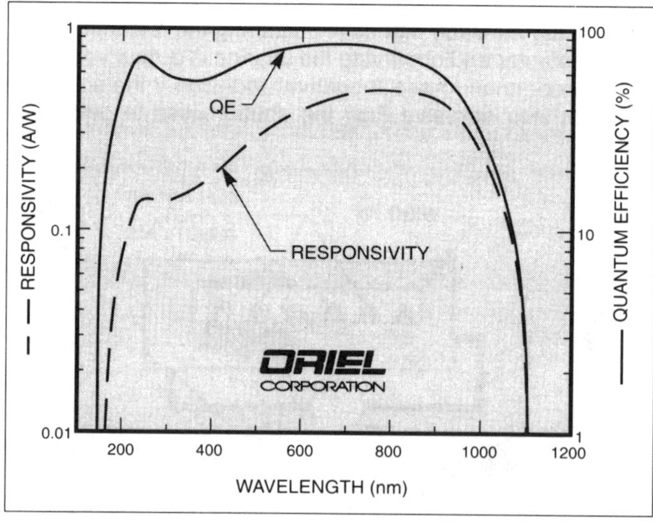

FIGURE 4.8. The responsivity and quantum efficiency of silicon. [Reprinted by permission from *ORIEL Instruments Catalog*, volume II (ORIEL Instruments, Stratford, CT, 1994).]

is most sensitive in the near infrared (0.8–1.7 μm) and HgCdZnTe for wavelengths of 2–12 μm. Later we will discuss optical immersion techniques to increase the active area of these devices. In the 1940s a popular photoconductive material for infrared solid-state detectors was introduced, lead sulfide (PbS); this material is still commonly used in the region of 1–4 μm [18].

p-i-n diode detectors have an approximate photopic response of 0.4 microamp/foot candle (μA/fc) and a radiometric response at 555 nm of 0.2 A/W [19]. *p-i-n* diode detectors are not very sensitive to temperature (−25 to +80 °C) or impact (compared to a photomultiplier tube, for example), making them an ideal choice for a field experiment. It is very important to keep the surface of any detector clean. This becomes an issue with *p-i-n* diode detectors because they are sufficiently rugged than they can be brought into field experiments that expose them to the natural elements. Buildup of dirt during an experiment should be avoided since it may bias the results.

Photomultiplier Tubes

The photomultiplier tube is a photoemissive device (see Fig. 4.9). The device consists of a vacuum tube with metal elements called dynodes under a high voltage. The first element, the photocathode, is negatively biased and will eject electrons when light is incident upon it, due to the photoelectric effect. The energy of the electrons is determined by the equation

$$\tfrac{1}{2}mv^2 = hf - \phi \tag{4.6}$$

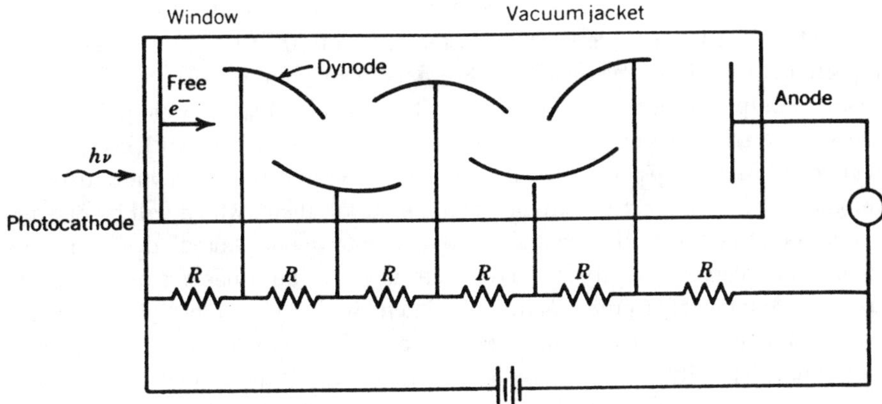

FIGURE 4.9. Photomultiplier tube schematic. (From E. L. Dereniak and D. G. Crowe, *Optical Radiation Detectors.* © 1984 John Wiley & Sons. Reprinted by permission of John Wiley & Sons, Inc.)

where m is the mass of the electron, v is its velocity, f is its frequency, ϕ is the work function (number of incident photons per second), and h is Planck's constant [20]. In other words, light interacts directly with the electrons in the detector material. An absorbed photon whose energy (hf) is greater than the work function of the photocathode frees an electron, with the surplus energy converted into kinetic energy of the electron. Electrons with enough kinetic energy escape the surface of the material and are known as photoelectrons. The voltage on the tube causes these ejected electrons to hit the next metal element at higher energy, due to the difference in potential of the elements, which causes several lower energy secondary electrons to be emitted. These electrons in turn are drawn to the next element, eventually causing a cascade effect that multiplies the effect of the original electron emitted from the photocathode (despite its name, this device does not "multiply" photons but electrons that are generated by the incident light). To achieve high amplification, kilovolts of biasing are not uncommon in PMTs; they are limited only by the breakdown voltage of the insulator between the dynodes. Fortunately, most PMTs are also low-current devices. Note that current flows only when photons below the critical wavelength are incident upon the detector. If no light is present, or only light which is not of the proper wavelength, then no current flows. The tube is often pumped down to vacuum, which is a good insulator; it will only force electrons off the cathode into the anode under a very high potential difference, and this is known as field emission. Some tubes are filled with gas to increase the breakdown threshold further. It is very important that the high-voltage supply be regulated to produce a stable bias voltage; otherwise, the gain may vary.

Photomultiplier tubes have spectral responses characterized by the photocathode: in general those used have low infared response and high ultraviolet response. The gain of a photomultiplier tube is determined by the voltage applied to it: since the device can take thousands of volts before the breakdown voltage between the insulating material and spacers are reached, amplifications of 10 to the 6 can be achieved [21]. The response speed is to the order of 10^{-10} s [22]. The response is proportional to light intensity over 6–8 orders of magnitude [23].

Photomultiplier tubes are seldom used outside of a laboratory setting, due to their high-voltage requirements (>1 kV bias), the voltage divider network consisting of hundreds of microamps, and their inherent fragility. In essense, they are made of a thin glass tube with precisely aligned metal plates in them, which will be damaged in transport if not carefully packed (as compared to p-i-n photodiodes). They also require more careful use: their cathodes are likely to burn out if exposed to high flux. An example of a problematic experiment would be a low-light experiment conducted in an experimental chamber, where the PMT would have little shuttering or filtering. Typically errors include forgetting to turn off the voltage to the detector before removing it from the experimental chamber or before turning on the room lights if the experiment is conducted on an unshielded optical bench. The photomultiplier tube is often used as a DC detector with low NEP, or a narrow-band AC detector with a lock-in amplifier (see Sec. 4.3.5.).

Charge-Coupled Devices

Multichannel detector arrays are desirable because of their two-dimensional features. Linear arrays of photodiodes can be used to form images; however, such arrays are susceptible to high readout noise which limits their detectivity. A more elegant alternative is the charge coupled detector. A CCD is a solid-state device composed of an array of metal–insulator–semiconductor capacitors. The major type of CCD available is composed of silicon wafer (semiconductor) with a silicon dioxide (insulator) surface grown on it and a matrix of metal squares deposited on top of the silicon dioxide layer, known as metal–oxide–silicon (MOS) technology (Fig. 4.10) [24]. When a photon of energy greater than the band gap is absorbed into the metal surface an electron–hole pair is formed. However, unlike p-i-n diodes, this array will store the charge, since it is composed of capacitors. Pulsed voltages will move the stored charges from their location to the next capacitor, and eventually to the digital electronics where they can be processed. This storage is significantly different from the real-time operation of single detector elements discussed previously. *Integration time* is now an important factor in setting up the experiment; with very long integration times dark current will build up. The elements may eventually saturate and the array loses all stored information; this concept is called saturation time, and it is also temperature dependent. Another issue is readout of the arrays; the most useful type is self scanning arrays, in which individual detector elements integrate the signal over time and read out serially. This limits the CCD dynamic range to only about 5 orders of magnitude, as compared with up to 10 orders of magnitude for individual detector elements. Linear arrays have pixels about 25 μm^2, spaced close together to provide high effective active area; however, each channel must be isolated to avoid cross-talk or ''blooming'' effects. The pixel width is narrow enough to give high resolution, and the height is large enough to provide adequate signal collection. This device is extremely powerful because it makes it possible to save images quickly and efficiently. However, it is limited by the speed of the electronics and the digitization levels—in the past, each element could save only 256 levels of light to a computer. CCD chips are often used for

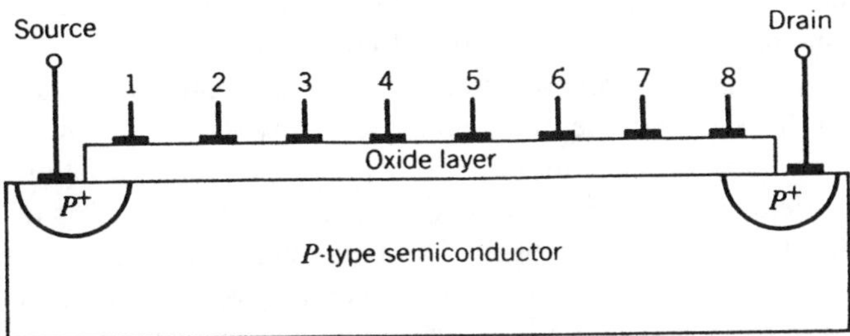

FIGURE 4.10. Metal–oxide–semiconductor field-effect transistor with eight gates. (From E. L. Dereniak and D. G. Crowe, *Optical Radiation Detectors.* © 1984 John Wiley & Sons. Reprinted by permission of John Wiley & Sons, Inc.)

detection in spectroradiometers, since a grating will spacially decompose incident light into wavelengths, and a CCD detector will automatically record the spatial position of incident light. A major advantage is multichannel scanning or imaging the entire slit of an imaging spectroradiometer.

CCD cameras can be used to record luminance information of images, becoming an imaging photometer. One commercially available system, CAPCALC, applies imaging photometry to vision research [25]. For this type of system it is important to realize that the linearity of response of the detector is altered by the acceptance of the optics. Each focal, aperture, and zoom setting has to be individually calibrated. Luminance detectors and imaging optics will be discussed below in Sec. 4.3.4. The CAPCALC system has been applied to lighting modelling, visibility through smoke and fog, and automotive window glazing [26]. It should be noted that most commercial CCD cameras automatically adjust the detector sensitivity to the light level incident on the camera (AGC), which must be disabled when designing this type of system.

Attempts have been made to convert color CCD video cameras to detectors that measure color luminance information. However, these color CCD chips use three CCD chips, each with a color filter. The color filters that are commercially available do not well approximate the CIE color coordinates and indeed have artificial features in the blue. In fact, most video cameras have traditionally not photographed properly in the blue, which resulted in Elizabeth Taylor having ''violet'' eyes, when her eyes were really blue.

Thermal and Photoconductive Detectors

Thermal detectors operate by converting incident radiation into a temperature change, which can be detected in several different ways such as monitoring the voltage generated at the junction of dissimilar metals, or by the pyroelectric effect. The bolometer was the first thermal detector invented (1880) and yet is still in use, despite the development of other thermal detectors. It measures the temperature change due the absorption of radiation by measuring a change in resistance in the material used to fabricate the bolometer. There are five types of bolometers—metal, thermistor, semiconductor, composite and superconducting. Of these, the thermistor and semiconductor bolometers are in most frequent use.

Thermistor is the most common bolometer used for popular applications, such as burglar alarms. This is because of its great durability—when hermetically sealed, this type of detector resists vibration, shock, temperature variations, and high humidity. The detection element of a thermistor bolometer is wafers of manganese cobalt and nickel oxides sintered together and mounted on a substrate like sapphire, which is electrically insulating but thermally conductive. A matched pair of these detectors are used in a bridge, with one detecting and one shielded, to account for ambient temperature variations (see Fig. 4.11) [27]. Sometimes thermistor bolometers are fabricated on lenses to increase their NEP, a technique known as optical immersion (see Fig. 4.12) [28]. If the detector is preceded by a hyperspherical lens

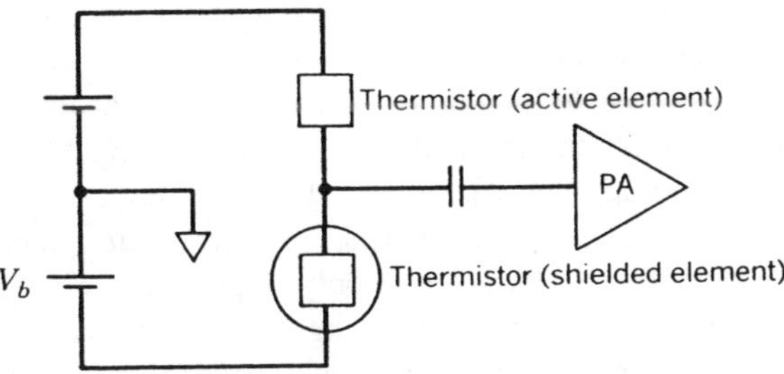

FIGURE 4.11. Biasing configuration for thermistor bolometer. (From E. L. Dereniak and D. G. Crowe, *Optical Radiation Detectors.* © 1984 John Wiley & Sons. Reprinted by permission of John Wiley & Sons, Inc.)

of refractive index n, as shown in this figure, the detector surface will look n squared times larger. Such detectors can be fabricated monolithically, with the detector element such as HgCdZnTe grown directly on the lens substrate. The spectral response of a typical thermistor bolometer is 0.4–10 μm micrometers, depending on window transmission and the emissivity of the absorber. Low-temperature semiconductor bolometers, in particular the germanium bolometer, are used for astronomical applications. They have a large spectral response, from 15 to 100 μm and are very useful when using only one high-performance detector is possible. These experiments are largely restricted to high altitudes or space platforms, due to atmospheric absorption [29].

Pyroelectric detectors are made of polarized dielectric materials, and when their

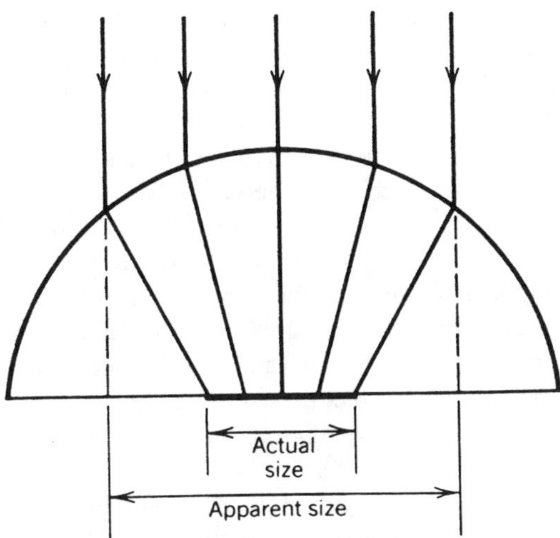

FIGURE 4.12. Immersion lens on thermistor. (From E. L. Dereniak and D. G. Crowe, *Optical Radiation Detectors.* © 1984 John Wiley & Sons. Reprinted by permission of John Wiley & Sons, Inc.)

temperature changes, this changes their polarization (due to an expansion of the crystal lattice spacing), and in turn causes a current to flow when connected to an external circuit. So, unlike bolometers, which measure the change in resistance when photons create a temperature change, pyroelectric detectors have a change in polarization with the temperature change, which creates a current flow. This is known as the *pyroelectric effect*. Since it is the temperature change that produces the current, pyroelectric detectors respond only to pulsed or chopped optical radiation; they are unaffected by steady-state background radiation. Small detectors can have extremely rapid response times, which depend on two time constants: the thermal time constant (determined by the thermal mass and packaging of the element) and the electrical time constant (the product of the shunt resistance and capacitance in the detector circuitry). The natural response frequency of these detectors is about 100–200 Hz, and the chopping frequency should be kept below this level for best noise performance; however, it can be extended electronically using lock-in amplifiers to around 1 kHz. The frequency response is lowered by adding a black coating, which provides uniform absorption over a wide spectral range; the detector may be only 25 μm thick, while the coating can be as much as 100 μm thick. These detectors are popular because they are sensitive at room temperature. A typical application is infared imaging. Pyroelectric detectors are beginning to be packaged in arrays [30].

Thermopiles are made by connecting a number of thermocouple junctions (typically 20–120) in series. Thermocouples are often used to measure temperature rather than simply being an optical instrument. When two dissimilar metals are connected in series (for example by spot welding) the temperature rise at the weld generates a voltage. When used with a reference junction at a known temperature (for example, dipped in ice water) the voltage will provide the temperature of the measuring junction. To detect radiation, one junction is blackened to absorb radiation. Typical materials used include bismuth and antimony, which have high thermoelectric coefficients. While one such detector would have a relatively low output voltage, connecting them into thermopiles increases the output voltage. Thermopiles are used for similar applications as pyroelectric detectors, only thermopiles are used for DC radiation, and pyroelectric detectors for pulsed or chopped radiation (see Sec. 4.3.5). Since they require no electrical bias, thermopiles have no flicker noise and excellent sensitivity in the infrared up to a few hertz (time constants are on the order of 50–200 ms). In fact, they are so sensitive care must be taken to stabilize the field of view by manually shuttering the radiation input and observing the change in output voltage. Nearly all room-temperature objects, including people, emit enough infrared to affect these devices; it is not uncommon to observe ''negative'' radiation if the source being measured is less emissive than the shutter used to obtain a ''zero'' reading [31].

In photoconductive materials absorbed incident photons change the conductivity of the materials without changing the temperature. This is done by the creation of free charge carriers. Photoconductive detectors are fabricated from PbS, PbSe, and HgCdZnTe. Optical immersion is also used with photoconductive detectors, and in general they are very similar to thermistors [32].

4.3.2 Filters

In this author's experience, the greatest difference between a high- and low-accuracy illuminance detector is the spectral correction. If any standard illuminance meter is waved under an incandescent lamp, it should read the same response within 3%. However, if the same detectors are waved in front of a surface illuminated by diffused red laser light, there can be as much as 100% difference. It is important to know how a detector manufacturer calculates spectral error, especially in the ''tails'' of the photopic curve. Some detectors measure their accuracy with only a white-light source, which can be misleading.

Illuminance and luminance detectors take a broadband measurement corrected by a photopic filter. Irradiance and radiance detectors also use filters to flatten spectral response, since the standard detection elements naturally have increasing response with wavelength. Other detectors, such as color meters, also use filters to match the CIE spectral response.

Filters are typically made of glass and plastic, and use ionic absorption and colloidal scattering to spectrally transmit light. Common compounds used in glass filters are nickel oxide, chromium oxide and ferric oxide, whose ions absorb according to Beer's law, and sulfur, cadmium sulfide, cadmium selenide, and gold, whose crystals embed in the glass and colloidally scatter [33].

Thin layer coatings can also use interference to eliminate narrow band passes, if the light is incident from a narrow solid angle. Care should be taken when using interference filters with uncollimated light; the transmission band is broader, with the peak shifted toward shorter wavelengths, and the resulting transmission curve is a weighted average over all the angles of incidence. Collimated light should always be used for this type of filter; if the filter is tilted with respect to the beam, the transmission peak shifts to shorter wavelengths (about 2%, or 12 nm for a 600-nm filter, is typical for tilts less than 20°). Polarization effects and band broadening are also problematic at angles greater than about 20° for interference type filters [34].

Filters can be used in series, rather than trying to combine different band passes into one filter. Since the light arriving at the filter is already diffused (see Sec. 4.3.3), sometimes placing additional filters side by side on the surface of your detector can create more accurate spectral correction. The principal behind this is that the best red filter absorbs too much blue to cover the whole signal, and the best blue filter absorbs too much red, but a little here and there can fine-tune the system. The result is called a *mosaic filter*. For this technique to work, high confidence in your diffusor and in the surface response from your detection element is required. Mosaic filters are used on more expensive detectors that require higher spectral accuracy [35].

Neutral-density filters are sometimes places over detectors to increase their dynamic range, when their full sensitivity is not required, and saturation is potentially a problem. There are two different ways that neutral-density filters are defined, in terms of optical density [$\log_{10}(I/I_0)$, where I/I_0 is the ratio of transmitted to incident light intensity] or f-stop, and they use similar symbols, so it is necessary to be careful when purchasing them. Spectral filters are available through several standard manufacturers, which can be found in the *Physics Today Annual Buyers Guide*, an American Institute of Physics publication. Filters can also be customized for your particular application, at a cost depending on your spectral requirements.

Colored glass filters can exhibit weak fluroescence and should not be positioned at the detector. Filters are sensitive to long-term UV exposure. Aging in filters can effect the accuracy, requiring recalibration. Filters are temperature sensitive, especially interference filters, whose layer intervals are particularly vulnerable to the temperature expansion and contraction. For this reason, care should be taken when mounting filters on high wattage lamps whose thermal properties may affect the filter; in this case, mount the filter on the detector instead. For example, if an interference filter is heated much above 80 °C for short periods of time, a small, nonreversible shift in center wavelength occurs at the shorter wavelengths. Optical cements used in laminated filters also tend to degrade at temperatures greater than 70 °C; they darken and increase filter absorption. Thermal shock can also result from abrupt localized heating of filters due to high power collimated beams of light. In general, filters should not be heated about 70 °C or exposed to shocks of greater than 5 °C per minute; if the filter must be mounted close to a high-power source, it is best to use a liquid filter or dichroic mirror between the source and filter.

4.3.3 Cosine Diffusors

It is straightforward to realize why a diffusor would be valuable in an illuminance meter; uniformly covering the detection element with signal eliminates concerns over the detector's surface response. It can be shown that the irradiance received by a flat surface in a collimated incident beam decreases with the cosine of the angle of incidence. Thus, the flux received by an ideal detector placed in such a beam should also show a cosine dependence; the detector signal should decrease with the cosine of incident angle. Such behavior is called *cosine correction*, because the reflectance of most materials increases with angle of incidence. This angular dependence on absorption causes the detector output to depart from Lambert's cosine law, and the detector must be corrected back to proper angular response. In general, the action of diffusing signal in an illuminance and irradiance detector is combined with cosine correction. In other words, the signal strength is weighted as to the solid angle of arrival. This cosine weighting is necessary due to the definition of illuminance and irradiance.

There are two ways of achieving this goal: using an integrating sphere or a transmitting diffusor. Integrating spheres are discussed more fully in Chap. 3. To summarize their relevant characteristics, they create almost perfect diffusing and cosine response, but only a small (1:100 in some cases) fixed ratio of signal reaches the detector, as well as their being expensive and bulky, which makes them unsuitable to most field experiments. In general they are used with laser experiments, whose extra collumination and coherence require their superior diffusing, as well as other laboratory-based measurements [36].

Transmitting diffusors are made from milk-white highly diffusing semitransparent materials having good conical-hemispherical spectral transmittance over the spectral range of good detector sensitivity [37]. For a transmitting diffusor cosine weighting is accomplished by means of the thickness of the diffusor, since the diffusor has some absorption. Typical detectors have cosine diffusors made of plastic. Opal glass is the traditional diffusing material (frosted opal glass is preferred). Opal glass can be purchased by certain standard laboratory suppliers, such as Oriel.

4.3.4 Imaging Optics

Luminance meters, like cameras, must image light on their sensing elements. The light is admitted from a restricted field of view, for example one degree. For this reason, the luminance meters' imaging optics is very important to their performance. Field of view is also critical for infrared detectors, because most objects at room temperature emit photons in the vicinity of 10 μm wavelength, which act as background noise. In Fig. 4.13 we see a typical imaging system for a luminance meter. View through optics is essential, because it is important to be certain the target is in the field of view. Since targets are imaged at different distances from the system, it is necessary to change the focal length of the lens without disrupting the baffling determining field of view. Designing such a system is outside the scope of this chapter; there are many fine "point and shoot" luminance meters commercially available. It is important to keep in mind that each optical element will affect the final values by their transmittance and vignetting, so calibration should be performed on the final unit, not the sensor inside.

Spectroradiometers use similar optics to luminance meters, with an added level of complexity—the grating which separates light by wavelength (see Fig. 4.14). Spectrometers or monochromators are like spectroradiometers without the focusing optics or detectors. Light must be focused into the spectrometer with external

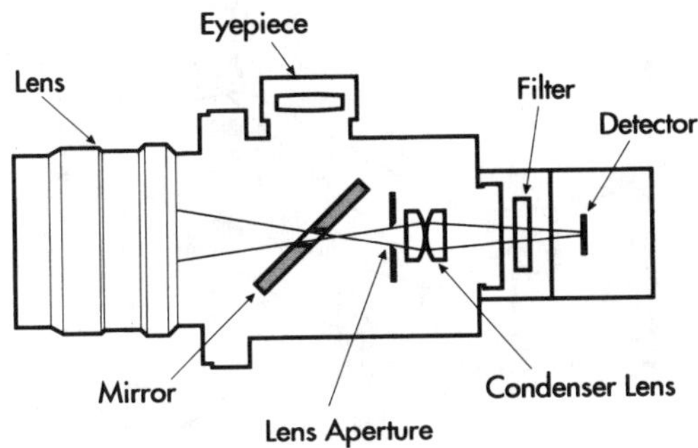

FIGURE 4.13. The optics of a luminance meter. For luminance measurements requiring small fields of view, a lens system with view-through optics is essential. [Reprinted by permission from *The Guide to Photometry* (Graseby Optronics, Orlando, FL, 1993), p. 6.]

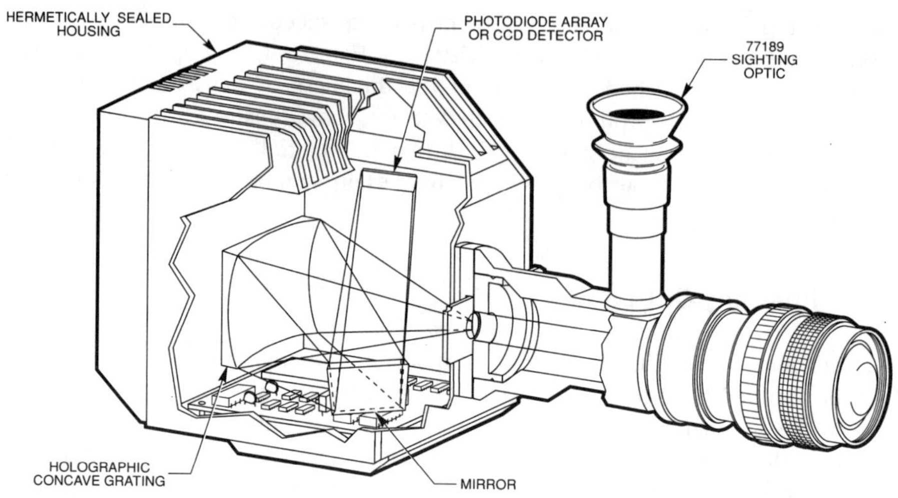

FIGURE 4.14. Cutaway diagram of a RadSpec spectroradiometer, with optional 77189 sighting optics. [Reprinted by permission from *ORIEL Instruments Catalog*, volume II (ORIEL Instruments, Stratford, CT, 1994).]

optics, and the light that comes out must be focused onto the detector (see Fig. 4.15): for this reason spectrometers can be used to spectrally calibrate a variety of detectors. A radiation budget is the calculation to determine the required sensitivity of a detector after the imaging optics. It is calculated by determining the approximate signal strength and the throughput of each optical element [38].

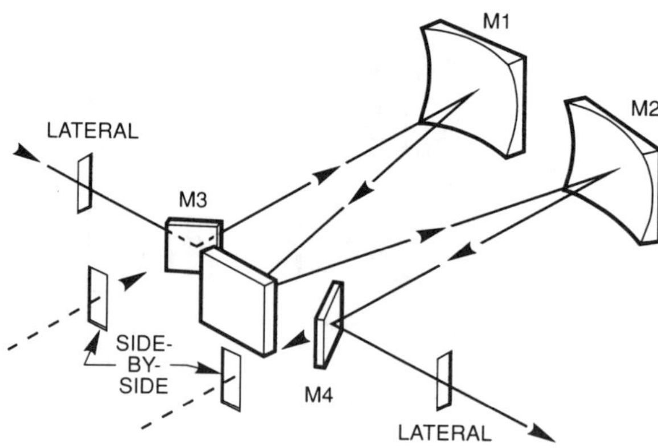

FIGURE 4.15. The optics of the 77200 1/4-m monochromator. [Reprinted by permission from *ORIEL Instruments Catalog*, volume II (ORIEL Instruments, Stratford, CT, 1994).]

4.3.5 Detector Electronics

When wiring an experiment, there are two kinds of electronics: detector powering and signal conditioning. They are not entirely independent: for example, for a PMT the high-voltage set on the detector will determine the detector sensitivity, as will the bias for a photodiode in the photoconductive mode. Since detector powering for each type of detector was discussed earlier, we will discuss here general signal conditioning. The first step of signal conditioning is preamplification, usually coupled with current to frequency (I/f) conversion. I/f conversion is changing a signal current to a signal voltage, which most electrical equipment can process; in other words, an I/f converter is a glorified resistor. In photomultiplier tubes, preamplification is inherent in the detector, but I/f conversion is not. Most preamplifiers designed for use with photodiodes implicitly I/f convert. After I/f conversion there are a number of options. If your signal is sufficiently strong, you can connect it directly to your readout. Otherwise, you want to move on to other forms of amplification. Most often, you will chose between electronic amplification and lock-in amplification. Electronic amplification is performed using solid-state electronics (operational amplifiers, etc.). Basically, it multiplies the signal (and noise) by a constant gain so that it is at the most sensitive part of the readout. It allows control over the part of the signal to which the readout is sensitive.

Lock-in amplification, or chopping, is an AC technique used to limit noise, although it can also be used to detect DC signals that are deliberately encoded with a known modulation. It is used only on very noisy signals. A mechanical light chopper (that looks somewhat like a fan) is placed between the signal and the detector. The amplifier is connected to the chopper, and so the amplifier ''knows'' from this reference signal when the signal to be detected is on and when it is off. As you can imagine, this makes for very precise background subtraction. However, that is not all the lock-in amplifier does.

The secret of the lock-in amplifier is that it narrows the bandwidth of the detector, and the narrower that bandwidth, the more precise the measurement. It does this by creating a weak periodic signal, and low pass filtering it (or narrowing the bandwidth). This helps eliminate $1/f$ noise (flicker noise, drifts, etc.; see ''Noise'' in Sec. 4.3.6). The signal you have after lock-in amplification is a differential signal, not a linear signal. It is good for determining changes in signal, not signal magnitude. Horowitz and Hill discuss lock-in amplification in *The Art of Electronics*, [39], p. 631.

> In order to illustrate the power of lock-in detection, we usually set up a small demonstration for our students. We use a lock-in to modulate a small LED [light-emitting diode] of the kind used for panel indicators, with a modulation rate of a kilohertz or so. The current is very low, and you can hardly see the LED glowing in normal room light. Six feet away a phototransistor looks in the general direction of the LED, with its output fed to the lock-in. With the room lights out, there's a tiny signal from the phototransistor at the modulating frequency (mixed with plenty of noise), and the lock-in easily detects it, using a time constant of a few seconds. Then we turn the room lights on (fluorescent), at which point the signal from the phototransistor becomes just a huge messy 120 Hz waveform, jumping in amplitude by 50 dB or more. The

situation looks hopeless on the oscilloscope, but the lock-in just sits there, unperturbed, calmly detecting the same LED signal at the same level. You can check that it's really working by sticking your hand in between the LED and detector. It's darned impressive.

Similar tricks to narrow the bandwidth are used in signal averaging, boxcar integration, multichannel scaling, pulse height analysis, and phase-sensitive detection. For example, boxcar averaging takes its name from gating the signal detection time into a repetitive train of N pulse intervals, or "boxes," during which the signal is present. Since noise which would have been accumulated during times when the gating is off is eliminated, this process improves the signal-to-noise ratio by a factor of the square root of N for white, Johnsson, or shot noise. (This is because the integrated signal contribution increases as N, while the noise contribution increases only as the square root of N [40].)

The amplified signal must then come to a form of readout, either analog, digital, or digital-computerized. Modern computers may have a card which performs analog to digital conversion of a voltage signal, and then inputs to the memory. One must be careful that the number of digital levels, for example, 128, is sufficient, and that your signal falls within the right voltage level, typical ± 10 V. This type of board can read the output of some preamplifiers for photodiodes quite well. Of course, if too big a voltage is sent to the computer that way, it can cause serious damage. Other ways some amplifiers are designed to output to a computer include RS232 serial and IEEE 488 parallel communication standards. If the experimental setup already requires computer control of motors, and the computer already has an appropriate communication bus, this can be very convenient. Since most forms of data analysis and reporting are already done on computer, designing a laboratory so that it reads directly to the computer can be valuable. Several companies provide software for computer instrument control and data analysis. High-end equipment designed for computer compatibility often comes with its own computer control software.

4.3.6 Detector System Fundamentals

In the following sections, we will consider some properties of the overall system that apply to many different kinds of detectors, namely, noise reduction and calibration.

Noise

A photometric system is subject to various types of noise. There can be noise in the signal, noise created by the detector, and noise in the electronics. A complete discussion of noise sources has already filled several good reference books; since this is a chapter on detectors, we will discuss briefly noise created by detectors. For a more complete discussion, the reader is referred to treatments by Dereniak and Crowe [41], who have categorized the major noise sources. It is often assumed that

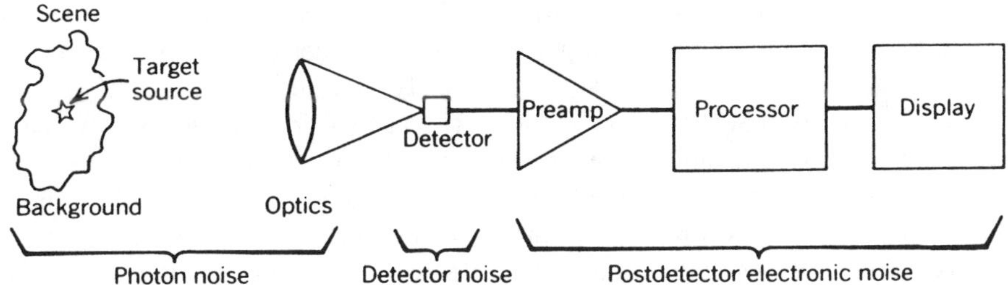

FIGURE 4.16. Detector system noise classification. (From E. L. Dereniak and D. G. Crowe, *Optical Radiation Detectors*. © 1984 John Wiley & Sons. Reprinted by permission of John Wiley & Sons, Inc.)

the noise in a detection system has a constant-frequency spectrum over the measurement range of interest; this is so-called white noise or Gaussian noise, and is often a combination of the effects in Fig. 4.16.

Shot noise occurs in all types of radiation detectors. It is due to the quantum nature of photoelectrons. Since individual photoelectrons are created by absorbed photons at random intervals, the resulting signal has some variation with time. The variation of detector current with time appears as noise; this can be due to either the desired signal photons or by background flux (in the latter case, the detector is said to operate in a background limited in performance, or BLIP, mode).

Generation-recombination, Johnson, and $1/f$ noises are particular to photoconductors. Absorbed photons can produce both positive and negative charge carriers, some of which may recombine before being collected. Generation-recombination noise is due to the randomness in the creation and cancellation of individual charge carriers. Johnson noise is the thermal noise caused by randomness in carriers generation and recombination due to thermal excitation in a conductor; it results in fluctuations in the detector's internal resistance, or in any resistance in series with the detector. Flicker or so-called $1/f$ noise is particular to biased conductors. Its cause is not well understood, but it is thought to be connected to the imperfect conductive contact at detector electrodes. It can be measured to follow a curve of $1/f^B$, where B is a constant that varies between 0.8 and 1.2.

Readout noise is particular to array detectors (such as CCDs) and is caused by the uncertainty in the large number of charge transfers that occur between storage registers on the way out of the detector. Microphonics is not a traditional source of noise, but can be listed with them because of the noiselike effect. It is due to vibrating or knocking about the detector and the detection electronics. It is most important in thermal detectors. This can change the electrical characteristics detector measurement circuit. Unlike the other noise sources mentioned above, this source can be eliminated by proper planning.

When designing an experiment, one must keep in mind likely sources of noise, their expected contribution, and how to best reduce them. Choose a detector so that the signal will be significantly larger than the detector's expected noise. Cooling some detectors will minimize the dark current; however, be sure to follow the manual. In general, it is far more important to keep the detector temperature stable than cool: Always make sure that a statistically significant amount of data is collected for analysis. Some AC techniques, such as lock-in amplification (which is discussed in Sec. 4.3.5), can help separate the signal from the noise. There are other rules of thumb that can be applied to specific detectors, as well. For example, although Johnson noise cannot be eliminated, it can be minimized. In a PMT, make the bias current shot noise the dominant noise source (at least 3× larger than the Johnson noise). In low-light situations, first choose the transimpedance gain as large as possible within your bandwidth limits; with zero bias, the AC noise at the PMT output is the Johnson noise. Then increase bias voltage until the variance is at least (3–4)× as large as the zero-bias case. The same rule of thumb applies to PbS- and PbSe-based detectors, though care should be taken not to exceed the maximum bias voltage of these devices or catastrophic breakdown will occur. In photodiodes, the shot noise is approximately 3× greater than Johnson noise if the DC voltage generated through a transimpedance amplifier is more than about 500 mV. In all cases, this results in a higher degree of linearity in the measurements while minimizing thermal noise.

Calibration

Calibration of detectors is the tracing of their response to a reference standard. How often a detector requires calibration is related to the care in use and storage, as well as the accuracy required for your application. However, if a laboratory uses the equipment for accredited standard measurements, the calibration intervals for illuminance and luminance meters is typically one year.

There are two basic philosophies in detector calibration: comparing the detector to a calibrated light source or a calibrated detector. In general, detectors travel better, but light sources remain stable over longer periods of time. Traditionally, accuracy was maintained by shipping lamps three at a time, presuming that only one would be changed in transport. However, there has been some discussion in the photometric community about using calibrated detectors to transfer standards between laboratories, while using calibrated light sources to maintain standards within a laboratory. NIST is leading this effort, and provides a service where silicon photodiode radiometers are rented from NIST and are calibrated for radiant power levels from 10^{-3} to 10^{-7} W in the 250–1064-nm spectral region. This measurement service has been documented in the NBS Special Publication No. SP250-17 [42].

The researcher who has only a few detectors is much better off having the manufacturer calibrate equipment than setting up a calibration facility, since this eliminates the need to purchase and maintain a traceable standard or set up a laboratory without light leakage. Measurement services are also provided to the public at several laboratories in the United States. The calibration and related measurement services provided by the U.S. national laboratory NIST are documented in a series of Special Publications, led off by the NIST Calibration Service

Users Guide, No. SP250. These documents can be obtained by writing the following addresses:
Superintendent of Documents
U.S. Government Printing Office
Washington, D.C. 20402-9325
(202) 783-3238

National Technical Information Service
5285 Port Royal Road
Springfield, VA 22161
(703) 487-4650

It is valuable to compare detectors calibrated by different manufacturers and laboratories before assuming that they are identical. Round-robin tests are a way of comparing different laboratories. The same lamps are sent from laboratory to laboratory, where they are measured, and careful records are kept as to burn time, etc. After the participating laboratories have completed their measurements, the lamps are sent to be remeasured at the first laboratory to ensure that the lamps were not changed in transport. This has proven to be a useful technique to maintain standard calibration between laboratories.

REFERENCES

1. *Oriel Catalog* (Oriel Corp., Stratford CT, 1994), Vol. II, Chap. 3, p. 3-15.
2. *The Photonics Directory Book 2: The Photonics Buyers Guide to Products and Manufacturers* (Laurin, Pittsfield, MA, 1993).
3. *Lasers & Optronics Technology and Industry Reference 1994* (Gordon, Morris Planes, NJ, 1993), Vol. 12, No. 10.
4. Exhibitor List, OSA 1994 Annual Meeting and Exhibit, October, 1994, held at the Loews Anatole Hotel, Dallas, Texas.
5. *Bulletin of the American Physical Society*, Dallas, Texas **34**, No. 3 (1989).
6. The United States National Committee of the International Commission of Illumination (USNC/CIE) publishes material that is available through CIE Publications, c/o TLA Lighting Consultants, Inc., 7 Pound Street, Salem, MA 01970. Phone: (508) 745-6870. FAX: (508) 741-4420.
7. Council for Optical Radiation Measurement (CORM) Secretary, 1043 Grand Ave #312, St. Paul, MN 55105.
8. *Optoelectronics Data Book* (Advanced Photonix, 1240 Avenida Acaso, Camarillo, CA 93012, 1993), pp. 36–37.
9. *Photomultiplier Tubes* (Hamamatsu Corp., Tokyo, 1990), p. 38.
10. CCD Video Camera C3077, Hamamatsu Corp., 1990, separate sheet.
11. Dereniak, E. L., and Crowe, D. G., *Optical Radiation Detectors* (Wiley, New York, 1984), Chap. 2, pp. 23–25.
12. Burns, G., *Solid State Physics* (Academic, Orlando, FL, 1985), Chap. 10, pp. 323–324.
13. Dereniak, E. L., and Crowe, D. G., *Optical Radiation Detectors*, Ref. 11, Chap. 2, p. 28.
14. *Oriel Catalog*, Ref. 1, Chap. 3, p. 3-10.
15. Burns, G., *Solid State Physics*, Ref. 12, Chap. 10, p. 330.
16. Dereniak, E. L., and Crowe, D. G., *Optical Radiation Detectors*, Ref. 11, Chap. 2, p. 32.
17. *Oriel Catalog*, Ref. 1, Chap. 3, pp. 3-8.
18. McCluney, W. R., *Introduction to Radiometry and Photometry* (Artec House, Boston, 1994), p. 205.
19. *General Purposes Detectors Technical Data Sheet*, Silicon Detectors Corporation, Camarillo, California, USA.

20. McCluney, W. R., *Introduction to Radiometry and Photometry*, Ref. 18, p. 193.
21. *Photomultiplier Tubes*, Ref. 9, p. 8.
22. McCluney, W. R., *Introduction to Radiometry and Photometry,* Ref. 18, p. 196.
23. *Oriel Catalog*, Ref. 1, Chap. 3, pp. 3-9.
24. Dereniak, E. L. and Crowe, D. G., *Optical Radiation Detectors*, Ref. 11, Chap. 9, p. 194.
25. Rea, M. S., and Jeffrey, I. G., ''A New Luminance and Image Analysis System for Lighting and Vision. I. Equipment and Calibration,'' J. IES, Winter, 64–72 (1990).
26. Bierman, A., Raffucci, J., Boyce, P. R., and DeCusatis, C., ''Exit-sign image degradation in smoke: A quantitative simulation,'' Int. J. Lighting Res. Tech. **28** (4), 177 (1996).
27. Dereniak, E. L., and Crowe, D. G., *Optical Radiation Detectors*, Ref. 11, Chap. 7, pp. 154–5.
28. Dereniak, E. L., and Crowe, D. G., *Optical Radiation Detectors*, Ref. 11, Chap. 7, p. 156.
29. Dereniak, E. L., and Crowe, D. G., *Optical Radiation Detectors*, Ref. 11, Chap. 7, pp. 163–5.
30. Dereniak, E. L. and Crowe, D. G., *Optical Radiation Detectors*, Ref. 11, Chap. 8.
31. *Oriel Catalog*, Ref. 1, Chap. 3, p. 3-7.
32. *Oriel Catalog*, Ref. 1, Chap. 3, p. 3-12.
33. *Oriel Catalog*, Ref. 1, Chap. 2, pp. 2-1 to 2-80.
34. *Oriel Catalog*, Ref. 1, Chap. 2, p. 2-15.
35. McCluney, W. R., *Introduction to Radiometry and Photometry*, Ref. 18, pp. 225–6.
36. *Optronics Laboratories Condensed Catalog*, Orlando, FL (1995), p. 5-6.
37. McCluney, W. R., *Introduction to Radiometry and Photometry*, Ref. 18, p. 221.
38. *Oriel Catalog*, Ref. 1, Chap. 2, p. 3-17.
39. Horowitz, P., and Hill., W., *The Art of Electronics* Cambridge, England: (Cambridge University Press, Cambridge, 1980), Chap. 14, pp. 628–631.
40. Horowitz, P. and Hill., W., *The Art of Electronics*, Ref. 39, Chap. 14, pp. 624–626.
41. Dereniak, E. L., and Crowe, D. G., *Optical Radiation Detectors*, Ref. 11, Chap. 2, pp. 36–43.
42. *The NBS Photodetector Spectral Response Calibration Transfer Program*, NBS Special Publication No. SP250-17.
43. *NIST Calibration Service Users Guide*, SP250; *Spectral Radiance Calibrations*, SP250-1; *Far Ultraviolet Detector Standards*, SP250-2; *Radiometric Standards in the Vacuum Ultraviolet*, SP250-3; *Regular Spectral Transmittance*, SP250-6; *Radiance Temperature Calibrations*, SP250-7; *Spectral Reflectance*, SP250-8; *Photometric Calibrations*, SP250-15; *The NBS Photodetector Spectral Response Calibration Transfer Program*, SP250-17; *Spectral Irradiance Calibrations*, SP250-20.

5

Measurement Procedures

Yoshi Ohno
NIST, A320/220, Gaithersburg, Maryland 20899

5.1 LUMINOUS-INTENSITY MEASUREMENT

In this section, the measurement of luminous-intensity transfer standard lamps are described. Luminous-intensity standard lamps, calibrated in one horizontal direction, provide the luminous-intensity unit (candela), and are required in the calibration facilities for light sources, illuminance meters, luminance meters, and other optical instruments. Luminous-intensity transfer standard lamps are calibrated either by measuring against luminous-intensity reference standard lamps (source-based method) or by using reference photometers (detector-based method). The luminous-intensity distributions of lamps and luminaires, measured by a goniophotometer, are discussed in Chap. 7.

5.1.1 Equipment

Photometric Bench

In order to calibrate or to use luminous-intensity standard lamps, a photometric bench is necessary, on which a standard lamp and a photometer head is precisely positioned and the lamp-to-photometer distance is varied freely and measured accurately. Figure 5.1 shows a schematic of a typical traditional photometric bench.

At one end of the bench, a luminous-intensity standard lamp is mounted on positioning stages that allow adjustment of the rotation, two horizontal positions, height, and the angle of the lamp. On the other side, a photometer head is mounted and aligned on the optical axis with the standard lamp. A photometric bench is often equipped with two telescopes, one perpendicular to the optical axis and another on the optical axis behind the lamp. The lamp position and orientation are aligned by using these telescopes. The center of the lamp filament is positioned at the origin of the distance scale, and the lamp-to-photometer distance is given by that scale. A He–Ne laser is also often used to represent the optical axis. Depending on the type of lamps, the lamp orientation can be adjusted by retroreflection of the laser beam from an alignment device. When the photometer is moved, the aperture screens are also moved accordingly.

In order to avoid stray-light errors, the entire photometric bench is usually

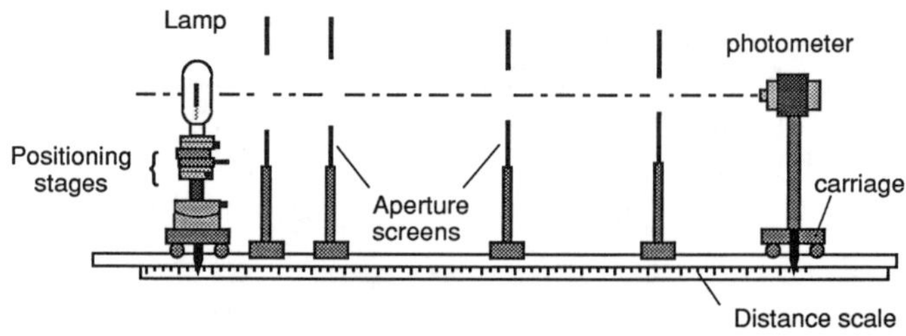

FIGURE 5.1. A traditional photometric bench.

installed in a darkroom where all the walls are painted black. When a darkroom is not available, a photometric bench can be installed in a light-tight box where all the walls are covered with a black velvet cloth. Figure 5.2 shows a typical layout of the components of a photometric bench. Aperture screens are arranged so that no light can directly enter the photometer from the surrounding walls and objects, and no light will leak out directly from the lamp to the surrounding space. The aperture screens are painted flat black or, for best results, covered with black velvet. Three or more aperture screens are normally used depending on the bench distance. The aperture of the screen has sharp edges to minimize errors by reflections. Such reflections can be visually checked by looking from the back of the photometer toward the lamp. If shiny rings are seen, the edges need to be sharpened. The photometer is preferably covered by an enclosure whose inner surfaces are covered with black velvet or painted flat black.

The lamp should be shielded with additional screens or a cover in order to minimize direct light spreading out in the room. Appropriate ventilation around the lamp should also be provided. Special attention sould be paid to the back of the lamp since any surface on the back of the lamp is viewed by the photometer. In order to minimize the reflection from the back surface, a black velvet cloth is normally placed more than 1 m away from the lamp. Otherwise, a light trap as described later in this section can be used.

The length of the bench is typically 3–10 m. A longer bench is preferred in order to produce a large range of illuminance from one standard lamp. Typical luminous-intensity standard lamps are not used at distances shorter than 1 m, where the inverse-square errors can be serious and the lamp-distance alignment becomes critical.

Any component, such as an automatic shutter or an additional aperture, should not be placed within about 10 cm in front of the photometer head unless it is permanently attached as part of the photometer. The reflections from the photometer front surface would reflect back strongly from objects at a short distance and can cause significant errors.

Figure 5.3 depicts another example of a photometric bench of a recent design [1,2]. The entire bench is in a light-tight box, the inside of which is covered with black velvet. Therefore, the bench can be installed in an ordinary laboratory room with bright walls or windows. The position of the carriage on the rails is measured

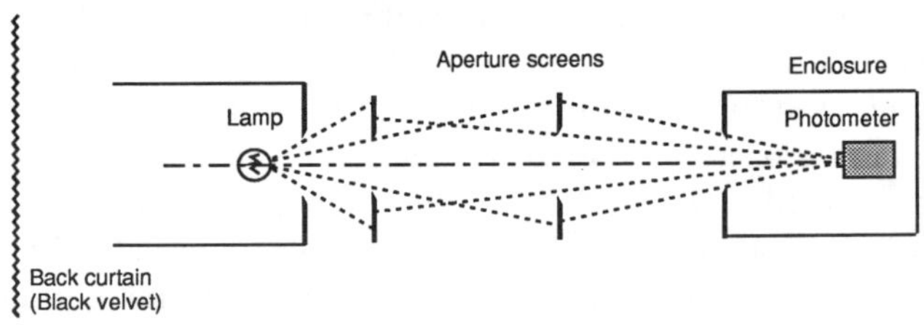

FIGURE 5.2. Layout of the components of a photometric bench.

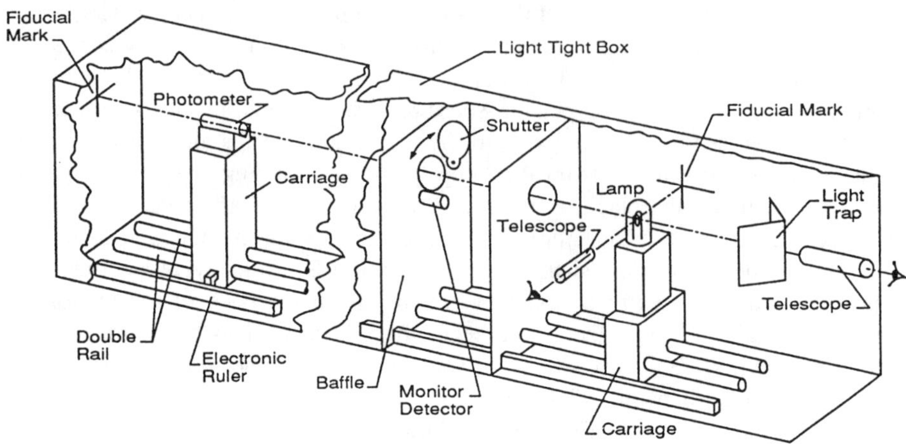

FIGURE 5.3. A photometric bench of a recent design.

by a linear encoder system that allows an accurate reading of the absolute position of the carriage by a computer. Two telescopes are used for alignment of the lamp. Variable diaphragms are used between compartments of the bench. Their aperture sizes can be adjusted according to the size of the lamp to minimize stray light on the photometer. A monitor detector beside the shutter is used to check the drift of the lamp continuously while the photometer is substituted with other photometers or moved to other positions. The light trap is made of a pair of black glass plates aligned at an appropriate angle. Light from the lamp is absorbed by many interreflections between the glass surfaces.

Photometer

A photometer is used on the photometric bench to measure the illuminance produced by the lamp. The photometer employs a silicon photodiode and a glass filter to match its spectral responsivity to the CIE $V(\lambda)$ function. Photometers employing a selenium cell should not be used due to their nonlinearity, fatigue, and instability. Refer to ''Requirements for Standard Photometers'' in Sec. 3.4.4 for recommended types of photometers for calibration purposes. The f_1' value (see ''Relative Spectral Responsivity'' in Sec. 5.2.4) of less than 3% is recommended for a photometer used for calibration purposes. Cosine correction over a wide angle range is not important here. It is rather preferred that the acceptance angle of the photometer be limited to about $\pm 30°$ in order to reduce stray-light errors.

The illuminance responsivity of a photometer head is usually 10^{-8}–10^{-9} A/lx. High quality silicon photodiodes (1 cm^2 area) normally exhibit a linear response up to 10^{-4} A and thus provide a large dynamic range. The photodiode signal is amplified by a current-to-voltage converter circuit having several gain ranges, e.g., 10^4–10^9 V/A, so the photometer can be measure illuminance over several orders of magnitude. If standard lamps are the primary standards in the laboratory (source-based method; see ''Source-Based Method of Luminous-Intensity Measurement'' in Sec. 5.1.2), the photometer is used only for relative measurements, and the absolute calibration including the amplifier gain is not necessary.

If the photometers are primary standards (detector-based method; see ''Detector-Based Method of Luminous-Intensity Measurement'' in Sec. 5.1.2), their absolute calibration and long-term stability are of prime importance. Normally a photometer together with an amplifier is calibrated for illuminance responsivity in V/lx at one of the gain ranges so that the absolute calibration of the amplifier gain is irrelevant. The relative gain factors of other ranges are determined either by using a constant-current source or by irradiating the photometer with a stable illumination. Gain factors at very high gain ranges ($>10^9$ V/lx) should be measured using illumination since the gain can change if the photodiode is replaced by a current source.

The photometer should be equipped with either a limiting aperture or a flat diffuser (see ''Requirements for Standard Photometers'' in Sec. 3.4.4). For calibration of sources other than 2856-K Planckian sources, the spectral mismatch correction factor, ccf* (T_c), of the photometer described in ''Spectral Mismatch of Photometers'' in Sec. 5.1.3 should be obtained.

Electrical Facility

Luminous-intensity standard lamps are normally operated with DC power. Regulated DC power supplies having short-term stability of better than 0.01% are used. The power supplies are usually operated on a constant-current mode, but for low currents, a constant-voltage mode can be used for better stability. The lamp current is measured as the voltage across a standard resistor (0.1–1 Ω), using a digital voltmeter (DVM) having high resolution (5 digits or more). An uncertainty on the order of 0.01% is normally required for current measurements since the light output of an incandescent lamp changes by about 7 times the change in lamp current. The DVM and the current shunt should be calibrated periodically.

It is most convenient to control the lamp current automatically with a computer-feedback system. In such a system, it is programmed so that the lamp current is ramped up and down slowly and maintains the current to within a control range during measurements. The power supply is operated in an external reference mode, and the reference voltage is supplied by a digital-to-analog converter (18 bits or more), which is interfaced to a computer.

A photometric bench is normally equipped with a socket (bipost or screw base) that has four separate contacts, two for the current supply and the other two for voltage measurements. This type of socket avoids errors in voltage measurement due to voltage drop at the contacts between the lamp and the socket.

Color-Temperature Measurement Facility

The color temperature of luminous-intensity standard lamps is important since none of the photometers is perfectly matched to the $V(\lambda)$ function, and their illuminance responsivity depends on the color temperature (spectral power distribution) of the source. For uniformity in photometric and colorimetric measurements and calculations, CIE specifies the reference spectral power distribution of an incandescent lamp, called the CIE Standard Illuminant A [3,4] which is a relative spectral power distribution of Planckian radiation at 2856 K. Thus, luminous intensity standard lamps are most often operated at 2856 K, but sometimes operated at lower color temperatures to reduce the aging rate. In any case, the color temperatures of luminous intensity standard lamps are always reported together with the luminous-intensity values.

Before the luminous intensity of a lamp is measured, the color temperature of the lamp is determined, or the current of the lamp is set to produce a certain color temperature. For this reason, a luminous-intensity calibration facility requires a capability for color-temperature measurement. Refer to Sec. 5.4 for the equipment and the procedures for color-temperature measurement and Sec. 3.7 for color temperature measurement.

Luminous-Intensity or Illuminance Reference Standards

For luminous-intensity measurements, either standard lamps or standard photometers can be used as reference standards. Traditionally, standard lamps have been most commonly used. Standard lamps should be reproducible and stable for repeated and long-term use, and be equipped with a special base or some other means to allow precise alignment. The lamps should be recalibrated at a certain interval of total burning time depending on the aging rate of the lamps. It is recommended that a group of more than three standard lamps be maintained as reference standards and that their scale be transferred to working standard lamps for routine calibration work. See Sec. 3.4.3 for the details of luminous-intensity transfer standard lamps.

Reference photometers can be maintained instead of luminous-intensity standard lamps as the laboratory's primary photometric standards. Reference photometers are calibrated for illuminance responsivity (A/lx or V/lx) normally for CIE Illuminant A. When used with a working lamp of known color temperature (normally 2856 K), the reference photometers provide the scale of illuminance over a large range based on the photometer's linearity. It should be noted that photometers do not transfer the luminous-intensity unit itself. In addition to the illuminance measurement, an accurate measurement of the lamp-to-photometer distance is necessary to determine the luminous intensity of a lamp. Reference photometers should be recalibrated periodically depending on their long-term stability. It is recommended that a group of three or more reference photometers be maintained as reference standards. See Sec. 3.4.4 for the details of reference photometers.

5.1.2 Measurement Procedures

Preparation and Handling of Test Lamps

Specially designed incandescent lamps are used for luminous-intensity transfer standards. Discharge lamps are not used for luminous-intensity standards because of their poor repeatability. The test lamps should be properly seasoned and screened before calibration. The procedures of lamp seasoning and screening are given in ''Lamp Seasoning'' and ''Lamp Characteristics and Screening'' in Sec. 3.4.3. The lamp identification number should be marked on each standard lamp.

The lamps should be handled carefully to avoid mechanical shocks to the filament. Before calibration, the bulb of the lamp is cleaned with a soft, lint-free cloth to remove dust accumulated from the packing material. The lamp bulb should not be touched with bare hands to avoid fingerprints. Gloves should be worn when handling the lamps. Special attention should be paid to quartz halogen lamps since water droplets or oily deposits on the bulb can cause permanent white spots on the quartz envelope after burning the lamp. Ethyl alcohol is used only when removing oily deposits such as fingerprints are to be removed. Lamps should be kept in a closed container when not in use.

Lamp Alignment

Luminous-intensity standard lamps are mounted in the base-down position, with the calibrated side of the lamp facing the photometer. As an example, Fig. 5.4 depicts the alignment of a bipost base lamp. The orientation of the socket is first aligned using an alignment device (a mirror mounted on a bipost base) and a laser beam. Using this method, the angle of the lamp can be reproduced to within $\pm 0.2°$. Once the socket is aligned, it need not be aligned again when the lamp is changed.

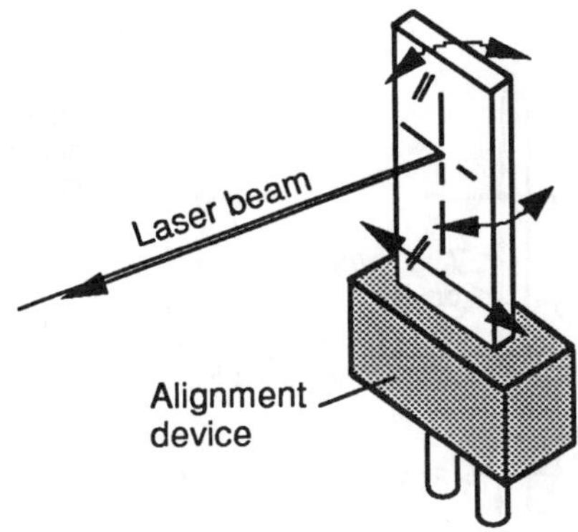

FIGURE 5.4. Alignment of the bipost base socket using an alignment device and a laser.

The alignment for the distance origin and the height of the lamp is performed using a side-viewing telescope as shown in Fig. 5.5. The distance origin of a clear-bulb lamp is adjusted to the center of the filament. For frosted lamps, the center line of the biposts is used. For clear-bulb lamps with a screw base (E26 and E39), orientation alignment is performed by viewing the filament with the telescopes.

Operation of Test Lamps

Luminous-intensity standard lamps are operated with DC power, and should be operated at a specified current rather than at a specified voltage because lamp voltage, in general, does not reproduce well due to the variation of sockets used among users. However, if the socket is designed well, the lamp voltage will reproduce fairly well on the same socket and is useful to monitor for changes in the lamp. The lamps should always be operated with the same electrical polarity since reversed polarity can affect the stability of the lamps.

The lamp current should be ramped up slowly to reach the specified value in ~30 s and ramped down in the same way. Turning the lamp on and off abruptly may cause the filament condition to change due to thermal shocks. After being turned on, the lamp is allowed to stabilize, typically for 10 min, while the lamp current is maintained at the specified value. The stabilization time depends on lamp type and each individual lamp. It is recommended that the lamp voltage and luminous intensity be monitored during stabilization. The lamp current is reported with an uncertainty statement, but the lamp voltage is often reported only for reference, without an uncertainty statement.

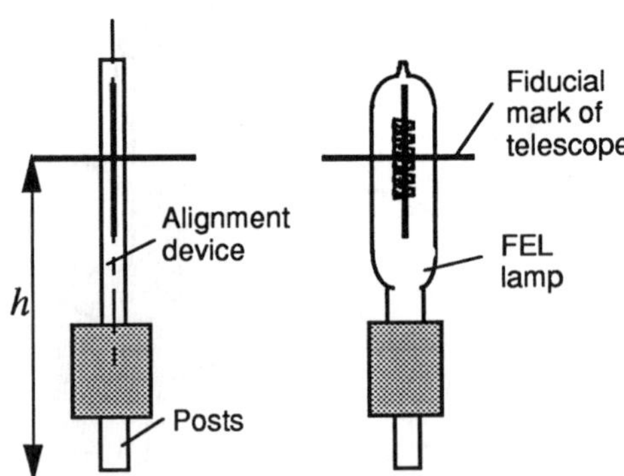

FIGURE 5.5. Alignment of the distance origin and height of the lamp for frosted-bulb lamp (left) and for clear-bulb lamp (right).

Measurement of Color Temperature

Before measuring luminous intensity, the color temperature of the lamp is measured for spectral mismatch correction. When a new lamp is calibrated for luminous intensity, the operating current of the lamp is first determined for a specified color temperature (normally 2856 K). The color temperature of the lamp is repeatedly measured, and the lamp current is adjusted so that the color temperature of the lamp equals the specified value.

When lamps are recalibrated, the original current is normally used, and the color temperature at that current is measured and reported. By using the same current, it is easier to keep a record of the lamps' repeatability and aging. The color temperature can shift due to the lamp aging, but as long as the change is in a small range (within ±20 K), it can be used as a "2856-K source" with practically no problem for luminous-intensity measurements. The procedures for color-temperature measurements are given in Sec. 5.4.

Detector-Based Method of Luminous-Intensity Measurement

Prior to measurement, the reference photometers and other electrical instruments are powered for at least half an hour for stabilization. The test lamp is mounted and aligned as described in "Lamp Alignment" in Sec. 5.1.2. The lamp current is ramped up slowly to the specified value, and allowed for the lamp luminous intensity and voltage to stabilize. Typically it takes 10 min but depends on lamp type and each individual lamp. The reference photometer is positioned at an appropriate distance (typically 3 m) from the test lamp, and the distance from the lamp is measured. First, the shutter in the bench is closed and the dark signal y_d of the photometer is taken, and then the shutter is opened and the signal y_s for the lamp is recorded. The luminous intensity I [in candelas, cd] of the test lamp is determined from

$$I = c_f d^2 (y_s - y_d)/R_v , \tag{5.1}$$

where d (in m) is the lamp-to-photometer distance, y (in V) is the photometer voltage, and s_v (in V/lx) is the illuminance responsivity of the photometer. c_f is a correction factor, such as the spectral mismatch correction and photometer temperature correction, applied as necessary. If no correction is made, c_f is unity. The details of corrections are described in Sec. 5.1.3.

The photometer signal is taken as an average of many samplings from a DVM in order to reduce noise components. The noise is often caused by microscopic vibration of lamp filaments and is more prominent with high-voltage lamps. When the measurement is finished, the lamp current is ramped down slowly and turned off. The burning time of the test lamp is recorded. In order to determine the repeatability of the test lamp, each test lamp is measured typically for three burnings. After each burning, the lamp is removed from the socket, completely cooled down, and mounted again. The mean value of the luminous intensity (and the lamp voltage) of the three burnings is reported. The standard deviation of the luminous-intensity values is also obtained to be used in the uncertainty budget (see Sec. 5.1.4). The room temperature and humidity during calibration are also recorded.

Source-Based Method of Luminous-Intensity Measurement

The source-based method is the traditional approach wherein the reference standards are maintained by a group of standard lamps, and the photometer is only used to compare one lamp to another. This method does not require photometers having superior long-term stability. This method is still predominantly used in industrial laboratories. Luminous-intensity standard lamps provide scales of luminous intensity and color temperature. Because lamps are subject to aging by use (see ''Lamp Characteristics and Screening'' in Sec. 3.4.3), the scale on the primary standard lamps is transferred to the working standard lamps to be used for routine calibration work.

Prior to measurement, the photometer and other electrical instruments are powered for at least half an hour for stabilization. First a standard lamp is mounted and aligned, and the lamp current is ramped up slowly to the specified value and allowed to stabilize. The photometer is positioned at an appropriate distance (typically 3 m) from the lamp, and its signal is taken. The responsivity s_I (in V/cd) of the photometer is obtained by

$$s_I = y_s / I_s, \tag{5.2}$$

where y_s is the photometer signal and I_s is the luminous intensity of the standard lamp. Two or three standard lamps are measured, and the responsivity of the photometer is obtained as an average for these standard lamps. In this occasion, the standard lamps are cross-checked with one another by comparing the responsivity value obtained from each lamp.

Then a test lamp is mounted and aligned so its lamp center (filament center for clear-bulb lamps) is exactly on the same position as the standard lamps. The photometer position should not be moved. The test lamp is operated at the specified current and electrical polarity. After stabilization, the photometer signal y_t is taken, and the luminous intensity I_t of the test lamp is obtained by

$$I_t = c_f y_t / s_I, \tag{5.3}$$

where c_f is a correction factor applied as necessary (see Sec. 5.1.3). The lamp-to-photometer distance is not necessary as long as the photometer position is fixed.

Each test lamp is normally measured for three burnings and the average luminous-intensity value is reported. Between measurements, the lamp should be removed from the socket, completely cooled down, remounted, and realigned. The standard deviation of the luminous-intensity values is also obtained to be used in the uncertainty budget (see Sec. 5.1.4). Standard lamps, whose repeatability is well known, are normally measured for one burning to minimize their burning time. The total burning time of each standard and test lamp should be recorded.

5.1.3 Sources of Errors and Corrections

Spectral Mismatch of Photometers

No photometer can be matched to the $V(\lambda)$ function perfectly, and an error occurs when a photometer measures a light source having a spectral power distribution different from the standard source for which the photometer is calibrated. In order to correct for such spectral mismatch errors, the photometer must be characterized for its relative spectral responsivity. The spectral mismatch correction factor ccf is given by

$$\mathrm{ccf}(S_t,S_s)=\frac{\int_\lambda S_s(\lambda)s(\lambda)_{\mathrm{rel}}d\lambda \int_\lambda S_t(\lambda)V(\lambda)d\lambda}{\int_\lambda S_s(\lambda)V(\lambda)d\lambda \int_\lambda S_t(\lambda)s(\lambda)_{\mathrm{rel}}d\lambda}, \tag{5.4}$$

where $S_t(\lambda)$ is the spectral power distribution of the test source, $S_s(\lambda)$ is the spectral power distribution of the standard source, $s(\lambda)_{\mathrm{rel}}$ is the relative spectral responsivity of the photometer, and $V(\lambda)$ is the spectral luminous efficiency. Using this equation, the photometer signal is corrected for any light source with known spectral power distribution. The photometer signal is multiplied by this factor to obtain a corrected result.

In the detector-based method, the illuminance responsivity of a photometer is normally obtained for the CIE Illuminant A. In this case, the normalized spectral mismatch correction factor ccf* is defined by

$$\mathrm{ccf}^*=\mathrm{ccf}(S_t,S_{\mathrm{A}}), \tag{5.5}$$

where S_{A} represents the spectral data of the CIE Illuminant A.

In the source-based method, if a photometer is calibrated for a light source S_1, which is different from the CIE Illuminant A, and measures another light source S_2, the spectral mismatch correction factor $\mathrm{ccf}(\mathrm{S}_2,\mathrm{S}_1)$ is given by

$$\mathrm{ccf}(\mathrm{S}_2,\mathrm{S}_1)=\mathrm{ccf}^*(\mathrm{S}_2)/\mathrm{ccf}^*(\mathrm{S}_1). \tag{5.6}$$

Equation (5.4) can be calculated over the visible region (380–780 nm) if the photometer is known to have negligible responsivity in the UV and IR regions. Otherwise, the spectral responsivity should be measured over the wavelength region 350–1100 nm (where silicon photodiodes are sensitive) so that the possible IR or UV leakage can also be corrected. The UV and IR sensitivities of photometers can be separately evaluated by the terms u and r as recommended by CIE [5].

For convenience, the ccf* of a photometer is expressed as a function of the color temperature T_c of the incandescent lamps to be measured. $\mathrm{ccf}^*(S_t)$ is calculated for Planckian radiation of several different color temperatures, and then the correction factors are fitted into a polynomial function. The ccf* is then given by

$$\mathrm{ccf}^*(T_c)=\sum_{j=0}^{n} a_j T_c^j, \tag{5.7}$$

where n can be 2 or 3. An example of this fitting is shown in Fig. 5.6. Once the polynomial constants are obtained for a photometer, the spectral mismatch correction factors for incandescent lamps are easily calculated from their color temperature.

Photometer Temperature Variation

The responsivities of photometers change depending on the temperature of their optical components. Measurement errors may occur if a photometer is used at an ambient temperature different from that at which it was calibrated. Unless the photometer is a temperature-controlled type (see ''Requirements for Standard Photometers'' in Sec. 3.4.4), or unless the laboratory temperature is always precisely controlled (within ±1 °C), the temperature dependence of the photometer should be evaluated and corrections should be made. The photometer temperature correction may not be necessary if the source-based method is used and measurements are done in a short period of time during which the laboratory temperature is constant to within ±1 °C.

The temperature coefficient (the responsivity drift per degree Celsius) of a photometer can be measured in a variable-temperature chamber. Before measurement, all the components inside the photometer must reach thermal equilibrium after the temperature of the chamber is changed (typically 1 h is required). An example of the temperature dependence of photometers is shown in Fig. 5.7.

When the photometer is used at temperature T_p, the temperature correction factor $k(T_p)$ is given by

$$k(T_p)=1-(T_p-T_0)c_p, \tag{5.8}$$

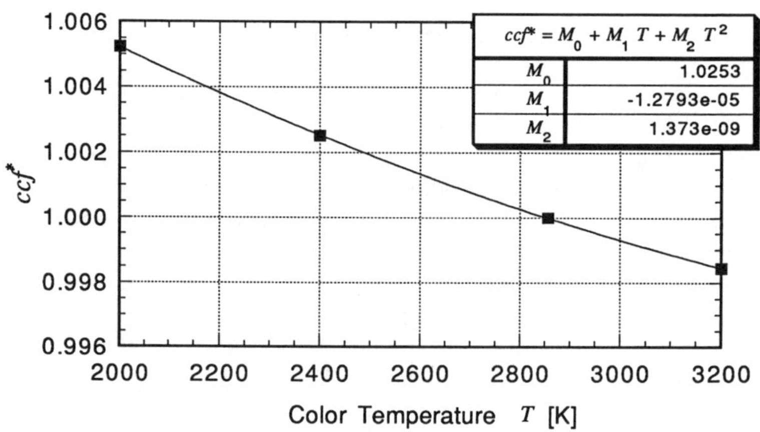

FIGURE 5.6. Polynomial fit for the spectral mismatch correction factor.

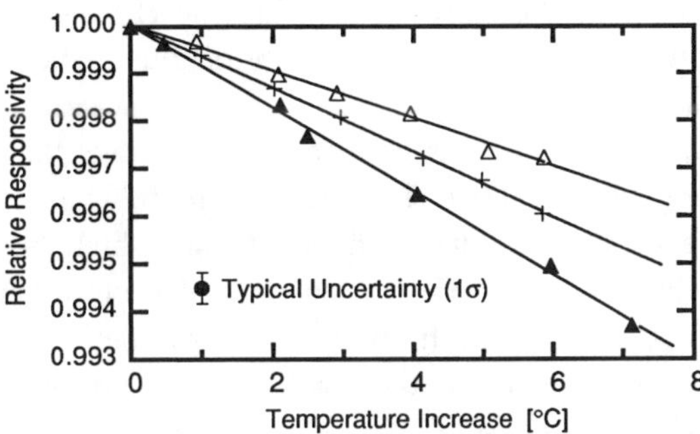

FIGURE 5.7. The temperature dependence of typical photometers.

where T_0 is the temperature at which the photometer has been calibrated and c_p is the temperature coefficient (relative change of responsivity per C°) of the photometer. The photometer signal is multiplied by $k(T_p)$). The photometer should be placed in the calibration laboratory with its power turned on for at least 1 h before use.

Stray Light

The photometer on the bench should measure light only from the lamp. Any other light reaching the photometer, e.g., from the surrounding walls, baffle screens, aperture edges, etc., is called stray light and causes measurement errors. Stray light can be more problematic with the detector-based method. In the source-based method, stray light tends to be canceled out when a test lamp is compared to a standard lamp. Stray light is also more influential when a photometer with a diffuser is used. Care should be taken to reduce stray light to a negligible level.

Stray light in a photometric bench can be evaluated and corrected as follows. A lamp is operated and the photometer signal is taken. Then the lamp is shielded by a small black plate with sharp edges, which is barely large enough to shield the entire bulb of the lamp from the photometer. The plate is placed near the photometer to avoid diffraction effects. The photometer views only stray light under this condition. Then, the lamp is turned off, and the photometer signal at complete darkness is taken for dark correction. The ratio of the photometer signal for the stray light (dark-corrected) versus the signal for the lamp is the stray light error. This value can be used for correction if it is not negligible. When stray light is to be checked, it is helpful to actually see the light from the photometer position. If shiny rings are seen on the baffle screens, they should be evaluated.

Departure from the Inverse Square Law

If the photometer is not properly aligned to the effective center of the lamp, the measured luminous intensity values vary with the lamp-to-photometer distance at calibration. This is usually not a problem if the lamp is a clear-bulb type and if the distance is measured from the center of the filament. However, frosted-type lamps are sometimes used when the spatial uniformity of illuminance is of prime importance.

For a frosted-type lamp, since its filament is not visible, the distance is usually measured from the geometrical center of the lamp bulb, which is unfortunately not the effective optical center of the lamp (the point from which the inverse square law holds). For example, a lamp with a frosted bulb of 6 cm diameter, the effective center of the lamp is probably 6 mm off from the geometrical center. If this lamp is calibrated at 3 m and used at 1 m, there will be a 0.8% error in the measurement. Frosted-bulb lamps with a considerable bulb size are not recommended as luminous-intensity standards. If such a lamp is to be used at various distances, the effective optical center of the lamp should be determined photometrically, and the lamp should be used based on the effective optical center of the lamp. The effective center of the lamp can be determined by measuring the luminous intensity at two very different distances and solving two simultaneous equations. Without determining the effective center of the lamp, frosted-type lamps should be used only at distances longer than 50 times the diameter of the bulb.

Linearity of the Photometer

Photometers of recent production employ high-quality silicon photodiodes and have a dynamic range of several orders of magnitude. The linearity is usually not a problem in luminous-intensity measurements at normal levels. However, the linearity of a photometer used in calibration work should be tested. Various methods are available for linearity measurements for detectors in general [6,7].

The linearity of a photometer, together with its amplifier, is normally measured using a 2856-K incandescent source overfilling the photometer head. A conventional, simple method is to use the inverse square law on a photometric bench. A small-size lamp with a clear bulb is used as a point source. Several lamps of different intensities are needed to cover several decades. The accuracy of this method is limited due to a departure from the inverse square law at shorter distances from the lamp. If a reference photometer with a known linearity is available, the linearity of other photometers can be measured simply by comparison to the reference photometer at various illuminance levels.

Figure 5.8 shows an example of the linearity of a well-designed photometer [1,2] having a built-in, low-noise amplifier with range settings from 10^4 to 10^{10} V/A, measured using a beam conjoiner device [7]. These data indicate that the photometer is linear over an output current range 10^{-10}–10^{-4} A. This corresponds to an illuminance range 10^{-2}–10^4 lx. The photometer can be linear at even higher illuminance levels, but typically there will be a temperature effect from the heat of the source on the $V(\lambda)$-correction filter at illuminance levels higher than 10^3 lx.

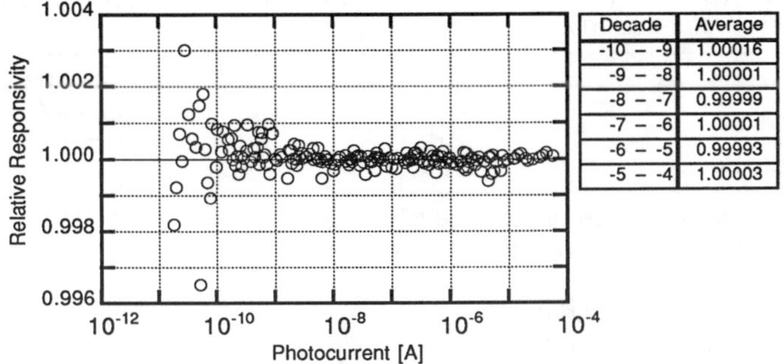

FIGURE 5.8. Linearity of a high-quality standard photometer.

Other Factors Affecting Measurement Results

The reproducibility of the lamp alignment can be a considerable error factor depending on the type of lamp and alignment method. It can be tested by measuring the luminous intensity of a stable standard lamp with repeated realignment and reburning. The variation of luminous intensity in these measurements indicates the reproducibility of lamp alignment if the lamp itself is reproducible to a negligible variation.

In the source-based method, the aging of the standard lamps is an important error factor to consider. Luminous-intensity standard lamps should be recalibrated at an appropriate total burning time. See ''Lamp Characteristics and Screening'' in Sec. 3.4.3 for the details of the characteristics of luminous-intensity standard lamps.

In the detector-based method, the long-term drift of the reference photometer is an important error factor. Reference photometers should be recalibrated at least once a year until the long-term stability data are established which indicate an appropriate calibration cycle. See ''Characterization of Standard Photometers'' in Sec. 3.4.4 for details.

5.1.4 Uncertainty Analysis

Uncertainty factors are categorized as type A and type B. Type-A uncertainty is of random nature and evaluated by statistical methods, and type-B uncertainty is of systematic nature and evaluated by other means. A detailed description for uncertainty analysis can be found in Ref. [8]. The guidelines given in this document are based on the recommendation by ISO [9]. These documents recommend the use of the terms *combined standard uncertainty* and *expanded uncertainty*. Combined standard uncertainty corresponds to the conventional 1σ uncertainty and is used only for uncertainty of fundamental units. For most other measurements and calibrations, expanded uncertainty with a coverage factor $(k=n)$ is used, which is n times the combined standard uncertainty. It is an internationally agreed custom to

TABLE 5.1. *An example of the uncertainty budget for the luminous-intensity calibrations (detector-based method).*

Uncertainty factor	Relative expanded uncertainty (k=2) (%)	
	Type A	Type B
Calibration of the reference photometers[a]		0.7
Long-term drift of the reference photometers between recalibrations[b]		0.5
Photometer temperature variation[c]	0.2	
Distance scale of the bench (0.5 mm in 3 m)[d]		0.03
Alignment of the lamp distance (1 mm in 3 m)[e]	0.07	
Spectral mismatch correction[f]		0.1
Lamp-current regulation (0.01%)[g]	0.07	
Lamp-current measurement uncertainty (0.01%)[h]		0.07
Stray light[i]		0.2
Random noise (lamp drift, etc.)[j]	0.1	
Repeatability of the test lamp (including alignment)[k]	0.5	
Overall uncertainty of test lamp calibration with respect to SI	**1.1**	

[a]Calibration of the reference photometers: The uncertainty of the reference photometer is stated in the calibration report issued by the national laboratory or the calibration laboratory that conducted the calibration.

[b]Long-term drift of the reference photometers: Estimated maximum drift of the reference photometer between calibrations. Refer to "Characterization of Standard Photometers" in Sec. 3.4.4 for details.

[c]Photometer temperature variation: If the photometer has no temperature controller or temperature sensor, and the laboratory temperature is kept to within $\pm n$ °C, the drift of the photometer responsivity is n times c_p, where c_p is the temperature coefficient of the photometer. c_p depends on each type of photometer and is typically on the order of 0.1%/°C. See "Photometer Temperature Variation" in Sec. 5.1.3 for further details. If corrections are made using photometer's built-in temperature sensor, or if the photometer is a temperature-controlled type, this uncertainty may be negligible.

[d]Distance scale of the bench (0.5 mm in 3 m): The uncertainty of calibration of the distance scale on the bench. The uncertainty in luminous intensity is twice the uncertainty of the distance due to the inverse square law.

[e]Alignment of the lamp distance: The reproducibility of the alignment of lamp position along the optical axis. This depends on the accuracy of the alignment tool used. The uncertainty in luminous intensity should be multiplied by 2.

[f]Spectral mismatch correction: This uncertainty depends on the accuracy of the relative spectral responsivity measurement for the photometer and the color temperature or the spectral distribution of the test lamp. A simple way to calculate this uncertainty is to multiply the uncertainty of the color temperature of the test lamp by the slope of the ccf* curve in Fig. 5.6.

[g]Lamp-current regulation: The range in which the lamp current is controlled during measurement. The uncertainty in luminous intensity should be obtained as 7 times the uncertainty of the current measurement.

[h]Lamp-current measurement: This factor is calculated from the calibration uncertainty of the current shunt and the DVM for lamp-current measurement. The uncertainty in luminous intensity should be obtained as 7 times the uncertainty of the current measurement.

[i]Stray light: Stray light can be evaluated and corrected with the procedure described in "Stray Light" in Sec. 5.1.3. However, the stray-light error may vary, depending on lamps and photometers. If the stray light is known to be less than a certain level (e.g., <0.1%), the maximum value can be included in the uncertainty budget instead of applying corrections to each test lamp.

[j]Random noise: Two times the standard deviation of the photometer signal when the test lamp is operating stably. This noise often comes from the lamp.

[k]Repeatability of test lamp: The repeatability of the test lamp should be taken into account in the uncertainty budget. 2 times the standard deviation of the luminous-intensity values in three lightings is normally used. If the repeatability of the particular test lamp is well known, the lamp can be measured in one or two lightings to reduce the lamp's burning time.

use expanded uncertainty with the coverage factor $k=2$, which corresponds to the conventional term 2σ uncertainty [8]. When an expanded uncertainty is expressed in percent, it is called *relative expanded uncertainty*.

Table 5.1 shows an example of the uncertainty budget for luminous-intensity calibrations using the detector-based method. These examples assume that the calibrations are performed at well-equipped commercial laboratories. The overall uncertainty U is calculated as the quadrature sum of all the factors as

$$U=\left(\sum_{i=1}^{n} U_i^2\right)^{1/2}. \tag{5.9}$$

When the overall uncertainty is used for analysis of other measurements, it is treated as a type-B uncertainty. Descriptions for each uncertainty factor are given in the table footnotes.

A possible large uncertainty factor that is not analyzed in this section is a change in the lamps during transportation. Lamps are subject to changes due to shocks to the filament during transportation, and there is no guarantee that they reproduce the photometric quantities precisely after shipping. When lamps must be shipped, a lot of packing material should be used to absorb shocks and containers should be handled with care. For best results, it is recommended that calibrated lamps be hand-carried.

Table 5.2 shows an example of the uncertainty budget for luminous-intensity calibrations using the source-based method. Only the first two factors are different from Table 5.1. Stray light is not listed since it is usually negligible with the source-based method.

5.1.5 Calibration Report

In calibration reports, the following items should be included:

1. Organization that performed calibration.
2. Customer's name.
3. Test lamp type and identifications.
4. Date of measurement.
5. Equipment used.
6. Standard lamps or reference photometers.

TABLE 5.2. *An example of the uncertainty budget for the luminous-intensity calibrations (source-based method).*

Uncertainty factor	Relative expanded uncertainty (k=2) (%) Type A	Type B
Calibration of the standard lamps[a]		1.0
Aging of the standard lamps between recalibrations[b]		1.0
Photometer temperature variation	0.2	
Distance scale of the bench (0.5 mm in 3 m)		0.03
Alignment of the lamp distance (1 mm in 3 m)	0.07	
Spectral mismatch correction		0.1
Lamp-current regulation (0.01%)	0.07	
Lamp-current measurement uncertainty (0.01%)		0.07
Random noise (scatter by dust, lamp drift, etc.)	0.1	
Repeatability of the test lamp (including alignment)	0.5	
Overall uncertainty of test lamp calibration with respect to SI	**1.5**	

[a]Calibration of the luminous-intensity standard lamps: The uncertainty of the luminous-intensity standard lamps is stated in the calibration report issued by the national laboratory or the calibration laboratory that conducted the calibration. This uncertainty normally includes the repeatability of the lamps.

[b]Aging of the standard lamps: This uncertainty is calculated from the aging rate of the standard lamps ("Lamp Characteristics and Screening" in Sec. 3.4.3) and their calibration intervals. For example, if the aging rate is 0.02%/h and the lamp is recalibrated every 25 h of its burning time, the uncertainty due to aging of the lamp will be 0.5%.

7. Burning position of the test lamp.
8. Alignment procedure of the test lamp.
9. Electrical polarity of the test lamp.
10. Stabilization time for the test lamp.
11. Corrections applied.
12. Number of burnings.
13. Lamp current (and voltage).
14. Uncertainty of the lamp current.
15. Luminous intensity (in cd).
16. Uncertainty of the luminous intensity.
17. Color temperature (if applicable).

5.2 ILLUMINANCE-METER CALIBRATION

The procedures for calibration of illuminance meters and standard photometers are described in this section. Illuminance meters and standard photometers are calibrated using luminous-intensity standard lamps (source-based method) or reference photometers (detector-based method).

The detector-based method is advantageous in illuminance-meter calibrations in that accurate distance measurement is not necessary and that many uncertainty factors tend to be canceled out in the substitution measurement. In the detector-based method, the aging of the lamp is not critical, and the lamp can be operated as long as necessary.

Using the source-based method, for illuminance-meter calibrations has its disadvantages. Distance measurement and lamp alignment are critical, and inverse-square-law errors can be problematic. The burning time of the lamp is restricted due to aging of the lamp.

5.2.1 Equipment

Basically the same equipment as required for luminous-intensity measurements (see Sec. 5.1.1) is used in illuminance calibrations. In the detector-based method, a working lamp, similar to luminous-intensity standard lamps (see ''Lamp Types'' in Sec. 3.4.3), is used. The requirements for the working lamp are not as strict as luminous-intensity standard lamps. The lamp alignment need not be precise, and only the short-term stability of luminous intensity during one burning is of concern.

Illuminance meters are normally calibrated over a large range (e.g., 20–2000 lx), and therefore two or three standard lamps of different power levels are usually used. Luminous-intensity standard lamps operated at 2856 K provide roughly the same number of candelas as their wattages. For example, a 20-W lamp provides ~20 cd (~2 lx at 3 m and ~20 lx at 1 m), a 200-W lamp provides ~200 cd (~20 lx at 3 m and ~200 lx at 1 m), and a 2-kW lamp provides ~2000 cd (~200 lx at 3 m and 2000 lx at 1 m), thus covering an illuminance range from 2 to 2000 lx on a 3-m photometric bench.

5.2.2 Measurement Procedures

Preparation of Test Items

Illuminance meters and standard photometers are calibrated for illuminance reading or illuminance responsivity (V/lx or A/lx). Prior to calibration, the diffuser surface of the illuminance-meter head or the photometer head under test (hereafter called *photometer head* to mean both) is cleaned with lens tissue to remove dust. If the photometer head is equipped with an aperture, it should not be touched since its edges are sharp and fragile. The surface inside the aperture (usually the filter surface) can be cleaned with air spray if dust particles are seen.

Before calibration starts, the reference plane of the photometer head under testing should be clarified. The reference plane is the plane from which the inverse square law holds and on which illuminance is measured by the photometer head. Figure 5.9 depicts examples of photometer heads and their reference planes. If the reference plane is erroneously defined, it can cause serious errors at close distances from a source. If the photometer head has an aperture or a flat diffuser, the reference plane is normally the surface of the aperture or the diffuser.

Some commercial photometer heads consist of only a silicon photodiode and a $V(\lambda)$-correction filter, and no aperture or diffuser is used. In this case, the photodiode surface is close to but not exactly the reference plane. Due to the refraction index of the $V(\lambda)$-correction filter, which is usually several millimeters in thickness, the effective reference plane may be several millimeters off from the photodiode surface. If the photometer head has a dome-shaped diffuser as shown in Fig. 5.9(a), the reference plane is not well defined, either. In such cases, the reference plane should be clearly specified by the manufacturer or the user of the instruments. If necessary, the standard plane can be experimentally determined using a photometric method (see ''Determination of the Reference Plane'' in Sec. 3.4.4). Other requirements for reference photometers are described in Sec. 3.4.4.

Prior to calibration, the photometer head under test is attached to a holder that can be mounted and aligned easily on the photometric bench. The position of the reference plane can be marked on the holder, or the distance offset from the holder edge is noted for alignment convenience. The illuminance meter and the photometer under test are powered and allowed to stabilize at the laboratory temperature for at least 1 h.

Detector-Based Method of Illuminance Calibration

In this method, illuminance meters and photometers are calibrated by direct comparison with reference photometers. First, a working lamp and a reference photometer are mounted on the bench and aligned to the optical axis using a laser

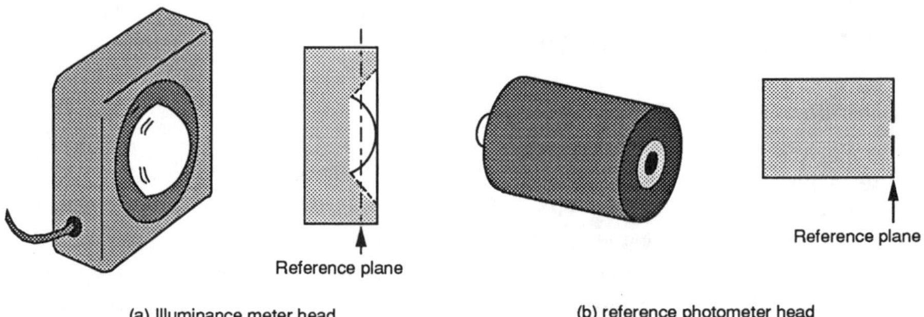

FIGURE 5.9. Illuminance-meter head, standard-photometer head, and their reference planes.

beam or a telescope. The orientation and distance alignment of the working lamp are not critical. The photometer head under test is then mounted in place of the reference photometer and checked if it is positioned on the optical axis and its surface is perpendicular to the optical axis. The photometer head must be aligned so that its reference plane is positioned exactly on the same position where the reference plane of the reference photometer was located.

The working lamp is operated at 2856 K and allowed to stabilize. Since the burning time of the working lamp is not of concern, the lamp can be stabilized long enough until its luminous intensity is perfectly stabilized (0.5–1 h). The reference photometer is positioned at an appropriate distance where a desired illuminance is provided, and the illuminance on its reference plane is determined. This illuminance is referred to as *set illuminance*. The reference photometer is then replaced by the test photometer head, with its reference plane adjusted exactly to that of the reference photometer. Then the reading of the test photometer or illuminance meter is taken. For illuminance meters, the ratio of its reading to the set illuminance is reported. For standard photometers, the illuminance responsivity (V/lx or A/lx) is obtained by the photometer signal divided by the set illuminance.

For illuminance meters, calibration is usually made for each range of the instrument. The lamp-to-photometer distance is changed to vary the illuminance level. However, the photometer head is placed no closer than 0.5 m from the lamp since the alignment error will be too critical at shorter distances. If the required illuminance range is too large for one lamp, another working lamp of a different power level is used.

For calibration of standard photometers, calibration at one illuminance level, at 2–4 m from the lamp, is normally sufficient. However, it is recommended that calibration be made at two illuminance levels (at different distances from the source) to check if the measurement system is working properly. The accuracy of the reference plane of the photometer can also be checked.

In the detector-based method of illuminance calibration, various forms of light sources can be used as the calibration source as long as the source produces a stable, uniform illuminance field over the receiving surface of the photometer head, having spectral power distribution similar to the CIE Illuminant A. For example, a lens system or an attenuator (neutral-density filters, etc.) can be combined to increase or reduce the illuminance levels. The light sources need not follow the inverse square law. The test source needs to be stable only during the calibration. Care should be taken to maintain the same color temperature and sufficient uniformity of illuminance.

The ambient temperature or the photometer temperature (for a temperature-monitored type) is an important factor. The temperature should be recorded at the same time the photometer signal is taken.

Source-Based Method of Illuminance Calibration

In this method, illuminance meters and photometers are calibrated based on the luminous intensity of a standard lamp and the lamp-to-photometer distance. First, the luminous-intensity standard lamp is mounted on the bench and aligned carefully as described in ''Lamp Alignment'' in Sec. 5.1.2. The photometer head under test is mounted so that the head is positioned on the optical axis and its receiving surface is perpendicular to the optical axis. The lamp-to-photometer distance for a desired set illuminance is calculated from the luminous intensity of the lamp, and the reference plane of the photometer under test is positioned at that distance.

The luminous-intensity standard lamp is operated at the specified current and allowed to stabilize. The reading of the test photometer or illuminance meter is taken. For an illuminance meter, the ratio of its reading to the set illuminance is reported. For standard photometers, the illuminance responsivity (V/lx or A/lx) is obtained by dividing the photometer signal by the set illuminance.

For illuminance meters, calibration is usually made for each range of the instrument. The lamp-to-photometer distance is changed to vary the illuminance level. The photometer head is placed no closer than 1 m from the lamp to avoid inverse-square-law errors. If the required illuminance range is too large for one lamp, two or more standard lamps of different power levels are used.

For calibration of standard photometers, it is recommended that calibration be made at two different distances from the source to check possible problems in photometer linearity or the lamp-to-photometer distance measurement.

The ambient temperature is an important factor for photometer calibrations, and should be recorded each time the photometer reading is taken.

Calibration for Light Sources Other than Illuminant A

For special purposes, some illuminance meters are used only to measure specific light sources other than the CIE Illuminant A. In this case, the calibration of the illuminance meter is first performed using the 2856-K source as described in ''Detector-Based Method of Illuminance Calibration'' or ''Source-Based Method of Illuminance Calibration'' in Sec. 5.2.2. Then, the relative spectral responsivity of the photometer head is measured, and the spectral mismatch correction factors for the specific light sources are obtained in the procedure given in ''Spectral Mismatch of Photometers'' in Sec. 5.1.3.

5.2.3 Sources of Errors and Corrections

Drift of the Working Lamp

During the substitution of the reference photometer and test photometers, the luminous intensity of the working lamp may drift and cause an error. To correct for this error, a $V(\lambda)$-corrected monitor detector can be installed off-axis in the photometry bench (see Fig. 5.3), and its signal recorded during the time the reference photometer and the test photometers are measured. If there is a significant drift, the monitor signal can be used for correction. If the bench does not have a monitor detector, the set illuminance should be measured with the reference photometer before and after measuring the test photometer(s). If the drift is small, the set illuminance is obtained as an average of the readings before and after the test-photometer measurement.

Ambient Temperature

The responsivity of illuminance meters and other photometers is affected by ambient temperature. Therefore, the ambient temperature during calibration should be measured and reported with the results. The recommended laboratory temperature is around 25 °C. In the detector-based method, refer to ''Photometer Temperature Variation'' on Sec. 5.1.3 regarding the correction for the temperature variations of the reference photometers.

Lamp-to-Photometer Distance

In the source-based method, the lamp-to-photometer distance is more critical than in the detector-based method because of the inverse-square-law errors associated with the standard lamp and the alignment errors mentioned above. Standard lamps with a clear bulb normally can be used at distances longer than 1 m without serious errors. If a frosted-bulb lamp is used, special care should be taken as described in the section titled ''Departure from the Inverse Square Law'' in Sec. 5.1.3.

In the detector-based method, although the inverse-square-law errors are less of a problem than in the source-based method, as the photometer comes closer to the lamp, the alignment (and the accuracy) of the photometer's reference plane will be more critical. For example, a 1-mm error of the photometer alignment at 50 cm from the lamp will lead to a 0.4% error in illuminance. Therefore, reference photometers under test should not be placed too close to the lamp.

With either method, the required illuminance range is usually too large for one lamp, and therefore two or more luminous-intensity lamps of different wattages are used as described in Sec. 5.2.1.

Reference Plane of a Photometer

If the reference plane of the photometer is not defined accurately, a measurement error occurs as the photometer is placed at varied distances from the source. Figure 5.10 is an example of a photometer having no aperture or a diffuser. In such a case, the true reference plane, denoted by P_0, is located near the silicon photodiode surface. Let us assume that the front surface of the photometer head P_1 is taken as its reference plane. When the photometer is placed at distance d from the source, the photometer signal is proportional to $1/(d+\Delta d)^2$ according to the inverse square law. If the photometer is calibrated at 3 m and measures an illuminance at 1 m from a light source, and if Δd is 1 cm, the relative measurement error will be -1.3%. Therefore it is important to define the reference plane of a photometer accurately. If

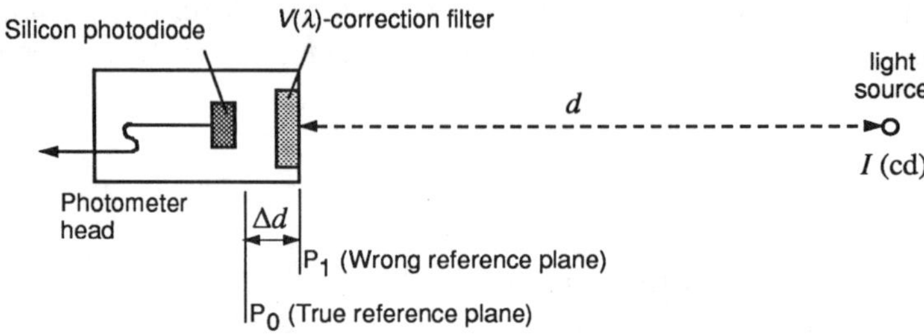

FIGURE 5.10. Measurement error caused by an inaccurate reference plane of a photometer.

the location of the reference plane of a photometer is not clearly known, it should be experimentally determined using the procedures described in ''Determination of the Reference Plane'' in Sec. 3.4.4.

Other Factors Affecting Measurement Results

Stray light in the photometric bench can cause significant errors in the calibration of illuminance meters and standard photometers. Stray light errors can be more problematic with the source-based method. In detector-based method, stray-light errors tend to be canceled out when the test photometer is measured by substitution with a reference photometer of the similar type. Refer to ''Photometric Bench'' in Sec. 5.1.1 and ''Stray Light'' in Sec. 5.1.3 for further information.

Photometers may have a linear response up to a level of 10^5 lx or higher, but the use of normal reference photometers should be limited to much lower levels to avoid possible errors caused by the heat from radiation. This effect is critical for the aperture-mode photometers with their $V(\lambda)$ filters exposed to the radiation. Use of reference photometers of this type should be normally limited to illuminance levels lower than 10^4 lx. Even if the photometer is a temperature-controlled type or temperature-monitored type, the heating of the filters by incoming radiation will not be eliminated or corrected if radiation is too high. If a reference photometer is to be used at illuminance levels higher than 10^4 lx, care should be taken so that the photometer is irradiated only for a minimum length of time necessary to take the reading.

5.2.4 Characterization of Illuminance Meters

Relative Spectral Responsivity

The relative spectral responsivity of a photometer is evaluated using the f_1' value as recommended by CIE [5]. f_1' indicates the degree of spectral mismatch of a photometer from the $V(\lambda)$ function. The spectral matching of a photometer or an illuminance meter is considered: high quality ($f_1' < 3\%$), medium quality ($3\% < f_1' < 8\%$), and poor quality ($f_1' > 8\%$). Some recent commercial photometers achieve f_1' values of less than 2%. f_1' is an evaluation index and cannot be used for corrections. f_1' is calculated from the relative spectral responsivity of the photometer according to

$$f_1' = \frac{\int_\lambda |s^*(\lambda)_{\rm rel} - V(\lambda)| d\lambda}{\int_\lambda V(\lambda) d\lambda} \times 100\ (\%). \tag{5.10}$$

$s^*(\lambda)_{\rm rel}$ is a normalized relative spectral responsivity of the photometer, as given by

$$s^*(\lambda)_{\rm rel} = \frac{\int_\lambda S(\lambda)_{\rm A} V(\lambda) d\lambda}{\int_\lambda S(\lambda)_{\rm A\ rel} d\lambda}, \tag{5.11}$$

where $S(\lambda)_{\rm A}$ is the spectral distribution data for CIE Illuminant A and $V(\lambda)$ is the spectral luminous efficiency function for photopic vision. The calculation can be done in the 380- to 780-nm region if the responsivity of the photometer in the UV and IR regions is negligible. The errors of photometers for various discharge lamps considered as ''white-light'' sources tend to be within the f_1' value of the photometer (there is no guarantee). However, the errors for colored-light sources such as traffic signals, cathode ray tube (CRT) displays, light-emitting diodes (LEDs), etc., can be much larger than the f_1' value.

The UV and IR responses should also be evaluated using u and r values defined in Ref. [5]. The u and r values can be obtained either by calculation from the relative spectral responsivity of the photometer measured in the 300–1100-nm region or by photometric measurements prescribed in the reference.

When a relative spectral responsivity is plotted on a graph, it is often normalized by its peak value or at 555 nm, which gives the wrong impression that the photometer has no error for the 555-nm monochromatic radiation. When the relative spectral responsivity is plotted, it is recommended to use the normalized function $s^*(\lambda)_{\rm rel}$ as given in Eq. (5.11). An example is shown in Fig. 5.11. When the $V(\lambda)$ function is plotted together, the difference between the two curves gives the actual errors for monochromatic radiation at each wavelength when the photometer is calibrated for the CIE Illuminant A.

Other Characteristics

Various other characteristics of illuminance meters or reference photometers can be evaluated using methods recommended in Ref. [5]. The degree of cosine correction of illuminance-meter heads is evaluated using a term f_2. The linearity of a photometer is evaluated using a term f_3. See also ''Linearity of the Photometer'' in Sec. 5.1.3 for linearity measurement. The temperature dependence of a photometer is described in the section titled ''Photometer Temperature Variation'' in Sec. 5.1.3. Characterization of standard photometers is described in Sec. 3.4.4.

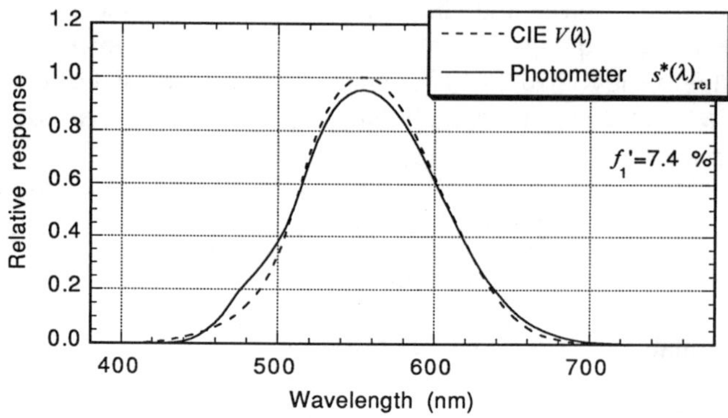

FIGURE 5.11. Relative spectral responsivity curve plotted as $s^*(\lambda)_{rel}$.

5.2.5 Uncertainty Analysis

Table 5.3 shows an example of the uncertainty budget for the calibration (detector-based method) of a typical illuminance meter in the illuminance range

TABLE 5.3. *An example of the uncertainty budget for the calibration of an illuminance meter (example).*

Uncertainty factor	Relative expanded uncertainty (k=2) (%)	
	Type A	Type B
Calibration of the primary reference photometers		0.7
Long-term drift of the reference photometers between calibrations		0.5
Photometer temperature variation	0.2	
Determination of the spectral mismatch correction factor		0.1
Random noise in the photometer measurements	0.1	
Illuminance nonuniformity		0.1
Stray light in the bench		0.2
Illuminance-meter head alignment (distance and angle)[a]	0.3	
Display resolution of the illuminance meter (1 in 199)[a]	0.5	
Inconsistency of the calibration factors at different levels[a]		0.5
Overall uncertainty of calibration with respect to SI	**1.2**	

[a]Uncertainty value depends on the test item.

300–3000 lx. The overall uncertainty value also depends on each individual illuminance meter and the illuminance levels.

Inconsistency of the calibration factors is the range of the calibration factors obtained at the illuminance levels tested. For example, if the calibration factor is 0.999 at 300 lx and 1.005 at 3000 lx, the inconsistency (0.6%) will be added. This inconsistency is most often caused by the incorrect definition of the reference plane of the illuminance-meter head since the calibration is conducted at different distances. It can also be caused by nonlinearity of the detector. See Sec. 5.1.4 for explanations of other factors.

5.2.6 Calibration Report

In calibration reports, the following items should be included:

1. Organization that performed calibration.
2. Date of measurement.
3. Customer's name.
4. Test photometer and its identification.
5. Reference plane of the photometer.
6. Equipment used.
7. Standard (working) lamps used.
8. Color temperature of the lamps.
9. Responsivity of the photometer or the ratios of measured illuminances to the set illuminances.
10. Uncertainty of the calibration.
11. Laboratory temperature.

5.3 LUMINOUS-FLUX MEASUREMENT

Luminous flux is the geometrically total light output from a light source and is probably the most important parameter for lamp products. The energy efficiency (lumen/watt) of light sources is determined from the total luminous flux and the electrical power consumption of the lamp. In this section, the measurement of luminous flux using an integrating-sphere photometer is described. The measurement method described here is based on the substitution method in which a test lamp is compared with a standard lamp in an integrating sphere. The luminous-flux measurements using a goniophotometer are discussed partly in Sec. 3.5.1. The realization of the luminous-flux unit is described in Sec. 3.5.

5.3.1 Test Items

In laboratories, limited types of luminous-flux standard lamps are measured and used as transfer standards or working standards. Gas-filled incandescent lamps ranging from 25 to 1000 W and some types of linear fluorescent lamps are often used as transfer standards (see ''Requirements for Standard Lamps'' in Sec. 3.5.3). Miniature lamps ranging from 1 to 30 W are also sometimes used. High-intensity discharge (HID) lamps are normally not used as transfer standards due to their poor repeatability.

In factories, all kinds of manufactured lamps including compact fluorescent lamps and HID lamps are measured in integrating spheres. In such cases, test lamps tend to be compared with different types of standard lamps due to the limited types of available standard lamps; thus substitution errors are most likely to occur. Ironically, corrections are most needed where they are most difficult to implement.

Before luminuous-flux standard lamps are submitted for calibration, the lamps must be seasoned and tested for stability. In general, new lamps have to be seasoned for at least 5% of their rated lifetime at the rated current. Incandescent lamps are normally seasoned on DC power for at least 50 h and fluorescent lamps for 500 h. The operating current, electrical polarity, and burning position must be specified for the calibration of the lamps (see also ''Seasoning and Screening'' in Sec. 3.5.3).

5.3.2 Equipment

Integrating-Sphere Photometer

Figure 5.12 shows the geometry for a typical integrating-sphere photometer for luminous-flux measurements. The integrating sphere is equipped with a photometer head, a baffle screen, an auxiliary lamp, and a lamp socket. Several important design criteria for an integrating sphere photometer are described below.

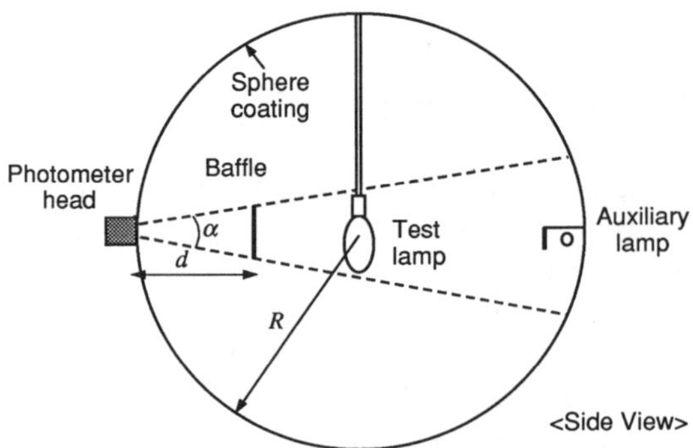

FIGURE 5.12. Geometry for a typical integrating-sphere photometer.

Sphere Size

The sphere should be large enough in three aspects. First, the size of the baffle is determined by the size of the test lamp. If the sphere is too small, the size of the baffle relative to the sphere size will be too large, and the spatial nonuniformity errors will be serious (see ''Baffle,'' p. 162). Second, if the sphere is too small, the sphere characteristics such as self-absorption by the lamp and source-size dependence of the sphere responsivity will be critical and difficult to make accurate corrections. Third, a lamp operated in the sphere generates considerable heat that affects the ambient temperature of lamp, sphere coating, and the responsivity of the photometer head. Especially for fluorescent lamps, the sphere should be large enough to maintain a stable temperature. CIE [10] recommends that the diameter of the sphere be larger than 10 times the diameter of compact lamps, and twice the length of tubular lamps. For luminous-flux measurement of general-purpose lamps (incandescent, linear fluorescent, compact fluorescent, and HID), integrating spheres of 1.5–3 m diameter are normally used. If the test lamps are small in size, a smaller sphere can be used as long as the temperature increase does not cause a problem.

Sphere Coating

Sphere coatings based on barium sulfate or polytetrafluoroethylene (PTFE) are commonly used. The reflectance of the coating can be controlled by mixing light-absorbing material into the coating. The reflectance of the coating is an important factor. Higher reflectance provides better spatial uniformity of the sphere responsivity, and the sphere is less subject to errors associated with the differences in angular luminous-intensity distributions. However, with higher reflectance, the sphere responsivity will be more sensitive to self-absorption by lamp, contamination by dust, spectral mismatch of the sphere response, etc. For this reason, CIE recommends 80% reflectance for sphere coatings [10]. However, in laboratories where these corrections are possible, higher reflectances (e.g., 98%) are preferred to reduce the spatial nonuniformity errors. If these corrections are not made, e.g., in factory facilities, lower reflectances (80–90%) may be advantageous. Higher reflectance is also advantageous for higher signals for low-flux sources and spectral flux measurements.

Photometer Head

The photometer head is a $V(\lambda)$-corrected silicon photodiode with a cosine-corrected angular responsivity. High-quality silicon photodiodes normally have a linear response over 7 orders of magnitude. Detectors other than silicon photodiodes are not recommended. Spectroradiometers are sometimes used in place of the photometer head, for color measurements, but for luminous-flux measurement, a photometer head with a single silicon photodiode normally provides a more linear response and reproducible results than spectroradiometers. (Photometers, of course, are subject to spectral mismatch errors.)

The spectral responsivity of the photometer head is matched to the $V(\lambda)$ function. However, it should be noted that the spectral responsivity of the total sphere system is also affected by the spectral throughput of the sphere, as determined by $\rho(\lambda)/[1-\rho(\lambda)]$. The overall spectral responsivity of the sphere system should be evaluated as described in ''Spectral Mismatch Correction'' in Sec. 5.3.4. It is recommended that the f_1' value of the total sphere system be less than 5%. The angular responsivity of the photometer should be cosine-corrected.

By utilizing a carefully designed amplifier (current-to-voltage converter) incorporated in the photometer head, a large dynamic range can be achieved [11]. Together with the wide linearity range of a silicon photodiode, a low-noise amplifier with several gain ranges (e.g., 10^4–10^9 V/A) allows luminous-flux measurements over several orders of magnitude using one integrating sphere. For example, 10^{-1}–10^5 lm can be measured with a 2-m integrating sphere [2,12].

The responsivity of a photometer head can change with ambient temperature or due to heat from the lamp in the sphere. The temperature-monitored photometer or the temperature-controlled photometer [13] can be employed for the most accurate measurement requirements.

Baffle

The baffle is needed to shield the photometer head from direct illumination by the lamp. However, the baffle is the main cause of the spatial nonuniformity of the sphere responsivity as described in ''Spatial Nonuniformity of the Sphere Response'' in Sec. 5.3.4. In order to minimize the spatial nonuniformity, the baffle should be located at $\frac{1}{3}$ to $\frac{1}{2}$ the sphere radius from the photometer head, and the size of the baffle should be the smallest possible to barely shield the largest lamp to be measured. CIE [10] recommends that the baffle be located $\frac{1}{3}$ of the sphere radius from the photometer head. The angle subtended by the baffle from the photometer head (α in Fig. 5.12) should be kept to a minimum, and is normally kept to less than 30°. This corresponds to a 30-cm baffle located at $\frac{1}{2}$ the radius in a 2-m sphere. It is recommended that two or more baffles of different sizes be prepared for a sphere, and a smaller baffle is normally used. The surface of the baffle facing the lamp should have the highest reflectance possible. The surface of the baffle facing the photometer can have a lower reflectance to improve the spatial uniformity of the sphere responsivity.

For linear fluorescent lamps, the shape of the baffle depends on the orientation of the lamp as shown in Fig. 5.13. The coaxial arrangement is advantageous in keeping the baffle size smaller. In the perpendicular arrangement, the baffle is a rectangular shape possibly with a circular part in the middle to shield the standard lamp.

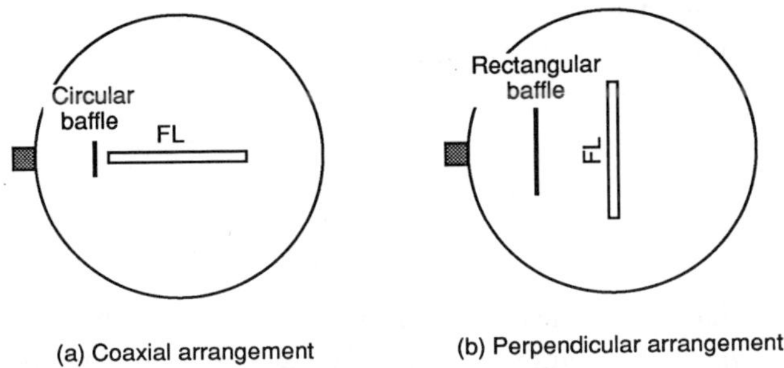

FIGURE 5.13. Arrangements for linear flourescent lamp (top view).

Auxiliary Lamp

Integrating-sphere photometers are equipped with an auxiliary lamp to allow measurement of self-absorption by lamps. The auxiliary lamp provides illumination to the entire sphere wall, but it must be shielded from the photometer head and the lamp to be measured. An auxiliary lamp is normally placed on the sphere wall opposite from the photometer head, but it can be placed elsewhere. The auxiliary lamp and its baffle should be small enough not to affect the performance of the integrating sphere. An automotive lamp or a miniature halogen lamp (~30 W) can be used. The self-absorption measurement is made with repeated opening and closing of the sphere to change the lamp inside the sphere. Therefore the auxiliary lamp should be stable against shocks resulting from opening and closing the sphere. The auxiliary lamp should be operated at a color temperature close to that of the test lamps.

Lamp Holder and Socket

The integrating sphere is equipped with an interchangeable lamp holder and socket. The lamp holder and socket should be painted with a high-reflectance sphere coating (e.g., barium sulfate paint of 98% reflectance). A screw-base socket (E26 and E39) is normally used for most of the luminous-flux standard lamps. It is desirable that the screw-base socket has four contacts, two for supplying the lamp current and two for measuring the lamp voltage.

It should be noted that any objects close to the lamp bulb can absorb light significantly. The top surface of the socket should be always kept white. If there is a gap on the socket, it should be filled with sphere coating. Special care should be taken when measuring miniature lamps because the size of sockets tends to be much larger relative to the size of the lamps, and the total flux can decrease significantly due to absorption by the socket surfaces. Sometimes miniature lamps are calibrated in combination with a particular socket, and in such cases, a combination of lamp and socket is required as a calibration artifact.

Temperature Sensor

If fluorescent lamps are measured, the lamp ambient temperature in the integrating sphere must be monitored. IES [14] specifies that a temperature sensor should be installed in the sphere, at the same height of the lamp and at a position closer than 0.9 m from the lamp, and should be shielded from direct illumination by the lamp [14]. A temperature sensor (such as a thermocouple) is often placed behind the baffle (on the detector side).

Electrical Facility

Incandescent Lamps

Luminous-flux standard lamps are operated using DC power, at a specified current, rather than a specified voltage, because lamp voltage, in general, does not reproduce well due to the variation of sockets used among users. However, if the socket is designed well, the lamp voltage reproduces fairly well on the same socket, and the lamp voltage is useful to monitor for changes in the lamp.

DC power supplies are usually operated on a constant-current mode, but for a very low current, a constant-voltage mode can be used with better stability. The lamp current is measured as the voltage across a standard shunt resistor (0.1–1 Ω), using a DVM having high resolution (5 digits or more). An uncertainty in the order of 0.01% is normally required for current measurements since the light output of an incandescent lamp changes by $\sim$7 times the change of lamp current. The DVM and the shunt resistor should be calibrated periodically.

Fluorescent Lamps

The electrical circuit for linear fluorescent lamp measurements is constructed using a reference ballast, the impedance of which is set to specified values for each type of lamp [15,16]. Figure 5.14 shows an example of the measurement circuit for rapid-start-type linear fluorescent lamps. A regulated AC power supply is used as the main power supply. The output voltage is transformed into higher voltage (400 V maximum) by using a step-up transformer, and fed into the reference ballast.

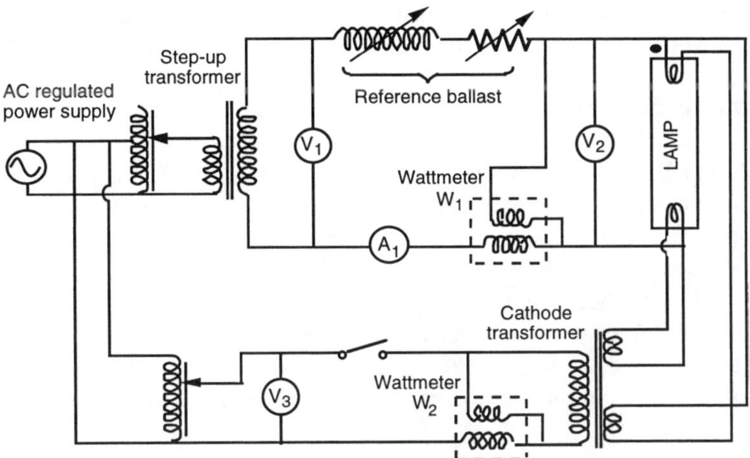

FIGURE 5.14. Measurement circuit for a rapid-start fluorescent lamp.

Prior to photometric measurements, the electrical parameters of the measurement circuit are set up according to the specifications of the lamp to be tested. The cathode transformer is calibrated, in advance, in terms of the primary voltage needed to provide the desired secondary voltage under normal load conditions and the power loss in the transformer under this normal load condition. The input voltage on voltmeter V_3 is adjusted to the calibrated value. The input power for the cathode transformer is read on wattmeter W_2. The impedance (reactance and resistance) of the reference ballast is set in accordance with the specifications of the lamp. The lamp is mounted on the socket and turned on, and the supply voltage V_1 is adjusted so that the lamp current A_1 equals the specified value. When the lamp current is adjusted, the cathode power is usually kept on. Measurements can be made with the cathode power off to simplify the electricial measurements. During photometric measurements, the supply voltage V_1, the lamp current A_1, the lamp voltage V_2, the lamp power W_1, and the cathode power W_2 are measured and recorded. If the cathode power is kept on during the measurements, the sum of W_1 and W_2 (corrected by the power loss of the cathode transformer) is reported as the total lamp power. Reference [14] describes the details for operation and electrical measurement of fluorescent lamps.

Luminous-Flux Standards

Standard lamps should be reproducible and stable for repeated and long-term use. The lamps should be recalibrated at an interval, specified in terms of the total burning time, which depends on the aging rate of the lamps. It is recommended that a group of more than three standard lamps be maintained as reference standards for a laboratory and that their scale be transferred to working standard lamps for routine calibration work. See Sec. 3.5.3 for the details of luminous-flux standard lamps.

Laboratory Environment

Fluorescent lamps are very sensitive to ambient temperature, and it is specified that measurements be made at an ambient temperature of 25±1 °C [14]. If fluorescent lamps are to be measured, the laboratory temperature should be controlled accordingly. It should be noted that the air temperature in the integrating sphere will increase slightly by heat from the lamp during stabilization.

Incandescent lamps are not sensitive to ambient temperature. However, stability of the total instrument, including the integrating sphere and the photometer head, requires that the laboratory temperature be maintained constant during measurements.

5.3.3 Measurement Procedures

Preparation of Test Lamps

The test lamps should be properly seasoned and screened before calibration. The procedures for lamp seasoning and screening are given in ''Seasoning and Screening'' in Sec. 3.5.3. Lamps should be handled carefully to avoid mechanical shocks to the filament, and the bulbs of lamps should be not be touched with bare hands. The lamp identification number should be permanently marked on each lamp. Prior to calibration, the bulbs of the lamps should be cleaned with a soft, lint-free cloth to remove dust from packing material. Lamps should be kept in a closed container when not in use.

Operation of Test Lamps

Incandescent standard lamps are operated using DC power at a specified current with specified electrical polarity. If the lamp current is not known, the lamp current is determined for a specified color temperature (normally 2856 K) as described in ''Measurement of Color Temperature'' in Sec. 5.1.2. It should be noted that some luminous-flux standard lamps are operated at a lower color temperature (e.g., 2700 K) in order to reduce the aging rate.

The lamp current is ramped up and down slowly (approximately 30 s). Photometric measurements should be made after the lamp has stabilized. It normally takes 5–10 min for incandescent lamps to stabilize. The lamp stabilization time should be a minimum necessary in order to reduce the total burning time of the lamp. Gas-filled incandescent lamps are sensitive to burning position. The luminous-flux lamps are normally calibrated in the base-up position and are always used in the same burning position.

Linear fluorescent lamps are operated in the horizontal position. The luminous flux of fluorescent lamps changes significantly with ambient temperature. The ambient temperature must be controlled within 25±1 °C. Before measurement starts, the lamp is normally stabilized for 15 min. If the ambient temperature is going to be too high, the integrating sphere can be opened during lamp stabilization. In this case, before measurement starts, the sphere should be closed very slowly to avoid drafts.

Total Luminous-Flux Measurement

The luminous flux of test lamps is determined by substitution with the standard lamps. Prior to the luminous-flux measurement, the correction measurements as described in Sec. 5.3.4 are conducted as necessary. After applicable correction factors are obtained, the luminous flux of test lamps is measured as follows.

First, two standard lamps are measured, then a group of test lamps are measured three times (remounted and relighted), and at the end, the same two standard lamps are measured. This sequence should be finished within a day. After the lamp is stabilized, the photometer signal y and the lamp's electrical data are taken. The lamp is turned off, and the photometer signal y_{dark} is taken for a dark reading. Then the "corrected signal" y' is obtained by

$$y' = \mathrm{ccf}^* \mathrm{scf}^* (y - y_{\mathrm{dark}}) / \alpha, \tag{5.12}$$

where α is the self-absorption factor (see "Self-Absorption Correction," in Sec. 5.3.4), ccf* is the spectral mismatch correction factor (see "Spectral Mismatch Correction," in Sec. 5.3.4), and scf* is the spatial nonuniformity correction factor (see "Spatial Nonuniformity of the Sphere Response," in Sec. 5.3.4). If corrections are not made, these correction factors will be unity. If applicable, the photometer temperature correction and photometer range correction are also made at this point.

The sphere responsivity R_s (V/lm) for each standard lamp measurement is obtained by

$$R_s = y' / \Phi_s, \tag{5.13}$$

where Φ_s is the luminous flux of the standard lamp. An average value for R_s is obtained from the four measurements of standard lamps. The luminous flux Φ_t of a test lamp is obtained by

$$\Phi_t = y' / R_s. \tag{5.14}$$

The average value for Φ_t is obtained from the three measurements of each test lamp. Two times the standard deviation of the three data sets for each test lamp is calculated and included in the uncertainty budget. When the photometer signal is taken, an average of several DVM readings is recorded together with the standard deviation. The standard deviation indicates the stability of lamp operation. Measurement of the standard lamps before and after the test lamp measurement ensures that the sphere condition has not changed during measurements. When measuring a large number of test lamps day by day, the sphere photometer can be calibrated periodically.

5.3.4 Correction Techniques

Self-Absorption Correction

A lamp in the sphere absorbs light emitted by itself and reflected back from the sphere wall, thus reducing the sphere responsivity. This effect is called *self-absorption*. The larger the lamp and the darker the color of the lamp, the more absorption the lamp exhibits. Unless the test lamps are exactly the same type as the standard lamps, the self-absorption should be measured and corrected. The following procedures should be used.

The auxiliary lamp in the integrating sphere is turned on and allowed to stabilize. The lamp is kept on during the self-absorption measurement. The photometer signal y_{01} is first taken for the *null* condition where the lamp and lamp holder are removed from the sphere. Then the photometer signal y_i is taken with each (standard and test) lamp $1,\ldots,i,\ldots,n$ (with an appropriate lamp holder) mounted in the sphere. These lamps are not operated. Then, the detector signal y_{02} is taken for the null condition again, and the self-absorption factor α_i of lamp i is obtained from

$$\alpha_i = y_i/[(y_{01}+y_{02})/2]. \tag{5.15}$$

Thus the self-absorption factor α is always given with respect to the null condition. When the luminous flux is measured, the measured photometer signal for lamp i is divided by α_i for correction. Normally the same self-absorption factors can be used for the same type of lamp used with the same lamp holder in the same sphere. However, the values will change as the sphere coating or the lamp holder becomes contaminated. The self-absorption is more critical with higher reflectance sphere coatings.

Spectral Mismatch Correction

The spectral responsivity of the integrating-sphere photometer is not perfectly matched to the $V(\lambda)$ function. An error occurs when a test lamp is compared with a standard lamp that has a different spectral power distribution from the test lamp. This error can be corrected by the spectral mismatch correction factor ccf as given by

$$\mathrm{ccf}(S_t, S_s) = \frac{\int_\lambda S_s(\lambda) R_s(\lambda) d\lambda \int_\lambda S_t(\lambda) V(\lambda) d\lambda}{\int_\lambda S_s(\lambda) V(\lambda) d\lambda \int_\lambda S_t(\lambda) R_s(\lambda) d\lambda}, \tag{5.16}$$

where $S_t(\lambda)$ is the spectral power distribution of the test lamp, $S_s(\lambda)$ is the spectral power distribution of the standard lamp, $R_s(\lambda)$ is the relative spectral responsivity of the sphere system, and $V(\lambda)$ is the spectral luminous efficiency. $R_s(\lambda)$ can be obtained by measuring the relative spectral responsivity of the detector $R_d(\lambda)$ and the relative spectral throughput of the integrating sphere $T_s(\lambda)$ as

$$R_s(\lambda) = R_d(\lambda) T_s(\lambda). \tag{5.17}$$

$T_s(\lambda)$ can be obtained by measuring the spectral irradiance of an incandescent standard lamp (with a clear bulb) operating inside and outside the sphere. $T_s(\lambda)$ is given as the ratios of the spectral irradiance at the sphere detector port and the spectral irradiance of the lamp measured on a photometric bench. The spectroradiometer used in this measurement should have a cosine response. An example of the spectral throughput and the relative spectral responsivity of an integrating sphere (with 97% reflectance barium sulfate coating) is shown in Fig. 5.15.

All the standard lamps used in a substitution measurement may not have the same color temperature. In this case, the spectral mismatch correction using Eq. (5.16) can be confusing. In order to simplify the correction, a normalized spectral mismatch correction factor

$$\mathrm{ccf}^* = \mathrm{ccf}(S_t, S_{\mathrm{A}}) \tag{5.18}$$

is introduced, where S_{A} represents the CIE Illuminant A. Then the correction is given by

$$\mathrm{ccf}(S_t, S_s) = \frac{\mathrm{ccf}^*(S_t)}{\mathrm{ccf}^*(S_s)}. \tag{5.19}$$

Then all the detector readings both for standard lamps and test lamps are simply multiplied by this factor for correction.

For convenience, the ccf* of a photometer is expressed as a function of the color temperature of the incandescent lamps to be measured and fitted to a polynomial function as described in ‘‘Spectral Mismatch of Photometers,’’ in Sec. 5.1.3.

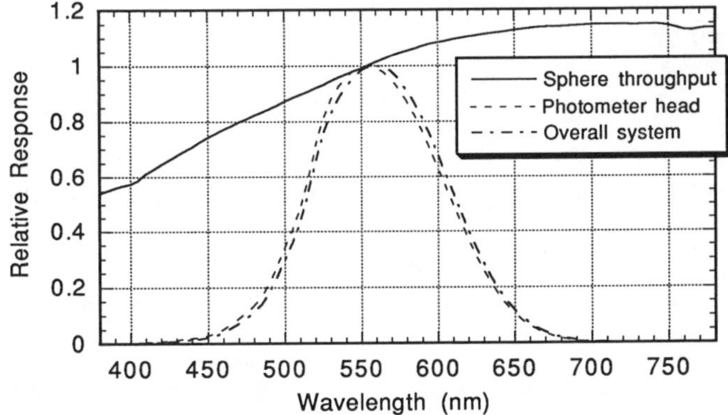

FIGURE 5.15. An example of the spectral throughput and the relative spectral responsivity of an integrating-sphere photometer.

Photometer Temperature

When a lamp is turned on in the integrating sphere, the photometer head on the sphere is heated up due to the heat from the lamp. For example, if a 1000-W incandescent lamp is measured in a 2-m integrating sphere several times, the photometer head temperature can increase by several degrees. Consequently, the responsivity of the photometer can drift by nearly half a percent. If the photometer head is a temperature-monitored type, correction can be made by the procedure given in ''Photometer Temperature Variation'' in Sec. 5.1.3. If the photometer head is a temperature-controlled type, the correction is not necessary.

Spatial Nonuniformity of the Sphere Response

The response of the sphere is not spatially uniform because of the baffle and other objects in the sphere, contamination of the sphere wall, gaps between hemispheres, etc. An error occurs when a test lamp is compared with a standard lamp that has different angular luminous-intensity distributions from those of the test lamp. Correction for this error has not been possible until recently, which significantly increased the measurement uncertainty of sphere photometry. A correction technique as described below was recently developed [2,17].

The spatial response distribution function (SRDF) $K(\theta,\phi)$ of the sphere is defined as the sphere response at a point (θ,ϕ) of the sphere wall or on a baffle surface, divided by the sphere response at (0,0). $K(\theta,\phi)$ can be obtained by measuring the detector signals while rotating a narrow beam inside the sphere. $K(\theta,\phi)$ is then normalized for the sphere response to an isotropic point source. The normalized SRDF, $K^*(\theta,\phi)$, is defined as

$$K^*(\theta,\phi)=4\pi K(\theta,\phi) \bigg/ \int_{\phi=0}^{2\pi}\int_{\theta=0}^{\pi} K(\theta,\phi)\sin\theta\, d\theta\, d\phi. \tag{5.20}$$

The spatial correction factor scf* for the test lamp with respect to an isotropic point source is given by

$$\mathrm{scf}^*=1 \bigg/ \int_{\phi=0}^{2\pi}\int_{\theta=0}^{\pi} I^*(\theta,\phi)\,K^*(\theta,\phi)\sin\theta\, d\theta\, d\phi, \tag{5.21}$$

where $I^*(\theta,\phi)$ is the normalized luminous-intensity distribution of the test lamp given by

$$I^*(\theta,\phi)=I_{\mathrm{rel}}(\theta,\phi) \bigg/ \int_{\phi=0}^{2\pi}\int_{\theta=0}^{\pi} I_{\mathrm{rel}}(\theta,\phi)\sin\theta\, d\theta\, d\phi, \tag{5.22}$$

where $I_{\mathrm{rel}}(\theta,\phi)$ is the luminous-intensity distribution of the internal source. However, a goniophotometer is not necessarily required to obtain $I_{\mathrm{rel}}(\theta,\phi)$. Only the relative angular intensity distribution is necessary, and its accuracy is not critical. For example, the data for a group of lamps of the same type can be represented by

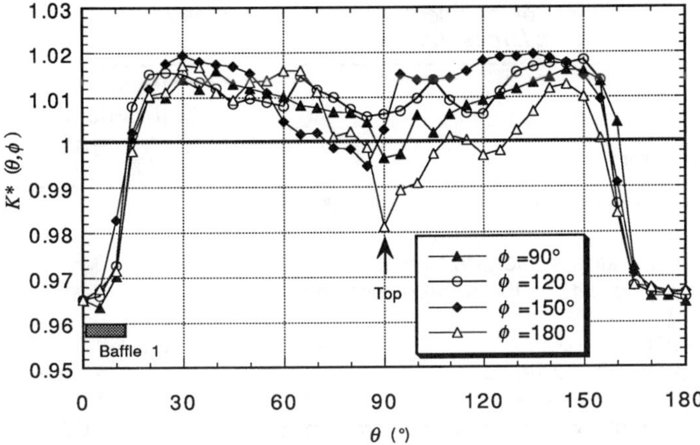

FIGURE 5.16. SRDF of a 2-m integrating sphere. (θ=0° is at the detector. ϕ=180° is the plane passing through the top of sphere.)

one lamp. Once the distribution data are taken, they are used for the lifetime of the lamps.

The SRDF, $K(\theta,\phi)$, is measured by rotating a beam source that is insensitive to burning position. Figure 5.16 shows an example of the normalized SRDF, $K^*(\theta,\phi)$, of a 2-m integrating sphere with an average sphere wall reflectance of 96%. The polar coordinates (θ,ϕ) in the graphs are referred to the position of the detector as illustrated in Fig. 5.12. ϕ=180° is the plane passing through the top of the sphere.

Other Factors Affecting Measurement Results

As with any other photometric measurements, the linearity of the photometer head affects the results (see ''Linearity of the Photometer'' in Sec. 5.1.3). If any object (such as a lamp holder) is facing too close to the lamp bulb, there will be light absorption that cannot be corrected by the self-absorption measurement. The lamp socket should always be kept clean with high reflectance. The relative error in lamp-current measurement for incandescent lamps affects luminous flux by 7 times.

5.3.5 Uncertainty Analysis

The uncertainty of luminous-flux measurements depends on what corrections are made. If any of the corrections described in Sec. 5.3.4 are not made, they should be accounted for in the uncertainty budget. Even if a correction is made, there is still an uncertainty for the correction. An example of an uncertainty budget for an incandescent lamp calibration is shown in Table 5.4.

A possible large uncertainty factor that is not analyzed in this section is any change in the lamps during transportation. Lamps are subject to changes due to

TABLE 5.4. *An example of the uncertainty budget for total luminous-flux calibrations of standard incandescent lamps (typical).*

Uncertainty factor	Relative expanded uncertainty (k=2) (%) Type A	Type B
Calibration of luminous-flux standard lamps[a]		1.0
Aging of the working standard lamps[b]		1.0
Transfer from working standards to test lamps		
Spatial nonuniformity of the sphere response[c] (not corrected)		0.8
Self-absorption correction[d]	0.2	
Spectral mismatch correction[e]		0.2
Repeatability of test lamps[f]	0.3	
Overall uncertainty of test lamp calibration with respect to SI	**1.7**	

[a]Calibration of the luminous-flux standard lamps: The uncertainty of the luminous-flux standard lamps is stated in the calibration report issued by the national laboratory or the calibration laboratory that performed the calibration. This uncertainty normally includes lamp's repeatability.
[b]Aging of the standard lamps: This uncertainty is calculated from the aging rate of the standard lamps (see "Lamps Characteristics and Screening" in Sec. 3.4.3) and their calibration intervals. For example, if the aging rate is 0.02%/h and the lamp is recalibrated every 50 h of its burning time, the uncertainty due to aging of the lamp is estimated to be 1.0%.
[c]Spatial nonuniformity of the sphere response: This uncertainty is associated with differences of the angular intensity distributions of the test lamps and the standard lamp.
[d]Self-absorption correction: Uncertainty of the determination of the correction factor.
[e]Spectral mismatch correction: Uncertainty of the determination of the correction factor.
[f]Repeatability of test lamps: Calculated as two times the standard deviation of the three measurements for each test lamp.

shocks to the filament during transportation, and there is no guarantee that they reproduce the photometric quantities precisely after shipping. When lamps must be shipped, a lot of packing material should be used to absorb shocks and the containers should be handled with care. For best results, it is recommended that calibrated lamps be hand-carried.

5.3.6 Calibration Report

In calibration reports, the following items should be included.

1. Organization that performed calibration.
2. Customer's name.
3. Test lamp type and identifications.
4. Date of measurement.
5. Measurement instruments used.
6. Standard lamps used.

7. Burning position of the test lamp.
8. Electrical polarity of the test lamp.
9. Stabilization time.
10. Corrections applied.
11. Number of burnings.
12. Lamp current (and voltage).
13. Luminous flux (in lm).
14. Color temperature.
15. Uncertainty of the luminous flux.
16. Uncertainty of the lamp current.

5.4 COLOR-TEMPERATURE MEASUREMENT

In this section, the phrase *color temperature* is used to represent a correlated color temperature (see Sec. 3.7.1). Note that, in some countries, distribution temperature (see Sec. 3.7.1) is commonly used for incandescent lamps. Color-temperature standard lamps are often needed for calibration of colorimeters and other color-measuring instruments. Luminous-intensity standard lamps are normally calibrated for color temperature also. The color temperatures of standard lamps are measured in one horizontal direction, most often at 2856 K, but sometimes in a range 2000–3200 K.

5.4.1 Equipment

A photometric bench as used in luminous-intensity measurements (see Sec. 5.1.2) is used in color-temperature measurements. The bench allows alignment and operation of a test lamp and controls stray light. Distance measurement is not necessary. Instead of a photometer head, a color-measuring device such as a spectroradiometer, a two-channel colorimeter, or a tristimulus colorimeter is used.

For the most accurate measurement of color temperature, a scanning-type spectroradiometer is used. In this case, it takes typically 10 min for one measurement. However, very often, the lamp current must be adjusted for a desired color temperature (e.g., 2856 K), and measurements must be repeated several times. In this case, a scanning-type spectroradiometer would be too slow. Therefore, either two-channel colorimeters or tristimulus colorimeters are normally used. Diode-array-type spectroradiometers are also used for fast measurement of color temperature.

Depending on the instrument to be used, spectral irradiance standard lamps or color-temperature standard lamps (see Sec. 3.7.3) are needed for calibration of the instrument.

5.4.2 Measurement Procedures

General Procedures

A test lamp is aligned in a similar manner as for the luminous-intensity measurements, but the lamp alignment is not critical. The lamp is turned on at a specified lamp current and allowed to stabilize (typically for 5 min) before the measurement begins. The color-temperature reading is recorded together with the lamp current and voltage. There are several methods available for color-temperature measurements. The details are described in the following sections. If the calibration is to be made for different currents, after changing the lamp currents, stabilize it for a while (typically 2 min is sufficient) before taking the reading. If the lamp current is to be set for a desired color temperature, the current is first set approximately, and measurements are repeated until the desired value is obtained. For color-temperature calibration, only one burning of the test lamp will normally suffice, but it is recommended to repeat the measurement to verify the calibration.

Red-to-Blue Ratio Method

This is a traditional method for fast measurement of color temperature. A two-channel colorimeter is used, which consists of two filtered detectors, each having a narrow-band spectral response in the blue region and red region, respectively. One way to use this colorimeter is a null comparison method. First, a standard lamp is operated, and the ratio of the signals from the two channels (red-to-blue ratio) is calibrated for the known color temperature (e.g., 2856 K) of the lamp. Then a test lamp is operated, and the lamp current is adjusted until the same red-to-blue ratio is obtained. The operating current of the test lamp is thus determined for the same color temperature as the standard lamp.

The red-to-blue ratio can be calibrated for many different color temperatures and fit to a polynomial function. Thus the color temperatures of the lamp over the fitted range can be measured. If the detectors are reproducible and stable for a long time, the colorimeter can be used without calibration for some period of time. It should be noted, however, that this method is liable to errors if different types of lamps (with different relative spectral power distributions) are compared. See Sec. 3.7.3 for further details. To reduce this error, the best peak wavelengths for the two channels are 460 and 680 nm.

Method Using a Tristimulus Colorimeter

This is another means for fast measurement of color temperature. A tristimulus colorimeter consists of three (or four) filtered detectors, each having relative spectral responsivity matched to the CIE color-matching functions $\bar{x}(\lambda)$, $\bar{y}(\lambda)$, and $\bar{z}(\lambda)$. A tristimulus colorimeter is normally calibrated for CIE Illuminant A (2856 K) and can directly indicate the color temperature of a test lamp. This is a convenient instrument, but if the color temperature (or the spectral power distribution) of the test lamp is different from that of the standard lamp, the instrument tends to cause an error due to the spectral mismatch of the detectors. With typical commercial instruments, the error may be negligible in the ±100-K range from the calibration point. The safest way to use this instrument is the null-comparison method as described in ''Red-to-Blue Ratio Method'' in Sec. 5.4.2.

Method Using a Spectroradiometer

By using a spectroradiometer, one can measure the spectral power distribution of a test lamp from which the color temperature of the lamp is calculated. This is the way in which the color-temperature standard lamps are calibrated. Measurements are based on spectral irradiance standards. The measurement uncertainty depends on the type of spectroradiometer. Mechanical-scanning-type spectroradiometers are generally reliable, but the measurements are time consuming.

Diode-array-type spectroradiometers are becoming more and more popular. The measurement can be as fast as tristimulus or two-channel colorimeters. This type of spectroradiometer, however, is generally not as accurate as a scanning-type spectroradiometer even though repeatability is good. When the spectroradiometer is calibrated with a spectral irradiance standard lamp (e.g., at 3100 K) and it measures lamps of different color temperatures, errors can occur just as with tristimulus colorimeters. It is recommended that this type of spectroradiometer be tested for accuracy in color-temperature measurements at several different color temperatures and corrections be made (see Sec. 3.7.3).

The color temperature of a lamp is often measured together with its luminous intensity. Figure 5.17 shows an example of a setup for measurement of color temperature and luminous intensity. When color temperature is measured, a PTFE plaque is placed on the optical axis, and the spectroradrometer views the diffuse reflection at 45°. The plaque is mounted on a kinematic base and is removed

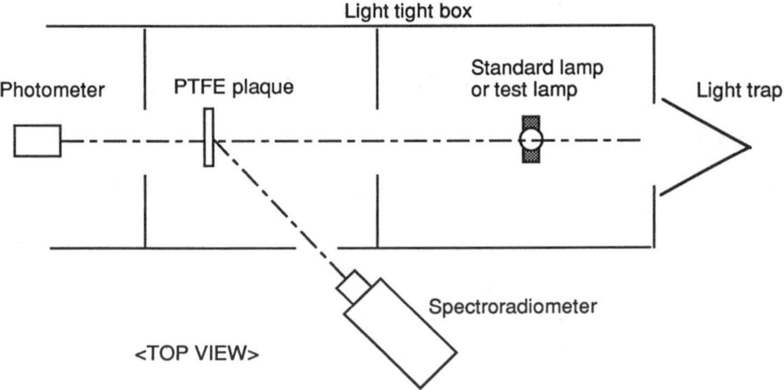

FIGURE 5.17. Setup for color-temperature and luminous-intensity measurements.

TABLE 5.5. *An example of the uncertainty budget for the color-temperature calibration..*

Factor	Expanded uncertainty, k=2 (K)				
	2000 K	2300 K	2600 K	2856 K	3200 K
Calibration of primary standard lamps	4	5	6	8	10
Aging of standard lamps between calibrations	2	2	3	4	5
Reproducibility of colorimeter	1	1	2	2	2
Reproducibility of test lamps	2	2	3	3	4
Overall uncertainty of test lamps	5	6	8	10	12

easily when the luminous intensity is measured. The spectroradiometer including the PTFE plaque is calibrated against a spectral irradiance standard lamp so that the spectral radiance factor of the plaque need not be known.

5.4.3 Uncertainty Analysis

Table 5.5 shows an example of an uncertainty budget for color-temperature calibration.

5.4.4 Calibration Report

In calibration reports, the following items should be included.

1. Organization that performed calibration.
2. Customer's name.
3. Test lamp type and identifications.
4. Date of measurement.
5. Measurement instrument used (including calibration record).
6. Standard lamp used (if applicable).
7. Alignment procedure for the test lamp.
8. Electrical polarity of the test lamp.
9. Stabilization time of the test lamp.
10. Number of burnings.

11. Current (and voltage) of the test lamp.
12. Color temperature (K).
13. Uncertainty of the color temperature.
14. Uncertainty of the lamp current.

ACKNOWLEDGMENT

This chapter was reviewed by Ms. Sally Bruce, Dr. Joseph L. Dehmer, Dr. Albert C. Parr, Dr. Jack J. Hsia, and Dr. Chris E. Kuyatt at National Institute of Standards and Technology. They spent long hours reviewing this chapter and gave precious comments and discussions which contributed to improving this chapter into its final form. The author is very grateful to them.

REFERENCES

1. Cromer, C. L., Eppeldauer, G., Hardis, J. E., Larason, T. C., and Parr, A. C., ''National Institute of Standards and Technology Detector-Based Photometric Scale,'' Appl. Opt. **32**, 2936–2948 (1993).
2. Ohno, Y., ''Photometric Calibrations,'' NIST Special Publication 250-37 (1996).
3. *Colorimetry*, 2nd ed., CIE Publication No. 15.2 (1986).
4. CIE Standard Colorimetric Illuminants, CIE/ISO 10526 (1991).
5. *Methods of Characterizing Illuminance Meters and Luminance Meters*, CIE Publication No. 69 (1987).
6. Budde, W., ''Multidecade Linearity Measurements on Si Photodiodes,'' Appl. Opt. **18**, 1555–1558 (1979).
7. Thompson, A., and Chen, H., ''Beamcon III, A Linearity Measurement Instrument for Optical Detectors,'' J. Res. NIST, **99**, 751–755 (1994).
8. Taylor, B. N., and Kuyatt, C. E., *Guidelines for Evaluating and Expressing the Uncertainty of NIST Measurement Results*, NIST Technical Note No. 1297 (1994).
9. *Guide to the Expression of Uncertainty in Measurements*, First ed., BIPM/IEC/IFCC/ISO/IUPAC/IUPAP/OIML (1993).
10. *Measurements of Luminous Flux*, CIE Publication No. 84 (1989).
11. Eppeldauer, G., and Hardis J. H., ''Fourteen-Decade Photocurrent Measurements with Large-Area Silicon Photodiodes at Room Temperature,'' Appl. Opt. **30**, 3091–3099 (1991).
12. Ohno, Y., Cromer, C. L., Hardis, J. E., and Eppeldauer, G., ''The Detector-Based Candela Scale and Related Photometric Calibration Procedures at NIST,'' J. IES, **23**, 88–98 (1994).
13. Eppeldauer, G., ''Temperature Monitored/Controlled Silicon Photodiodes for Standardization,'' in *Surveillance Technologies*, SPIE Proc. **1479**, 71–77 (1991).
14. IES Approved Method for the Electrical and Photometric Measurements of Fluorescent Lamps, IES LM-9-1988 (1988).
15. Rapid Start Types Dimensional and Electrical Characteristics, ANSI C78.1-1991 (1991).
16. Fluorescent Lamp Ballasts—Methods of Measurement, ANSI C82.2-1984 (1984).
17. Ohno, Y., ''Realization of NIST 1995 Luminous Flux Scale Using Integrating Sphere Method,'' J. IES **25** (5), 13–22 (1996).
18. Ohno, Y., ''Integrating Sphere Simulation—Application to Total Flux Scale Realization,'' Appl. Opt. **33** (13), 2637–2647 (1994).
19. Ohno, Y., Lindemann, M., and Sauter, G., ''Analysis of Integrating Sphere Errors for Lamps Having Different Angular Intensity Distributions'', J. IES **26** (1) (1996).

6

Measurement Procedures and Case Studies

David C. Gross

Test and Measurement Laboratories, Osram Sylvania, Inc., Beverly, Massachusetts 01915

An expert is someone who knows some of the worst mistakes that can be made in his subject and how to avoid them.

—W. Heisenberg

6.1 INTRODUCTION

Measurement of a system returns a value that is related to the system's actual performance. The ability of the measured value to represent the actual parameter is subject to many confounding mechanisms. Indicated values may be in error due to the measurement system inaccuracies. The measurement may not be completely and exclusively of the intended feature. The device under test may not be operating in the expected fashion. In these three ways the value reported will be compromised and may lead to erroneous decisions. We must keep in mind the use of the results we generate. The purpose of a test will direct which measurements are performed and set the required level of accuracy and confidence.

In the measurement of visible light, many light sources are challenging in their operation and prone to subtle problems. The current state of measurement technology will provide highly accurate results, only with close vigilance and a comprehensive understanding of the measurement process.

This chapter will strive to provide introductory guidance for making photopic measurements, those related to the eyes response. Recommendations for further guidance [1–12] are discussed in Sec. 6.10. Existing standards [13–24] indicate how measurements should be made but may fall short for some readers. The methods to be used are specified but often not described in detail. The users of these standards are expected to be experienced practitioners of photometry. Fortunately the situation is not truly bleak for the novice. The science of light measurement is comprised, to a large part, of common sense. With a faith in cause and effect, care and some guidance any technically trained investigator can make accurate and reliable photopic measurements.

Limitations in equipment, time, and care lead to lower accuracy. This natural trade-off may be the key to photometry because of the inherent uncertainty of all measurements [25]. Prior knowledge of the result's use is vital to the judgment of how much accuracy, and therefore effort, is required.

6.2 ROOTS OF PROBLEMS

Measurement of light is primarily a matter of avoiding mistakes. It is actually very easy to make a measurement; aim a meter at your source and read the display. The result you get may be good enough or it may be in error by an amount which would lead you to the wrong decision. Errors and uncertainty [25] are unavoidable but must be controlled to a level that allows for valid conclusions. In this chapter I will concentrate on the sources of errors. Reduction of these errors will allow you to bring your measurement accuracy to the level needed.

The roots of error in lighting measurement will be put in three categories: the measurement equipment, the environment, and the light source and its operation. Please note that the measurement method is not seen as a source of error, but that the method used will extend or limit the errors arising from these three categories.

6.2.1 Measurement Equipment Error

The photopic measurement of light provides a value that is proportional to the stimulus produced by a light source. This is accomplished by modifying the spectral response of the detector to match the spectral response of the eye. This may be done by filtering or by dispersing the light to measure the colors separately. Section 6.3 will address this fundamental choice of equipment.

The common parameters of interest, luminous flux and illuminance, require input optics that have specific spatial throughput. This is difficult to accomplish in both cases and may color the incoming light distorting the spectral response.

To measure flux an integrating sphere is the most practical method [10,11,26–29]. With the source in the sphere, a baffled port is used to collect light for the measuring instrument. In practice the sphere is calibrated with a source of different spectral, spatial, and directional output. When the test source is measured the sphere's imperfections (ports and baffles, and the paint's color, uniformity, and reflectance) lead to errors.

Illuminance can be as troublesome to measure as flux, due to the requirement of measuring the light with cosine response. To measure all the light through a plane the response in all directions must be equal and only diminish as the exposed area is foreshortened. This is difficult to accomplish and is again best done using an integrating sphere with an input port.

6.2.2 Measurement Environment

Environmental troubles are separated from equipment problems for this discussion to highlight the importance of outside influences on the measurement process. The equipment used for measuring light is sensitive by its nature. As the environment changes, the equipment's response will change. If the change happens between calibration and measurement, then the results will be compromised. An example is the changing moisture content of barium sulfate sphere paint. The high reflectance of the paint is due in part to the high surface area. The amount of water makes a dramatic difference because it reduces the surface area by matching the index of refraction between the paint and the air. The moisture content may change due to the humidity in the air or heating from a high powered source. The throughput of a sphere is proportional to $\rho/(1-\rho)$, where ρ is the average wall reflectance [11]. If the majority of the sphere wall, 97%, is painted and the reflectance changes from 95% to 95.1%, the throughput changes from 11.74 to 11.90. This 1.4% increase was seen as a 1000-W calibration lamp dried our sphere paint. This indicated that the sphere needed a larger warming lamp when not in use.

Some types of light sources are very sensitive to temperature and will vary in their operation as the environment changes. Ambient temperature for these lamps must be held to the standard test temperature while preventing the source from changing the air temp. This must be done with still air, as moving air will cool the source and lead to behavior significantly different from stable operation.

The question of what type of measurements are to be made now arises. If the source is being measured for comparison to other sources, then a reference standard test may be appropriate [17,19–23].

Measurements directed toward an application should consider the use of the lamp. The amount of light delivered in actual use should be measured, but this will undoubtedly lead to trouble. The operation of measurement equipment in the field, even an air-conditioned office, can suffer from environmental factors. Light pollution from displays, windows, and other lighting systems, may not be possible to control. The temperature of the air and its velocity are not easily controlled and strongly affect fluorescent lamp systems. An actual site, if not specially prepared, will provide uncontrolled operation of the light sources. All three categories of error exacerbate in application testing and require monitoring as control may be impossible. As a minimum the background light, input voltage and the ambient lamp temperature should be measured during testing.

6.2.3 Source and Source Operation

Artificial light sources would appear to be easier to measure than natural light because they are controllable, but they have problems of their own. Incandescent lights are a relatively stable source, and for this reason are used as standards. Operating as a glowing resistor the filament light bulb provides little challenge to stabilize or operate. The incandescent lamp's efficiency problem leads to problems in measurements. About 90% of the energy is wasted, coming out as heat and infrared radiation. The heat leads to warming of the surrounding air, with all the disturbance that brings. The infrared radiation passes in all directions and heats all that it strikes. If baffles are put in place to block light, they may heat and form thermal drafts. The instrumentation is subject to heating and will vary in all possible ways: mechanical (wavelength), electrical (bias supply), chemical (detector response), and optical (coatings drying). These problems and others are present with discharge sources. Due to the nature of their operation the arc-based discharge lamps have electric input that is not resistive or even inductive or capacitive. The resulting wave forms for current and voltage include high harmonic content and are a challenge to control and measure. The internal chemistry of the lamps is thermally controlled and will lead to operations that are not stable until the thermal gradients have diffused all the controlling chemicals, Hg, salts, etc., to their lowest energy. The lowest energy for the controlling chemical may not be the lowest energy for the system and oscillations may be perpetual. Though these oscillations are not common I will present an example.

Mercury Rain. In a fluorescent lamp the current is carried on ionized mercury vapor whose pressure is controlled by condensate at the coldest spot in the lamps. If this cold spot is at the top of the lamp and enough mercury condenses to form a large liquid ball it may fall or roll down through the arc, dramatically raising the Hg vapor pressure. The photometric and electrical parameters jump significantly, though the lamp has been stabilized for hundreds of hours.

The only protection against these problems is a clear and thorough understanding of the light source that you are testing. This understanding extends to the electrical equipment that is driving and measuring the lamps. An example of the problems that operating equipment can introduce is the use of a power amplifier as a line conditioner. To provide clean, stable, 60-Hz voltage to the lighting system, an engineer used a wideband amplifier with 60-Hz input. The system under test was a transformer type ballast and fluorescent lamps. The lighting system ran at low (~50%) output and the power amplifier's output voltage dropped to 85%. The reason was that the lamps, during startup, experience anode oscillations at ~5 kHz. The wideband power amplifier sensed this signal and attempted to follow it and maintain its zero source impedance. A high-frequency ballast can also trick some voltage supplies into distorting their output. This is a prime example of the need for an oscilloscope to monitor electrical wave forms during all electrical work. Knowledge of the amount of electricity (current, voltage, or power) is not sufficient to tell you if there is trouble.

6.3 CELL OR SPECTROMETER?

The measurement of a visible light relies on the defined and accepted photopic response curve, $V(\lambda)$. This curve, discussed in Chap. 2, is the basis for comparable measurements between instruments and laboratories. Instruments achieve this response in one of two ways.

1. Cell-based photometry uses a detector with photopic correction. The detector incorporates filters to provide the necessary relative response across the spectrum.

2. Visible spectroradiometry relies on separating the radiation into its parts and measuring the spectra wavelength by wavelength. The resulting power spectrum is weighted to get a photometric result.

The two systems differ greatly in their usefulness and cost. A photocell filtered to give a value that relates to illuminance can be purchased at modest cost [30]. Spectrometers often exact a price that limits their availability to part-time users. The price of a spectrometer goes beyond monetary costs. The time and effort needed to operate and maintain a spectroradiometer are usually beyond the commitment of a part-time user. This is a generalization, and technology could eliminate this distinction.

The results from the two types of systems differ to a similar degree. The readings from a cell can be in error due to the inability to exactly match the response of the standard curve. Changes to the filter and detector can lead to degradation of the response curve. The detector's accuracy can be further compromised when input optics are used to direct light to the detector [31]. Unless the optics have an even transmission across the spectra, the spectral response is compromised. The most common case is the use of a cell on an integrating sphere. The spheres throughput must be spectrally flat to maintain the photopic response of the cell.

The spectrometer system is calibrated for throughput at each wavelength separately, taking into account all changes and any spectral weighting from input systems. This ability makes the spectrometer my choice for use on an integrating sphere. The multiple reflections in a sphere will multiply the slight spectral differences of paint that has high reflectance. Cells are often used on spheres, but testing sources that differ from the calibration source require corrections for the spectra. As an indicator of change a sphere photocell excels. During warmup and while a scan is being taken the monitoring of the photocell will indicate trouble due to instabilities or warn that the lamp has not yet reached equilibrium.

6.3.1 Are Cells Really that Bad?

Let us take a look at the published specifications of a handheld photometer that costs $\sim \frac{1}{20}$ of a starting engineer's salary. This photocell meter is designed for illuminance measurements and therefore can be evaluated directly without input optics.

Figure 6.1 approximates the published specification of the detector. The data were retrieved from a plot in the sales literature. As you can see, the response deviates from the weighting used to calculate the lumen by several percent. More important, for an area where the weighting function is small, the error can become proportionally very large. If used to measure a He–Ne laser at 632 nm, after calibration with a blackbody, the total error is $\sim 10\%$! Using these data, lumens of several types of sources can be calculated and compared to the standard weighting

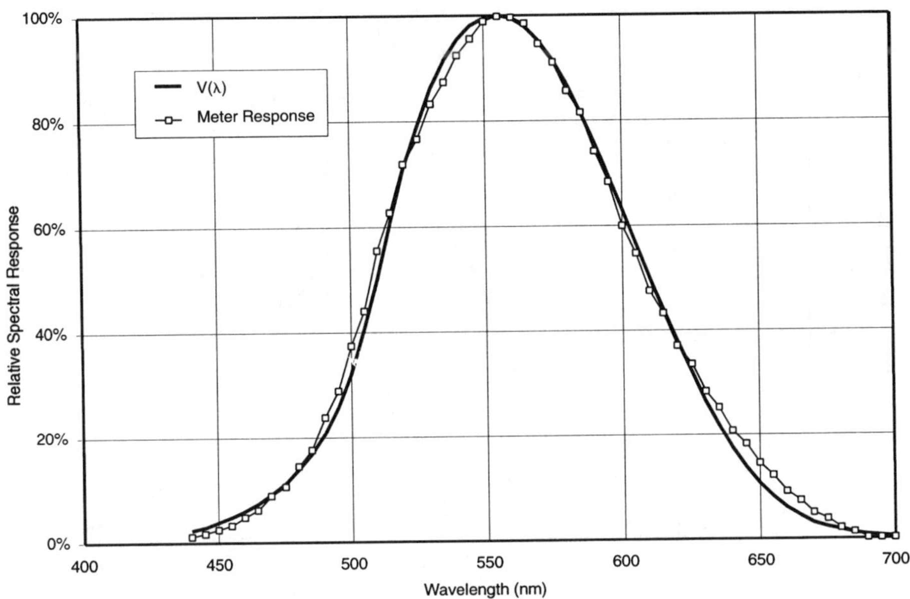

FIGURE 6.1. Illuminance meter's published response.

curve. The results are surprisingly good for illumination light sources. This can be attributed to the need for some balance across the spectrum to achieve white light. If a source contained only one wavelength, then the source would not be white and we would probably not test it for lumens. Lumens of a He–Ne laser are not a useful measure. On the other hand, light from a light-emitting diode (LED) needs to be quantified relative to the eye's response. To address this, detectors that are optimized for accuracy in the red region are available for LED testing above 600 nm.

So cells are not that bad? Think again! Actual spectral errors may be quite large or the spectral response may have changed over time. Using a monochromator configured to measure detector's response and a calibrated detector as a standard, a handheld illuminance meter was characterized. This meter corresponds to the above discussion of published data. The meter was ''experienced,'' older than 5 years, and had physical wear indicating use. The results shown below are quite poor but are representative of several meters from this manufacturer.

Figure 6.2 shows data scaled to give the proper [the same as $V(\lambda)$] output with a 2856 K blackbody calibration source. This is then the response of this meter relative to an ideal meter, $V(\lambda)$, when calibrated with source A. The impact of this on measuring other sources is calculated by weighting a spectrum with the proper $V(\lambda)$ and with the ''calibrated'' meter's response. See Chap. 3 and Ref. [14].

Figure 6.3 shows the change in reading for an incandescent lamp (blackbody) at

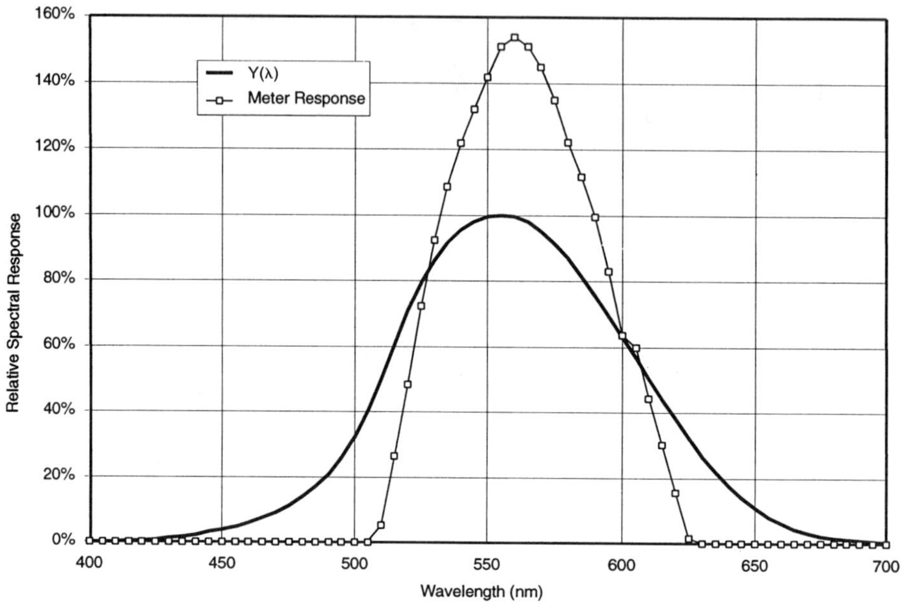

FIGURE 6.2. Illuminance-meter measured response (magnitude set using source A).

various temperatures. The error is not significant until the light source is quite red because of the smooth nature of the blackbody spectra.

The measurement of discharge lighting is a tougher case because these spectra contain separate emissions at various wavelengths. Broad spectra tend to balance the errors for over- and underreporting, leading to reasonable results from systems with very distorted spectral response. The weighting of the spectra from a fluorescent, Fig. 6.4, and a mercury halide lamp, Fig. 6.5, show $>10\%$ error from this meter because of the deviations from the standard $V(\lambda)$ response.

The results of the above calculations or empirical tests with lamps of known output may be used to correct the response of a cell for a particular source. These corrections are only as good as the correspondence between the test source and the spectral adjustment source. This spectral adjustment is in fact a calibration, but due to the complexity the process often appears to be a fudge factor.

Above is the main problem of cell photometry: A cell may introduce errors that are difficult to detect. Remember that recalibrating the level of a cell does not measure the change in the cell's spectral characteristics. If a cell is used with input optics, then the spectrum of the input system will introduce changes that can only be understood by evaluation of the input optic's spectral transmittance, along with the source, detector, and calibration source's spectral characteristics.

The advantages of cells are compelling for many. Moderate cost, ease of use, reliability, portability, and sensitivity are all advantages over spectrometers. For

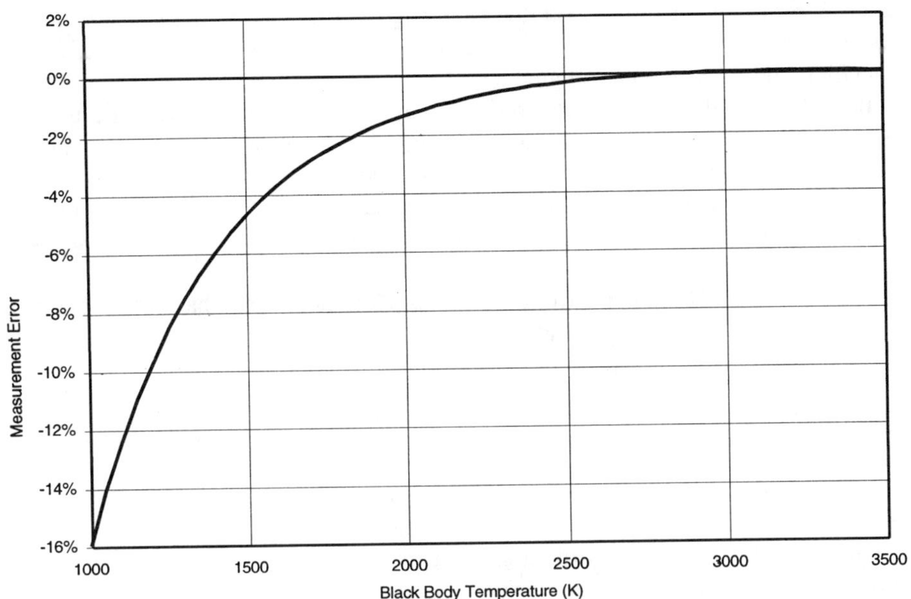

FIGURE 6.3. Meter's error (F_1) vs temperature of blackbody.

these reasons cell-based photometry is often the reasoned choice even in the face of spectral challenge.

If cost and portability limit the choice of instrumentation, then characterize the response of your ''photopic'' cell.

6.3.2 The Down Side of Visible Spectroradiometers

A spectroradiometer is an instrument that demands much from its user. First the cost of a spectrometer for photometry will be about 10 times that of a comparable cell photometer. At present the available systems require environmental considerations regarding temperature humidity and shock. This may change, but for now they are precision optical instruments requiring care. The dispersion of light lowers the light level at the detector, making the signal detection that much harder.

There are two types of spectrometers, monochromators and polychromators, which sweep the spectra or present a rainbow to an array of detectors, respectively. The joys of moving optical parts are detailed in Chap. 8 and will not be recounted here. Suffice to say that normal use can lead to troubles that test the operator as well as the source. The array-based polychromator systems require some special care in the setting of the wavelength. The spread of wavelength across the array is never even and requires corrections. The determination of the wavelengths across the spectra is hampered by the limited emission lines and their positions. If a line, such as the 546-nm line, is directly on the wavelength, then determining if it is measured correctly is simplified. Few lines in nature are so convenient, requiring careful matching and often human judgment.

In both cases the spectrometer only runs if the software runs. Using purchased software may uncover needs that were never considered by the software's author. For example, a lumen time series may be needed. Writing your own software will give you complete control and intimate knowledge of the system, but is not practical except in extreme cases. In choosing a spectrometer system, test the software against your present and projected needs.

In summary, cell- and spectrometer-based photometry each have their place. In the table below the relative disadvantages of each system are shown:

Cell-based	Spectrometer-based
Spectral accuracy	High price
Unable to correct for input optics	Slow
	Fragile
Single value (no spectral information)	Upkeep
No color quality (color rendering index)	Large size
	Optical inefficiency
	Cooling and electrical requirements

In keeping with the spirit of this chapter only the disadvantages are shown, but they can easily be translated into positive points. For example, results from spectrometers allow for the calculation of the color rendering index (CRI) [21].

6.4 CASE STUDY: MY NEW SPECTRORADIOMETER

Use of a measurement system requires that you have confidence in its operation and can rely on its results. A new spectrometer, purchased, built, or inherited, can be tested on each of its parts to isolate specific functions. Checks are combined into a suite of qualifying tests for instrument acceptance. The order in which the tests are performed is nearly irrelevant because of their independence: linearity, input spatial response, wavelength accuracy, and time response.

6.4.1 Linearity

Linearity can be assumed for the optical parts of the system because of light's superposition. Mirrors, gratings, and diffusers are not optically active so their properties will be the same for one photon as for many. Gradual thermal changes due to hot sources are always a hazard but should be considered environmental rather than linearity problems.

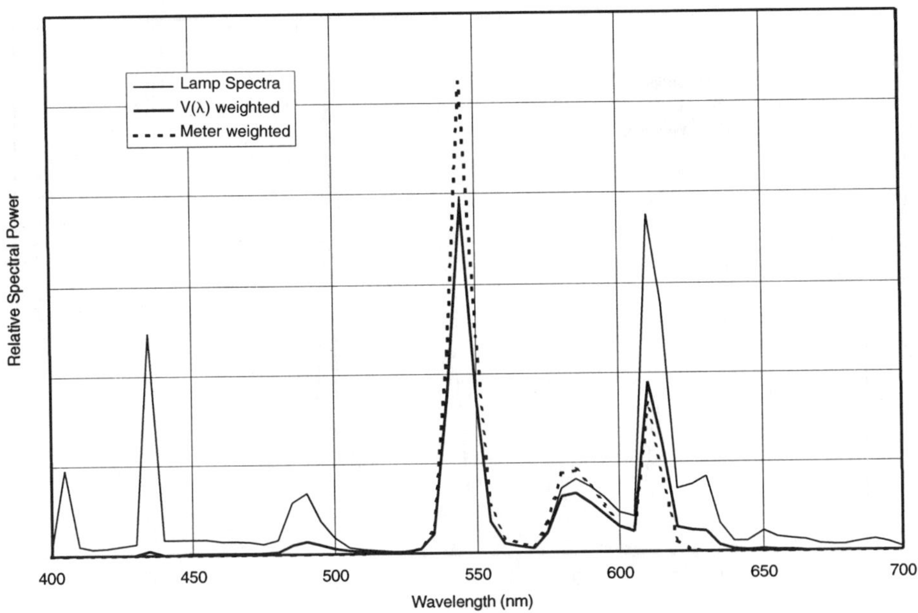

FIGURE 6.4. 3200-K fluorescent lamp.

Data collection for spectrometers require a complex detection system that starts with a detector and often ends with capture software running on a non-real-time computer. The details of how the signal gets measured are not important at this point, and total understanding is probably not possible when software is part of the system. What is of importance is that the recorded value is linearly proportional to the signal.

Three methods for measuring linearity will be described here, but because this is such a fundamental concept many more can be devised.

Neutral-Density Filters

Filters that provide constant attenuation throughout the spectrum are a convenient tool for lowering input signals. A set of filters can provide varying levels of input with which to test the systems response. The amount of light passed by the filters can be very accurately determined by a dual-beam spectrometer. This instrument is usually a turn-key operation and will measure the ratio of light through the filter and straight to its detector. Once characterized these filters will be seen to be not quite spectrally flat (see Fig. 6.18). Using the transmittance of the filter at specific wavelengths, we can compare various input levels to the reported values.

A calibration lamp is set up in its normal configuration with one change. The path

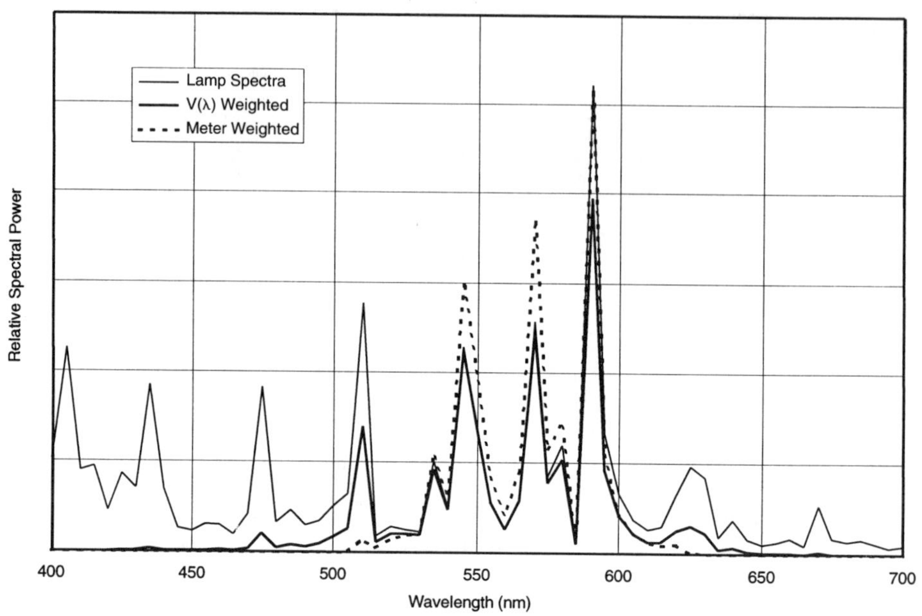

FIGURE 6.5. 5400-K metal halide.

of light to the system input must be apertured, as shown in Fig. 6.6, so that all passed light can be filtered. The measurement is taken with no filter in place and recorded. This scan is repeated with each filter in place and then again with no filter. Readings are used at three wavelengths for the comparison. The incandescent calibration lamp has a varying output that changes by an order of magnitude from the blue to the red. This can be used to extend the range of this test by using a short wavelength and a long wavelength. Readings at 400, 550, and 700 nm are tabulated with the ratios for the filters at those wavelengths. Though the filters are not flat, they change so slowly that a slight wavelength error in the system under test should

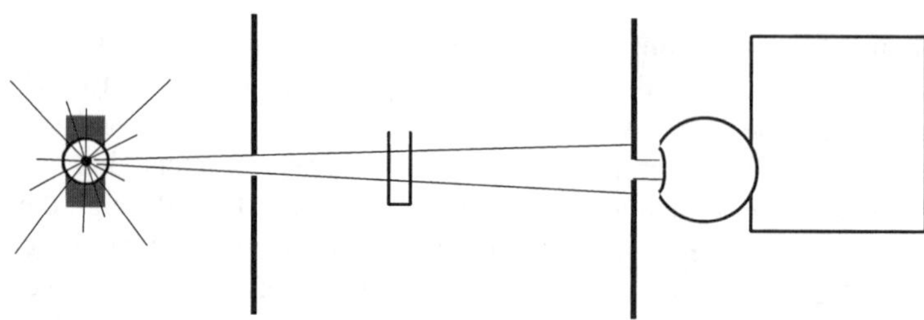

FIGURE 6.6. Setup for transmission measurement.

not introduce significant errors. It is clear that this method is simple, fast, and easy but requires a range of filters to cover the full range of the instrument. In addition, the dual-beam spectrometer may not have the range to give significant numbers for low transmission filters. The use of limited filters with the next method can extend and speed the testing.

Troubles to avoid are tilt in the filter and placing the filter near the source or input aperture. During characterization and use the light should be normal for the filter to maintain the same path length within the filter. Reflections from the filter surfaces will not be part of the assigned transmission and must be avoided in use.

Distance

This method relies on the fact that transfer of radiation will drop by the reciprocal of the distance squared, all other things being equal. Keeping all things equal is aided by having a rail or bar on which you can slide sources in and out. These rails permit the distance to vary without changing the direction or orientation. If a rail is available, use it, if not, be very careful. The use of a laser to keep the system in line can help but will not replace a bar. The laser is aligned normal to the input by bouncing the beam back from the middle of a piece of glass placed flat on the face of the input optics. This beam is the optical axis of the system and can be used to keep the source on line. As the source is moved the beam must be kept on the lamp and an alignment rod placed behind the lamp on the cart. The change of signal seen by changing distance is equal across the spectrum because air is transparent and the input optics are filled from a point source. Distance measurements are not needed for a spectrometer because the full spectrum can be compared. (This method is used for cell photometers by careful distance measurement and the inverse square law.) If there is nonlinearity, then the ratio of one spectrum to another will not be a constant flat line. For example, if the detector system saturates at a high level, the ratio of two spectra will drop off at the wavelengths where the detector signal is highest. For my spectrometer and an incandescent lamp, the detector signal is highest at 550 nm. A ratio of the bright to dim signal will be constant up to the point where the signal is overloading the detector where the ratio will drop. Knowledge of these details is not necessary because a constant ratio is the indicator of proper linearity. This test is especially good for the electronic and software systems, as the signal levels varies continuously across the spectra testing for discontinuities at range changes.

Combined Inputs

The first two methods requires the availability of filters or space to move the source. In addition the proof of linearity may seem abstract. The method of combining two signals is very straightforward and requires readily available tools [32,33]. The mixing of light obeys the law of superposition, except in special cases such as with a laser. If two sources shine on a detector, the detector will respond to each separately. By arranging the sources to be measured separately and then together the ability of the detector system to linearly measure a signal can be tested.

Starting with a calibration lamp as a source, set up two paths to the detector.

The input aperture is tilted (Fig. 6.7) so that the two beams are nearly equal at the input to the system. Baffles are placed to reduced the stray light to the system and allow for mounting of filters and stops. The levels can be varied by using variable apertures or a simple block part way across the baffle opening. Test the baffles system and stops by blocking both paths of light to the detector. Take a full spectra as a background reading. If the scan looks like an incandescent lamp, continue to place baffles in the paths to the detector. A baffle around the detector input may be very effective. Open one stop, and let the light to the system input from one path; then take a reading. Open the second path so that both paths are illuminating the input; then record a spectra. Close the first path so that you are measuring the unmeasured light path. This order is suggested so that you have the least chance of bumping the alignment between readings. The reading with both paths open should equal the sum of the readings for each path measured separately.

Continue to test by restricting the light through the apertures until you reach the lowest level that you will possibly use or down to the level of background. The steps should overlap so as to provide complete coverage of the ranges used. By using the red as the upper measure and the blue end of the spectra as the lower level, the overlap will be quite large.

6.4.2 Input Spatial Response

Illuminance

The measurement of illuminance is the determination of light incident on a planar surface element. The ideal is that all light, regardless of its direction, will be

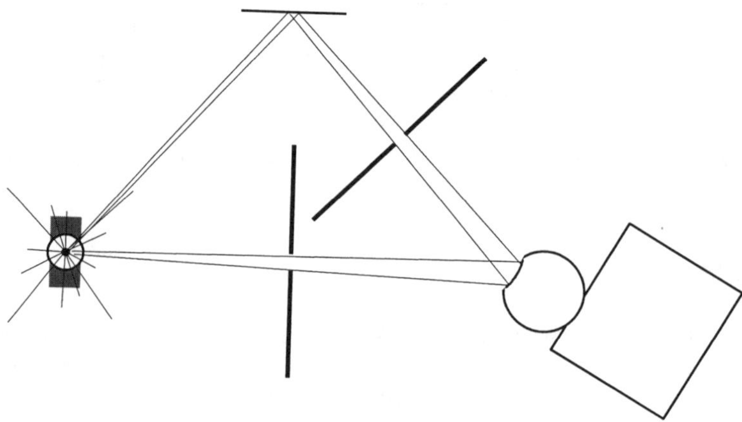

FIGURE 6.7. Linearity test: combined signals.

measured if it strikes the surface element. This is the same, as all light that goes through a flat aperture will be equally measured. When in a constant beam of light that fills the aperture, the amount of light that goes through the aperture will be a function of the angle that the beam of light makes to the normal direction of the flat aperture. In the normal direction, 0°, full measure of the light gets through. At 90° to the surface of the aperture, none of the light gets through. At angles in between, the cosine of the angle will be the ratio between the light through and the amount from the normal direction. By mounting any detector on a rotation stage and in a uniform beam from a distance, the angular response can be measured. Mount the aperture or diffuser face on the axis of rotation, so the distance will not change. It may be impractical to mount the spectroradiometer on a stage, particularly off axis. A cell may be used as a surrogate, taking the place of the radiometer at the sphere's exit aperture. Mount a strong constant source at a distance 30 times larger than the detector or source. At this distance, the light will be parallel for practical purposes. Block all but the beam that illuminates the detector. The detector has a wide acceptance angle so you will measure your shirt, the light fixtures, and many other things if there is light near the detector. Two rooms, one with the source and one with the detector, with a hole between them is the best way to get the detector to a dark place. Rotate the detector and record measurements for varying angles. Set +90° and −90° by observing the light across the face of the detector or its bezel if the detector is curved. Compare the response with a cosine curve, after normalized to unity at 0°. The discrepancies that you see between the cosine and the response are the inaccuracies that you will get when you calibrate with a small source on axis and measure light from all directions, for example room lighting.

Flux

The integrating sphere is a wonderful way to gather the total light from a source and measure it at a single stroke. But for it to be accurate, the source should be small compared to the diameter of the sphere. Failing that, the calibration source needs to be of the same physical form and have the same spatial output distribution as the tested source. For fluorescent lamps this is not practical. To be large as a 4 ft lamp, the sphere should be ~15 ft in diameter, which is much larger than it sounds.

To measure the response of a sphere for a sources off axis, a simple setup can be used. String two uninsulated wires across the sphere. Fluorescent-lamp sockets are best for attaching the wires because the test area then comprises the space the fluorescent lamp normally occupies. The wires are electrically connected to a supply that is used for calibration lamps giving a fixed drive current for the test lamps. This test lamp is fashioned from a small frosted incandescent lamp with three legs attached (Fig. 6.8). The three legs allow it to rest on the wires, with two on one wire fixed to the base and other riding the second wire and completing the circuit. The legs should be bent to hold the lamp in the same position as the normal fluorescent lamp under test. The sphere is used to measure the output of the lamp in

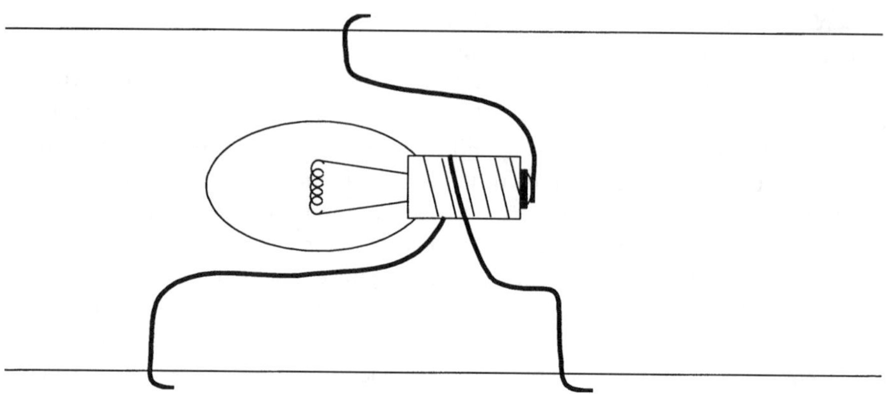

FIGURE 6.8. Device for testing source position uniformity in spheres.

various positions at a fixed current and therefore output. If the lamp is mounted parallel to the wires, then the shadow effect of the base can be determined by performing measurements in each orientation.

A comparison of the sphere throughput as a function of source position will indicate if the magnitude of errors is for large sources. It is not suggested that a correction be applied to readings, rather institute improvements to the sphere's coating and configuration.

Measurements of directional uniformity have been performed with the use of a small spotlight mounted in the sphere at the lamp position and adjustable in direction across the sphere's interior [34]. Care must be taken to prevent disturbance of the sphere's throughput with moving parts or changing of the lamps output. If the mounting and lamp system changes as the aim is changed, then secondary effects may impact your results. Be aware of changes in the filament's orientation with respect to gravity that may cause significant variations in the output.

6.4.3 Wavelength Accuracy

Fluorescent room lights make for a good quick check of wavelength. Unfortunately with improvements in phosphors the spectra is getting crowded with other lines. For now you can only count on the 436-nm mercury line from a fluorescent lamp. For a complete check use a low-pressure Hg pen lamp, germicidal, blacklight, or EPROM eraser lamp. (*Caution*: The germicidal and pen lamps present health hazards.) These lamps will give you the atomic emissions of mercury that you can identify. Table 6.1 lists the strong mercury emission lines and their relative intensity from a pen-type lamp.

If your system is not variable, i.e., someone else's software or fixed pixel positions, then determining where you have measured the line can be tricky. A typical photometry system will have a band pass of 5 nm and report every 2 nm. The

TABLE 6.1. *Mercury "pen" lamp output.*

Wavelength (nm)	Relative intensity
404.66	43
435.83	100
546.07	78

strongest wavelength is not the center of the line. An emission line whose center falls between the reported wavelengths should not be considered as being at the wavelength with the highest response. Figure 6.9 shows the 404.656-nm line as measured with a 5-nm Gaussian bandwidth. The highest reading at 404 nm is not the position of the emission line.

The Hg emission line at 546 nm is helpful in this case because it is nearly on the even nanometer wavelength. If all is aligned correctly, the 546 nm wavelength will have the largest signal with symmetrical shape, and the 544 and 548 nm wavelengths will be equal in magnitude. For other lines, model with a Gaussian or triangular function and use the center of that fit [35–37].

6.4.4 Time Response

In scanning spectrometers the detector signal may vary by orders of magnitude as emission lines are crossed. This is tempered in photometry because it is known that

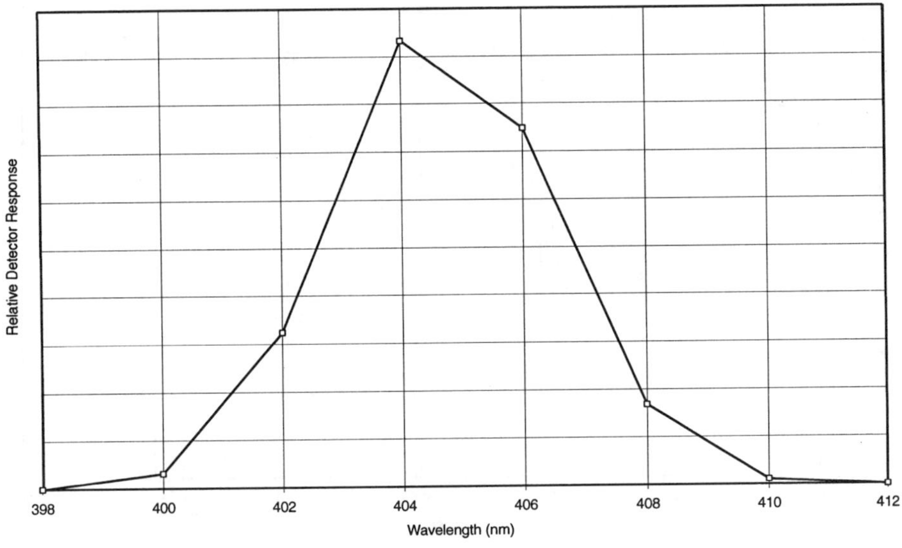

FIGURE 6.9. Hg emission line at 404.656 nm.

a 5-nm bandwidth will provide acceptable colorimetric results and lower peak heights [38,39]. Nevertheless, a rapid change in signal may result in a detector response that is different from that which was seen during calibration with a incandescent lamp.

Rapid changes in signal levels may cause hysteresis or overshoot in detector systems. To test for this problem, change the signal level at a rate that is quicker than the scan step speed. If the change in level is seen in more than one data point, this might be the result of detector system lag.

Figure 6.10 shows the normal scan of an incandescent lamp and an interrupted scan of that lamp. Note how the signal drops all the way to zero and returns without overshoot to the proper level. This shows that the electrical and software time constants and delays are sufficient for this instrument at this scan speed for these levels of illumination. If the scan speed is raised or other parameters change, you must recheck.

6.5 LIST OF INSTRUMENT PROBLEMS

There are a large number of ways for the equipment to have difficulties. The trouble may occur quickly, between calibration and use, or slowly so that the change is attributed to other causes. Below is a list with short descriptions of the causes and effects. This is in no way complete and should be used when trouble is

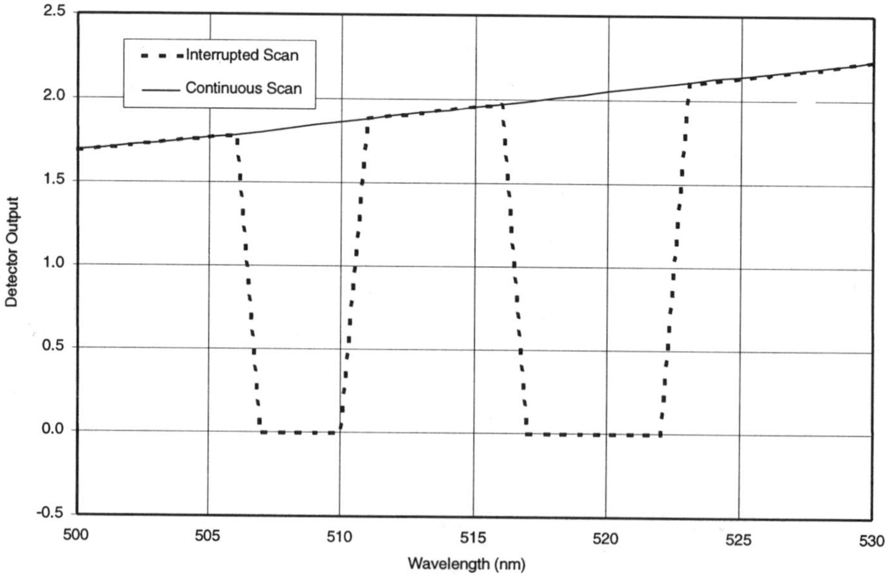

FIGURE 6.10. Check of detector hysteresis.

suspected, and not as a checklist. You should always suspect trouble. The wide variety and subtlety of these problems should be a spur to make checking a way of life.

6.5.1 Filter Fade

Absorption filters work by the chemical makeup of material. Exposure to radiation—how else can you use it—will break down the material, leading to a changed absorption. This is normally seen as a change in transmission. For glass, solarization decreases transmission in the blue with exposure to UV. Plastics turn brown through oxidation of broken polymer chains, first seen as yellowing. Plastic filters will show increased transmission as their dyes bleach.

6.5.2 Heat Changing Spectrometer Mechanically

A spectrometer is a very precise optical instrument where beam alignment determines the wavelength of light passed and the amount of light transmitted. A slight temperature gradient, such as the heating of one side by an incandescent calibration lamp, may lead to a change in the optical throughput. A calibration that is true when the spectrometer has one side hot may not be true after the optics cool down.

6.5.3 Sphere Dirt

An integrating sphere works by getting the light from all directions equally to the detector. The top half of the sphere will not have dust falling on it so that the top half will almost always be cleaner and brighter than the bottom half. A spotlight pointing up in a sphere will usually produce a higher reading than the same source pointing down.

6.5.4 Sphere Moisture

Spheres that use high reflectance coatings of barium sulfate or magnesium sulfate rely on the small particle size to get reflections near 100%. If there is moisture in the paint, the amount of surface area is reduced and the reflection is lowered. A common problem is that the sphere will dry during the calibration routine and render the calibration invalid. It is necessary to always have a lamp burning in the sphere to maintain the temperature and limit the absorption of water.

6.5.5 Dust on Optics: Scatter and Fluorescence

The presence of dust on any optical surface will lead to scatter, absorption, and fluorescence. As light is moving through your optical instrument it is directed by mirrors and lenses to specific places. If a dust mote is in the light's path, the light may be absorbed or scattered. Scattered light rays may pass the spectral selection, raising the background signal level. If absorbed, it can be changed to heat and lost to the measurement or changed to a different wavelength. A particularly difficult version of this is dust fluorescing in the sphere coating when the source has ultraviolet emissions.

6.5.6 Detector Hysteresis

As discussed in Chap. 4, the overloading of a detector may change the response for a time or permanently. This problem is detected when successive calibrations are compared.

6.5.7 Detector-Temperature Changes

It may not be practical to hold the temperature of the detector constant, in which case its response will vary. The dark noise level is a good indicator of the detector temperature and should be monitored.

6.5.8 Fluorescence

When measuring sources that also produce UV, the optics of the instrumentation may introduce visible signals through fluorescence. Filtering of the UV before the input optics and prior to calibration will prevent this. Beware of filters that block UV but, in doing so, fluoresce.

6.5.9 UV Damage

The damage of optical parts due to radiation is energy related and may have threshold frequency. Therefore any problem seen with visible light may be worse with UV, and some problems only occur with the higher photon energy of UV.

6.5.10 Heating of Electronics

The heating of electronic components can change their values, leading to changes in the circuits behavior. Amplifier gain and photomultiplier tube high voltage must remain constant for the calibration to be valid.

6.6 CASE STUDY: LIGHT BULB EFFICIENCY WHEN DIMMED

The production and measurement of light is largely driven by economics because of the large amount of light needed and its consumption of energy resources. This first case study will address this cost/light relationship and its pitfalls.

Customer's question: Can we save money using dimmers?

Translation: Measure the efficacy, usable light per power, versus dimming for a dimming incandescent lamp.

Constraints: Use what you have: digital multimeters and a cell-based illuminance meter.

6.6.1 Step 1: Is the Cell Any Good?

The light meter needs to be linear to measure the changes in light from the light bulb. If half the light is incident on the meter, will the reported value be reduced by one half? Methods for testing this property of the meter are detailed in Sec. 6.4.1. At this point we will assume that the meter's response is linear. The spectral response of the meter should match the $V(\lambda)$ photopic response to give meaningful information regarding levels of visible light. Checking this requires special equipment, and we will make the wrong assumption that because the measurements are relative, the spectral response will not matter. This step should include actual testing, but for this example we will naively trust our light meter.

6.6.2 Step 2: Setup

The ''system under test'' is an incandescent light bulb operated by a dimmer. The system is supplied with electrical power by a line conditioner to provide a stable voltage source. For this study the line conditioner is not vital, but it is good practice to have a stable supply as local electrical loads may quickly change laboratory line voltage. If the voltage changes between the time you record the electrical values and the noting of the photometric data, then any correlation is corrupted. The wiring of the system includes two digital multimeters (DMM) set to record the voltage and current at the light bulb. The incandescent light is a resistor, the current waveform matches the voltage waveform, and therefore the power consumed is the product of the voltage and current.

The detector is mounted at a distance such that undimmed, the indicated value is near the top of the meter's range. The meter used in this test has $3\frac{1}{2}$ digits and we will chose the 0–2000 range. The range chosen is sensitive enough so that the distance needed to get ~1500 counts is far enough away from the source that the meter is not heating up during the test. The detector is baffled by a black plate with a hole in it. The baffle is set between the detector and the lamp to prevent reflections from reaching the detector. It may seem that reflections may not be able to lead to error because they will vary with the lamp's output. Constant reflections that do not change or color the light will not cause problems, but with people moving about in the room and reflective surfaces it is important that the detector view is limited to the source.

6.6.3 Step 3: Testing the Setup

The detector system may have an offset that is significant and therefore should be subtracted from each reading. To find the offset the detector is shuttered so that no light may get in. The best way is a cap that covers the detector tightly and completely. This offset value will vary with temperature and should be measured before and after the testing to see if heating has affected the detector.

The determination of stray light is done with the systems on and the direct light path blocked. Where the light is blocked is important because blocking the signal path may also stop the path of stray light. Generally stopping all direct light from the entering the meters aperture at each baffle is a good way to find the stray light value. The determination of stray light is critical for accuracy, but for these relative measurements it is probably not critical.

Finally if the room is not dark you are not measuring the source, you are measuring the source and other light. With the source off and the room in the state you would use for testing, all equipment on, record the detector reading. If this value is significant, then the value may be subtracted from all readings, but this is poor practice as this background light may vary as people move and equipment displays change.

6.6.4 Step 4: Measurement

After all systems have been on for a time, monitor all parameters for changes. Record values as a time series to be assured that the variations are truly random noise and not a slow drift covered in noise. When the systems are stable to 1% for the time period which the test should run, 15 min, then the testing may begin. 1% is set as an arbitrary level of stability. This level must be chosen based on the precision needed. Record the values of voltage, current, and the detector reading at one end of the testing regime. In this case start at the high setting so the lamp is stable. Decrease the lamp voltage by some even value using the dimmer. The size of each step can be judged by balancing the needs of the number of data points with the possible drift and operator error if too many readings are taken. Five data points are likely to be sufficient to determine the dimming curve, so twice that many would give a good smooth curve. This may not be the case, and sufficiency will be known only after the testing. In this case a quick run will show the behavior of the curve and indicate exactly how many points to measure on the second run.

Important principle: You need to know how the system you are testing will perform before you test it! This requires you to pretest a new subject system to determine levels and behavior before the test plan can be defined. The testing of a new source should result in two findings, first how to test and second the source's performance.

Starting at 115 V and going down at 10 V increments will provide 12 data points. At zero the run is continued back up to 115 V with the same steps to check for cooling and other hysteresis problems.

6.6.5 Step 5: Data Analysis

The graph of relative lumens and watts (Fig. 6.11) is easily constructed by normalizing the wattage and light to the maximum setting. The graph may be constructed as relative efficacy (illuminance/power) versus dimmer position. This graph shows a gain in efficacy as the bulb is dimmed which may not be expected. Knowledge of incandescent lamps suggests that there is something very wrong with this test.

Faults

This case study was performed to highlight the problems that will occur in naively performing measurements without consideration of the system under test and the measurement system.

First, there is an intentional problem in the procedure. The system is the combination of the lamp and dimmer and therefore the input power to the dimmer and lamp must be used, not the lamps power. In this test the amount of power that the dimmer added to the system is ~0.5 W and not significant. Measuring the lamp in this way points out the fallacy that electrical measurements are easy. Not in this case. The dimmer and lamp is not a linear system, and therefore the measurements must be done with an eye to various problems. Evaluation with an oscilloscope

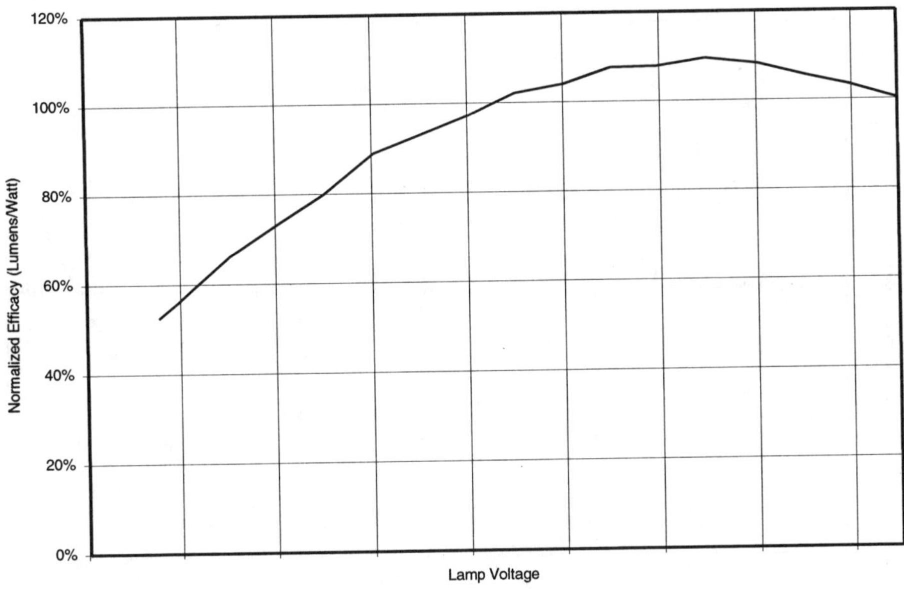

FIGURE 6.11. Dimmer efficacy.

shows in Fig. 6.12 that the current waveform is distorted and will contain high-frequency components.

The power can only be measured with a wideband wattmeter. After seeing the waveform the need for a line conditioner is clear. Distortion of the supply voltage caused by the sharp current waveform could further muddle this study.

The second problem is the use of DMMs for the voltage and current measurements. Assuming the dimmer draws no appreciable power still leaves the bandwidth limitations of the meters. The use of 60-Hz meters will lead to a mismeasurement of the power, even with a resistive load because of higher-frequency components in distorted waveforms. Figure 6.13 shows the difference between appropriate and insufficient electrical measurement methods.

The measurement of the light is distorted by the use of a nominally photopic cell with an erroneous response curve. The spectral response of the meter used is shown in Fig. 6.14. In comparison with a spectrometers measurement (Fig. 6.15) the relative output is not poor until the lamp is quite dim and therefore at a low color temperature.

Now let us look at the final graph of system efficiency (Fig. 6.16) as measured by the naive way and with a spectrometer and wideband wattmeter.

At this point we should think again about how the results of our test will be used. The broad shape of the curve tells us if energy is being wasted. The handheld multimeter errors have resulted in a curve that shows a more efficient use of energy as the lamp dims. This error far overshadows the photocell light measurements. For

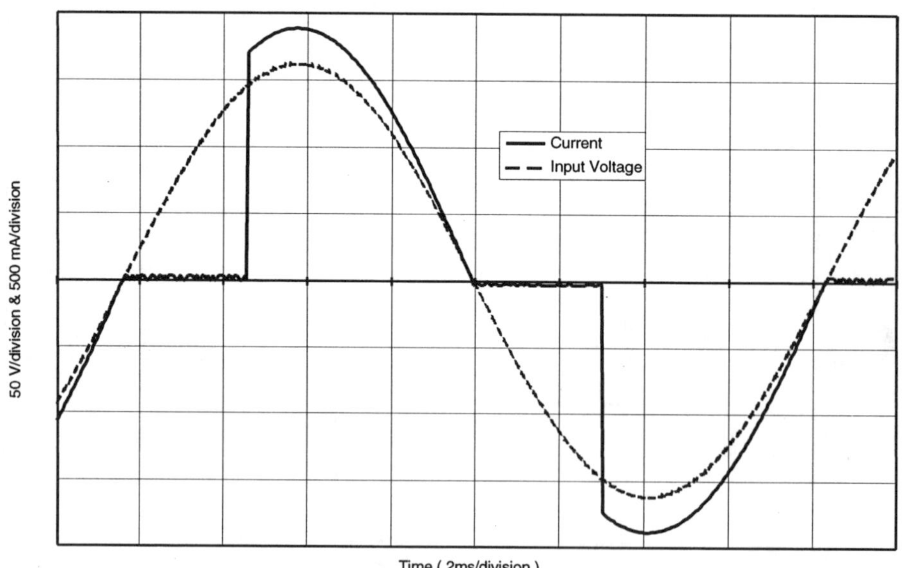

FIGURE 6.12. Wave shapes at 2 ms/div.

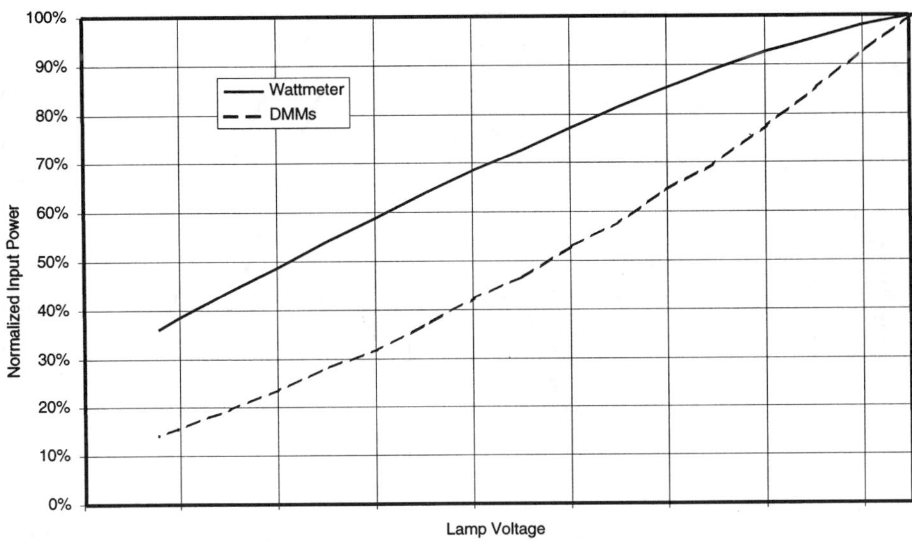

FIGURE 6.13. Power measurement.

this test the photocell would give you sufficiently accurate results if correlated with the wattmeter's readings. The results would then show the drastic drop in efficacy that is known to occur [40] if an incandescent lamp is operated below its designed voltage.

Advice

The question posed for this case study has several problems. The question is: Can we save money using dimmers?

The clear understanding of the purpose of a test starts with a critical review. This review should include attacks on the question. Several should have come to mind for knowledgeable photometrists.

- The efficacy of incandescent lamps decreases greatly at lower powers [40]. Are you expecting a "miracle"?
- If you are concerned with costs, why are you using incandescent lamps?
- If the installation is large enough to warrant power reductions, can you turn off some sources and thereby maintain efficacy?
- Will favorable test results lead to banks of dimmers?

These questions should be diplomatically raised to the customer, the last question in particular. What decision turns on the result is vital for determining the equipment, effort, and time needed.

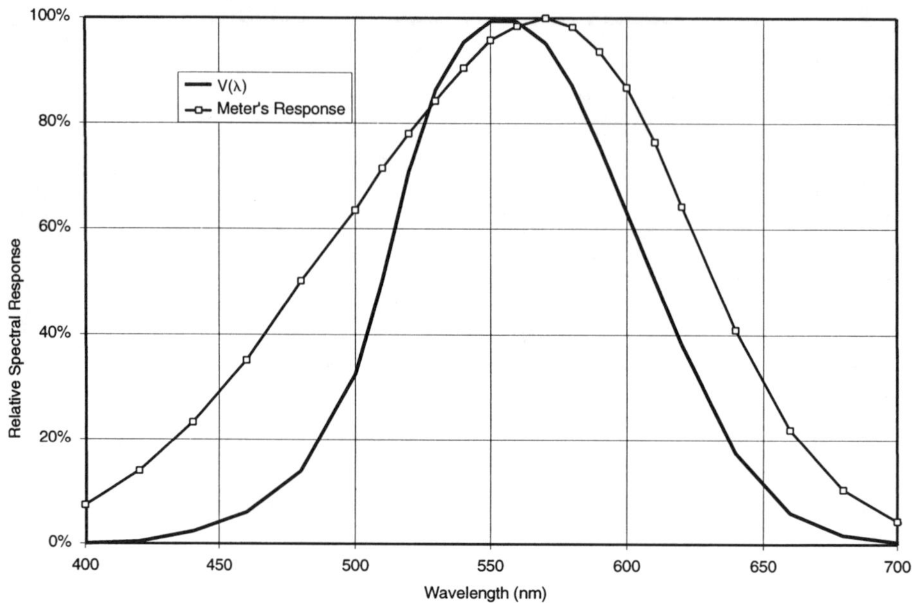

FIGURE 6.14. Meter's spectral response.

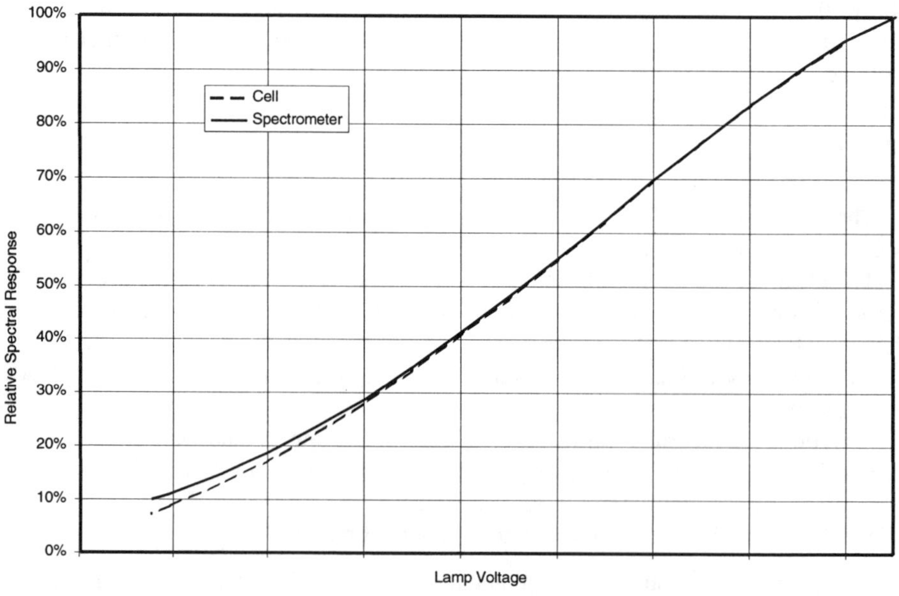

FIGURE 6.15. Output measurements.

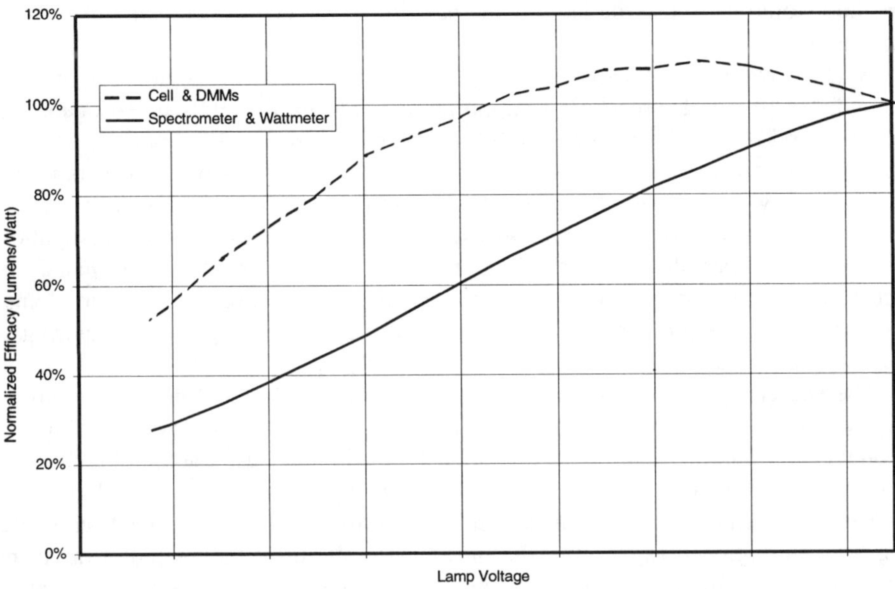

FIGURE 6.16. Measurement results.

6.7 CASE STUDY: LOOKING FOR A 0.3% CHANGE

Lighting systems improve in an evolutionary fashion with small improvements adding up over the years. How can you know if you have improved if the change is less than 1%? From the inherent uncertainty in the calibration source alone [6], it is clear that the change must be measured on a relative basis.

This example is chosen to allow for easy relative measurements. The simplicity arises from the fact that the source modification is on the outside and may be accomplished while the source is operating. Quantifying internal changes to the source, chemistry, etc., requires comparing distinct samples. These comparison tests cannot be finessed; using statistics of a sufficiently large number of samples is the only way.

A standard fluorescent lamp is constructed with end caps upon which the electrical and mechanical connections are made. The glass envelope may allow light out, which would then be absorbed by the end caps. Is this loss significant? Are we shipping lumens that the customer does not receive? Based on theoretical studies there could be 3% trapped light. A large-scale test was planned to measure lamps with and without end caps. This method would yield results only on a large scale because of the variability in the manufacture and operation of fluorescent lamps. Slight variations in lamp construction will lead to differing operation. Confounding is the fact that upon relighting a fluorescent lamp, it may change from its last operation significantly due to the electrode's condition and temperature. A simpler method would be to test the lamp with and without the end caps and not disturb its operation.

A lamp without end caps is seasoned to provide a stable source. Again the knowledge of this source [17] is important, telling us that it is unreasonable to expect good results during the first 100 h of lamp operation. For common sources this type of information can be found in books or from experienced workers in the field. For new sources or when in doubt the source's behavior can be characterized in a similar fashion to evaluate the measurement system. This type of investigation is invaluable especially before the start of a large test program. The critical analysis after a large test program on a new source can uncover most pathological problems. Unfortunately the problems may have invalidated the readings, thus requiring a repeat of the test.

The end-cap pins normally support the lamp in a socket and provide electrical connections. The electrical connections are easily accomplished by extending the lead wires. Mounting in the sphere must be done with the least amount of additional structures minimizing optical effect or heat sinking at the bulb wall. Thin white thread can easily hold the lamp and minimizes disturbances. The lead wires are threaded through end caps so that the caps may be slid up to the lamp or away from the lamps by a distance of several inches. The lamp is operated at a fixed current to maximize stability. The room temperature is stable at the specified 25 °C, but opening the sphere will provide drafts that cannot be prevented. The system used features a diode array spectrometer which makes a measurement in ~2 s. An average of 10 readings is recorded for this test to improve the precision.

A measurement is taken with the end caps against the lamp ends. The end caps are then moved as far as possible from the lamp and another reading is taken. The difference should be the light that the end caps absorb. But even with special care, the amount of noise and drift in the readings will be nearly equal to the difference (Fig. 6.17). So the cycle is repeated with one reading at the lamp and one pulled back, and again, and again. On the spot, comparison showed a slight difference, but a graph of the data against time shows a drift that can be compensated for. The true difference is ~0.3%.

What about the end caps in the sphere? Moving the end caps away from the lamp and into the path of the light may be enough to lower the sphere throughput and mask the extra light. This is checked by introducing two additional end caps into the sphere and measuring the change in light.

6.8 CASE STUDY: TOO MUCH LIGHT

Equipment to measure radiation and light is usually designed for maximum sensitivity. The signal will still be too low in some cases. High sensitivity is often achieved by using photomultiplier tube (PMT) detectors that have a response of amps per lumen. These tubes run at high voltages and would melt if amps of current were run through them. In practice most PMTs need to be limited to 1×10^{-6} A

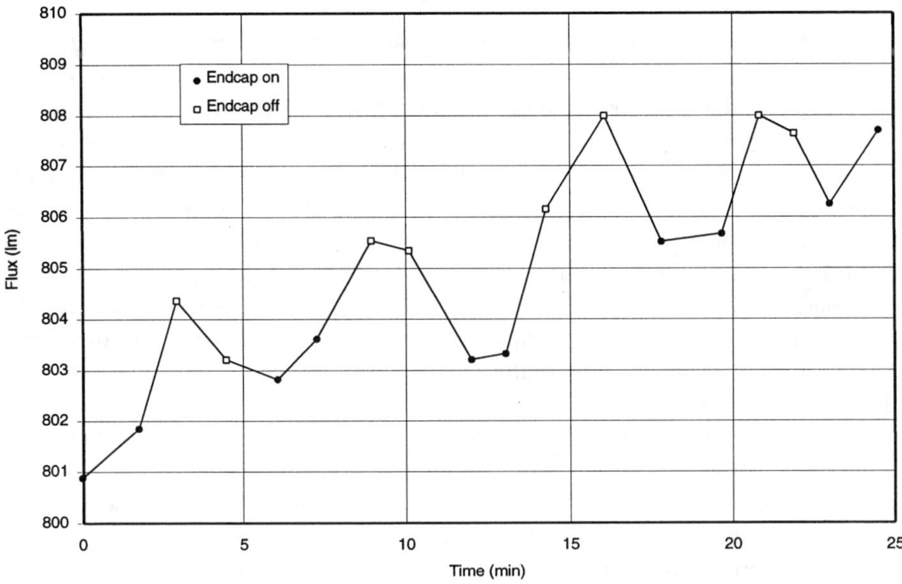

FIGURE 6.17. Total flux with and without end caps.

(1μA) to prevent problems. The spectrometer that selects the light for the PMT usually lowers the incoming light well below this limit. The measurement of focused light such as a headlight can raise the input levels well above the normal illumination levels. A source that is monochromatic has the chance to put most of its energy through the system at one monochromator position. For these reasons it is necessary to have techniques for reducing the input signal.

It may be that the system has enough dynamic range, so the system can lower its sensitivity to make the readings at high intensities and still be calibrated. In this case, anything that reduces signal is okay because the spectral changes will be calibrated out. This will be the best solution because the linearity across the ranges is the only concern and systems are optimized for reading low signals.

6.8.1 Illumination from Point Sources

If the device tested is a "point source," with uniform illuminance at the detector port, then you can just back away. The inverse square law works well for this type of measurement. Calibration can be done at the normal distance, and then the test sample can be moved out to a range that eliminates detector overloads. Though this technique has limited use it is the cleanest and most accurate because there is only one additional error source. The signal needs to be scaled for the actual distance as compared to the calibration distance.

6.8.2 Illumination from Extended Sources

Extended sources are those that will not scale as the distance is increased. If the illumination is needed at a distance of 50 cm, then the sources over 5 cm in their longest dimension will not accurately scale. This is the rule of $\frac{1}{10}$ for the inverse square law. Infinitesimal source and detector will have exact inverse square law behavior but what about real objects? If all the light goes from the farthest two points between a source and detector, that change will be the bound for errors. With a 1-cm-diam source and detector, separated by 10 cm, the longest path is 10.05 cm. From the inverse square law this worst case path reduces the signal by 1%.

A source with a reflector will also cause problems if the distance is changed. The light will not expand to an even sphere for these lamps until distances become large. If you can get back far enough, scaling is possible but you may not have that much distance to work with.

The option then becomes to attenuate the light at the instrument. We have gone to the trouble of getting cosine-corrected input optics. If we put anything in front of the input the response will no longer be cosine and the quantity that we measure will not be illumination. Inside the input optics there can be made a place for a filter and/or an aperture. If a sphere is used to provide cosine reception and mix the polarization, then a limit can be put on the output of the sphere to reduce the signal. The use of a filter will require that the spectral changes from the filter must be determined. It can be seen in Fig. 6.18 that the changes through the spectrum for this neutral-density filter would drastically change the measurements if not full accounted for.

Readings of a calibration lamp with and without the filter in place are placed in ratio form to find the change that the filter introduces. These reading should be done with care and with an eye to the signal levels. The filter may reduce the signal to the point of noise problems. Quantization of signal is seen in Fig. 6.18 due to the large range that must be covered to characterize this $\frac{1}{1000}$ 3 OD, neutral-density filter. The ratio of the spectra should match the transmission of the filter when read on a dual-beam spectrometer. There may be problems if the filter is placed at the sphere exit because the efficiency of the sphere will change if the exit port reflectance changes. The calibration spectra is adjusted to provide the proper response when the filter is in place. This method of adjusting a normal calibration is used because the calibration signal levels are assumed to be within the normal range of the equipment. Therefore the calibration and the test source are of different levels and the use of the filter would lower the signal of the calibration source. The calibration source, being an incandescent lamp, varies over a range of 1:10 or more. If the calibration signal is substantially reduced, then the amount of noise in the calibration can become significant. It may be practical with some systems to reduce the throughput and then do a calibration. The standard calibration with the use of known measured transmission spectra should give you better results in terms of noise levels. Calibration and readings will then both be in the same range, where the instrument's performance is optimum

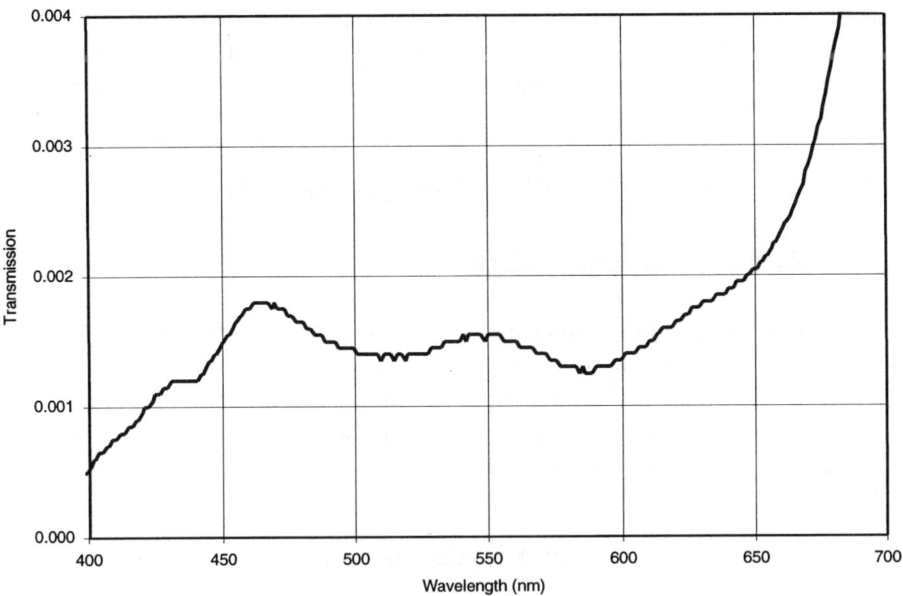

FIGURE 6.18. Transmission of neutral density filter.

The use of an aperture should be better for spectral flatness but this is not always the case. If the aperture is at a sphere, then the color of the paint will change the response of the sphere and the system. Black paint works best to reduce the spectral changes to the sphere's throughput. If the aperture is placed in the spectrometer the results can depend on the exact positioning and therefore will not be as repeatable. One good way to reduce the signal is to have an aperture between two spheres so that the signal path is not affected and the attenuation at all wavelengths is equal. The trade-off for better attenuation is lower throughput. As always check carefully.

6.9 LABORATORY OPERATIONS

The measurement of light requires several systems. Needed are the following: mechanical, optical, and electrical systems to make the measurements, systems to operate the source if artificial, calibration systems, and finally a system for performing the measurement task is needed. Dependable accuracy, the actual *truth* of the measurement, relies on its precursor precision, the repeatability of the measurement. Though a moderate level of accuracy may be the best obtainable, the level of precision should far exceed the level of accuracy. Precision allows for the comparative use of measurements beyond the accuracy of the measurements.

A controlled procedure of operation is the backbone of precision. The inevitable changes can be tracked, limited, and their effects quantified by the use of a consistent and constant method of operation. Below is an outline of the required parts of the laboratory's operation followed by an example detailing daily operation.

6.9.1 Requirement for a Photometry Laboratory

Calibration of Nonoptical Equipment

The setting of conditions, voltage, temperature, etc., must be accurate and traceable, so that variations in conditions can be quantified and limits to operational uncertainty determined. A calibration service or an in-house calibration department that tracks the performance and understands the calibration cycle of the particular equipment is helpful in maintaining accuracy (ISO Guide 25).

Calibration of the Photometer Using Standards

I advocate the in-house calibration of the photometric tools to promote an understanding of their operation and limits. If you do your own calibration, you will have a better feel for the instrument's capabilities and limitations. This may be impractical, in which case the calibration must be done in a proper manner using a system that traces deviations from specified performance and tracks uncertainty.

Environmental Control

The best calibration is for naught if the equipment is subjected to environmental conditions outside the range specified by the manufacturer.

The temperature can have a tremendous impact on detector response, artificial sources, and input optics. Other less dramatic effects from temperature may appear if you cure the symptoms and not the source. See Secs. 6.4.2 and 6.4.10.

Limited Access

The laboratory used as a smoking room or the loan of equipment are dramatic cases of human effects on the laboratory's operation. *Note:* The calibration system must require recalibration after any break in the chain of custody (i.e., a loan). A casual interruption of operations can have a more subtle and difficult to trace impact on results. Access to the laboratory should be limited to provide a stable environment and promote calm uniform operations.

Formal Operating Documentation

The operation of the laboratory must be detailed by formal written procedures that are true to the actual operation of the laboratory. This documentation is for the purpose of recording what is done, not a plan for how you would like to operate.

Consistent Operations

The operations of the laboratory should be regular, even boring. The repeatability of measurements relies on doing things the same way each time in order to provide for precision and uncover inevitable problems. Selection and maintenance of the human parts of the system should include consideration of the potential effects of monotony on measurement results.

Documentation of Results

The collection and communication of results should be optimized to make the data transmission as lossless as possible. Good data with clerical errors are the same as poor data with the additional loss of face. At the time of measurement, remember that it is impossible to go back and get data later; therefore record everything, e.g., time of day. All parameters that can conceivably be recorded should be so as to give the customer as much information as possible to analyze. All data that are collected should be retained for the eventual questions. Where possible, supply the customer with all data, as this allows for comparison and analysis that was not initially contemplated.

6.9.2 How I Work in My Laboratory

My laboratory was designed and constructed for the measurement of spectral irradiance and therefore may include features that are impractical to achieve otherwise.

All is black: the ceiling, walls, floor, equipment, curtains, wires, laboratory fixtures, lighting fixtures, wall switches, etc.—you get the picture. The black is flat to reflect as little light as possible directly back. To help keep the floor black and not turn dirt gray, a clean room sticky mat is placed outside the door.

The floor plan (Fig. 6.19) shows the division of the room into a measurement area and the space for the operator. The curtains separating the two sections are opaque and drag on the floor blocking all light. At the ceiling the track is covered by a valance 6 in. tall, blocking all light in that direction. There is sufficient curtain width so it laps against the walls and has a generous overlap at the split.

Because the sources to be measured include arc discharges, the electrical supply system includes circuits that are supplied by line conditioners that are known to

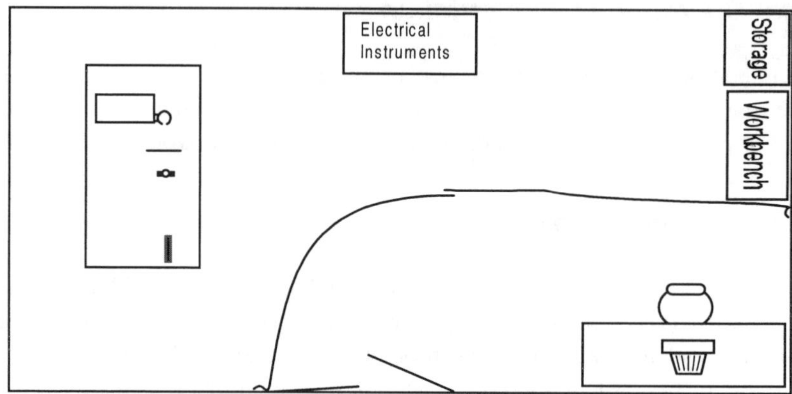

FIGURE 6.19. Spectroradiometry laboratory.

handle the distorted current requirements without introducing harmonics to the voltage waveform.

Because the measurement instrumentation is able to filter its input power, conditioned power is only needed for the lamp operation circuit. Separate circuits for lamps and instrumentation allow the line conditioner to be sized slightly lower and protects the instruments from restabilization or even damage if a circuit breaker is tripped.

The customers for my measurements include many product lines and researchers. With an eye to flexibility the spectroradiometer is mounted on a table (Fig. 6.20) that may be wheeled anywhere within the room. At one end of the table is an optical breadboard, a 2 ft×3 ft steel plate and honeycomb with the surface drilled and taped every inch. This breadboard allows me to mount anything anywhere and know that it will stay there. An alignment laser and the fixture for a calibration lamp are usually bolted in place so that the setup for calibration can be verified and used without delay.

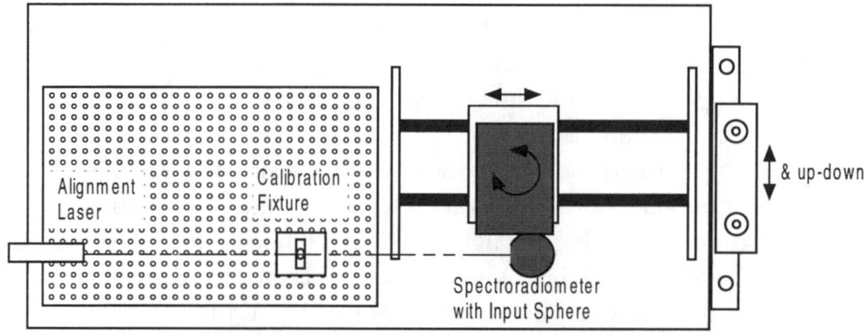

FIGURE 6.20. Spectroradiometer table top view.

In the middle of the table is a track and carriage with locking bolts. The spectroradiometer is mounted on the carriage, that which allows rotation of exactly 180°. The spectrometer's input apertures normal, the optical axis, is aligned parallel to the single direction of movement on the track. Movement is then always along the optical axis when facing either direction.

The other end of the table includes a stage that moves vertically and horizontally, giving me three degrees of freedom between a source mounted on the stage and the spectrometer input optics.

The control of the spectroradiometer is accomplished at the desk by a computer connected with a data cable dropped from the ceiling. The computer is connected to the company network simplifying transfer of data and backup to another location of the day's work.

At the computer is a *laboratory notebook*. The laboratory notebook is a hardbound book with its own serial number and every page is numbered, including the book's serial number. The pages have grids for drawing and include spaces for signatures and witnesses. As per the instructions in front, all entries are written with a pen and nothing is erased or covered up. Errors are marked with a single line through the entry. This is actually very important in photometry, as an entry error is often caused by instability and therefore the ''bad'' data are very useful and must not be lost. The notebook is used primarily to document measurements and their results rather than complete descriptions of lamp and measurement setups. In our system there is one book for each photometer. Additionally there is a single book that details our use of standards and creation of working standards.

The entries in the notebook are made such that each measurement gets a serial number. The date and reading number are combined to form the serial number as in 960211-0, the first reading on February 11, 1996. The computer data file references this number in the description field of the data and uses it for the file name. The file name for this reading is 9602110.DAT or 9602110.CAL depending on the measurement type. The first line of an entry is the serial number followed by the test description including the customer's request number. The following lines contain test parameters such as distance, spectral range, and rate, circuit, and electrical parameters. Finally, as a check, the computer file name is recorded, mainly to serve as an aid to ensure correct entry into the computer.

Typical entries are shown below:

960211-0 DAILY CAL FEL 94485 $D=50$ cm
350–800 @1 nm $I=8.2000$
$V=104.535$
std file used 94485.STD
file saved as 9602110.CAL

960211-1, CHECK LAMP FEL 94484 $D=50$ cm
350–800 @10 nm $I=8.2000$
$V=105.445$
file saved as 9602111.DAT

960211-2 TR964987 F40/CW #1 $D=10$ cm
$V_{in}=236$ V $V_{lamp}=104.3$ V
$W_{in}=43.1$ W $W_{lamp}=39.7$ W
$I=427.1$ mA

file saved as 9602112.DAT

6.9.3 A Day of Spectroradiometry

Beginning of the day check:
Ambient temperature.
Detector temperature.
Continuity of computer and equipment power.

Calibration:
Set up calibration fixture or check alignment.
Install and light calibration lamp.
Wait 20 min, adjust current, check voltage.
Record electrical parameters in notebook.
Select lamp file (.STD file) for calibration lamp being used.
Calibrate spectroradiometer.
Check electrical values to detect drift during calibration.
Record calibration in notebook, save computer data as .CAL file.
Compare calibration with past calibrations.

Check calibration:
Install and light check lamp.
Wait 20 min, adjust current, check voltage.
Record electrical parameters in notebook.
Select calibration to use (.CAL file).
Measure, may be less than full resolution.
Record in notebook, save computer data as .DAT file.
Compare spectra with lamp's assigned values (.STD file).

Measure source:
Setup lamp fixture.
Setup lamp circuit.
Light lamp, check for proper operation.
Wait for the time that is specified for this lamp type.
Monitor for stability compare electrical with past readings and/or experience.
Record all parameters in notebook.
Select calibration to use (.CAL file).
Measure spectral irradiance.
Record in notebook, save computer data as .AT file.
Compare readings with past values.
If the power of the lamp increases the ambient temperature, then check detector temperature.

6.10 RECOMMENDATIONS

My primary recommendation for further guidance is Keitz [1]. Though difficult to obtain, this text is worth the effort for students and engineers because of its practical nature. The usefulness is not diminished by time or the inclusion of visual photometry.

The standard text, Walsh [2], is comprehensive but difficult to use and thin on practical help.

Color Science [3] is an in-depth reference of color vision and all related calculations. The inclusion of comprehensive data tables makes this a one-stop resource for color calculations.

Stimson [4] and McCluney [5] provide an up-to-date treatment of photometry and include radiometry. McCluney is particularly thorough in discussions of optical instrumentation and materials.

NIST (formerly NBS) provides specific guides [6–8], detailing their calibration sources and how to use them. The Self-Study Manuals [9] cover their subjects with a thoroughness that requires extensive study and work to use.

Notable past publications from the NBS (now NIST) are listed in Refs. [10–12]. These papers are useful, though dated.

REFERENCES

1. Keitz, H. A. E., *Light Calculations and Measurement* (Cleaver Hume, London, 1955).
2. Walsh, J. W. T., *Photometry* (Constable, London, 1958).
3. Wyszecki, G., and Stiles, W. S., *Color Science* (Wiley, New York, 1967).
4. Stimson, A., *Photometry and Radiometry for Engineers* (Wiley, New York, 1974).
5. McCluney, R., *Introduction to Radiometry and Photometry* (Artech House, Boston, 1994).
6. Walker J. H., *et al., NBS Measurement Services: Spectral Irradiance Calibrations*, NBS (now NIST) Special Publication No. 250-20 (1987).
7. Brooker, R. L., *et al., NBS Measurement Services: Photometric Calibration*, NBS (now NIST) Special Publication No. 250-15 (1987).
8. Walker, J. H., *et al., NBS Measurement Services: Spectral Radiance Calibration*, NBS (now NIST) Special Publication No. 250-1 (1987).
9. *NBS Self-Study Manual on Optical Radiation Measurements*: 910-1, Radiance, Spectral Distribution; 910-2, Distribution of Radiation, Measurement Equation; 910-3, Polarization; 910-4, Slit scattering, deconvolution, apertures; 910-5, UV Solar measurement; 910-6, Coherence; 910-7, Linearity; 910-8, Black Body.
10. Rosa, E. B., *Theory, Construction and Use of the Photometric Integrating Sphere*, J. Res. Nat. Bur. Stand. **18** (447) (1922).
11. Fussell, W. B., *Approximate Theory of the Photometric Integrating Sphere*, NBS Technical Note No. 594-7 (1974).
12. Gibson, K. S., *Spectrophotometry*, NBS Circular No. 484 (1949).
13. CIE, *Color Rendering*, Publication No. 13.3 (Commission Internationale de l'Eclairage, Vienna, 1995).
14. CIE, *Method of Characterizing the Performance of Radiometers and Photometers*, 53 (Commission Internationale de l'Eclairage, Vienna, 1985).
15. CIE, *Spectroradiometry*, 63 (Commission Internationale de l'Eclairage, Vienna, 1984).
16. CIE, *Method of Characterizing Illuminance Meters & Luminance Meters*, Publication No. 69: (Commission Internationale de l'Eclairage, Vienna, 1987).
17. IESNA, *IES Approved Method for the Electrical and Photometric Measurements of Fluorescent Lamps*, IES LM-9-1988 (Illuminating Engineering Society of North America, New York, 1988).
18. IESNA, *IES Practical Guide to Colorimetry of Light Sources*, IES LM-16-1984 (Illuminating

Engineering Society of North America, New York, 1984).

19. IESNA, *IESNA Approved Method for Photometric Testing of Reflector-Type Lamps*, IES LM-20-1994 (Illuminating Engineering Society of North America, New York, 1994).
20. IESNA, *IES Approved Method for Photometric Testing of Indoor Fluorescent Luminaires*, IES LM-41-1995 (Illuminating Engineering Society of North America, New York, 1995).
21. IESNA, *IES Approved Method for the Electrical and Photometric Measurements of General Service Incandescent Filament Lamps*, IES LM-45-1991 (Illuminating Engineering Society of North America, New York, 1991).
22. IESNA, *IES Approved Method for the Electrical and Photometric Measurements of High Intensity Discharge Lamps*, IES LM-51-1984 (Illuminating Engineering Society of North America, New York, 1984).
23. IESNA, *IESNA Guide to Spectroradiometric Measurements*, IES LM-58-1994 (Illuminating Engineering Society of North America, New York, 1994).
24. IESNA, *IES Approved Method for the Electrical and Photometric Measurements of Single-Ended Compact Fluorescent Lamps*, IES LM-66-1991 (Illuminating Engineering Society of North America, New York, 1991).
25. Taylor, B. N., and Kuyatt, C. E., *Guideline for Evaluating and Expressing the Uncertainty of NIST Measurement Results*, NIST Technical No. 1297 (1994).
26. Ohno, Y., ''Integrating Sphere Simulations Application to Total Flux Scale Realization,'' Appl. Opt. **33**, 2637 (1994).
27. Spears, G. R., ''Fluorescent Lamps—Improved Photometric Integrating Sphere,'' J. IES **5**, 165 (1976).
28. Mohan, K., Schaefer, A. R., and Zalewski, E. F., ''Measurement of Geometrically Total Spectral Radiant Power,'' Appl. Opt. **14**, 1035 (1975).
29. Houis, W. A., and Knoll, J. S., ''Characteristics of an Internally Illuminated Calibration Sphere,'' Appl. Opt. **22**, 4004 (1983).
30. ''1994 IESNA Survey of Illuminance and Luminance Meters,'' IESNA Lighting Design and Application, June, 31 (1994).
31. Collins, R. G., ''A Comparison of Spectroradiametric and Broadband Measurement Methods for the Determination of Chromaticity of Discharge Lamps,'' J. IES, **25**, 81 (1995).
32. Saunders, R. D., and Shumaker, J. B., ''An Automated Radiometric Linearity Tester,'' Appl. Opt. **23**, 3504 (1984).
33. Coslovi, L., and Righini, F., ''Fast Determination of the Nonlinearity of Photodetectors,'' Appl. Opt. **19**, 3200 (1980).
34. Ohno, Y., ''New Method for the Realizing a Luminous Flux Scale Using an Integrated Sphere with an External Source, J. IES **25**, xxx (1995).
35. Tseng, C., *et al.*, ''Wavelength Calibration of a Multichannel Spectrometer,'' Appl. Spectrosc. **47**, 1808 (1993).
36. Brownrigg, J. T., ''Wavelength Calibration Method for Low-Resolution Photodiode Array, Spectrometers,'' Appl. Spectrosc. **47**, 1007 (1993).
37. Berlot, P. E., and Locascio, G. A., ''Ultraviolet-Visible Photodiode Array Spectrophotometer Wavelength Calibration Method. A Practical Computer Algorithm,'' Analyst **116**, 313 (1991).
38. Jerome, C. W., and Eby, J., ''Chromaticity Computation,'' Illuminating Eng. **65**, 54 (1970).
39. Venable, W. H., ''Accurate Tristimulus Values from Spectral Data,'' Color Res. Appl. **14** (5), 260 (1989).
40. *IESNA Lighting Handbook*, 8th ed. (Illuminating Engineering Society of North America, New York, 1993).

7

Photometric Reports

Randall P. Bergin
Independent Testing Laboratories, Inc., 3386 Longhorn Road, Boulder, Colorado 80302

7.1 OVERVIEW: PHOTOMETRIC TESTING OF LUMINAIRES

Long before a commercial construction project is built, the lighting levels have already been established, the types and quantities of luminaires have been determined, and the locations indicated on the blueprints. For this to be possible, the performance of the luminaires must be known in order to predict the lighting levels and uniformities. The proper distribution and lamp type must be selected to achieve these goals.

Photometric testing of the luminaires must be conducted to determine the performance so that the data can be used in lighting calculations. Methods and procedures for the testing must be followed in order for results from different labs to be consistent, repeatable, and reliable. The format must be similar enough that reports from different labs may be compared without causing confusion. A standard computer file format [1] makes transfer and use of the test results in a wide variety of lighting application programs possible without the need for keying the data in every time.

The procedures for testing and reporting the data are contained in guides published by the Illuminating Engineering Society of North America (IESNA) for North America, and the Commission Internationale de l'Eclairage (CIE) in Europe. While the specific details vary between the IESNA and CIE format reports, both describe the photometric distribution of the luminaire in candelas (intensity of the light, in a particular direction). Most of the commonly used performance attributes of a luminaire can be taken from or calculated from the candela distribution.

The performance of the lamps themselves may also be evaluated. This may be limited to total flux (lumen) output only or may describe the photometric distribution of the source or the color or quality of light produced.

Test procedures and reports also exist for a variety of other specialized equipment, generally where conformity, safety, or minimum performance requirements are established, such as signal lights used at traffic intersections, and warning and emergency lights of all types. The test procedures may come from a variety of different groups or agencies including Underwriters Laboratories, Inc. (UL), Institute of Traffic Engineers (ITE), American Society for Testing and Materials (ASTM), and the Society of Automotive Engineers (SAE). Various governmental agencies from around the world also publish requirements for test procedures.

7.1.1 Laboratory Procedures

An introduction to photometric testing and the procedures used can be found in the IESNA *Lighting Handbook* [2] and in IESNA publications LM6 [3], *General Guide to Photometry*, and LM36 [4], *Practical Guide to Photometry*. Photometric tests are conducted under controlled conditions, within temperature limits specified for the type of test being performed. Test equipment consists of the goniometer, the photometer, electrical measuring equipment, and a thermometer.

The goniometer is a device for measuring angles in the spherical coordinate system. The coordinate system can be visualized by using the latitudes and longitudes on a globe for reference. The luminaire would be mounted at the center of the globe, and the candela value is recorded at the intersection of each latitude and longitude. Depending on the type of test, the polar axis may be either vertical, called type A by the IESNA, or horizontal, called type B. Angular measurement should be accurate to within 0.25°.

The photometer is a device for measuring intensity of the light. The photometer should have a photopic correction filter so that the spectral response of the system corresponds to the CIE 1931 Standard Observer, i.e., the typical daytime response of human vision. The photometer also must have a linear response over several decades of intensity. Since the intensity of a luminaire may range from fractions of a candela to millions of candelas, it is important that the photometer be capable of accurately measuring changes in intensity of this magnitude.

Electrical measurements are required to make sure that the lamps or ballasts are operated at their correct characteristics for the test. Generally, volts, amperes, and watts are recorded. For incandescent lamps, tests are generally performed at 90% of rated input voltage with the current held constant. High-intensity discharge (HID) lamps are operated at rated lamp wattage. Fluorescent lamps are operated at rated input voltage to the ballast. Most photometric laboratories currently use multimeters that measure volts, amps, and watts. For more information, see IESNA publication LM28 [5] on the selection and use of electrical instruments. The stability and quality of the electricity supplied is also important. Voltage should be held constant to within ±0.1%, with voltage total harmonic distortion of less than 3%. Careful selection of a voltage regulator is needed to meet these requirements.

Temperature must be monitored to keep the ambient temperature within the test limits. Fluorescent lamps must be tested in a 25±1 °C free air ambient, with the lamps operated on the ballast the luminaire is to be supplied with, at rated input voltage to the ballast. Fluorescent lamps are very sensitive to ambient temperature and air movement; therefore these must be carefully monitored to ensure compliance. Changing lamp or ballast types will affect the performance of the luminaire.

HID-type lamps, including mercury, metal halide, or high-pressure sodium lamps, are operated at rated lamp watts. The current should be held constant when incandescent sources are tested. HID and incandescent lamps are less sensitive to temperature variations and the ambient temperature must be maintained at 25±5 °C. All testing is performed in a draft-free environment.

Before any testing is done, appropriate lamps must be obtained and seasoned for testing. IESNA publication LM54 [6], *IES Guide to Lamp Seasoning*, covers the required time any lamp must be seasoned before they are used for testing. This is important since electrical characteristics and lumen output can change drastically during the first hours they are operated. Fluorescent lamps must be tested for symmetry to ensure that the output variation around the lamp does not exceed 4% [7]. If symmetry is not checked there is the possibility that the resulting efficiency will be incorrect.

7.1.2 Absolute versus Relative Photometry

Photometric testing falls into one of two categories: absolute and relative testing. *Absolute testing* is just what the name implies, the actual intensity distribution of the luminaire is measured for the particular lamp and ballast used. If the total efficiency is desired, the total lumen output of the lamp must be measured, and the efficiency is calculated by dividing the total lumen output of the luminaire by the total lumen output of the lamp.

For *relative photometry*, the report is issued as if the lamp used produced rated lumens. This is accomplished by generating a correction factor and multiplying the measured candlepower for the luminaire by this factor. The correction factor is calculated by taking the rated lamp lumens and dividing by the measured total light output of the lamp or lamps used for the test in the luminaire. For incandescent, HID, and single-ended fluorescent lamps, an exploration of the distribution of the lamp is made and the data are averaged to one plane. Zonal lumens are calculated by multiplying the candelas at each vertical angle by the appropriate zonal constant, and the zonal lumens are summed to get the total lumen output [2]. For linear fluorescent lamps, the total lumen output can be calculated from one candela reading taken perpendicular to the lamp or lamps and multiplying by a mathematically derived constant that varies with the length of the lamp and the test distance, approaches 9.15 as a limit [7,8], and is referred to as the lumen to candela ratio. Most luminaire photometry is done using relative photometry because it makes comparison of reports on similar luminaires easier, and since an accurate lumen output of the lamp is not required because the report is generated using a ratio, constant calibration of the photometric range is not required.

7.2 PHOTOMETRIC REPORTS FOR COMMERCIAL LUMINAIRES

In the field of commercial lighting, where luminaires are used for general area lighting, there are three basic types of photometric reports. These are indoor, roadway, and floodlight. Each report type is designed to give the end user of the report the performance characteristics of the luminaire within the context of the end use of the product. Each type includes not only the distribution of the luminaire in candelas, but also information which can be used to aid a designer in choosing the proper luminaire for a given task.

7.2.1 Indoor Reports

Photometric reports for luminaires primarily designed for interior spaces will generally contain a candela tabulation, polar candela plot, zonal lumens and summary, total efficiency, luminaire spacing criterion (SC), average luminance tables, and zonal cavity coefficients of utilization. Other information that may be included are paint reflectance [9], shielding angles, a sketch of the luminaire, maximum luminaire luminance, and a visual comfort probability (VCP) table.

Front Page of an Indoor Report

The following photometric report is for a typical recessed linear fluorescent luminaire of the type commonly found in office ceilings. The same general type of report is used for all indoor-type reports [7,10]. Generally, the front page (Fig. 7.1) contains (a) the name of the laboratory that performed the testing, (b) a report or reference number identifying the particular test, (c) the date the test was performed, (d) the manufacturer of the luminaire, (e) the catalog number of the luminaire, (f) a description of the luminaire, (g) a description of the lamp or lamps operating in the luminaire during the test, and (h) the manufacturer and catalog number of the ballast used, if applicable. The standard way the luminaire is to be mounted (i) is indicated. This may be surface, recessed, pendant, track, or some other word that describes the way the luminaire is to be used.

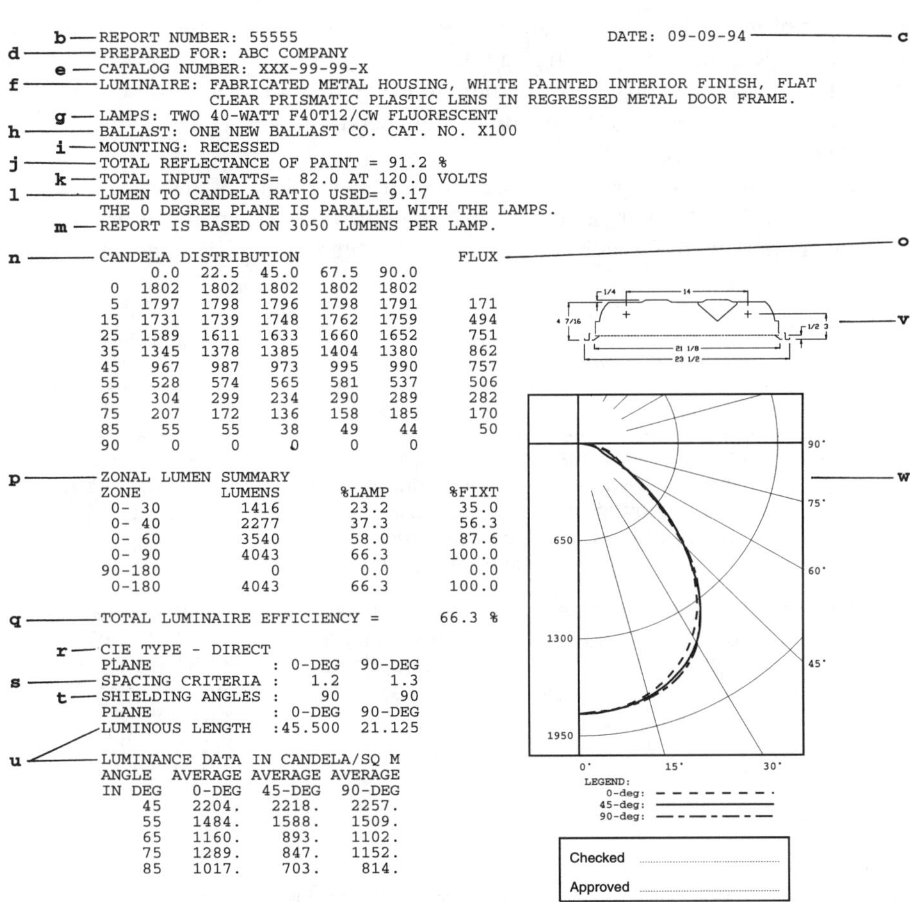

a — *THE TESTING LABORATORY*

b — REPORT NUMBER: 55555 DATE: 09-09-94 — c
d — PREPARED FOR: ABC COMPANY
e — CATALOG NUMBER: XXX-99-99-X
f — LUMINAIRE: FABRICATED METAL HOUSING, WHITE PAINTED INTERIOR FINISH, FLAT CLEAR PRISMATIC PLASTIC LENS IN REGRESSED METAL DOOR FRAME.
g — LAMPS: TWO 40-WATT F40T12/CW FLUORESCENT
h — BALLAST: ONE NEW BALLAST CO. CAT. NO. X100
i — MOUNTING: RECESSED
j — TOTAL REFLECTANCE OF PAINT = 91.2 %
k — TOTAL INPUT WATTS= 82.0 AT 120.0 VOLTS
l — LUMEN TO CANDELA RATIO USED= 9.17
THE 0 DEGREE PLANE IS PARALLEL WITH THE LAMPS.
m — REPORT IS BASED ON 3050 LUMENS PER LAMP.

n — CANDELA DISTRIBUTION FLUX — o

	0.0	22.5	45.0	67.5	90.0	FLUX
0	1802	1802	1802	1802	1802	
5	1797	1798	1796	1798	1791	171
15	1731	1739	1748	1762	1759	494
25	1589	1611	1633	1660	1652	751
35	1345	1378	1385	1404	1380	862
45	967	987	973	995	990	757
55	528	574	565	581	537	506
65	304	299	234	290	289	282
75	207	172	136	158	185	170
85	55	55	38	49	44	50
90	0	0	0	0	0	

p — ZONAL LUMEN SUMMARY

ZONE	LUMENS	%LAMP	%FIXT
0- 30	1416	23.2	35.0
0- 40	2277	37.3	56.3
0- 60	3540	58.0	87.6
0- 90	4043	66.3	100.0
90-180	0	0.0	0.0
0-180	4043	66.3	100.0

q — TOTAL LUMINAIRE EFFICIENCY = 66.3 %

r — CIE TYPE - DIRECT

PLANE	: 0-DEG	90-DEG
s — SPACING CRITERIA	: 1.2	1.3
t — SHIELDING ANGLES	: 90	90
PLANE	: 0-DEG	90-DEG
u — LUMINOUS LENGTH	:45.500	21.125

LUMINANCE DATA IN CANDELA/SQ M

ANGLE IN DEG	AVERAGE 0-DEG	AVERAGE 45-DEG	AVERAGE 90-DEG
45	2204.	2218.	2257.
55	1484.	1588.	1509.
65	1160.	893.	1102.
75	1289.	847.	1152.
85	1017.	703.	814.

FIGURE 7.1. Front page of an indoor report.

If the luminaire contains a painted reflector, the total reflectance of the paint is reported (j). Input electrical characteristics (k) are reported for fluorescent luminaires, along with the lumen to candela ratio (l) used to calculate lamp lumen output. The lumens per lamp (m) for which the report is based must be indicated.

Next on the report is (n) the candela distribution. This is usually a condensed exploration, with more complete data shown on one of the following pages. Indoor photometric data are presented in the type A angular coordinate system, where the polar axis of the angular measurement system is vertical.

The coordinate system can be visualized by using the latitudes and longitudes on a globe for reference. Most IESNA indoor reports use 22.5° lateral planes, so the longitudes on the globe would be oriented 22.5° apart, for a total of 16 planes of data. The vertical increment during the exploration is usually 2.5° or 5°, so there would be latitudes every 2.5° or 5° vertically. The luminaire would be mounted at the center of the globe, and the candela value is recorded at the intersection of each latitude and longitude. Depending on the symmetry of the distribution of the luminaire, the data are averaged so that the entire distribution can be described by a fewer number of planes. Indoor reports will generally fall into one of four types of symmetry, listed in Table 7.1.

Indoor photometric reports are shown with each plane of data in a vertical column, with the lateral angle above each column. The vertical angles are listed in the first column, with 0° (nadir) at the top and the highest vertical angle, in this case 90°, at the bottom.

TABLE 7.1. *Classes of symmetry.*

Type	Description	Type of luminaire	Planes reported
Symmetric	The luminaire has the same vertical distribution in all lateral planes.	Spun reflectors with axially mounted lamp, such as low-bay, high-bay, and recessed symmetric downlights.	All planes (minimum 10) averaged to one plane.
Quadrilateral	The luminaire has each quadrant symmetric.	Most linear fluorescent luminaires, such as 2×4 troffers, "strip" fluorescents, etc.	Five planes reported (0°, 22.5°, 45°, 67.5°, 90°) by averaging 0° & 180°; 22.5°, 157.5°, 202.5° & 337.5°; 45°, 135°, 225° & 315°; and 67.5°, 112.5°, 247.5° & 292.5° planes.
Bilateral	The luminaire is symmetric about one plane only.	Most with incandescent, HID, and single-ended compact fluorescent lamps in a horizontal position, and most wall wash and unidirectional types.	Nine planes reported (0°, 22.5°, 45°, 67.5°, 90°, 112.5°, 135°, 157.5°, 180°) by averaging 22.5° & 337.5°; 157.5° & 202.5°; 45° & 315°; 135° & 225°; 67.5° & 292.5°; 112.5° & 247.5°; and 90° & 270° planes.
Asymmetric	The luminaire exhibits no symmetry; one plane only.	Directional step lights, tunnel lights, and any unit that does not fit one of the other categories.	16 planes reported, no planes are averaged.

There may be a final column (o) that tabulates the flux in lumens for each solid angle centered on each of the vertical angles in the first column. The lumens are calculated by taking a weighted average of the candela values for each vertical angle and multiplying the result by the zonal constant for the solid angle defined. Tabulations of these zonal constants may be found in several IESNA documents, or may be calculated from

$$K=2\pi(\cos\theta_1-\cos\theta_2), \tag{7.1}$$

where θ_1 is the lower angle of the cone and θ_2 is the upper angle of the cone.

The zonal lumen summary (p) indicates where the output of the luminaire is going. The first column shows the vertical zones, in degrees, the second column shows the total lumens in each zone, and the third and fourth columns show the percentage of lamp output and percentage of luminaire output.

The total luminaire efficiency (q) is the percent of lumens that are emitted by the luminaire.

The CIE classification [2] (r) describes how much of the light is below horizontal and how much is above horizontal relative to the total luminaire output. These are direct (90% or more down), semidirect (60%–90% down), general diffuse and direct–indirect (40%–60% up and down), semi-indirect (60%–90% up), and indirect (90% or greater up).

The luminaire spacing criterion [11,12] (s) is derived from a formula that gives the greatest distance apart luminaires should be spaced relative to their mounting height above the task plane.

The shielding angle [2] (t) is the angle below horizontal at which the bare fluorescent lamp first becomes visible, measured in degrees. For lensed luminaires, the shielding angle is always 90°, since the bare lamp is never visible.

Average luminance [13] data (u) are calculated by taking the candela at a given vertical and lateral angle and dividing by the projected area of the luminous portions of the luminaire at that viewing angle. The vertical angles are in the first column, the lateral angles are above each column. Maximum luminance may also be shown. Maximum luminance [14] must be measured by scanning the luminous surface or surfaces of the luminaire to locate the highest 645 mm^2 (1 $in.^2$) area, at each vertical and lateral angle.

This report also shows a sketch of the luminaire (v) and a polar candela plot (w) showing the candela curves for three planes of data.

Coefficient of Utilization Table for Indoor Reports

The coefficient of utilization (CU) table (Fig. 7.2) gives the percent of lumens produced by the lamp or lamps that makes it to the task plane. The line labeled RC (a) is the effective ceiling reflectance, the line labeled RW (b) is the wall reflectance, and the first column is the room cavity ratio (RCR) (c). The tabulation shows the percent of utilized lumens for the ceiling and wall reflectances and room cavity ratios.

THE TESTING LABORATORY

REPORT NUMBER: 55555 DATE: 09-09-94

PREPARED FOR: ABC COMPANY

COEFFICIENTS OF UTILIZATION - ZONAL CAVITY METHOD

EFFECTIVE FLOOR CAVITY REFLECTANCE 0.20

a	RC	80				70				50			30			10			0
b	RW	70	50	30	10	70	50	30	10	50	30	10	50	30	10	50	30	10	0
c	0	79	79	79	79	77	77	77	77	74	74	74	71	71	71	68	68	68	66
	1	73	70	68	66	71	69	67	65	66	64	62	63	62	61	61	60	59	57
	2	67	62	58	55	66	61	58	54	59	56	53	57	54	52	55	53	51	50
	3	62	56	51	47	60	55	50	47	53	49	46	51	48	45	49	47	44	43
	4	57	50	45	41	56	49	44	41	48	43	40	46	42	39	45	42	39	38
	5	53	45	40	36	52	45	39	36	43	39	35	42	38	35	41	37	34	33
	6	49	41	36	32	48	40	35	32	39	35	31	38	34	31	37	34	31	30
	7	46	38	32	28	45	37	32	28	36	31	28	35	31	28	34	30	28	26
	8	43	34	29	26	42	34	29	25	33	29	25	32	28	25	32	28	25	24
	9	40	32	27	23	39	31	26	23	31	26	23	30	26	23	29	26	23	22
	10	38	29	24	21	37	29	24	21	28	24	21	28	24	21	27	24	21	20

ALL CANDELA, LUMENS, LUMINANCE, COEFFICIENT OF UTILIZATION AND VCP VALUES IN THIS REPORT ARE BASED ON RELATIVE PHOTOMETRY WHICH ASSUMES A BALLAST FACTOR OF 1.000. ANY CALCULATIONS PREPARED FROM THESE DATA SHOULD INCLUDE AN APPROPRIATE BALLAST FACTOR.

FIGURE 7.2. Coefficient of utilization table for indoor reports.

The room cavity ratio is calculated using the room dimensions from the following:

$$\mathrm{RCR} = 5h(L+W)/(LW), \tag{7.2}$$

where h is the mounting height of the luminaires above the work plane, L is the room length, and W is the room width.

The height is measured from the task plane to the luminaire mounting height. For an office space using suspended luminaires the distance would be from the top of the desk (76 cm or 30 in. above the floor) to the luminaire. The reflectances of the surfaces must be accurate for the procedure to give dependable results. In rooms where the luminaires are suspended, an effective ceiling cavity reflectance must be determined. First, a ceiling cavity ratio must be calculated using formula (7.2), but with the distance from the major light opening of the luminaire to the ceiling as the height. Using the actual ceiling and wall reflectances, the effective ceiling cavity reflectance must be taken from a table such as Fig. 9-22, [2] "Percent Effective

Ceiling or Floor Cavity Reflectances for Various Reflectance Combinations,'' in the *IESNA Lighting Handbook* [2]. Effective floor cavity reflectances that are significantly different than 20% may also be calculated using the same method and table. For surface- and recess-mounted luminaires the actual ceiling reflectance should be used.

Once the room cavity ratio has been calculated, the coefficient of utilization can be taken from the CU table using the proper effective cavity or surface reflectances. The average illuminance level may be calculated using

$$\text{Average illuminance} = \frac{\begin{pmatrix}\text{lumens}\\ \text{per lamp}\end{pmatrix} \times \begin{pmatrix}\text{no. of lamps}\\ \text{in luminaire}\end{pmatrix} \times \begin{pmatrix}\text{coefficient}\\ \text{of utilization}\end{pmatrix} \times \begin{pmatrix}\text{light-loss}\\ \text{factor}\end{pmatrix}}{(\text{floor area})}, \tag{7.3}$$

where the lumens per lamp value is the manufacturer's lumen rating, the no. of lamps in luminaire is the total number of lamps operating in each luminaire, the coefficient of utilization is taken from the coefficient of utilization table, expressed as a decimal, the light-loss factor is the total depreciation due to ballast factor, lamp lumen depreciation due to age, luminaire dirt depreciation, temperature effects, etc., the floor area is the total floor area (if expressed in meters the illuminance will be in lux; if expressed in feet, the illuminance will be in footcandles).

Visual Comfort Probability Table for Indoor Reports

The visual comfort probability (VCP) is a calculation based upon substantial empirical data [15] representing the probability that a normal observer will not find a given lighting system uncomfortable due to discomfort glare from a defined viewing position. The empirical data were collected using lensed direct-only fluorescent luminaires in a uniform symmetric pattern; therefore any evaluation based on VCP should be made with similar-type luminaires. Its validity with other lighting systems has not been substantiated.

Variables that can affect the VCP values include lamp lumen rating, illuminance level in the room, room surface reflectances, and light-loss factors. The table (Fig. 7.3) should state the conditions used for the calculation.

The first two columns specify the room dimensions (a). Two groups of numbers are presented, the first group is labeled ''LUMINAIRES 0 DEGREE PLANE'' (b). These calculations are based on viewing the lighting system parallel with the 0° plane. For most fluorescent photometric tests conducted in accordance with IESNA practices, this would be parallel with the lamps. The second group is labeled ''LUMINAIRES 90 DEGREE PLANE'' (c). These calculations are based on viewing the lighting system parallel with the 90° plane. For most fluorescent photometric tests conducted in accordance with IESNA practices, this would be perpendicular to the lamps.

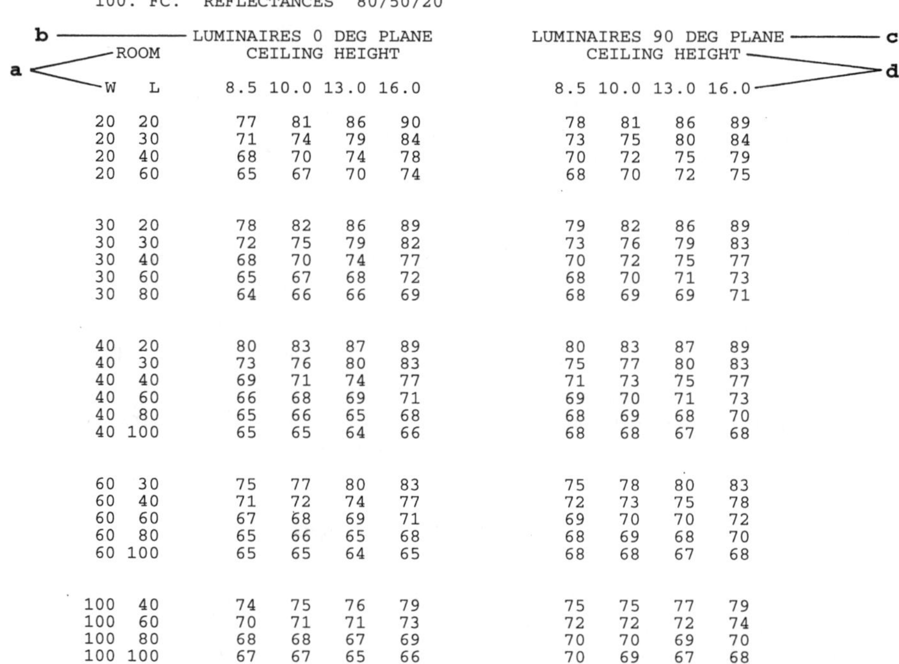

THE TESTING LABORATORY

REPORT NUMBER: 55555 DATE: 09-09-94
PREPARED FOR: ABC COMPANY

VISUAL COMFORT PROBABILITY TABLE

RATED LUMENS PER LAMP 3050.

100. FC. REFLECTANCES 80/50/20

ROOM		LUMINAIRES 0 DEG PLANE CEILING HEIGHT				LUMINAIRES 90 DEG PLANE CEILING HEIGHT			
W	L	8.5	10.0	13.0	16.0	8.5	10.0	13.0	16.0
20	20	77	81	86	90	78	81	86	89
20	30	71	74	79	84	73	75	80	84
20	40	68	70	74	78	70	72	75	79
20	60	65	67	70	74	68	70	72	75
30	20	78	82	86	89	79	82	86	89
30	30	72	75	79	82	73	76	79	83
30	40	68	70	74	77	70	72	75	77
30	60	65	67	68	72	68	70	71	73
30	80	64	66	66	69	68	69	69	71
40	20	80	83	87	89	80	83	87	89
40	30	73	76	80	83	75	77	80	83
40	40	69	71	74	77	71	73	75	77
40	60	66	68	69	71	69	70	71	73
40	80	65	66	65	68	68	69	68	70
40	100	65	65	64	66	68	68	67	68
60	30	75	77	80	83	75	78	80	83
60	40	71	72	74	77	72	73	75	78
60	60	67	68	69	71	69	70	70	72
60	80	65	66	65	68	68	69	68	70
60	100	65	65	64	65	68	68	67	68
100	40	74	75	76	79	75	75	77	79
100	60	70	71	71	73	72	72	72	74
100	80	68	68	67	69	70	70	69	70
100	100	67	67	65	66	70	69	67	68

FIGURE 7.3. Visual comfort probability table for indoor reports.

The next column headings describe the various ceiling heights for which VCP is calculated (d), followed by the VCP probabilities for each of the room conditions tabulated.

To use the table, first ensure that the lumen rating, illuminance level, and reflectances are correct for the design lighting system. Pick the room dimensions that most closely approximates the room in question, and read the VCP value.

7.2.2 Roadway Reports

Photometric reports for luminaires primarily designed for roadway and parking areas will generally contain an isoilluminance plot, candela tabulation, polar candela plot, flux distribution and summary, total efficiency, and a roadway coefficient of utilization plot. Other information that may be included are a sketch of the luminaire, an isocandela diagram or polar candela plot, or a roadway isoluminance and luminance yield diagram.

Isoilluminance Diagram for Roadway Reports

The following photometric report is for a typical roadway-type luminaire utilizing a HID lamp. The same general type of report is used for all roadway- and area-type luminaires [16]. Generally, the front page (Fig. 7.4) contains (a) the name

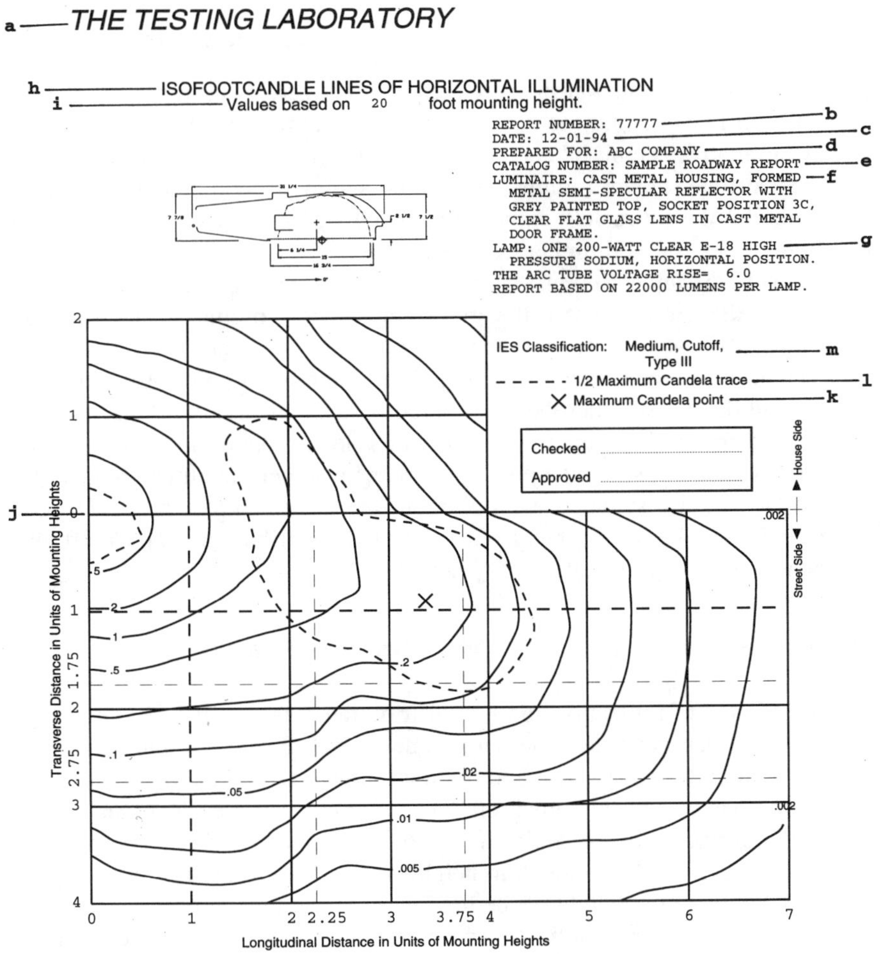

FIGURE 7.4. Isoilluminance diagram for roadway reports.

of the laboratory that performed the testing, (b) a report or reference number identifying the particular test, (c) the date the test was performed, (d) the manufacturer of the luminaire, (e) the catalog number of the luminaire, (f) a description of the luminaire, and (g) a description of the lamp or lamps operating in the luminaire during the test (g).

The plot may be in lux or footcandles, the units will be stated on the report (h). The plot is generally produced for a given mounting height (i), or may have a table for a number of mounting heights. The grid showing the transverse and longitudinal coordinates are in units of mounting heights so that the plot may be used for other mounting heights. The isoilluminance values shown may be corrected to other mounting heights by multiplying by the mounting height correction factor:

$$\text{mounting height correction factor}$$
$$= (\text{plotted mounting height})^2/(\text{desired mounting height})^2. \qquad (7.4)$$

The luminaire is located at the zero longitudinal, zero transverse point (j), with the 0° plane oriented along the zero longitudinal line toward the street side. Most roadway luminaires exhibit bilateral symmetry; therefore only one half of the distribution needs to be shown. Luminaires with no symmetry will have isoilluminance lines plotted for the entire 360° exploration. Axially symmetric luminaires will have concentric circles for isoilluminance lines.

The maximum candela point (k) and $\frac{1}{2}$ maximum candela trace (l) indicate the locations on the ground where the maximum candela produced and the isoline representing 50% of maximum candela would fall. These are used to aid in the classification of the distribution (m) [17]. The longitudinal point of the maximum candela point defines whether the distribution is short, medium, or long. The transverse limit of the half maximum plot within the longitudinal limits defined by the maximum candela classification point determines whether the luminaire is a type I, II, III, or IV. Axially symmetric distributions are classified as type V. The cutoff classification must be determined from the candela distribution [17]. Cutoff classifications are cutoff, semicutoff, and noncutoff.

Coefficient of Utilization Diagram for Roadway Reports

The coefficient of utilization and flux distribution diagram (Fig. 7.5) is an application aid that displays where the lamp lumens are projected by the luminaire, in both graphical and tabular form, and the total luminaire efficiency.

The graph has two lines plotted, one labeled (a) street side, the other (b) house side. The coefficients of utilization (c) is the decimal factor of the lamp lumen output. The street width to mounting height ratio (d) gives the transverse distance relative to the mounting height of the luminaire, so the diagram can be easily used for any mounting height. Both the street and house side curves are plotted for installations where contributions may be coming from either the front or back of the luminaire.

THE TESTING LABORATORY

REPORT NUMBER: 77777 DATE: 12-01-94
PREPARED FOR: ABC COMPANY

COEFFICIENTS OF UTILIZATION AND FLUX DISTRIBUTION

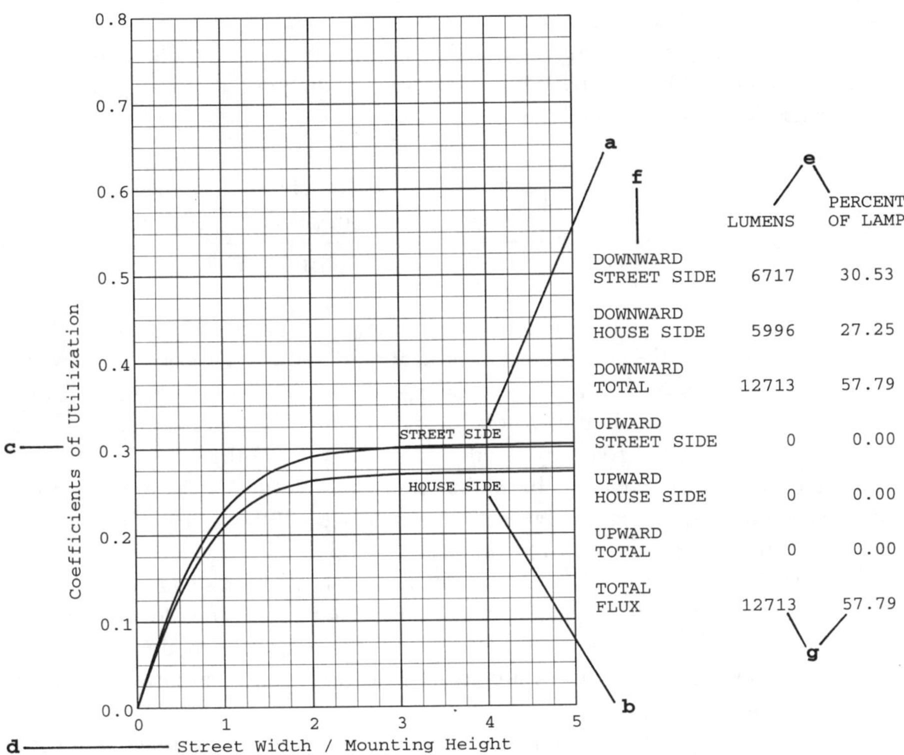

	LUMENS	PERCENT OF LAMP
DOWNWARD STREET SIDE	6717	30.53
DOWNWARD HOUSE SIDE	5996	27.25
DOWNWARD TOTAL	12713	57.79
UPWARD STREET SIDE	0	0.00
UPWARD HOUSE SIDE	0	0.00
UPWARD TOTAL	0	0.00
TOTAL FLUX	12713	57.79

ALL CANDELA, LUMENS AND COEFFICIENT OF UTILIZATION VALUES IN THIS REPORT ARE BASED ON RELATIVE PHOTOMETRY WHICH ASSUMES A BALLAST FACTOR OF 1.00. ANY CALCULATIONS PREPARED FROM THESE DATA SHOULD INCLUDE AN APPROPRIATE BALLAST FACTOR.

FIGURE 7.5. Coefficient of utilization diagram for roadway reports.

The diagram is used to calculate the average illuminance level on a roadway. The first step is to divide the distance from the luminaire to the opposite curb line by the mounting height. Any overhang or setback of the luminaire location must be considered. The coefficient of utilization is taken from the diagram, using the street side curve if the luminaire is oriented toward the opposite curb. Second, the setback or overhang distance to the near curb line is divided by the mounting height. If there is an overhang and the luminaire is oriented toward the opposite curb line, the house side CU curve is utilized, and the coefficients of utilization numbers are added. If there is a setback from the near curb line and the luminaire is oriented toward the opposite curb line, the street side CU curve is utilized, and the coefficients of utilization numbers are subtracted, since the lumens from the luminaire location to the near curb line are not utilized. If the luminaire is above the near curb line, the second step is ignored. If there is contribution from other luminaires with different geometric orientations, the above process is repeated, taking into account the orientation and setback or overhang of the other luminaires. An example would be two luminaires mounted back to back on poles in the median of a roadway. The average illuminance level is calculated using

$$\text{Average illuminance} = \frac{\begin{pmatrix}\text{lumens}\\ \text{per lamp}\end{pmatrix} \times \begin{pmatrix}\text{coefficient of}\\ \text{utilization}\end{pmatrix} \times \begin{pmatrix}\text{light-loss}\\ \text{factor}\end{pmatrix}}{(\text{road width}) \times (\text{pole spacing})}, \quad (7.5)$$

where the lumens per lamp value is the manufacturer's lumen rating, the coefficient of utilization is from the coefficients of utilization diagram, expressed as a decimal, the light-loss factor is the total depreciation due to ballast factor, lamp lumen depreciation due to age, luminaire dirt depreciation, temperature effects, etc., the road width is the actual roadway width, curb to curb, and the pole spacing is the distance between pole locations in the longitudinal direction. If the road width and pole spacing are expressed in meters, the illuminance will be in lux; if expressed in feet, the illuminance will be in footcandles.

The flux distribution is tabulated in lumens and in percent of lamp lumens (e), and is tabulated (f) for four quadrispheres: downward street side, downward house side, upward street side, and upward house side. These are summed into two hemispheres: downward total and upward total. The total lumen output and luminaire efficiency is also tabulated (g).

A roadway luminance diagram may also be included that is similar to the isoilluminance diagram, but includes the effect of the roadway characteristics and driver location. The isolines would be in candelas per square meter. The roadway pavement classification [17] and calculation method, either IESNA [17] or CIE [18], will be specified. The point relative to the maximum veiling luminance (disability glare) and its value may be shown. A luminance yield diagram may be shown on the roadway luminance diagram, or as a separate plot. It is used just like the coefficients of utilization diagram, except the results will be in luminance, as candelas per square foot if the dimensions are in feet or as candelas per square meter if meters are used.

Candela Tabulation for Roadway Reports

Roadway photometric data are measured in the type-A angular coordinate system, where the polar axis of the angular measurement system is vertical. Most

IESNA roadway reports use 10° lateral planes from 5° to 175° lateral, plus the 0°, 90°, 180° (a), and the plane containing the maximum candela value, for a total of 22 tabulated planes for a luminaire with bilateral symmetry (b) (Fig. 7.6).

The vertical angles reported are usually 2.5° increments from 0° to 90° vertical, then 10° from 95° to 175°, and 180° for luminaires with light above horizontal (c).

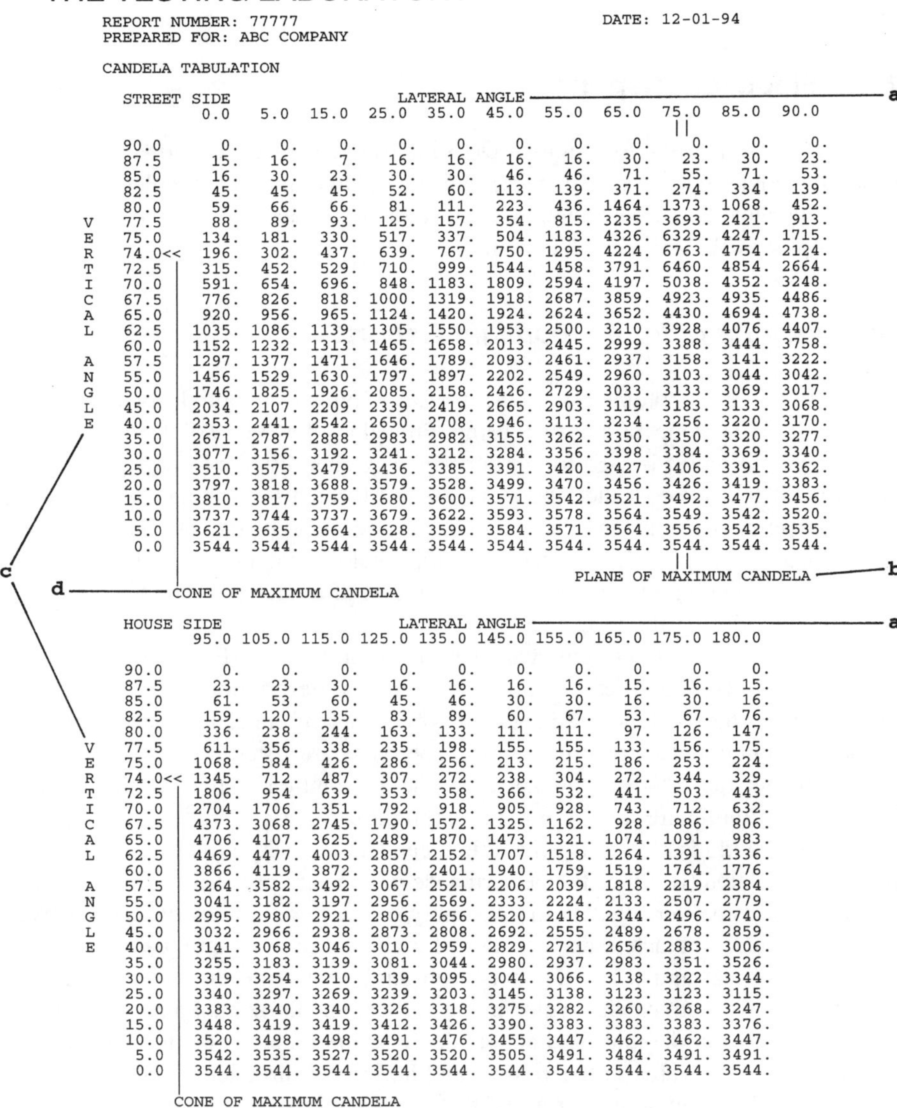

THE TESTING LABORATORY

REPORT NUMBER: 77777 DATE: 12-01-94
PREPARED FOR: ABC COMPANY

CANDELA TABULATION

STREET SIDE — LATERAL ANGLE (a)

VERTICAL ANGLE	0.0	5.0	15.0	25.0	35.0	45.0	55.0	65.0	75.0	85.0	90.0
90.0	0.	0.	0.	0.	0.	0.	0.	0.	0.	0.	0.
87.5	15.	16.	7.	16.	16.	16.	16.	30.	23.	30.	23.
85.0	16.	30.	23.	30.	30.	46.	46.	71.	55.	71.	53.
82.5	45.	45.	45.	52.	60.	113.	139.	371.	274.	334.	139.
80.0	59.	66.	66.	81.	111.	223.	436.	1464.	1373.	1068.	452.
77.5	88.	89.	93.	125.	157.	354.	815.	3235.	3693.	2421.	913.
75.0	134.	181.	330.	517.	337.	504.	1183.	4326.	6329.	4247.	1715.
74.0<<	196.	302.	437.	639.	767.	750.	1295.	4224.	6763.	4754.	2124.
72.5	315.	452.	529.	710.	999.	1544.	1458.	3791.	6460.	4854.	2664.
70.0	591.	654.	696.	848.	1183.	1809.	2594.	4197.	5038.	4352.	3248.
67.5	776.	826.	818.	1000.	1319.	1918.	2687.	3859.	4923.	4935.	4486.
65.0	920.	956.	965.	1124.	1420.	1924.	2624.	3652.	4430.	4694.	4738.
62.5	1035.	1086.	1139.	1305.	1550.	1953.	2500.	3210.	3928.	4076.	4407.
60.0	1152.	1232.	1313.	1465.	1658.	2013.	2445.	2999.	3388.	3444.	3758.
57.5	1297.	1377.	1471.	1646.	1789.	2093.	2461.	2937.	3158.	3141.	3222.
55.0	1456.	1529.	1630.	1797.	1897.	2202.	2549.	2960.	3103.	3044.	3042.
50.0	1746.	1825.	1926.	2085.	2158.	2426.	2729.	3033.	3133.	3069.	3017.
45.0	2034.	2107.	2209.	2339.	2419.	2665.	2903.	3119.	3183.	3133.	3068.
40.0	2353.	2441.	2542.	2650.	2708.	2946.	3113.	3234.	3256.	3220.	3170.
35.0	2671.	2787.	2888.	2983.	2982.	3155.	3262.	3350.	3350.	3320.	3277.
30.0	3077.	3156.	3192.	3241.	3212.	3284.	3356.	3398.	3384.	3369.	3340.
25.0	3510.	3575.	3479.	3436.	3385.	3391.	3420.	3427.	3406.	3391.	3362.
20.0	3797.	3818.	3688.	3579.	3528.	3499.	3470.	3456.	3426.	3419.	3383.
15.0	3810.	3817.	3759.	3680.	3600.	3571.	3542.	3521.	3492.	3477.	3456.
10.0	3737.	3744.	3737.	3679.	3622.	3593.	3578.	3564.	3549.	3542.	3520.
5.0	3621.	3635.	3664.	3628.	3599.	3584.	3571.	3564.	3556.	3542.	3535.
0.0	3544.	3544.	3544.	3544.	3544.	3544.	3544.	3544.	3544.	3544.	3544.

PLANE OF MAXIMUM CANDELA (b)

c

(d) CONE OF MAXIMUM CANDELA

HOUSE SIDE — LATERAL ANGLE (a)

VERTICAL ANGLE	95.0	105.0	115.0	125.0	135.0	145.0	155.0	165.0	175.0	180.0
90.0	0.	0.	0.	0.	0.	0.	0.	0.	0.	0.
87.5	23.	23.	30.	16.	16.	16.	16.	15.	16.	15.
85.0	61.	53.	60.	45.	46.	30.	30.	16.	30.	16.
82.5	159.	120.	135.	83.	89.	60.	67.	53.	67.	76.
80.0	336.	238.	244.	163.	133.	111.	111.	97.	126.	147.
77.5	611.	356.	338.	235.	198.	155.	155.	133.	156.	175.
75.0	1068.	584.	426.	286.	256.	213.	215.	186.	253.	224.
74.0<<	1345.	712.	487.	307.	272.	238.	304.	272.	344.	329.
72.5	1806.	954.	639.	353.	358.	366.	532.	441.	503.	443.
70.0	2704.	1706.	1351.	792.	918.	905.	928.	743.	712.	632.
67.5	4373.	3068.	2745.	1790.	1572.	1325.	1162.	928.	886.	806.
65.0	4706.	4107.	3625.	2489.	1870.	1473.	1321.	1074.	1091.	983.
62.5	4469.	4477.	4003.	2857.	2152.	1707.	1518.	1264.	1391.	1336.
60.0	3866.	4119.	3872.	3080.	2401.	1940.	1759.	1519.	1764.	1776.
57.5	3264.	3582.	3492.	3067.	2521.	2206.	2039.	1818.	2219.	2384.
55.0	3041.	3182.	3197.	2956.	2569.	2333.	2224.	2133.	2507.	2779.
50.0	2995.	2980.	2921.	2806.	2656.	2520.	2418.	2344.	2496.	2740.
45.0	3032.	2966.	2938.	2873.	2808.	2692.	2555.	2489.	2678.	2859.
40.0	3141.	3068.	3046.	3010.	2959.	2829.	2721.	2656.	2883.	3006.
35.0	3255.	3183.	3139.	3081.	3044.	2980.	2937.	2983.	3351.	3526.
30.0	3319.	3254.	3210.	3139.	3095.	3044.	3066.	3138.	3222.	3344.
25.0	3340.	3297.	3269.	3239.	3203.	3145.	3138.	3123.	3123.	3115.
20.0	3383.	3340.	3340.	3326.	3318.	3275.	3282.	3260.	3268.	3247.
15.0	3448.	3419.	3419.	3412.	3426.	3390.	3383.	3383.	3383.	3376.
10.0	3520.	3498.	3498.	3491.	3476.	3455.	3447.	3462.	3462.	3447.
5.0	3542.	3535.	3527.	3520.	3520.	3505.	3491.	3484.	3491.	3491.
0.0	3544.	3544.	3544.	3544.	3544.	3544.	3544.	3544.	3544.	3544.

CONE OF MAXIMUM CANDELA

FIGURE 7.6. Candela tabulation for roadway reports.

In addition, the cone of data at the vertical angle containing the maximum candela value is reported (d).

Roadway photometric reports are shown with each plane of data in a vertical column, with the lateral angle above each column. The vertical angles are listed in the first column, with 0° (nadir) at the bottom and the highest vertical angle tabulated at the top.

Some reports may include more candela data, often in a separate tabulation, around the point of maximum candela to better define the distribution.

7.2.3 Floodlight Reports

Photometric reports for luminaires primarily designed for floodlighting will generally contain a summary page, candela tabulation, flux tabulation, and an isocandela plot. Other information could include a sketch of the luminaire or a candela plot of the axial candela planes.

Summary Page for a Floodlight Report

The following photometric report is for a typical floodlight-type luminaire utilizing a HID lamp and tested using Procedure LM35-1989 [19]. The same general type of report is used for all floodlights. Generally, the front page (Fig. 7.7) contains (a) the name of the laboratory that performed the testing, (b) a report or reference number identifying the particular test, (c) the date the test was performed, (d) the manufacturer of the luminaire, (e) the catalog number of the luminaire, (f) a description of the luminaire, and (g) a description of the lamp or lamps operating in the luminaire during the test.

A summary of the floodlight characteristics follows. The IES type [19] (h), also called the NEMA Classification, provides a general description of the distribution. This describes the total included angle of 10% of maximum candlepower (candelas) in the horizontal and vertical directions. The types are listed in Table 7.2.

The smaller the IES type, the narrower the distribution. If the field angle is less than 10°, the luminaire falls into the searchlight category.

The value and location of the maximum candela intensity produced by the floodlight are tabulated (i). Since the performance parameters of the floodlight are based on this value, small changes in the assembly of the floodlight can severely change the performance characteristics because of a change in maximum candela.

The (j) vertical and (k) horizontal beam angles describe the included angle to 50% of the maximum candela value in the vertical and horizontal directions. The (l) vertical and (m) horizontal field angles describe the included angle to 10% of the maximum candela value in the vertical and horizontal directions. The field lumens (n) are calculated by summing all of the zonal lumens whose candela values are greater than 10% of the maximum. The field efficiency (o) is calculated by dividing field lumens by total lamp lumens.

a —— *THE TESTING LABORATORY*

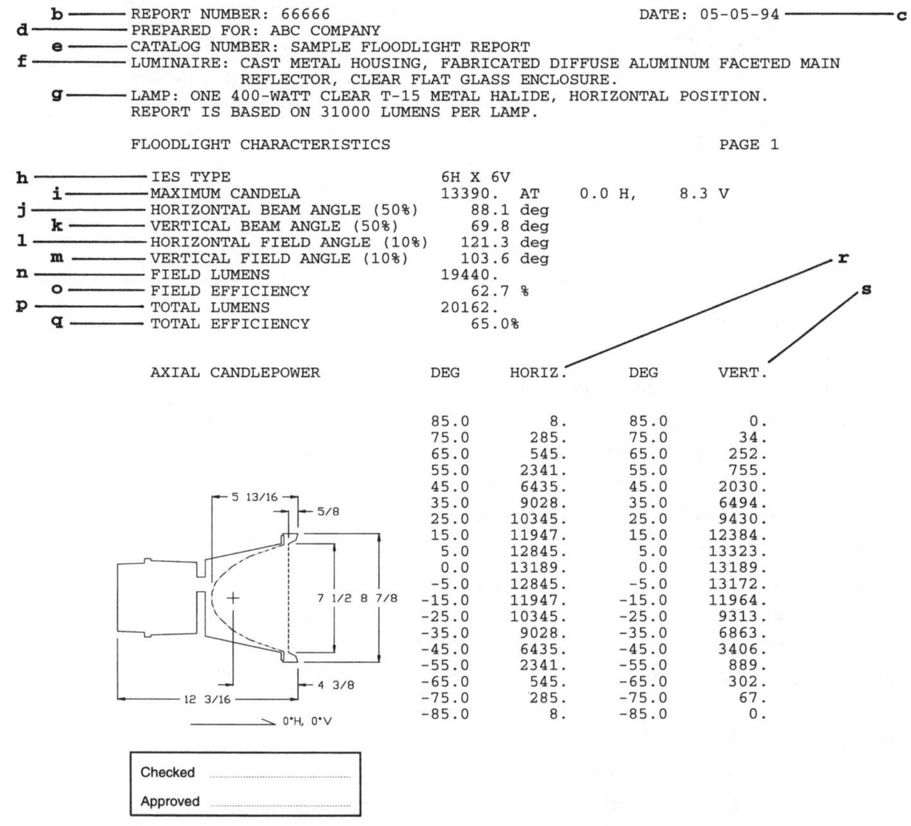

b — REPORT NUMBER: 66666 DATE: 05-05-94 — c
d — PREPARED FOR: ABC COMPANY
e — CATALOG NUMBER: SAMPLE FLOODLIGHT REPORT
f — LUMINAIRE: CAST METAL HOUSING, FABRICATED DIFFUSE ALUMINUM FACETED MAIN REFLECTOR, CLEAR FLAT GLASS ENCLOSURE.
g — LAMP: ONE 400-WATT CLEAR T-15 METAL HALIDE, HORIZONTAL POSITION.
REPORT IS BASED ON 31000 LUMENS PER LAMP.

FLOODLIGHT CHARACTERISTICS PAGE 1

h	IES TYPE	6H X 6V
i	MAXIMUM CANDELA	13390. AT 0.0 H, 8.3 V
j	HORIZONTAL BEAM ANGLE (50%)	88.1 deg
k	VERTICAL BEAM ANGLE (50%)	69.8 deg
l	HORIZONTAL FIELD ANGLE (10%)	121.3 deg
m	VERTICAL FIELD ANGLE (10%)	103.6 deg
n	FIELD LUMENS	19440.
o	FIELD EFFICIENCY	62.7 %
p	TOTAL LUMENS	20162.
q	TOTAL EFFICIENCY	65.0%

AXIAL CANDLEPOWER

DEG	HORIZ. (r)	DEG	VERT. (s)
85.0	8.	85.0	0.
75.0	285.	75.0	34.
65.0	545.	65.0	252.
55.0	2341.	55.0	755.
45.0	6435.	45.0	2030.
35.0	9028.	35.0	6494.
25.0	10345.	25.0	9430.
15.0	11947.	15.0	12384.
5.0	12845.	5.0	13323.
0.0	13189.	0.0	13189.
-5.0	12845.	-5.0	13172.
-15.0	11947.	-15.0	11964.
-25.0	10345.	-25.0	9313.
-35.0	9028.	-35.0	6863.
-45.0	6435.	-45.0	3406.
-55.0	2341.	-55.0	889.
-65.0	545.	-65.0	302.
-75.0	285.	-75.0	67.
-85.0	8.	-85.0	0.

Checked ....
Approved ....

ALL CANDELA AND LUMEN VALUES IN THIS REPORT ARE BASED ON RELATIVE PHOTOMETRY WHICH ASSUMES A BALLAST FACTOR= 1.000. ANY CALCULATIONS PREPARED FROM THESE DATA SHOULD INCLUDE AN APPROPRIATE BALLAST FACTOR.

FIGURE 7.7. Summary page for a floodlight report.

The total lumens (p) are calculated by multiplying all of the zone center candela values by the appropriate zonal constant [19] and summing all of the lumens. The total efficiency (q) is calculated by dividing the total lumens by the total lamp lumens.

A tabulation of the candela values in the (r) horizontal and (s) vertical gives an overview of the distribution. This may be shown graphically instead.

Candela Tabulation for Floodlight Reports

Floodlight photometric data are presented in the type-B angular coordinate system, where the polar axis of the angular measurement system is horizontal. The

TABLE 7.2. *IES types (NEMA classification).*

Type	Total angle to 10% of maximum candela (field angle)
1	10°–18°
2	18°–29°
3	29°–46°
4	46°–70°
5	70°–100°
6	100°–130°
7	130° and up

coordinate system can be visualized by using the latitudes and longitudes on a globe for reference. For the type-B system, the globe would be turned 90° so that the polar axis is horizontal and the equator is vertical. The floodlight would be mounted at the center of the globe and aimed at a point on the equator. The aiming point is defined as 0° horizontal, 0° vertical. The equator is the vertical axis, and the longitude through the aiming point is the horizontal axis. The vertical and horizontal angular increments are defined by the IES type [19].

There may be up to three candela tabulations (Fig. 7.8) for one report. One showing field candelas over the entire field exploration, one showing field-only candelas, and one showing the total exploration on 10° by 10° increments starting 5° off the major axis plus the major axes. Most floodlights exhibit bilateral symmetry, so that the horizontal angles would start at 0° and increment to 90°, while the vertical angles would start at −90° and end at +90°. For IES types 6 and 7, the field angle increment is 10°, so only one candela tabulation is required.

The first column contains the vertical angles (a), while the horizontal angles (b) are listed in the first row. The candela values (c) are tabulated for each horizontal and vertical angle.

Lumen Tabulation for Floodlight Reports

There may be up to three lumen tabulations (Fig. 7.9) for one report: one showing field lumens over the entire field exploration, one showing lumens for those zones whose candela is greater than 10% of the maximum candela, and one showing the total exploration on 10° by 10° increments with the zone centers starting 5° off the major axes. The zonal lumens are calculated by multiplying the candela values at the zone centers by the appropriate zonal constant [19] and are tabulated in the same manner as the candela values, only the angles of the zone edges are tabulated instead of the zone center angles.

The first column contains the vertical angles (a), while the horizontal angles (b) are listed in the first row. The lumens (c) are tabulated for each zone.

THE TESTING LABORATORY

REPORT NUMBER: 66666 DATE: 05-05-94
PREPARED FOR: ABC COMPANY

CANDELA TABULATION -- ENTIRE EXPLORATION PAGE 2

AVERAGE OF RIGHT AND LEFT SIDES

VERTICAL ANGLES

b	0.0	5.0	15.0	25.0	35.0	45.0	55.0	65.0	75.0	85.0	
85.0	0.	0.	0.	0.	0.	0.	0.	0.	0.	0.	
75.0	34.	50.	50.	42.	34.	25.	17.	8.	0.	0.	
65.0	252.	252.	252.	235.	176.	84.	59.	42.	17.	0.	
55.0	755.	763.	738.	663.	503.	235.	109.	84.	50.	0.	
45.0	2030.	2014.	1963.	1896.	1888.	1510.	638.	143.	109.	0.	
35.0	6494.	6460.	6116.	5596.	4908.	3549.	1351.	285.	134.	0.	
25.0	9430.	9363.	8642.	7668.	6720.	4858.	2022.	352.	151.	0.	
15.0	12384.	12174.	11209.	9464.	8256.	5974.	2458.	461.	168.	8.	
5.0	13323.	12963.	11989.	10328.	9028.	6435.	2391.	529.	210.	8.	
a — 0.0	13189.	12845.	11947.	10345.	9028.	6435.	2341.	545.	285.	8.	c
-5.0	13172.	12845.	11922.	10362.	9053.	6444.	2307.	537.	302.	8.	
-15.0	11964.	11788.	11075.	9514.	8239.	5579.	1846.	419.	176.	8.	
-25.0	9313.	9254.	8566.	7543.	6544.	4321.	1435.	319.	151.	0.	
-35.0	6863.	6846.	6469.	5865.	5026.	3247.	1099.	277.	134.	0.	
-45.0	3406.	3322.	3264.	2995.	2618.	1728.	663.	168.	101.	0.	
-55.0	889.	889.	873.	789.	612.	285.	134.	92.	59.	0.	
-65.0	302.	302.	302.	260.	210.	109.	67.	42.	17.	0.	
-75.0	67.	59.	59.	59.	42.	25.	8.	0.	0.	0.	
-85.0	0.	0.	8.	0.	0.	0.	0.	0.	0.	0.	

BEAM EDGE = 6695. CP = 50% OF MAX.
FIELD EDGE = 1339. CP = 10% OF MAX.

FIGURE 7.8. Candela tabulation for floodlight reports.

Isocandela Diagram for Floodlight Reports

Isocandela lines representing the candela distribution are plotted on a rectangular grid as a visual reference of the performance of the floodlight (Fig. 7.10). Vertical angles (a) are shown on the vertical axis, and the horizontal angles (b) are shown at the top and bottom of the plot. The isocandela lines and their values are plotted.

Testing and Reporting of Floodlight Data

Currently, the IES LM35-1989 guide specifies that data be reported in the type-B coordinate system. Almost universally, floodlights are tested on equipment in the

THE TESTING LABORATORY

REPORT NUMBER: 66666 DATE: 05-05-94
PREPARED FOR: ABC COMPANY

LUMEN TABULATION PAGE 3

AVERAGE OF RIGHT AND LEFT SIDES

VERTICAL ANGLES	b — 0	10	20	30	40	50	60	70	80	90	TOTAL
90											
	0.	0.	0.	0.	0.	0.	0.	0.	0.		0.
80											
	2.	1.	1.	1.	1.	0.	0.	0.	0.		6.
70											
	8.	7.	6.	4.	2.	1.	1.	0.	0.		29.
60											
	23.	22.	18.	13.	5.	2.	1.	0.	0.		84.
50											
	61.	58.	52.	47.	32.	11.	2.	1.	0.		264.
40											
	196.	180.	154.	122.	76.	24.	4.	1.	0.		757.
30											
	284.	254.	211.	167.	105.	35.	5.	1.	0.		1062.
20											
	369.	329.	261.	206.	129.	43.	6.	1.	0.		1344.
10											
	393.	352.	285.	225.	138.	42.	7.	2.	0.		1444.
a — 0											
	389.	350.	286.	226.	139.	40.	7.	2.	0.		1439. — c
-10											
	357.	325.	262.	205.	120.	32.	5.	1.	0.		1309.
-20											
	280.	252.	208.	163.	93.	25.	4.	1.	0.		1027.
-30											
	207.	190.	162.	125.	70.	19.	4.	1.	0.		778.
-40											
	101.	96.	83.	65.	37.	12.	2.	1.	0.		396.
-50											
	27.	26.	22.	15.	6.	2.	1.	0.	0.		100.
-60											
	9.	9.	7.	5.	2.	1.	1.	0.	0.		35.
-70											
	2.	2.	2.	1.	1.	0.	0.	0.	0.		7.
-80											
	0.	0.	0.	0.	0.	0.	0.	0.	0.		0.
-90											
	2708.	2454.	2020.	1591.	955.	290.	48.	14.	0.		10081.

LUMEN TOTAL THIS PAGE = 10081.

FIGURE 7.9. Lumen tabulation for floodlight reports.

type-A coordinate system and converted to type B. Originally, the type-B system was used to ease manual calculations using the data. Since most lighting calculations are now done using computer programs, and since the type-B system is more difficult to implement in these programs, future versions of LM35 will implement testing and reporting in the type-A coordinate system. This will also allow more data to be utilized, with better resolution of the distribution.

7.3 SUMMARY

In this chapter the major types of photometric tests and reports for most commercial lighting equipment based on IESNA procedures were explored. The relevant

THE TESTING LABORATORY

REPORT NUMBER: 66666 DATE: 05-05-94
PREPARED FOR: ABC COMPANY

ISOCANDELA CURVES PAGE 4

AVERAGE OF RIGHT AND LEFT SIDES

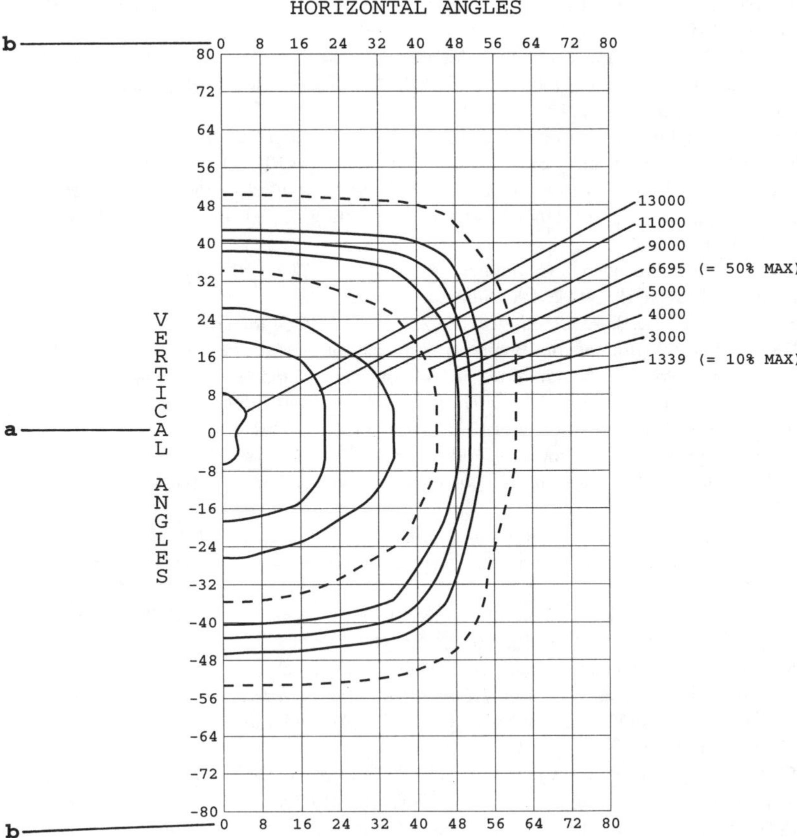

FIGURE 7.10. Isocandela diagram for floodlight reports.

IESNA guides should be reviewed for more detail. As discussed, the CIE has procedures that differ in detail from those of the IESNA. Relevant CIE documents should be reviewed for more details.

REFERENCES

1. IESNA Testing Procedures Committee, *IES Standard File Format for Electronic Transfer of Photometric Data and Related Information*, IESNA LM-63-95 (Illuminating Engineering Society of North America, New York, 1995).
2. Rea, M. S., *IESNA Lighting Handbook*, 8th ed. (Illuminating Engineering Society of North America, New York, 1993)
3. IESNA Testing Procedures Committee, *IES General Guide to Photometry*, IESNA LM-6, J. IES **50**, No. 3, 147–152; No. 4, 201–210 (1955).
4. IESNA Testing Procedures Committee, *IES Practical Guide to Photometry*, IESNA LM-36, J. IES **1**, No. 1, 73–96 (1971).
5. IESNA Testing Procedures Committee, *IESNA Guide for the Selection, Care, and Use of Electrical Instruments in the Photometric Laboratory*, IESNA LM-28-89 New York (Illuminating Engineering Society of North America, New York, 1989).
6. IESNA Testing Procedures Committee, *IESNA Guide to Lamp Seasoning*, IESNA LM-54-91 (Illuminating Engineering Society of North America, New York, 1991).
7. IESNA Testing Procedures Committee, *IESNA Approved Method for Photometric Testing of Indoor Fluorescent Luminaires*, IESNA LM-41-85 (Illuminating Engineering Society of North America, New York, 1987).
8. IESNA Testing Procedures Committee, *Appendix G—The Calculation of Factor K*, IESNA LM-10A, J. IESNA **19**, No. 2, 190, Summer (1990).
9. IESNA Testing Procedures Committee, *IES Approved Method for Total and Diffuse Reflectometry*, IESNA LM-44-90 (Illuminating Engineering Society of North America, New York, 1990).
10. IESNA Testing Procedures Committee, *IESNA Approved Method for Photometric Testing of Indoor Luminaires Using High Intensity Discharge or Incandescent Filament Lamps*, IESNA LM-46-92 (Illuminating Engineering Society of North America, New York, 1992).
11. IESNA Design Practice Committee, *Recommended Practice for Classification of Interior Luminaires by Distribution: Luminaires Spacing Criterion*, Lighting Design Appl. **7**, No. 8, 20–21 (1977).
12. Levin, R. E., *Revision of the S/MH Concept*, Lighting Design Appl. **7**, No. 8, 22–25 (1977).
13. IESNA Testing Procedures Committee, Determination of Average Luminance of Luminaires, J. IES **1**, No. 2, 181–184 (1972).
14. IESNA Testing Procedures Committee, *IES Guide for Measurement of Photometric Brightness (Luminance)*, Illuminating Eng. **56**, No. 7, 457–462 (1961).
15. IES Committee on Recommendations of Quality and Quantity of Illumination, *Outline of a Standard Procedure for Computing Visual Comfort Ratings for Interior Lighting*—RQQ Report No. 2 (1972), ES LM-42-1972, J. IES **2**, No. 3, 328 (1973); Appendix to RQQ Report No. 2 (1972), J. IES **2**, No. 4, 504 (1973).
16. IESNA Testing Procedures Committee, *IESNA Approved Method for Photometric Testing of Roadway Luminaires Using Incandescent Filament and High Intensity Discharge Lamps*, IESNA LM-31-88 (Illuminating Engineering Society of North America, New York, 1989).
17. IESNA Roadway Lighting Committee, *American National Standard Practice for Roadway Lighting*, ANSI/IESNA RP-8-83 (Illuminating Engineering Society of North America, New York, 1983) (reaffirmed 1993).
18. *Calculation and Measurement of Luminance and Illuminance in Roadway Lighting*, CIE Publication No. 30-2 (CT-4.6) 1982 (Commission Internationale de L'Eclairage, Paris, 1982).
19. IESNA Testing Procedures Committee, *IESNA Approved Method for Photometric Testing of Floodlights Using Incandescent Filament for Discharge Lamps*, IESNA LM-35-89 (Illuminating Engineering Society of North America, New York, 1989).

8

Spectroradiometry Methods

William E. Schneider and Richard Young

Optronics Laboratories, 4470 35th St., Orlando, Florida 32811

8.1 SPECTRORADIOMETRIC VERSUS PHOTOMETRIC QUANTITIES: DEFINITIONS AND UNITS

Radiometry is the science and technology of the measurement of electromagnetic radiant energy. It is more commonly referred to simply as "the measurement of optical radiation." Whereas radiometry involves the measurement of the total radiant energy emitted by the radiating source over the entire optical spectrum (1 nm–1000 μm) and spectroradiometry is concerned with the spectral content of the radiating source, photometry is only concerned with that portion of the optical spectrum to which the human eye is sensitive (380–780 nm). More specifically, photometry relates to the measurement of radiant energy in the "visible" spectrum as perceived by the standard photometric observer. Loosely, the standard photometric observer can be thought of as the "average" human.

A number of radiometric, spectroradiometric, and photometric quantities are used to describe a radiating source. Although there has been some disagreement in the past with respect to these definitions, those described herein are those most commonly used at the present time.

8.1.1 Radiometric Quantities

The radiometric quantities listed below are those most frequently used in the measurement of optical radiation [1,2].

Radiant energy is the total energy emitted from a radiating source (in J).

Radiant energy density is the radiant energy per unit volume (in J m^{-3}).

Radiant power or flux is the radiant energy per unit time (in J s^{-1} or W).

Radiant exitance is the total radiant flux emitted by a source divided by the surface area of the source (in W m^{-2}).

Irradiance is the total radiant flux incident on an element of surface divided by the surface area of that element (in W m^{-2}).

Radiant intensity is the total radiant flux emitted by a source per unit solid angle in a given direction (in W s^{-1}).

Radiance is the radiant intensity of a source divided by the area of the source (in W sr^{-1} m^{-2}). Figure 8.1 shows the geometry for defining radiance. Note that a steradian is defined as the solid angle that, having its vertex in the center of a sphere, cuts off an area of the surface of the sphere equal to that of a square with sides of length equal to the radius of the sphere.

Emissivity is the ratio of the radiant flux density of a source to that of a blackbody radiator at the same temperature.

Pure physical quantities for which radiant energy is evaluated in energy units are defined in Table 8.1.

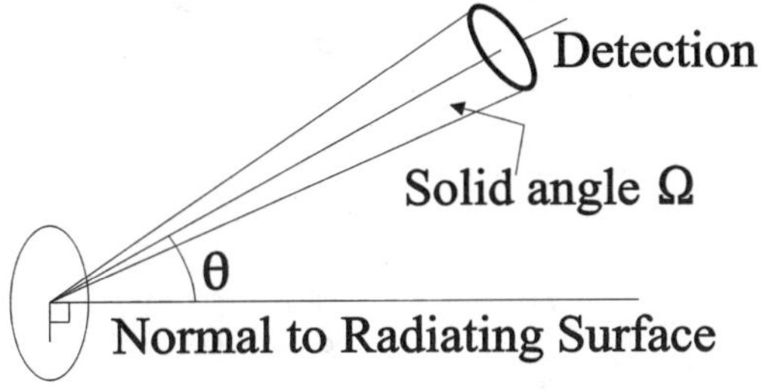

FIGURE 8.1. Geometry for defining radiance.

TABLE 8.1. *Fundamental radiometric quantities.**

Quantity	Symbol	Defining equation	Units
Radiant energy	Q, Q_e		J (joule)
Radiant energy density	w, w_e	dQ/dV	J m^{-3}
Radiant power or flux	Φ, Φ_e	dQ/dt	J s^{-1} or W (watt)
Radiant exitance	M, M_e	$d\Phi/dA_{source}$	W m^{-2}
Irradiance	E, E_e	$d\Phi/dA_{surface}$	W m^{-2}
Radiant intensity	I, I_e	$d\Phi/d\Omega$	W sr^{-1}
Radiance	L, L_e	$d^2\Phi/d\Omega(dA \cos\theta)$ $dI/dA \cos\theta$	W m^{-2} sr^{-1}
Emissivity	ϵ	$M/M_{blackbody}$	

*From Ref. [1], with permission.

8.1.2 Photometric Quantities

When the radiometric quantities listed in Table 8.1 are evaluated by means of a standard photometric observer, they correspond to an analogous photometric quantity (see Table 8.2). Each pair of quantities, radiometric and photometric, are represented by the same principal symbol (save emissivity and luminous efficacy) and are distinguished only by the subscript. The subscript e (or no subscript) is used in the case of physical (radiometric) quantities and the subscript v is used for photometric quantities.

Although there are many common terms used to define photometric light output, the basic unit of measurement of light is the lumen [3]. All other photometric quantities involve the lumen. The eight fundamental photometric quantities concerned with the measurement of light are defined below.

Luminous energy is the total energy as perceived by a standard 2° observer (in lm s).

Luminous energy density is the luminous energy per unit volume (in lm s m^{-3}).

TABLE 8.2. *Fundamental photometric quantities.**

Quantity	Symbol	Defining equation	Units
Luminous energy	Q_v	$K_m \int V(\lambda) Q(\lambda) d\lambda$	lm s
Luminous energy density	w_v	dQ_v/dV	lm s m^{-3}
Luminous flux	Φ_v	dQ_v/dt	lm (lumen)
luminous exitance	M_v	$d\Phi_v/dA_{source}$	lm m^{-2}
Illuminance	E_v	$d\Phi_v/dA_{surface}$	lm m^{-2}
Luminous intensity	I_v	$d\Phi_v/d\Omega$	lm s^{-1} or cd (candela)
Luminance	L_v	$d^2\Phi_v/d\Omega(dA \cos\theta)$ $dI_v/dA \cos\theta$	cd m^{-2}
Luminous efficacy	K	Φ_v/Φ	lm W^{-1}

*From Ref. [1], with permission.

Luminous flux is the luminous energy per unit time (in lm).

Luminous exitance is the ratio of the luminous flux emitted to the surface area of the source (in lm m^{-2}).

Illuminance is the luminous flux per unit area incident on a surface. It is the luminous flux divided by the area of the surface when the surface is uniformly irradiated (in lm m^{-2}).

Luminous intensity is the luminous flux per unit solid angle in the direction in question (in lm sr^{-1} or candela).

Luminance is the ratio of the luminous intensity to the area of the source (in cd m^{-2}).

Luminous efficacy is the ratio of the total luminous flux to the total radiant flux (in lm W^{-1}).

8.1.3 Spectroradiometric Quantities

When radiant energy, or any related quantity, is measured in terms of its monochromatic components it becomes a function of wavelength. Therefore, the designations for these quantities must be preceded by the adjective *spectral*, as in *spectral irradiance*. The symbol itself, for each quantity, is followed by the symbol for wavelength (λ). For example, spectral irradiance has the symbol $E(\lambda)$ or $E_e(\lambda)$.

If the spectral distribution of the source is known, the following relationship between the lumen and the watt can be used to convert from one to the other:

$$\Phi_v = 683 \int_\lambda \Phi(\lambda) V(\lambda) d\lambda \quad (\text{lm}),$$

where $\Phi(\lambda)$ is the spectroradiometric power distribution of the light source (expressed in "watts per unit wavelength interval"), $V(\lambda)$ is the relative photopic luminous efficacy function (normalized at 555 nm), and λ is the wavelength (usually expressed in nanometers). The value of 683 lm W^{-1} is the absolute luminous efficacy at 555 nm.

When a photometric or radiometric measurement of a light source is made, it is *not* possible to convert from photometric to radiometric units or vice versa, *unless* the spectral distribution of the source is precisely known. In the special case of a monochromatic light source, such as a laser, the equation simplifies as shown below. For example, in the case of a 1-mW He–Ne laser whose output is at 633 nm, the luminous flux is

$$\begin{aligned}\Phi_v &= (683 \ \text{nm})\Phi(633 \ \text{nm})V(633 \ \text{nm}) \\ &= (683 \ \text{nm} \times 10^{-3} \times 0.235\,334\,4 = 0.1607 \ \text{lm}).\end{aligned}$$

In general, a measurement of the spectrometric output of a light source will provide the most accurate photometric data.

8.1.4 Transmittance

Transmittance is the ratio of the transmitted radiant or luminous flux to the incident radiant or luminous flux [4].

$$\text{Luminous transmittance} \quad \tau_v = \Phi_v^t / \Phi_v^i;$$

$$\text{Spectral transmittance} \quad \tau(\lambda) = \Phi^t(\lambda) / \Phi^i(\lambda).$$

Here the superscript t refers to the transmitted flux and superscript i to the incident flux.

The total transmittance (τ) of a medium (or object) consists of two parts, regular transmittance (τ_r) and diffuse transmittance (τ_d), where

$$\tau = \tau_r + \tau_d.$$

If radiant or luminous flux travels through a sample such that the exit angle may be predicted from the entry angle according to Snell's refraction law [5], the transmittance is referred to as *regular*. When the flux is scattered as it travels through the sample, or on a macroscopic scale Snell's law no longer applies because of the roughness of the surface, the transmittance is referred to as *diffuse*.

The luminous (or photometric) transmittance of the medium is dependent on the spectral composition of the radiating source. Accordingly, the nature of the radiating source must be specified when determining the photometric transmittance of a medium. For example, the photometric transmittance of a blue filter will be considerably higher when the radiating source is a xenon arc lamp than when the radiating source is a tungsten lamp operating at a color temperature of 2856 K.

The photometric transmittance of a medium for a specified radiating source can be determined photometrically or spectroradiometrically. However, the specified radiating source (or a source with the same spectral distribution as the specified source) must be used when employing the photometric technique. Users commonly fail to account for source distributions when performing spectroradiometric calculations. While this is still valid—the photometric transmittance of a medium is equivalent to specifying ''an equal energy source,'' i.e., a source having equal energy at all wavelengths over the visible spectrum—it should be borne in mind that this is not a practical source for comparative photometric measurements.

Photometric transmittance (τ_v) can be computed from a knowledge of the spectral transmittance [$\tau(\lambda)$] and the relative spectral distribution of the specified source [$\Phi(\lambda)$] as follows:

$$\tau_v = \int_\lambda \tau(\lambda)\Phi(\lambda)V(\lambda)d\lambda,$$

where $V(\lambda)$ is the relative photopic luminous efficacy.

8.1.5 Reflectance

Reflectance is the ratio of the reflected radiant or luminous flux to the incident radiant or luminous flux:

$$\text{Luminous reflectance} \quad \rho_v = \Phi_v^r / \Phi_v^i;$$

$$\text{Spectral reflectance} \quad \rho(\lambda) = \Phi^r(\lambda) / \Phi^i(\lambda).$$

Here the superscript r refers to reflected flux and the superscript i to incident flux.

The total reflectance of an object (ρ) is divided into two parts: specular reflectance (ρ_s) and diffuse reflectance (ρ_d) where

$$\rho = \rho_s + \rho_d.$$

Specular reflectance consists of reflection of radiant or luminous flux without scattering or diffusing in accordance with the laws of optical reflection as in a mirror. Diffuse reflectance consists of scattering of the reflected flux in all directions.

As with photometric transmittance, the photometric reflectance of an object is dependent on the spectral composition of the radiating source and spectral composition of the radiating source must be known or specified when determining photometric reflectance. Also, as with photometric transmittance, photometric reflectance can be determined photometrically or spectroradiometrically.

8.1.6 Spectral Responsivity

Spectral responsivity $R(\lambda)$ generally refers to the electrical signal generated by a photodetector $[s(\lambda)]$ when irradiated with a known radiant flux of a specific wavelength $[\Phi(\lambda)]$ and is determined using the relationship

$$R(\lambda) = s(\lambda) / \Phi(\lambda).$$

The output signal of the detector can be in amperes, volts, counts/seconds, etc.

The spectral responsivity of a photodetector can be either a power response or an irradiance response. A power response generally involves under filling the detector with monochromatic flux, whereas an irradiance response involves uniformly overfilling the detector with monochromatic flux. It is possible to convert from one type of response to another if the area of the receiver (sensitive portion of the photodetector) is known and if the receiver is uniform in sensitivity.

8.2 SPECTRORADIOMETRIC STANDARDS

The accurate measurement of optical radiation involves not only the use of a stable, well-characterized photometer, radiometer, or spectroradiometer, but also, somewhere along the line, the use of a standard. This standard can be in the form of a radiating source whose radiant flux output and geometrical properties are accurately known, or a detector whose response is accurately known. For most spectroradiometric applications, a standard source should be used to calibrate the measurement system if the system is to be used to measure the spectral output of sources. A standard detector should be used to calibrate the measurement system if the system will be used to measure the spectral response of detectors. This section will describe the ''basic'' spectroradiometric standards that have been set up by National Institute of Standards and Technology (NIST) and that are available through commercial calibration laboratories. With the exception of blackbody standards, only those spectroradiometric standards suitable for use over the visible spectrum are covered.

In many instances, a ''basic'' spectroradiometric standard is not suitable for calibrating a measurement system. For example, it is extremely difficult to use a ''basic'' spectroradiometric standard that has a nominal spectral irradiance of 0.1 $\mathrm{W\,m^{-2}\,nm^{-1}}$ to calibrate a measurement system that will be measuring irradiance levels that are 6–10 decades less. What is needed is a ''special-purpose,'' low-light-level standard that has an irradiance level comparable to the source to be measured. Accordingly, various special-purpose spectroradiometric standards whose calibrations are based on the ''basic'' spectroradiometric standards will also be described.

8.2.1 Blackbody Standards

Since most of the spectroradiometric standards available for use over the visible spectrum are based on the spectral radiance of a blackbody as defined by the Planck radiation law, a short discussion on blackbody radiation is in order. A blackbody is a theoretical source whose surface absorbs all of the radiant flux incident on it regardless of wavelength or angle of incidence, i.e., it is perfectly black [6,7]. Planck's law defines the spectral radiance L_λ of a blackbody as

$$L_\lambda = \frac{c_1}{n^2\lambda^5(e^{c_2/n\lambda T}-1)}\ \mathrm{W\ cm^3\ sr},$$

where c_1 is the first radiation constant (3.7418×10^{-12} W $\mathrm{cm^2}$), c_2 is the second radiation constant (1.4388 cm K), n is the refractive index of air (1.000 27), λ is the wavelength in air (in cm), and T is the thermodynamic temperature in (K).

When radiant flux is incident on an object, it is either reflected (ρ), transmitted (τ), or absorbed (α). Thus,

$$\rho+\tau+\alpha=1.$$

If all of the energy incident on the object is absorbed, the absorptance is unity and, according to Kirchoff's law [8], the emissivity of the object is also unity [9]. In reality, a perfect absorber over all wavelengths and temperatures does not exist. There are a number of black paints, oxidized metals, and evaporated blacks that have an absorptance of 0.9 or better, but nothing close to unity. However, a near-perfect absorber or emitter can be formed by placing a small hole in the wall of a hollow, isothermal enclosure whose interior surface has a high absorptance ($\geqslant 0.9$). Radiant flux incident on the opening of the enclosure is subject to the following:

1. Approximately 90% of the radiant flux incident on the surface is absorbed.
2. The remaining 10% is diffusely reflected (provided that the interior surface of the sphere has a diffuse reflectance).
3. A negligible portion of the radiant flux will escape through the small opening on the first order reflectance.
4. Significantly smaller and smaller portions of the incident flux will escape through the opening for the higher-order reflectances.

By careful design of geometry, choice of materials, and method of heating, a device can be constructed whose absorptance or emissivity is very close to unity. A number of methods exist for computing the effective emissivity of various radiating enclosures [10,11].

There are many commercially available blackbodies having various geometrical shapes and constructed of different materials for use over different temperature ranges. Some of these blackbodies are field-portable and some are elaborate laboratory infrared standards that operate at the freezing temperatures of different metals. However, most of these blackbodies are used primarily in the infrared at wavelengths above about 1000 nm. Blackbodies suitable for use in the visible spectrum must operate at temperatures of 2500 K or higher. These are extremely expensive devices and are not practical for normal laboratory calibrations.

8.2.2 Basic Spectroradiometric Standards

NIST Standard of Spectral Radiance

With the exception of a blackbody radiator, there were no convenient spectroradiometric standards prior to about 1960 [12]. Although the blackbody was and still is the primary standard used for most infrared calibrations, its use in UV, visible, and near-IR is very limited. The NIST scale of spectral radiance consists of a tungsten-ribbon-filament lamp whose calibration is based on the radiant flux emitted by a blackbody of known temperature as determined from the Planck radiation equation [13–15]. The originally lamp chosen by NIST was the GE30A/T24/3. It has a mogul bipost base, and a nominal rating of 30 A at 6 V. Radiant energy is emitted from the flat strip filament through a 1.25 in. fused silica window located in the lamp envelope. The window is parallel to and at a distance of about 3–4 in. from the plane of the filament. Spectral radiance values are reported over the wavelength range 225–2400 nm. The estimated rms uncertainty varies with wavelength from 1.0% at 225 nm to 0.3% at 2400 nm and is about 0.6% throughout the visible spectrum. Typical spectral radiance values are shown in Table 8.3.

Tungsten-ribbon-filament lamp standards of spectral radiance have found wide use in the calibration of spectroradiometric and other instrumentation used to measure the spectral radiance of a small area. However, the use of these radiance standards is limited by the small area that can be calibrated and by the low radiance that the standards provide at lower wavelengths. These standards are useful when measuring the spectral radiance of plasmas, furnaces, or other small area radiating sources. These radiance standards can also be used as an irradiance standard by carefully imaging the filament onto a small, precision slit with an accurately measured opening. The irradiance at a distance from the opening can then be calculated. However, a number of difficulties exist: the source area is quite small, which limits the irradiance level; the effective transmittance or reflectance of the imaging optics must be measured; and the angular field is limited.

NIST Standard of Spectral Irradiance

NIST established quartz halogen lamps as standards of spectral irradiance in 1963 [16] in order to eliminate the problems associated with using the tungsten-ribbon-filament lamp standards of spectral radiance. A GE 200-W quartz iodine

TABLE 8.3. *Measured values of spectral radiance and blackbody temperature of a tungsten-ribbon-filament lamp.*

λ (nm)	L_λ (W cm^{-3} sr^{-1})	T_{bb} (K)	λ (nm)	L_λ (W cm^{-3} sr^{-1})	T_{bb} (K)
225	8.805	2677.3	575	97 830	2534.3
230	13.23	2677.1	600	113 700	2521.5
240	28.11	2676.3	654.6	147 300	2493.2
250	55.41	2674.7	675	158 600	2481.9
260	102.7	2672.9	700	171 800	2468.4
270	180.5	2671.0	750	194 000	2440.6
280	302.8	2669.1	800	211 800	2414.1
290	480.7	2665.4	900	229 000	2356.7
300	738.4	2662.2	1050	223 400	2269.4
325	1 867	2653.0	1200	198 100	2183.0
350	4 003	2643.0	1300	177 000	2125.9
375	7 540	2632.4	1550	127 300	1991.8
400	12 770	2620.8	1700	102 600	1916.3
450	29 100	2596.9	2000	66 390	1778.6
475	40 160	2584.6	2100	57 480	1736.0
500	52 880	2572.0	2300	42 780	1650.3
550	82 060	2546.8	2400	37 740	1618.1

lamp was examined and found to have acceptable characteristics for use as a standard of spectral irradiance. It is a rugged lamp in a small quartz envelope of relatively good optical quality. The small size of the lamp envelope together with the small area of the filament yields an approximate point source irradiance field at fairly close distances, thus permitting placing the lamp within 0.5 m of the spectroradiometer. The tungsten–halogen cycle permits operating the lamps at color temperatures as high as 3100 K, thus providing significantly higher irradiance levels in the ultraviolet spectral region. These new standards were calibrated over the wavelength range 250–2500 nm. A similar 1000-W lamp was set up by NIST in 1965 relative to the 200-W standards. This standard had an irradiance level approximately 5 times that of the original 200-W standard. In 1975, NIST switched from the 1000-W DXW-type lamp to a 1000-W FEL type lamp [17,18]. These FEL lamps were converted to a medium bipost base that enabled more convenient use with a kinematic lamp holder, allowing the lamps to be removed and replaced exactly in the same position (see Fig. 8.2).

Calibration of the FEL Standards of Spectral Irradiance are based on the NIST Standards of Spectral Radiance and are calibrated over the wavelength range 250–2400 nm. Typical spectral irradiance values are given in Table 8.4. The estimated rms uncertainty varies with wavelength from 2.23% at 250 nm to 6.51% at 2400 nm and has an average uncertainty of about 1% in the visible region.

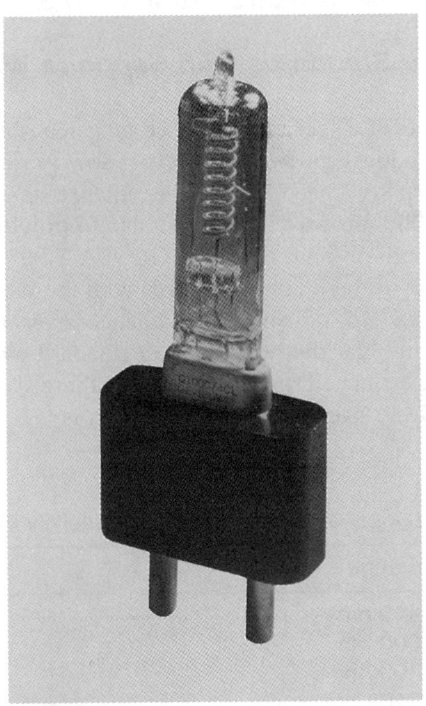

FIGURE 8.2. FEL spectral irradiance standard lamp and bipost system.

TABLE 8.4. *Spectral irradiance of a 1000-W FEL-type lamp at 50 cm.*

λ (nm)	E_λ (W cm^{-3})	λ (nm)	E_λ (W cm^{-3})	λ (nm)	E_λ (W cm^{-3})	λ (nm)	E_λ (W cm^{-3})	λ (nm)	E_λ (W cm^{-3})
250	0.182	320	3.549	390	18.31	700	190.3	1540	126.0
260	0.320	330	4.752	400	21.77	800	221.9	1600	166.9
270	0.531	340	6.223	450	44.06	900	231.6	1700	101.6
280	0.833	350	7.998	500	73.12	1050	220.7	2000	67.1
290	1.258	360	10.05	555	108.7	1150	203.0	2100	59.7
300	1.830	370	12.46	600	137.5	1200	193.1	2300	46.0
310	2.592	380	15.22	654.6	168.6	1300	173.3	2400	40.0

NIST Standard Detector

NIST has established an absolute spectral responsivity scale based on a high-accuracy cryogenic radiometer. Table 8.5 gives the estimated uncertainties assigned to selected silicon photodetectors calibrated relative to the NIST Scale. NIST also provides responsivity uniformity plots at specific wavelengths.

8.2.3 Special-Purpose Spectroradiometric Standards

Spectral Radiance Standards with Sapphire Windows

This standard consists of a specially modified tungsten ribbon filament lamp (GE 18A/T10/2P) with an optical grade, sapphire window [19]. These standards were developed in order to satisfy a need for a single radiance standard that could be used over the entire 250–6000 nm wavelength and also to provide an alternative to the more expensive and more difficult to obtain GE 30A/T24/3 lamp.

The spectral radiance of these lamp standards with the sapphire window is traceable to the NIST Standard of Spectral Radiance over the wavelength range 250–2400 nm and to the blackbody calibration standard over the 2400–6000 nm wavelength range. The estimated rms uncertainty of these special purpose standards relative to the NIST Scale over the visible spectrum is 2%.

TABLE 8.5. *Estimated uncertainty in absolute responsivity measurements.*[a]

Wavelength range	Uncertainty (%)
400 nm$\leq\lambda\leq$440 nm	±0.7
440 nm$\leq\lambda\leq$900 nm	±0.22
900 nm$\leq\lambda\leq$1000 nm	±0.3
1000 nm$\leq\lambda\leq$1100 nm	±0.7

[a]This is a relative expanded uncertainty (k=2).

Spectral Irradiance Standards

A series of tungsten halogen lamps having wattages of 1000, 200, and 45 W has been set up as special-purpose standards of spectral irradiance [20]. These standards are directly traceable to the NIST FEL Standard of Spectral Irradiance over the wavelength range 250–2400 nm and to a blackbody calibration standard for wavelengths above 2400 nm. Whereas the spectral irradiance of the 1000-W DXW standard is similar to that of the FEL 1000-W standards, the 200-W and 45-W standards have irradiance levels of about 5 times and 20 times less than the 1000-W FEL standard, respectively. However, all of these standards are calibrated when operating at a color temperature of about 3000 K; thus, the relative spectral distribution is approximately the same for all of the lamps. The estimated rms uncertainty relative to the NIST Scale is on the order of 1% over the visible spectrum.

Plug-In Tungsten-Lamp Standards of Spectral Irradiance

Plug-in, pre-aligned irradiance standards [21] are available for accurately calibrating various spectroradiometers for spectral irradiance response (see Fig. 8.3). These standards consist of a compact, 200-W tungsten halogen lamp operating at a color temperature of about 3000 K. The short working distance of about 13 cm provides irradiance levels significantly higher than that normally obtained with higher wattage standards. The combination of greater precision in optical alignment and higher irradiance levels provides for a more accurate calibration of the spectroradiometer. The ''plug-in pre-aligned'' concept also eliminates tedious and time-consuming setup and alignment that is normally associated with spectroradiometric standards as they merely attach to the integrating sphere input optics portion of the spectroradiometer. These standards have an estimated rms uncertainty relative to the NIST Scale of 1% over the visible spectrum.

FIGURE 8.3. Plug-in pre-aligned standards of spectral irradiance.

Integrating-Sphere Calibration Standards

All of the spectroradiometric standards described herein are calibrated for either spectral radiance or spectral irradiance, and they are only calibrated for one set of conditions, i.e., one specified lamp current, one distance, etc. In addition, all of these standards, with the exception of the ''plug-in pre-aligned'' standards, operate in the open air and cannot easily be attenuated. In many instances, there is a need for a large-area, uniformly radiating source that is accurately calibrated for spectral radiance and, in some cases, calibrated for spectral irradiance. In addition, accurate attenuation of the radiance or irradiance of the source may be desirable. A carefully designed integrating-sphere–tungsten-lamp combination meets the above criteria.

Integrating-sphere calibration standards generally consist of two parts: a source module with an optics head and a separate electronic controller. The source module for a typical sphere source is shown in Fig. 8.4. This unit incorporates a 150-W tungsten halogen, reflectorized lamp with a micrometer-controlled variable aperture between the lamp and the integrating sphere in order to vary the flux input to the sphere. The integrating sphere is coated with a highly reflecting, diffusely reflecting material such as polytetrafluoroethylene (PTFE) or $BaSO_4$. These materials have reflectances above 99% throughout the visible spectrum. The variable aperture at the entrance port of the sphere provides for continuous adjustment of the sphere radiance over a range of more than 10^6. A precision silicon detector–filter combination with an accurate photopic response is mounted in the sphere wall and monitors the sphere luminance. The electronic controller contains the lamp power supply and the photometer amplifier. The source module with optics head is designed such that it can be configured with different sized integrating spheres; thus, the diameters

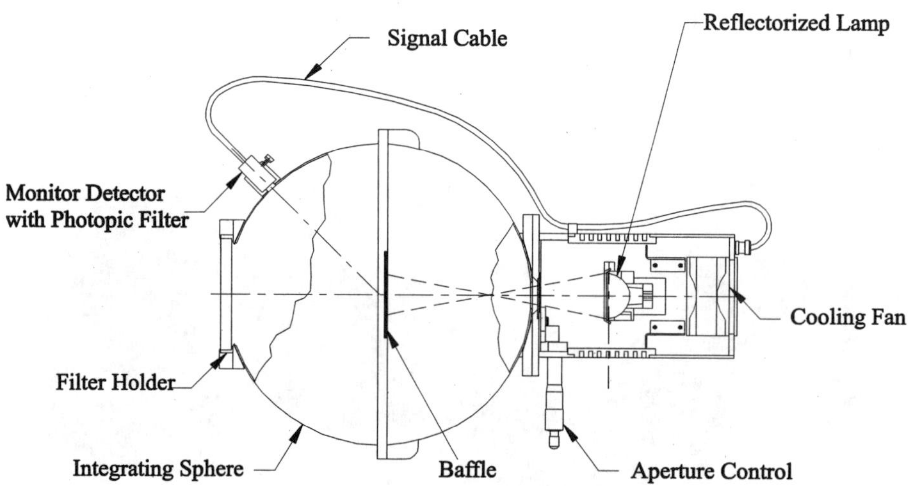

FIGURE 8.4. Sectional view of an integrating-sphere calibration standard.

of the exit (radiating) port can be made progressively larger as the sphere diameter increases. A 4:1 ratio of sphere diameter to exit port diameter will generally provide a uniformity in the radiance at the exit port of ±0.5%.

Since integrating-sphere sources are quite uniform in radiance and have well-defined radiating areas, the spectral irradiance can be computed once the source has been calibrated for spectral radiance. It should be noted that these integrating-sphere sources are also used quite extensively as photometric standards. In general, the detector or photopic filter that serves as a monitor in the sphere wall is calibrated such that the luminance of the sphere is displayed on the electronic controller.

A sphere source of this design that incorporates a 4-in.-diam integrating sphere will typically have a 1-in.-diam radiating port with an adjustable luminance from 100 000 to 0.001 cd/m^2. A 12-in.-diam version will have a 3-in.-diam radiating port with an adjustable luminance from 12 000 to 0.001 cd/m^2. Typical estimated uncertainties are 2% rms relative to the NIST Scale.

For measurement application requiring even lower output levels than that obtainable with the sphere source described above, an extremely low-light-level integrating-sphere source such as that shown in Fig. 8.5 is available. This source is similar to that described above, but two methods are used to attenuate the sphere radiance without changing the spectral distribution of the source. A low-wattage tungsten halogen lamp is mounted on a moveable track such that the distance from the lamp to the entrance port of the sphere can be varied from 5 to 30 cm. Immediately in front of the entrance port of the sphere is a six-position aperture wheel containing precision apertures having diameters from 28 to 0.15 mm. Thus, the combination of varying the lamp–sphere distance and inserting different sized apertures over the entrance port enables precise setting of the radiance or luminance of the radiating port. This integrating-sphere source is also designed such that the optics head can accommodate different size spheres. Luminance levels as low as 10^{-5} cd m^{-2} can be obtained with this version.

FIGURE 8.5. A low-light-level integrating-sphere calibration standard.

In some cases, it is desirable to have even larger radiating areas than those obtainable with the integrating-sphere sources described above. In these instances, integrating spheres having diameters of 30 in. or greater are available. These sphere sources generally have one or more lamps mounted inside the sphere. Attenuation is accomplished by turning one or more of the lamps off.

In general, spectral radiance values for integrating-sphere calibration standards are reported for a specified lamp current. Varying the operating current not only changes the magnitude of the spectral radiance, it also changes the spectral distribution. Thus, it is imperative that the lamp be operated at the specified current when used as a spectroradiometric standard. However, this is not the case when these integrating-sphere sources are used as photometric standards. The silicon detector–photopic filter monitor will measure luminance, and the accuracy of the luminance display is not dependent on the lamp current provided that the detector-filter is accurately matched to the Commission Internationale de l'Eclairage (CIE) photopic efficacy function.

Detector Spectral Response Calibration Standards

Silicon photodiodes are available having spectral responsivity calibrations in terms of power response or irradiance response over the wavelength range 200–1100 nm [22–25]. These calibrated detectors generally have a 1-cm^2 active area and are mounted in black anodized aluminum housings with a convenient BNC connector. They may also include a removable precision aperture. Figure 8.6 shows a typical spectral responsivity plot (power response in A/W).

The photodetectors are normally calibrated with zero bias voltage (short-circuit mode). The transfer uncertainty in the calibration of these detectors relative to the NIST scale is on the order of 0.5% over the visible spectrum.

8.3 SPECTRORADIOMETRIC INSTRUMENTATION—GENERAL FEATURES

At the heart of any spectroradiometer is some mechanism for separating the optical radiation into its spectral components [26,27]. By far the best and most common mechanism is a monochromator. This section will review the components of the monochromator, their role in determining the performance of the system as a whole, and the typical specifications that users should look for when selecting a system.

A typical monochromator, such as those shown in Figs. 8.13, 8.14, and 8.16–8.18, consists of entrance and exit slits, collimating and focusing optics, and a wavelength-dispersing element such as a grating or prism. Additional mechanisms such as optical choppers or filter wheels may be included and are often mounted inside the monochromator.

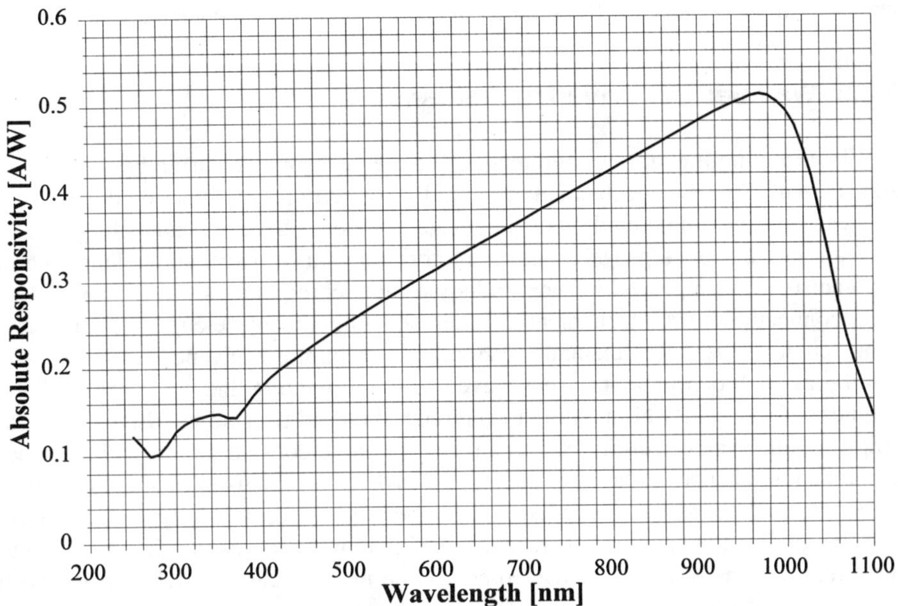

FIGURE 8.6. Spectral responsivity of a typical silicon photodetector.

8.3.1 Wavelength-Dispersing Elements

Most modern monochromators use diffraction gratings, but a few of the older prism-based monochromators are still in use. Diffraction gratings have a few disadvantages when compared to prisms (mainly the multiple-order effects covered later), but their greater versatility, ease of use, wavelength range, and more constant dispersion with wavelength means that grating monochromators are used almost exclusively in spectroradiometry. Since grating monochromators are by far the most widely used, further discussion will exclude the other types.

8.3.2 Collimating and Focusing Optics

The first optical element of a monochromator is usually a collimating optic (typically a concave mirror), which alters the diverging beam coming from the entrance slit into a collimated beam directed at the grating. The grating acts on this incident collimated beam to create a series of collimated diffracted beams, each at a different angle that depends on wavelength. By rotating the grating, each wavelength in turn will strike the focusing optic, creating an image of the entrance slit at the exit slit position. In some monochromators, the use of collimating and focusing optics is eliminated by employing curved gratings (as shown in Fig. 8.11). These are generally more compact than plane-grating monochromators, but are limited in wavelength range because the gratings are no longer interchangeable.

8.3.3 Wavelength Drive Mechanisms

When rotating the grating, two main drive mechanisms are currently used: direct and sine-bar. Sine-bar mechanisms essentially convert the sine function dependence between grating angle and wavelength to a linear drive mechanism giving a constant number of steps (of a stepper motor) per unit of wavelength. This type of mechanism was employed in most old-style monochromators since it provided an easy connection to a mechanical counter showing the wavelength selected and hence could be used in ''manual'' systems. As technology advanced, reliable self-calibrating monochromators were developed, eliminating the need to read the counter before use. By eliminating the counter, the need for sine-bar drives was also removed, allowing direct-drive mechanisms (giving a constant number of steps per grating angle change) to be introduced. Direct-drive mechanisms can be thought of as having a ''theoretical'' sine-bar, where the angle necessary to give the correct wavelength is calculated and then selected. The elimination of the sine-bar mechanism removes many of the associated errors and in some cases allows grating turret and wavelength drive mechanisms to be combined.

8.3.4 Stray Light

Although monochromators are used to isolate a particular spectral component from all other wavelengths, any practical monochromator will transmit residual ''out-of-band'' wavelengths; this is known as stray light. For most visible applications, this effect is virtually negligible but for certain usage, such as measurements for night-vision systems, this stray light can swamp the real spectral components being analyzed. In such cases, a double monochromator is used to reduce the typical 10^{-4} stray-light levels to 10^{-8} (expressed as the ratio of detected stray radiation to the total detectable radiation entering the monochromator). A double monochromator is essentially two identical single monochromators where the output of the first is the input of the second. Naturally, for this to work both monochromators have to be set to the same wavelength, and experience has shown that the only way to achieve reliable results is to have both monochromators sharing the same base plate and drive mechanism. Attempts at bolting two single monochromators together have been tried many times in the past (and sometimes at present) with poor results, and this approach should be avoided.

8.3.5 Blocking Filters

Since gratings disperse light by diffraction, any wavelength of light may be diffracted in several directions (orders) as shown in Fig. 8.7. In practice, this means

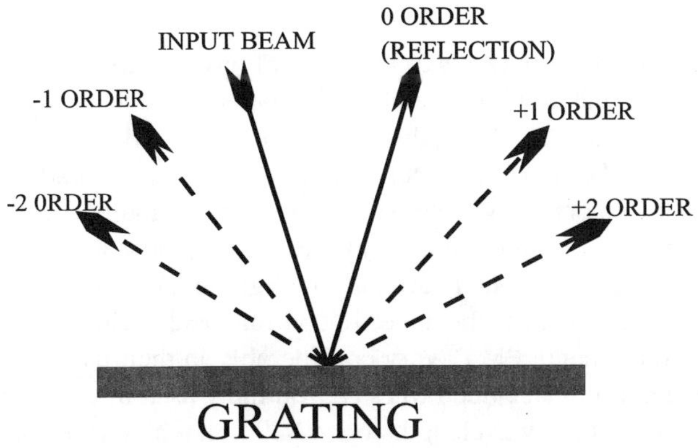

FIGURE 8.7. The various orders of diffraction from a grating.

that the first order of 900 nm is diffracted at the same angle as the second order of 450 nm and the third order of 300 nm. When measuring light at 900 nm it is therefore essential to remove the 450- and 300-nm components, and this is done by blocking filters. Blocking filters absorb short wavelengths while transmitting long wavelengths. For example, a 550-nm blocking filter would absorb both the 450- and 300-nm components while leaving the 900 nm essentially unaffected. In visible applications, at least two blocking filters are generally used: the first to block any UV components and the second to prevent 380-nm light recurring at 760 nm. These filters are placed in the beam at the appropriate wavelengths during a scan and are often placed within a filter wheel and selected automatically, making the entire process ''transparent'' to user.

8.3.6 Grating Optimization

The grating consists of many finely spaced lines (or grooves) etched into a surface. The density of these grooves, expressed in grooves per millimeter (groove mm^{-1}) determines the angular separation of wavelengths (dispersion) in any monochromator. Also, by altering the shape of the groove profile, it is possible to make certain angles of diffraction more efficient. This process is known as *blazing* and is used to increase the throughput of the monochromator in desired spectral regions. However, this process also decreases the throughput outside of this region, giving the normal ''rule-of-thumb'' that the usable wavelength range of a grating is two-thirds to twice the blaze wavelength. For example, a grating blazed at 500 nm should be usable from 330 to 1000 nm, and in practice this is almost always selected for visible work. Other gratings can be chosen to increase the accuracy for specific sources (very blue or red), but the close agreement of maximum efficiency with the peak of the photopic response function makes the 500-nm blaze grating the best choice for general applications.

8.3.7 Detectors

Although the monochromator is at the heart of any spectroradiometer, a detector is always present in the system. The choice of detector is generally dictated by the light levels to be measured, the stability required, and any nonvisible application that the system may be required to perform. Almost all spectroradiometers use one of two types of detectors for visible work: a silicon photodiode or photomultiplier tube (PMT). The silicon photodiode may be used in most medium-to-high light-level applications, and the PMT is used at low light levels. Silicon photodiodes can vary in their usable wavelength range, but virtually all respond to visible wavelengths. On the other hand, PMTs vary considerably in their usable range, and only trialkali (S20) or gallium arsenide PMTs are routinely used in photopic applications. The S20 PMTs respond to wavelengths up to 830 nm, which is adequate for normal photopic applications, whereas gallium arsenide PMTs are chosen for night-vision tests because of their longer wavelength response of up to 930 nm.

8.3.8 Signal Detection Systems

In many cases, the electronics to amplify and process the signals from the detector, commonly referred to as signal detection systems, influence the performance of the system. For instance, DC amplifiers able to resolve currents from silicon detectors of about 10^{-12} A (1 picoamp) are readily available from most manufacturers, but some suppliers offer much superior performance (to 10^{-15} A, or 1 femtoamp). Since the inherent noise of a silicon detector at room temperature is about 10^{-15} A, anything but the best amplifiers will both decrease the sensitivity range and increase the noise of the detection system. The noise and sensitivity of a PMT depend both on the PMT voltage and the type of signal detection system. In the DC amplification mode, the dark current of an S20 PMT is, at best, in the picoamp range at room temperature. The amplifier does not therefore require sensitivity ranges less than this, but should still be of good quality to prevent adding noise unnecessarily.

Lock-in amplification, also referred to as AC amplification, can also be used with both PMTs and silicon detectors. Here, an optical chopper spins to alternately transmit and block light at frequencies of tens to hundreds (or even thousands) of cycles per second (hertz). An amplifier, locked in to the chopping frequency, gives a signal proportional to the difference between light and dark phases of the cycle. This technique can be used to eliminate those components that are not at the chopping frequency, such as most of the noise, and compensates for dark-current drifts in the system. However, like any other amplifier, lock-in amplifiers can also add noise to the signal, and are most useful when used with infrared detectors rather than the relatively noiseless visible detectors such as silicon detectors and PMTs.

Photon counting (PC) signal-detection systems provide greater sensitivity than DC or AC modes, but are limited to PMTs. PC detection systems for silicon photodiodes have been developed [28] and are commercially available, but can only be used with detectors too small for practical spectroradiometric use. In the PC mode, each photon hitting the photocathode of the PMT produces a current pulse. These pulses are separated from the large number of smaller pulses from the rest of the PMT and counted. The result is a system that actually counts the individual photons that are absorbed by the PMT. Since a photon is the smallest quantity of light (equivalent to an atom of an element) this represents the ultimate theoretical sensitivity achievable.

8.3.9 Monochromator Throughput and Calibration Factors

When making spectroradiometric measurements, the factors relating the intensity of light at each wavelength to the signal observed must be determined. In theory, if the contribution of each component in the system is known, these factors can be calculated. However, in practice this yields no more than "a rough estimate." To determine the exact factors, a calibration standard (where the intensity of light is known at each wavelength) is used. These calibration standards should be traceable to NIST or other national standards laboratory to ensure correct results.

8.3.10 Slits and Aperture Selection

Selection of appropriate slits and apertures is critical in obtaining correct spectroradiometric results, and yet remains a subject fogged in mystery for most users. This section aims to help dispel that fog, giving the user a clear insight as to which slits or apertures should be selected for their particular application.

Monochromator slits are rectangular, generally much taller than they are wide, and are positioned so that the long side is normal to the plane of the monochromator (i.e., usually vertical). An aperture may be any shape, though it is usually circular, and is used in place of a slit in certain applications. For the purpose of this discussion, circular apertures will be assumed, since these are the most common and represent a very different shape than that of a slit.

Input or exit optics will frequently require a circular aperture to define a field of view (telescopes, microscopes, and other imaging optics), the beam convergence or divergence and uniformity (collimating optics), or the size of an image (some reflectance, transmittance, or detector response accessories). Generally, the only accessories that do not require apertures are nonimaging types such as integrating spheres. The requirements of the input-exit optics always determine the selection of an aperture or slit. If the accessory attached to the entrance requires an aperture, then it is installed in place of the entrance slit; if the accessory attached to the exit requires an aperture, then it is installed in place of the exit slit.

Theoretically, two apertures may be installed: at both the entrance and exit slits. However, such a configuration requires exact alignment of heights and is subject to large changes in throughput with small changes in temperature, flatness of benches, etc. This inherently unstable arrangement is therefore rarely used, except in special applications, and generally whichever side of the monochromator that is opposite the accessory should always have a slit installed.

Limiting Apertures

For large sources or nonimaging optics, the entrance aperture (or slit) limits the size and distribution of light entering the monochromator. However, for certain other applications, the size and distribution may be limited by other factors, making the selection (or indeed the presence) of the slit or aperture irrelevant.

In certain applications, e.g., spectral radiant intensity measurements, the aperture must be underfilled. This means that the aperture no longer defines the size or shape of the image entering the monochromator. An ''equivalent aperture,'' having the same size and shape as the image, should then be used to determine the expected behavior of the system.

Similarly, when using fiber optics (even slit-shaped fibers) the slit width or aperture only applies when smaller than the fiber. If the fiber is smaller then the slit (or aperture) it is the fiber that determines the expected behavior of the system. When using a light source that forms an image of the source onto the entrance slit, the same considerations apply.

The concept of limiting apertures can also be applied at the exit of a monochromator, since detectors may be small, e.g., a 5-mm slit is the limiting aperture when using a 10 mm×10 mm silicon detector, but not when using a 3 mm×3 mm PbS detector. Even if the detector is coupled to the monochromator using imaging optics, it may still define a limiting aperture once magnification effects have been considered.

Slit–Slit Bandpass and Slit Function

When using a monochromatic source, the monochromator forms an image of the entrance slit at the exit. The exit slit therefore acts as a mask, defining the portion of the image that reaches the detector. As the wavelength is altered, the image moves across the exit slit, and a scan of detector signal versus wavelength is called the *slit function* and may be used to find the full width at half maximum (FWHM) or as it is more commonly called, the bandpass.

The slit function, and hence bandpass, can be calculated quite simply for an ideal instrument. Two possible shapes exist: one where the entrance and exit slits are the same, and the other where they are different. If the detector responds equally to all light passing through the exit slit then, as illustrated in Fig. 8.8, the signal is proportional to the area of overlap between the image of the entrance slit and the mask formed by the exit slit. This gives a triangular slit function for equal slits, and a flat-topped function for different slits. In the case of different slits, the image could be wider than the exit slit and still give the same result, so the slit function is independent of which slit is the entrance and which is the exit.

Although these two shapes exist, in fact only one is sensible for spectroradiometry: equal slits. This is because with different slits the throughput (and hence

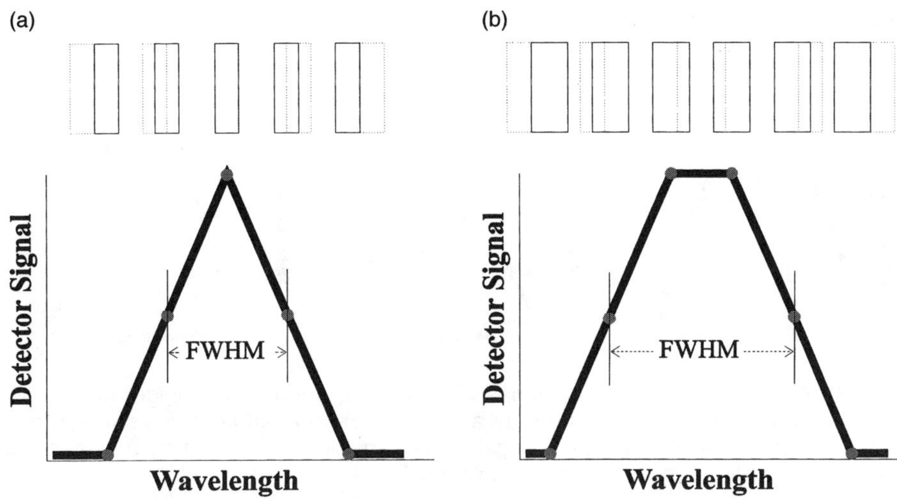

FIGURE 8.8. Slit function and bandpass for (a) equal slits and (b) different slits. In each case, the detector signal is proportional to the overlap between the image (dashed line) and the exit slit (solid line).

signal) is limited by the smaller slit while losing resolution to the wider slit, and severe sampling errors can arise in measurements of sharp spectral features with normal scan intervals. On the other hand, equal slits provide the maximum signal at any bandpass and give accurate peak areas with most scan intervals less than the FWHM.

In a real system, the triangular function will have a rounded top and baseline intercept. Also, if the slits are very narrow, the function may often resemble a skew-Gaussian curve rather than a triangle. These are due to normal aberrations found within any monochromator system and do not affect the general principles outlined.

Aperture–Slit Bandpass and Slit Function

As with slit–slit systems, an image of the entrance aperture is formed at the exit slit, though the resulting slit function is not nearly so obvious. The important difference for aperture–slit calculations is that the shapes are sufficiently mismatched to create three possibilities for the slit function, as shown in Fig. 8.9. The first (a), where the aperture is much larger than the slit, gives a cosine-curve-shaped slit function. The second (b), where they are the same width, gives an almost triangular shape (except the sides are S shaped) with essentially the same bandpass as the slit–slit equivalent. The third option (c), where the aperture is much smaller than the slit, gives a flat-topped profile with S-shaped sides.

For exactly the same reasons as those applying to slit–slit configurations, best

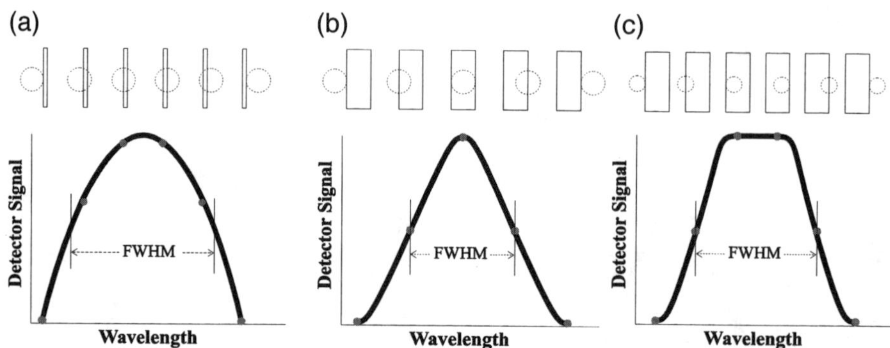

FIGURE 8.9. Slit function and bandpass for (a) an aperture much bigger than a slit, (b) an aperture with the same width as a slit, and (c) an aperture much smaller than a slit. In each case, the detector signal is proportional to the overlap between the image (dashed line) and the exit slit (solid line).

results are obtained by matching the slit and aperture widths as closely as possible. However, since the size of the aperture and slit are normally determined by additional factors such as field of view and sensitivity, it is not unusual to make measurements with the slit larger than the aperture. There are *no* circumstances where slits much smaller than the aperture give better results than matched combinations.

Although the above-noted description used an aperture entrance and slit exit, the same results would be obtained if they were reversed since we are treating the monochromator as ideal. In real systems, very slight differences may exist between the two configurations since aberrations will distort the images of slits and apertures differently.

Dispersion, Bandpass, and Limiting Resolution

The previous section dealt with the shape of the slit function. However, to put actual values to the band pass the dispersion must also be known. The dispersion (or more correctly inverse linear dispersion) is the wavelength region (in nm) in 1 mm distance in the plane of the slit. It varies with the focal length of the monochromator and the grating groove density (grooves per millimeter). For any particular monochromator, if the dispersion with a 1200-grooves/mm grating is known (this is usually available from the manufacturer), the band pass with any grating or slit width can be calculated using

$$B = \frac{1200DS}{G},$$

where B is the bandpass (in nm), D is the dispersion (in nm/mm) with a 1200-groove/mm grating, G is the groove density of the grating used (in grooves/mm), and S is the slit width (in mm).

Thus, if a 600 groove/mm grating is used with a monochromator of 4 nm/mm dispersion and 1.25-mm slits, the bandpass will be 10 nm. Table 8.6 shows the bandpasses of various slit–slit combinations (for a single monochromator with 4 nm/mm dispersion), highlighting the recommended (equal entrance and exit) selections.

The equation above is based on two basic assumptions: that the dispersion remains constant with wavelength and is perfectly linear at all slit widths. Real systems have a variable dispersion with wavelength, though good designs can optimize this to just a few percent, and aberrations and alignment errors generally limit the bandpass at small slit widths. This limit at small slit width is called the limiting optical resolution of the system. Spectroradiometric measurements are generally made at bandpasses well above the limiting optical resolution of the system to ensure that the slit function is reasonably constant at all wavelengths.

Because the slit function changes with the relative size of the aperture–slit combinations, mismatches (even to smaller widths) can lead to increased bandwidths. This means that the above formula for slit–slit combinations will not apply to aperture–slit sizes. Table 8.7 shows the bandwidths of various aperture–slit combinations (for a single monochromator with 4 nm/mm dispersion), with the recommended configuration highlighted.

8.3.11 Throughput versus Bandpass

Optimizing practical measurements often involve a trade-off between smallest bandpass and highest signals. If there are fixed parameters, such as field of view, that must be observed, the slit–aperture configuration will also be fixed. However, if several combinations are possible, then the system may be optimized. Assuming that the slits and apertures are matched, as recommended, and the entrance slit (or aperture) is uniformly illuminated, the signal increases with width. The magnitude

TABLE 8.6. *Bandpasses (in nm) for various slit–slit combinations for a single monochromator with 4 nm/mm dispersion. (Recommended combinations are in bold face.)*

Entrance slit (mm)	Exit slit (mm) 0.25	0.5	1.25	2.5	5.0
0.25	**1.0**	2.0	5.0	10.0	20.0
0.5	2.0	**2.0**	5.0	10.0	20.0
1.25	5.0	5.0	**5.0**	10.0	20.0
2.5	10.0	10.0	10.0	**10.0**	20.0
5.0	20.0	20.0	20.0	20.0	**20.0**

TABLE 8.7. *Bandpasses (in nm) for various aperture–slit combinations for a single monochromator with 4 nm/mm dispersion. (Recommended combinations in bold face.)*

Aperture diameter (mm)	Exit slit (mm)				
	0.25	0.5	1.25	2.5	5.0
1.5	3.8	3.6	**4.0**	7.5	15.0
3.0	7.7	7.6	7.3	**8.1**	15.0
5.0	13.0	12.9	12.3	12.2	**15.0**

of this increase varies with the type of source and whether slit–slit or aperture–slit combinations are used.

If the source is a broadband type, such as a tungsten lamp, doubling the slit–slit sizes would double both the intensity entering the monochromator and the bandpass of the system: giving a fourfold increase in signal. Under the same conditions, an eightfold increase would be seen for aperture–slit combinations, since both width and height of the aperture are doubled.

When a monochromatic source such as a mercury lamp is used, only a twofold and fourfold increase is seen for the respective slit–slit and aperture–slit combinations because only one wavelength component exists. This difference between monochromatic and broadband sources accounts for the observed changes in the spectra of mixed sources, such as fluorescent lamps, as the bandpass is altered. In such a case, as the bandpass is decreased, the peaks due to the monochromatic lines become much "taller" in proportion to the broad emission of the phosphors.

Double Monochromators and Middle Slits

The sections above deal with entrance and exit slits and apertures for a single monochromator. The basic principles apply also to double monochromators, but with modifications. Essentially, a double monochromator is two identical single monochromators working in series, but using a single drive mechanism and housing. Theoretically, and sometimes practically, double monochromators have four slits—an entrance and exit for each of the single monochromator components. More generally, the exit of one is the entrance of the other and hence only three slits are used: the entrance, middle, and exit. The influence, and hence selection, of the middle slit largely depends on whether the double monochromator is additive or subtractive.

The majority, by far, of double monochromators in use are additive. *Additive* means that light that is dispersed by the first monochromator component is further dispersed by the second. Thus, if a single monochromator has a dispersion of 4 nm/mm, then a double additive monochromator of the same basic design would have a dispersion of 2 nm/mm. When using Tables 8.6 and 8.7 for the equivalent additive double-monochromator systems, the band pass calculated should be divided by 2. However, even though both monochromator components share the same drive and housing, links, and gears can mean that each part may not transmit exactly the same wavelength (this is commonly referred to as ''coordination'' of the monochromators). If this happens, the throughput of the system will suffer, and if this wavelength mismatch is temperature dependent (as it is in all practical systems), then it is inherently unstable and unsuitable for spectroradiometry. Compensating for this is actually quite easy: the middle slit should be at least 1 to 2 mm larger than the entrance and exit whenever possible. Apertures are *never* used in the middle slit position.

Subtractive monochromators have two main applications. The first of these is performing extremely fast measurements, of the order of picoseconds. Here, the difference in path lengths through an additive monochromator might spread out a pulse because of the finite speed of light, but the symmetrical arrangement of a subtractive double monochromator provides a constant length for all paths. The second application, for measurements such as detector spectral response, is more common. In additive double monochromators, the small difference in wavelength between the left and right edges of the slit can cause serious errors if the detectors are not uniform. However, the subtractive double monochromator has no residual dispersion at the exit slit, eliminating this source of error. The lack of residual dispersion is because the second component monochromator operates, not to further disperse like an additive system, but to combine (the opposite of disperse in this context) any light dispersed by the first component monochromator. This means dispersion effectively only occurs in the first component monochromator, and this system should be treated as a single monochromator using the entrance and middle slits as far as slit function, bandpass, etc., is concerned. However, the coordination of the two components are still important, and since there is no effective movement of the image at the exit slit, it is prudent to:

1. have the entrance slit slightly larger than the exit slit (or aperture);
2. have the middle slit slightly larger than the entrance slit.

These conditions are to optimize stability, which is essential for good spectroradiometry, but they do compromise the narrowest bandpasses and triangular slit function. If a triangular slit function is required, equal entrance and middle slits should be chosen. If the narrowest bandpass is also desired, then the entrance and middle slits should be equal to the exit slit or aperture (but beware of the inherent instability of such an arrangement).

8.4 PERFORMANCE SPECIFICATIONS

Having described the function and use of each component of a spectroradiometer in Sec. 8.3, it is worthwhile considering typical specifications of a system and how these affect results.

8.4.1 *f*-Number

The *f*-number of any focusing system is the focal distance divided by the limiting aperture. It is therefore inversely related to the solid angle encompassed by, and hence the "light-gathering power" of, the optical system. In other words, the smaller the *f*-number, the greater the throughput of a monochromator. However, there are practical limits to the *f*-number for monochromators. Since spherical mirrors are generally used in monochromators, instead of the ideal off-axis paraboloid shape, at *f*-numbers less than 3 or 4 the aberrations destroy the advantages of decreased *f*-number. Also, monochromators are almost always used with accessories appropriate to the type of measurement, and these accessories generally have larger *f*-numbers than the monochromator, thus limiting the overall system. For most applications, the best compromise is obtained by selecting a monochromator with an *f*-number of about 4.

8.4.2 Wavelength Accuracy Resolution and Repeatability

When a monochromator selects a wavelength, there may be a difference between the actual wavelength transmitted and that reported by the computer or wavelength counter. The wavelength accuracy required largely depends on the application and how fast the signal changes with wavelength, but for general photopic applications the monochromator should always be within ± 0.2 nm of the true wavelength. Some manufacturers now use a polynomial fit (also known as a cubic spline) to improve the accuracy, and this should be adopted whenever available.

As the slit widths are reduced, eventually a limiting bandpass is reached. This is the limiting optical resolution of the monochromator. For photopic, and indeed most other spectroradiometric measurements, bandpasses of less than about 0.5 nm are rarely used. This is readily achievable with a 250-mm focal length monochromator, which is probably the most usual instrument found in spectroradiometric laboratories. To scan sources containing monochromatic emission or absorption lines, such as fluorescent lamps, greatest accuracies are achieved with bandpasses of 1 nm or less. However, when scanning at these small bandpasses it should be borne in mind that at least two (and preferably five or more) data points per bandpass should be obtained. The monochromator should therefore have a step resolution (the smallest increment of wavelength movement) of one-fifth or less of the limiting optical resolution to accommodate all possibilities, and the user should realize that accurate measurements of line sources take more data points than broadband sources.

Obviously, to obtain accurate results the monochromator must transmit the same wavelength each time it is selected. In practice, small pseudorandom variations are found. To be reliable enough for spectroradiometric measurements, this variation—the reproducibility—needs to be about ± 0.1 nm or less.

8.4.3 Bandpass

To allow for the various applications and sources encountered in spectroradiometric measurements, the monochromator should normally allow bandpasses of 0.5–20 nm (or more) to be selected. The bandpass depends on the monochromator dispersion, and for a value of 4 nm/mm this corresponds to slits of 0.125–5 mm.

8.4.4 Sensitivity and Dynamic Range

The $V(\lambda)$ function values vary by over 6 decades, depending on wavelength. In order that a source is properly represented, this then is the minimum required dynamic range of a spectroradiometer for photopic measurements. Many broadband sources, such as tungsten lamps, may not require measurements over a wide range of signals, but others such as red light-emitting diodes (LEDs) certainly do. As an illustration, consider an extreme example of a LED with a maximum intensity emission at 830 nm. The $V(\lambda)$ value at 830 nm is 4.52×10^{-7}, which means that any emission at 555 nm from this LED will contribute over 2 million times that of the peak value towards the final photopic result. If the dynamic range of the instrument is unable to resolve this level of emission, the user will be unable to provide an accurate photopic value. To resolve this the instrument must be sensitive enough to detect any light at 555 nm, and yet still accommodate the peak intensity nearly 7 decades stronger.

The dynamic range of an instrument, being the ratio of the largest measurable signal to the smallest, is a relative measure of performance. It is not, in itself, an indication of quality. For instance, in the above example a range 10^{-10}–10^{-3} A would not be sufficient for accurate measurements if the signal at 830 nm was 10^{-7} A, since the higher ranges are redundant.

The smallest measurable signal is generally determined by digitization or noise, whichever is greater. Noise is a random variation of signal with time whereas the digitization limit is the difference between 0 and 1 of the analog-to-digital or similar converter (also often referred to as the signal resolution). Although this noise is electrical, it can be expressed in terms of an equivalent intensity of light. Noise equivalent power (NEP), noise equivalent irradiance (NEI), noise equivalent radiance (NER), etc., are used to specify the smallest signals measurable for most spectroradiometric systems. This parameter not only serves to place the dynamic range on an absolute scale, but also indicates the overall performance of the system since it depends on throughput of the monochromator and the quality of the detector and electronics. However, care should be exercised in comparing specifications from two manufacturers since this parameter varies with wavelength and conditions of measurement.

8.4.5 Stray Light

Stray light can be thought of as a limit to measurements of weak spectral components in the presence of strong spectral components. For instance, in the LED example above, if the stray light level was 10^{-4}, then the signal measured at 555 nm may be only two or three decades below that at 830 nm—even if there is no light at 555 nm. In such a case, the photopic value calculated would be grossly in error, despite the signal being well above the noise equivalent input of the detector and within the dynamic range of the system.

As already pointed out, many photopic applications require stray light levels of around 10^{-4} to achieve accurate results. Some applications, e.g., NVIS or the LED example above, require much lower stray-light levels than this. Alternatively, if other measurements are performed with the same instrument, such as UV hazard assessment, this may dictate this selection of the instrument. For these more stringent applications a double monochromator, with stray light levels of $\leqslant 10^{-8}$ are used. It is always acceptable to use a monochromator with low stray light for less stringent applications, but never vice versa.

8.4.6 Scanning Speeds

It is not unnatural for users to want faster and faster scanning. However, the correct balance between speed and accuracy must be found. Certainly, scans may be sped up by not selecting blocking filters, not changing gain ranges, and integrating signals for a short time, but results may become meaningless with this approach. Also, one should not confuse slow scans with slow systems. Often a slower instrument will move the monochromator, change filters, and gain ranges just as fast as a faster one—the difference is the time and care spent in measurement.

8.4.7 Stability

To be effective, a spectroradiometer must be stable over the time between calibration and measurement. Although some drift in the electronics would normally be expected, much of this is removed periodically by such routines as self-calibration and dark-level subtraction. The largest influence on the stability of spectroradiometers is the environment and treatment of the instrument. In fact, the interval between recalibrations is often more dependent on the environment than on the instrument. Movement and temperature changes should be kept to a minimum to maintain calibrations for as long as possible. Also, regular checks on the wavelength and throughput (using appropriate standards) should be made to track changes in calibration factors and hence predict when recalibrations are necessary to maintain the desired accuracy.

8.4.8 Software and Automation

Although this is not a specification as such, in that it only has a secondary influence on results, the degree of automation found in current systems results in the software, and not the user, often determining the methods of measurement and calculation. The user should therefore verify that the techniques dictated by the software are appropriate.

8.5 SPECTRORADIOMETRIC MEASUREMENT SYSTEMS

The components essential to a spectroradiometer have been discussed in Sec. 8.3. In addition, front-end or input optics that collect and transfer the incident optical radiation into the measurement system are needed when measuring the spectral output of various light sources, and both input and exit optics are required when measuring spectral transmittance, spectral reflectance, or spectral response of photodetectors. However, in all cases, proper calibration of the measurement system with the appropriate spectroradiometric standard is essential. This section will cover

1. the selection and use of various input/exit optic modules and
2. the selection and use of the appropriate standard to calibrate the integrated measurement system.

8.5.1 Source Measurements, Input Optics, and System Calibration

Spectral Irradiance and Radiant Flux: Cosine Collectors

Cosine collectors sample radiant flux according to the cosine of the incident angle. These devices accept radiation from the entire hemisphere. Two general types of cosine collectors are available: transmitting- or reflecting-type diffusers. Reflecting-type diffusers are vastly superior with respect to both cosine collection and wavelength range of usefulness. The predominant reflecting cosine collector used for optical radiation measurements is the integrating sphere.

A properly designed and coated integrating sphere (see Fig. 8.10) is extremely useful for many photometric, radiometric, and spectroradiometric measurements. Integrating spheres can be obtained with various geometries and with different coatings. A prerequisite for a good integrating sphere is a diffuse, highly reflective coating. The sphere geometry and type of coating used depend on the measurement application. $BaSO_4$- and PTFE-based coatings have reflectances approaching 100% in the visible spectrum.

Integrating spheres are particularly useful when measuring the spectral irradiance of large or irregularly shaped sources and are essential when measuring the irradiance of sunlight, fluorescent lamps, or any other large area or extended source. Integrating spheres are also quite useful when measuring the luminous, radiant, or spectral radiant flux of diverging or diffusely radiating sources. In such cases, all of the flux emitted by the source must be collected at the entrance port of the sphere. Figure 8.11 shows an integrating-sphere cosine collector mounted at the entrance slit of a double grating monochromator.

Calibration of a spectroradiometer with an integrating-sphere input is accom-

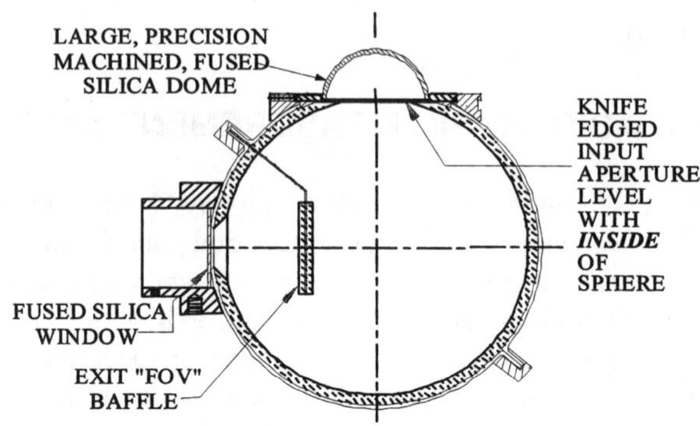

FIGURE 8.10. An integrating-sphere cosine receptor.

plished using a standard of spectral irradiance. The ''plug-in pre-aligned'' standard of spectral irradiance described in Sec. 8.3.1 is particularly well suited for this application. The spectral irradiance response function, $\Psi(\lambda)$ of a system is determined from the relationship

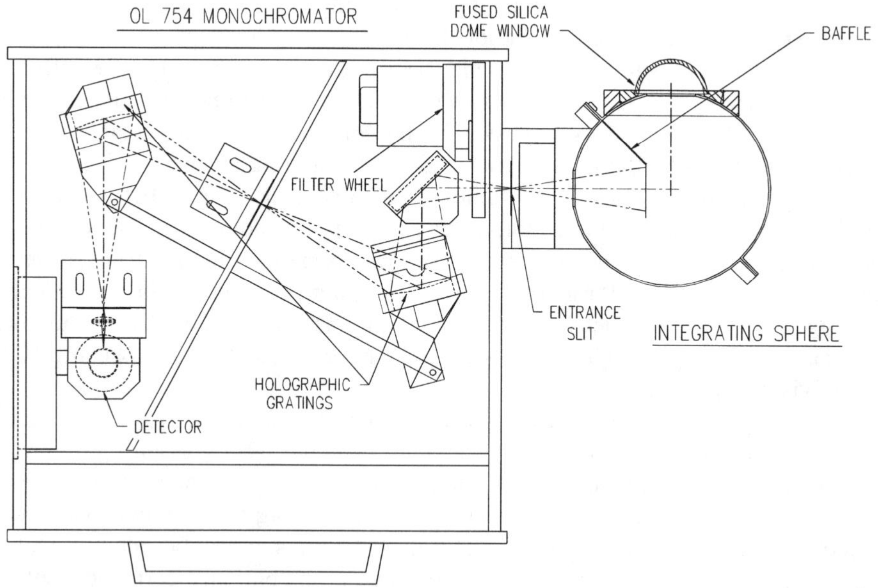

FIGURE 8.11. A spectroradiometer with an integrating-sphere cosine receptor.

$$\Psi(\lambda)=\frac{E(\lambda)}{s(\lambda)},$$

where $E(\lambda)$ is the spectral irradiance of calibration standard and $s(\lambda)$ is the photodetector signal at each wavelength.

Once the spectroradiometer has been calibrated for spectral irradiance response over the wavelength range of interest, it can be used to measure the spectral irradiance of the test source, $E^t(\lambda)$, using the relationship

$$E^t(\lambda)=\Psi(\lambda)s^t(\lambda),$$

where $s^t(\lambda)$ is the photodetector signal at each wavelength.

The spectral irradiance response calibration factors for the system can be converted to spectral radiant power simply by multiplying the calibration factors by the area of the entrance port of the integrating sphere. Thus, the spectroradiometer can now be used to measure the spectral radiant power of a source provided all of the flux emitted by the source is collected by the integrating sphere.

Spectral Irradiance: No Input Optics

Under certain measurement conditions, no special input optics are needed. For example, input optics are not required when measuring the spectral irradiance of point sources or collimated sources if the spectroradiometer responds uniformly over the angular field that the source subtends with the monochromator (or other optical dispersing element). Both the spatial transmission of the monochromator and the uniformity of the photodetector over its sensing area contribute to the spectroradiometer's overall spatial uniformity.

Calibration of a spectroradiometer having no input optics for spectral irradiance response can be accomplished using the 1000-, 200-, or 45-W lamp standards of spectral irradiance. The ''plug-in pre-aligned'' standards are not recommended in this case since the source does not approximate a point source because of the relatively short source–monochromator distance.

Spectral Radiance: Field-of-View Baffle Attachment

A field-of-view (FOV) baffle attachment, as shown in Fig. 8.12, is a mechanical device that limits the acceptance angle of the detector or monochromator. The FOV baffle attachment must therefore have a larger f-number than the monochromator.

Such baffles enable the spectroradiometer to measure the spectral radiance of large area, uniformly radiating sources. The source must overfill the FOV of the attachment. The distance from the entrance slit of the monochromator to the source is not critical when making measurements over the visible spectrum as long as the

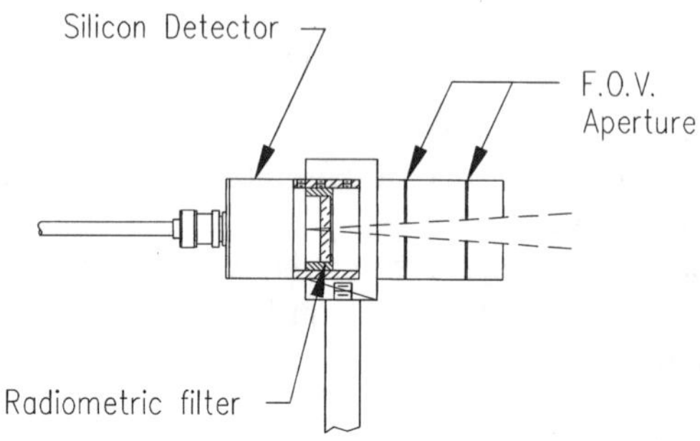

FIGURE 8.12. A field-of-view baffle attachment.

source maintains this overfill condition. An appropriate integrating-sphere calibration standard should be used to calibrate the system for spectral radiance response.

Spectral Radiance, Radiant Intensity, and Irradiance: Telescope Input Optics

Telescope input optics are used when measuring sources at large distances from the measurement system. In terms of nomenclature, telescope input optics convert the spectroradiometer into a telespectroradiometer. A telescope enables the system to measure spectral radiant intensity, spectral radiance, and spectral irradiance.

When measuring spectral radiant intensity or spectral irradiance with a telespectroradiometer, the source to be measured should underfill the field of view of the telescope. On the other hand, when measuring spectral radiance, the source should overfill the field of view of the telescope. In general, it is not essential to know the distance from the telespectroradiometer to the source when making spectral radiance measurements. However, the distance from the calibration source to the telespectroradiometer must be accurately known when calibrating the system for spectral irradiance or radiant intensity response.

Calibration of a telespectroradiometer for spectral irradiance response can be accomplished by positioning a 1000-, 200-, or 45-W lamp standard at a specified distance from the measurement system. The distance is dependent on the optical characteristics of the telescope. The spectral irradiance standards can also be used to calibrate the telespectroradiometer for spectral radiant intensity response as the spectral irradiance of the standards can be converted to spectral radiant intensity simply by multiplying the irradiance by the distance squared.

Calibration of a telespectroradiometer for spectral radiance response can be accomplished using an integrating-sphere calibration standard. However, as stated above, the radiating area of the calibration standard must overfill the field of view of the telescope.

Spectral Radiance: Microscope Input Optics

Mounting a microscope to the entrance port of the monochromator converts the system into a microspectroradiometer. The system can then measure the spectral radiance of small radiating sources. It is essential that the microscope have an accurate viewing system in order to image the source on the entrance aperture of the monochromator properly. In general, when using microscope input optics, the entrance slit of the monochromator is replaced with a small, circular, entrance aperture.

Input microscopes can be obtained with various optics and objective lenses. Calibration of a microspectroradiometer for spectral radiance response is easily accomplished using an integrating-sphere calibration standard, but may be difficult with other sources due to uniformity and working distance considerations.

Spectral Irradiance and Radiance: Fiber-Optic Probes

Fiber-optic probes can also be coupled directly to the monochromator. They are particularly useful when positioning or aligning the measuring device with respect to the source is difficult. A probe can be used without additional input optics or in combination with input optics such as an integrating sphere (for spectral irradiance, or spectral radiant flux measurements), a telescope (for spectral radiance, spectral irradiance, or spectral radiant intensity measurements), or a microscope (for spectral radiance measurements). For most applications involving a microscope, it is generally more convenient to couple the microscope to the monochromator using a fiber-optic probe as shown in Fig. 8.13.

In all cases where the input optics is coupled to the monochromator via a fiber,

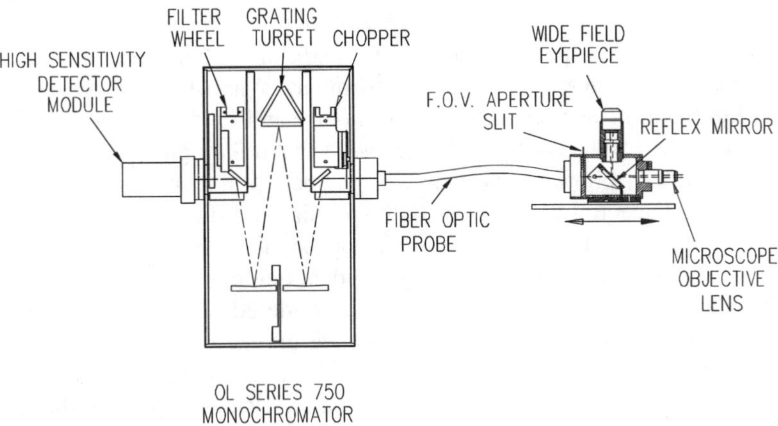

FIGURE 8.13. A spectroradiometer with fiber-optic probe and microscope input optics.

the appropriate calibration source is the same as that recommended above for the various input optic modules.

Spectral Radiance: Imaging Optics

Relatively simple imaging optics consisting of lenses or mirrors can be used for forming an image of the source on the entrance port of the monochromator. If the imaging optics does not have a viewing system, either the position of the source or the measuring device should be adjusted until a sharp image is visually observed at the entrance port of the monochromator. Imaging optics are used when measuring spectral radiance. Either the tungsten-ribbon-filament lamp standards of spectral radiance (described in Sec. 3.3) or the integrating-sphere calibration standards can be used to calibrate the measurement system for spectral radiance response.

8.5.2 Spectral Transmittance

Regular spectral transmittance measurements are relatively straightforward. The simplest setup for measuring regular spectral transmittance involves positioning a light source normal to the entrance slit of the monochromator and recording the detector signal with and without the object to be measured in the optical path. The ratio of the signals is the regular transmittance of the object at the wavelength setting of the monochromator.

An instrument dedicated primarily for spectral transmittance measurements is called a *spectrophotometer*. Spectrophotometers generally have all essential components contained in a single enclosure and quite frequently use a "double-beam" optical design. In a double-beam spectrophotometer, the light source is split into two fairly equal light paths and recombines at the photodetector. The sample to be measured is inserted in one path and the signal-detection system measures the ratio of the "test" signal to the "100%" signal. Although spectrophotometers capable of measuring spectral transmittances as low as 0.01% are readily available, the versatility of spectrophotometers are somewhat limited. Whereas most spectrophotometers are limited to measurements of transmittance and reflectance, a well-designed spectroradiometer can be configured to measure spectral output of light sources and spectral response of photodetectors as well as spectral transmittance (both regular and diffuse) and spectral reflectance (both specular and diffuse).

The spectral transmittance of an object is dependent on the nature of the radiating flux, i.e., the transmittance is effected by incident angle, polarization, temperature, etc. Also, the effect of fluorescence must be considered when measuring transmittance. For highest accuracy, it is desirable to measure the transmittance of an object in exactly the same manner in which it will be used. This can be more nearly achieved using a more flexible spectroradiometer. Using a spectroradiometer for transmittance measurements requires a light-source module and a transmittance attachment. The versatility of a spectroradiometer enables configuration of the system for broadband ("white-light") illumination of the sample or for monochro-

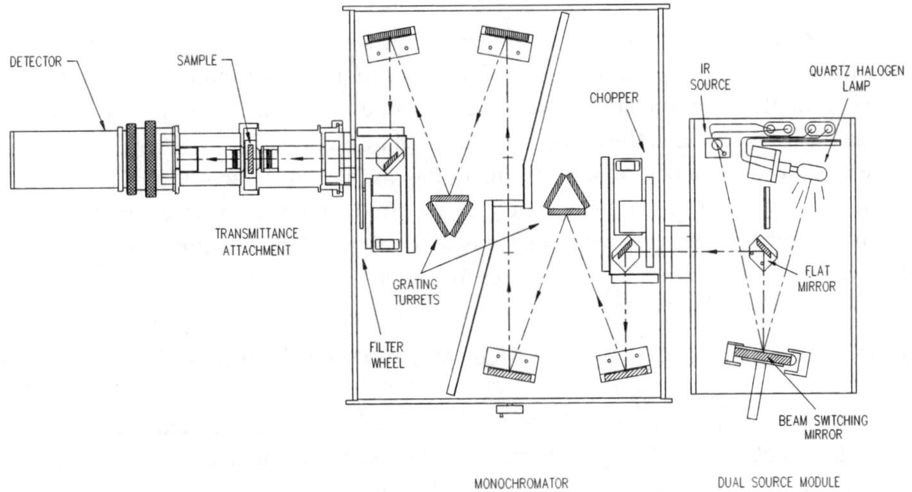

FIGURE 8.14. A spectroradiometer configured for measuring regular spectral transmittance.

matic illumination. Figure 8.14 shows a spectroradiometer configured for measuring regular spectral transmittance with monochromatic flux incident on the sample.

Diffuse transmittance requires an integrating-sphere attachment similar to that shown in Fig. 8.15 for collecting all of the transmitted radiant flux. This design works well with double-beam spectrophotometers. A modified version of this attachment can be used either at the input or output of a spectroradiometer. The transmittance measurements made with this attachment measures the total spectral transmittance (sum of regular and diffuse).

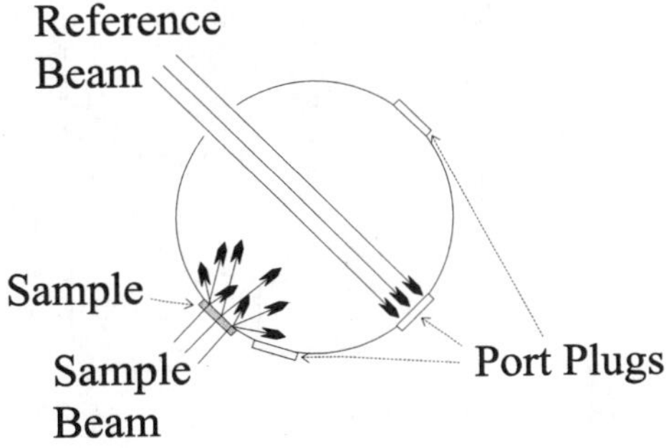

FIGURE 8.15. Components of an integrating-sphere attachment for diffuse transmittance measurements.

8.5.3 Spectral Reflectance

The spectral reflectance of an object is also dependent on the nature of the radiating flux, and whether or not the incident flux is broadband or monochromatic is of particular importance. Fluorescence is quite common when an object is irradiated with broadband flux. The effect of fluorescence is eliminated when the incident flux is monochromatic. In many cases, the reflectance of an object when irradiated by a particular light source is required. As with spectral transmittance measurements, spectroradiometers can be configured with source modules and reflectance attachments to suit the measurement requirement.

Specular Reflectance with Monochromatic Incident Flux

Figure 8.16 shows the optical layout of a spectroradiometric system configured to measure spectral specular reflectance as a function of the incident angle with monochromatic flux incident on the sample. In this case, a dual-source module is mounted at the input to a double grating monochromator. A variable-angle, all-mirror, specular reflectance attachment is mounted at the exit port of the monochromator. An important feature of this design is the ''self-calibration'' capability. Accurate measurements can be made without the use of an auxiliary standard (calibrated mirror). The flexible receiver design enables the detector to be positioned at the 0° incident angle position for a 100% reading. Specular reflectance measurements can then be made for various angles of incidence.

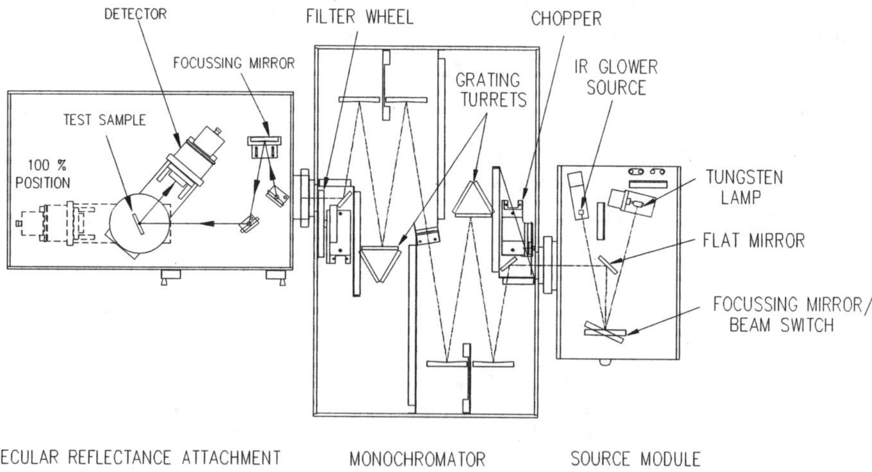

FIGURE 8.16. A spectroradiometer configured for measuring specular reflectance with monochromatic incident flux.

Specular Reflectance with Broadband Incident Flux

Measurements of spectral specular reflectance with broadband flux incident on the sample can be made simply by mounting the reflectance attachment between the source module and the monochromator. For these measurements, the detector is now mounted at the exit port of the monochromator and a fiber optic probe couples the reflectance attachment to the entrance port of the monochromator.

Diffuse Reflectance with Monochromatic Incident Flux

Figure 8.17 shows the optical layout of a spectroradiometric measurement system configured for measuring diffuse reflectance with monochromatic incident flux. The integrating sphere reflectance attachment uses an accurate, double-beam design with an automatic or manually controlled optical beam switching mirror. The double-beam design enables the user to use the significantly more accurate *comparison method* of measuring diffuse reflectance. This particular version incorporates a sample compartment directly in front of the entrance port of the integrating sphere that enables the user to measure the transmittance of regular or diffusely transmitting samples. A removable specular light trap enables the user to make diffuse reflectance measurements with or without the specular component included. The optical design also enables the user to measure specular reflectance at a fixed angle of incidence.

For highest accuracy, a calibrated, diffuse reflectance plaque should be used as the reference standard when making diffuse reflectance measurements over the

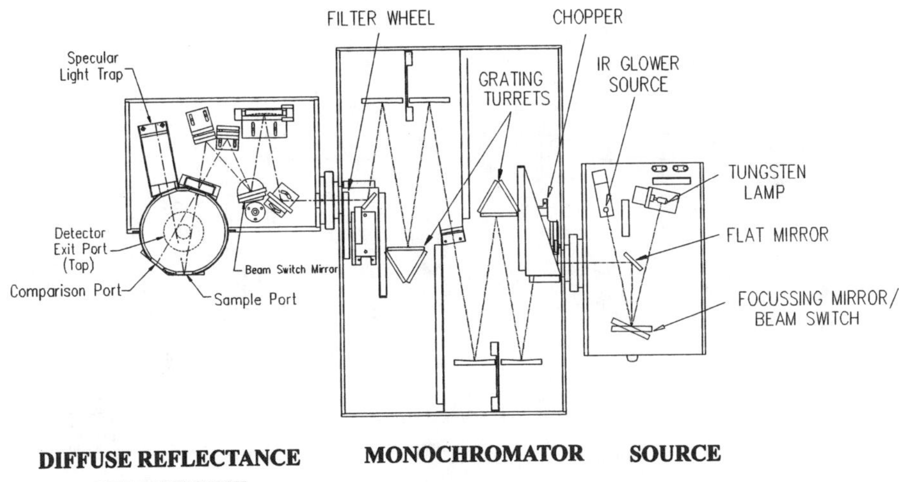

FIGURE 8.17. A spectroradiometer configured for measuring diffuse reflectance with monochromatic incident flux.

visible spectrum. A plaque coated with PTFE is generally considered superior to other coatings. PTFE has a diffusely reflecting surface with a reflectance above 99% over the visible spectrum.

Diffuse Reflectance with Broadband Incident Flux

Measurements of spectral diffuse reflectance with broadband flux incident on the sample can be made simply by mounting the integrating-sphere reflectance attachment between the source module and the monochromator. For these measurements, the detector is now mounted at the exit port of the monochromator and a fiber-optic probe couples the reflectance attachment to the entrance port of the monochromator.

8.5.4 Spectral Responsivity

Figure 8.18 shows the optical layout of a spectroradiometric system configured to measure the spectral response of photodetectors. In this case, the input optics consists of a source module mounted at the entrance port of a double grating monochromator and a reflective collimating exit optics module mounted at the exit port of the monochromator. The detector, either standard or test, is mounted to the exit optics module.

Measuring power or irradiance response of a photodetector is a two-step procedure. The first step involves positioning the standard detector in the collimated beam and measuring the monochromatic flux or irradiance. Monochromatic flux can be determined from

$$\Phi(\lambda)=s^{s}(\lambda)/R^{s}(\lambda),$$

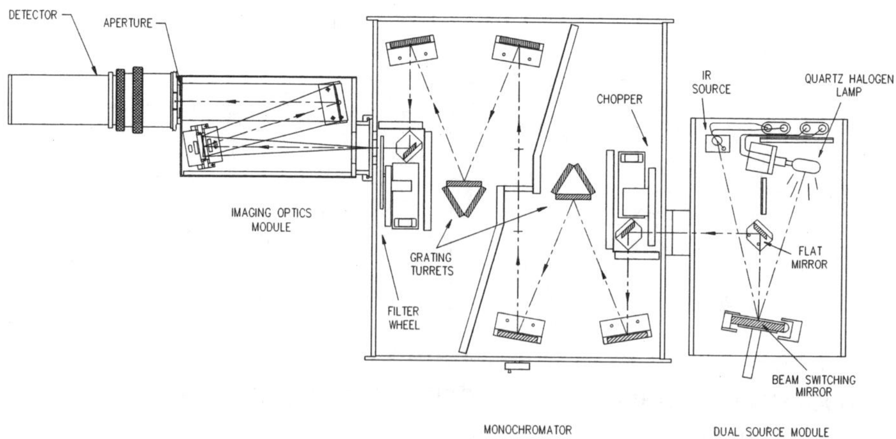

FIGURE 8.18. A spectroradimeter configured for measuring detector spectral response.

where $R^s(\lambda)$ is the spectral power response of standard detector, and $s^s(\lambda)$ is the standard detector signal at each wavelength. The standard detector is replaced by the test detector and the spectral power response of the test detector, $R^t(\lambda)$, is determined from

$$R' = \Phi(\lambda) s'(\lambda),$$

where $s^t(\lambda)$ is the test detector signal at a specified wavelength.

The irradiance response is measured in the same manner as the power response except that the entire detector (both standard and test) are uniformly irradiated and the irradiance response of the standard detector is used in the computation as opposed to the power response.

Measurements over the visible spectrum generally employ a NIST traceable standard silicon detector. Since most silicon detectors are quite uniform in sensitivity over the visible spectrum, and in many instances the area of the standard detector is exactly 1 cm^2, the power response (in A W^{-1}) and irradiance response (in A W^{-1} cm^2) are identical. Typical uncertainties in measuring the monochromatic flux or irradiance over the visible spectrum are on the order of 1%.

8.6 CALCULATING PHOTOMETRIC AND COLORIMETRIC PARAMETERS

8.6.1 Tristimulus Values

Chromaticity values may be calculated for sources (data file in radiometric units) or objects (data file in fractional transmittance or reflectance). For objects, values represent the color as seen under a standard illuminant (A, B, C, or D65).

Sources

The X, Y, and Z tristimulus values for sources are calculated using the CIE 1931 spectral tristimulus values, $\bar{x}(\lambda)$, $\bar{y}(\lambda)$, and $\bar{z}(\lambda)$ as follows:

$$X = \sum_{\lambda=380\text{ nm}}^{780\text{ nm}} \bar{E}(\lambda)\bar{x}(\lambda)\Delta\lambda,$$

$$Y = \sum_{\lambda=380\text{ nm}}^{780\text{ nm}} \bar{E}(\lambda)\bar{y}(\lambda)\Delta\lambda,$$

$$Z = \sum_{\lambda=380\text{ nm}}^{780\text{ nm}} \bar{E}(\lambda)\bar{z}(\lambda)\Delta\lambda,$$

where $E(\lambda)$ are spectral values of the data file, $\bar{x}(\lambda)$, $\bar{y}(\lambda)$, and $\bar{z}(\lambda)$ are the CIE 1931 spectral tristimulus values, and λ is the wavelength interval of the data file (nm).

Objects

The X, Y, and Z tristimulus values for objects are calculated using the CIE 1931 spectral tristimulus values, $\bar{x}(\lambda)$, $\bar{y}(\lambda)$, and $\bar{z}(\lambda)$ as follows:

$$X=k\sum_{\lambda=380\text{ nm}}^{780\text{ nm}}\bar{E}(\lambda)\Gamma(\lambda)\bar{x}(\lambda)\Delta\lambda,$$

$$Y=k\sum_{\lambda=380\text{ nm}}^{780\text{ nm}}\bar{E}(\lambda)\Gamma(\lambda)\bar{y}(\lambda)\Delta\lambda,$$

$$Z=k\sum_{\lambda=380\text{ nm}}^{780\text{ nm}}\bar{E}(\lambda)\Gamma(\lambda)\bar{z}(\lambda)\Delta\lambda,$$

$$k=\frac{100}{\Sigma\bar{E}(\lambda)\bar{y}(\lambda)\Delta\lambda},$$

where $\bar{E}(\lambda)$ is the relative spectral power of an illuminant, $\Gamma(\lambda)$ is the spectral reflectance or transmittance data, $\bar{x}(\lambda)$, $\bar{y}(\lambda)$, and $\bar{z}(\lambda)$ are CIE 1931 spectral tristimulus values, and $\Delta\lambda$ is the wavelength interval of the data file (in nm).

The $\bar{x}(\lambda)$, $y(\lambda)$, and $\bar{z}(\lambda)$ values used in the equations above refer to a 2° field-of-view observer. Equivalent X_{10}, Y_{10}, and Z_{10} values for a 10° field of view can be similarly calculated using the CIE 1964 supplementary spectral tristimulus values.

8.6.2 Photometric Output Calculations

For sources, the photometric output (in this case illuminance, E_v) is calculated by

$$E_v(\text{lm cm}^{-2})=Y[W\text{ cm}^{-2}](683\text{ lm W}^{-1}).$$

For objects, the value of Y is expressed in percent and is the photometric transmittance or reflectance.

8.6.3 CIE 1931 Chromaticity Calculations

The x, y, and z chromaticity coordinates of the data file are calculated from the tristimulus values X, Y, and Z as follows:

$$x=\frac{X}{X+Y+Z},$$

$$y=\frac{Y}{X+Y+Z},$$

$$z=\frac{Z}{X+Y+Z}.$$

8.6.4 UCS 1976 *u*, *v*, *u*′, and *v*′ Coordinates Calculations

The UCS 1960 (u,v) coordinates are calculated:

$$v=\frac{6y}{12y-12x+3}=\frac{2}{3}v',$$

$$u=\frac{4x}{12y-2x+3}=u'.$$

The UCS 1976 u' and v' coordinates are calculated:

$$u'=\frac{4x}{12y-2x+3}=u,$$

$$v'=\frac{9y}{12y-2x+3}=\frac{3}{2}v.$$

8.6.5 Correlated Color-Temperature Calculations

Correlated-color-temperature calculations are based on Robertson's method [29] using a table of 30 isotemperature lines. Robertson's successive approximation method should be accurate to calculate correlated color temperatures to within 0.1 μrd [the micro-reciprocal-degree (μrd)$=10^6/T$, where T is the temperature in kelvins). The maximum error from 1600 to 3000 K should be less than 0.2 K plus the measurement uncertainty. Robertson states that the errors may be larger for sources with chromaticities farther than 0.01 from the Planckian locus. However, the concept of correlated color temperature has little meaning outside the immediate vicinity of the Planckian locus.

8.6.6 CIE LAB/LUV Color Space Calculations

CIE LAB Color Space Calculations are performed as per the recommended 1976 CIE formulas. For source computations, X, Y, and Z are normalized equally such that $Y=100$. The equations are as follows:

$$L^*=116\left(\frac{Y}{Y_n}\right)^{1/3}-16,$$

$$a^* = 500\left[\left(\frac{X}{X_n}\right)^{1/3} - \left(\frac{Y}{Y_n}\right)^{1/3}\right],$$

$$b^* = 200\left[\left(\frac{Y}{Y_n}\right)^{1/3} - \left(\frac{Z}{Z_n}\right)^{1/3}\right].$$

Here X_n, Y_n, and Z_n are tristimulus values of the reference white: The equations above are modified slightly when X/X_n, Y/Y_n or Z/Z_n is less than 0.01. The modified equations are shown below:

$$L^* = 116\left[f\left(\frac{Y}{Y_n}\right) - \left(\frac{16}{116}\right)\right]$$

$$a^* = 500\left[f\left(\frac{X}{X_n}\right)^{1/3} - f\left(\frac{Y}{Y_n}\right)^{1/3}\right],$$

$$b^* = 200\left[f\left(\frac{Y}{Y_n}\right)^{1/3} - f\left(\frac{Z}{Z_n}\right)^{1/3}\right],$$

where $f(Y/Y_n) = (Y/Y_n)^{1/3}$ for Y/Y_n greater than 0.008 856 and $f(Y/Y_n) = 7.787(Y/Y_n) + 16/116$ for Y/Y_n less than or equal to 0.008856; $f(X/X_n)$ and $f(Z/Z_n)$ are similarly defined:

$$u^* = 13L^*(u' - u'_n),$$

$$v^* = 13L^*(v' - v'_n).$$

For sources, values represent a comparison to a standard illuminant for an ideal white object. For objects, values represent a comparison to an ideal white object under a given standard illuminant.

8.6.7 Color-Difference and Color-Rendering Calculations

Detailed calculations of color differences and color-rendering indices are beyond the scope of this section, but are reviewed elsewhere [30]. However, these are routinely used commercially as indicators of lamp performance and a brief discussion of their significance is in order.

Two sources may have the same chromaticities or correlated color temperature and yet be spectrally different. When a transmitting or reflecting sample is viewed under these sources, it is therefore possible that different chromaticities result for each source. Differences in chromaticities are normally calculated in some uniform color space (e.g., LAB, LUV, or CIE 1964 WUV). The agreement (expressed in percent) between any source and a reference illuminant is termed *color rendering.*

The CIE has specified the standard method for assessing the color-rendering properties of sources [31–33]. The method consists of a series of "special color-rendering indices" and an average "general color-rendering index." Each of the special color-rendering indices represents CIE 1964 color differences between the chromaticities for 14 specified samples under the test source and a specified reference illuminant. The first eight of these specific color-rendering indices are averaged to give the general color-rendering index. The closer the general color-rendering index is to 100%, the more the test source is thought to resemble the reference illuminant.

8.6.8 Detector Photometric Parameters

Several photometric parameters, specified by CIE, may be used to reflect the performance of a photopic detector. Essentially these values represent to agreement between the response of a detector and the CIE standard observer. By far the most frequently quoted, and hence measured, of these parameters is the f_1' value. This parameter represents an overall ''goodness of fit'' rather than an accuracy at a specific wavelength, and is given by

$$f_1' = \frac{\int |\bar{R}(\lambda) - V(\lambda)|}{\int V(\lambda)} \times 100 \ (\%),$$

where $\bar{R}(\lambda)$ is the detector response and $V(\lambda)$ is the CIE standard observer response.

This parameter is expressed in percent, with lower values indicating a better fit to the standard observer. As this value approaches zero, the errors introduced by calibrating with a lamp of one spectral distribution and measuring a lamp of another spectral distribution rapidly diminish. Opinions differ on the range of values that are acceptable for photopic measurements, largely because it depends so much on the source, but most experienced users agree that if the calibration and test source differ considerably, then a fit to within 2% is required. This is generally only achievable by measuring the unfiltered detector response and ''tailoring'' the filter to match. For less demanding applications, values of less than 5% are generally acceptable.

8.6.9 Limits in Using Photometric and Colorimetric Calculations

When calculating correlated color temperature or dependent parameters such as the color-rendering index, it should be borne in mind that these assume a blackbody or similar spectral distribution. Many sources in common use, e.g., fluorescent lamps or LEDs, have spectral distributions very different from that of a blackbody. Although results may be obtained, and indeed the lighting industry often uses these parameters to characterize their lamps, results should be interpreted with caution.

8.7 ACCURACY AND ERRORS

8.7.1 Random, Systematic, and Periodic Errors

These three types of error are the fundamental components limiting the accuracy of any measurement. Each is, at least theoretically, distinct and separable but in practical situations a huge amount of work is required to isolate them from each other.

Figure 8.19 illustrates these basic types of error:

Random errors are variations about the mean that, if sampled sufficiently, would form a Gaussian or other similar statistical function. These errors decrease with increased sampling, either by longer integration times or by multiple scans.

Systematic errors are offsets to the "true" value. These are, by definition, constant and do not change by multiple sampling. Often these errors are due to basic assumptions not being realized in practice.

Periodic errors are those that arise from periodic or pseudoperiodic events. Such examples include variations due to air conditioning changing the local temperature or AC interference from sources or line voltages.

The statistical distributions of these different types of error are given in Fig. 8.20.

When specifying the accuracies of a measurement, these individual errors must be estimated and combined. Most standards laboratories quote errors to be 2 or 3 standard deviations (approximately 95% or 99% confidence levels, respectively),

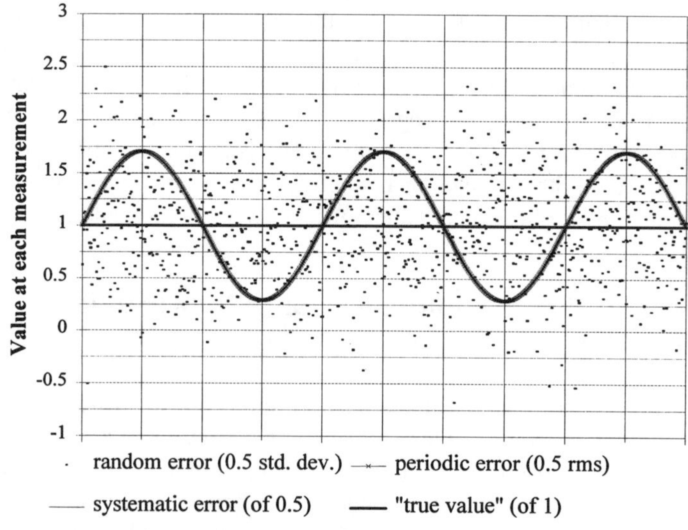

FIGURE 8.19. Types of error in measurements.

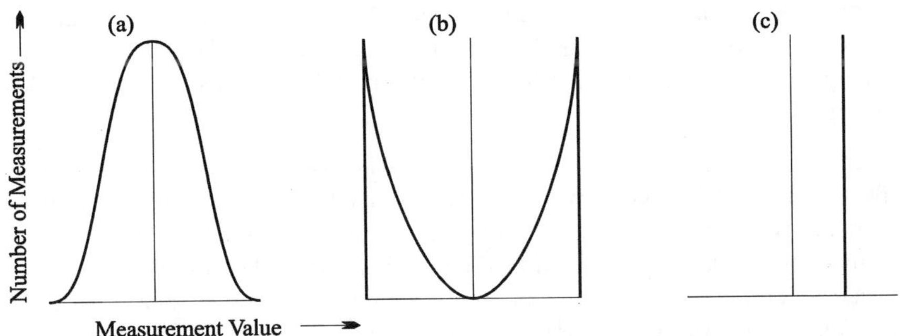

FIGURE 8.20. Statistical distributions about a central "true" value of (a) random errors, (b) simple periodic errors, and (c) systematic errors.

but since only one type of error—random—has any significance expressed in this way, the interpretation of these values may be difficult for the user.

8.7.2 Error-Source Photometry and Spectroradiometry

To illustrate the effects of these errors on an actual measurement, consider the use of a photometer in the determination of the illuminance from a light source at some specific distance. In this measurement we would expect the following.

- Random noise from:
  the detector;
  electronics;
  the light source.
- Systematic errors from:
  the measurement of the distance;
  any error in the calibration factor of the photometer, including errors associated with the calibration light source;
  noncosine collection of light;
  differences between the detector relative spectral response and the $V(\lambda)$ function;
  differences between the calibration source spectral distribution and the test source spectral distribution;
  stray light from reflections off walls and objects and emission from other sources, e.g., computer screens, in the laboratory;
  nonlinearity of the detector–amplifier combination with intensity;
  errors associated with setting the true dark level of the system (i.e., the average of the dark noise).
- Periodic errors from:
  temperature, humidity, and air-movement variations affecting the response and spectral characteristics of the photometer or the intensity of the light source;

drift of zero levels or gain in the detector or amplifier;
stray-light variations due to movement of objects and personnel, changing computer screen displays, and daylight ''leaks'' into the laboratory;
sampling errors and ''beating'' when measuring AC sources or if AC ''pickup'' is significant.

From the list above, it should be apparent that just because close agreement of results are obtained between several measurements, this does not imply that they are accurate. Confidence in the accuracy of results is only achieved when all sources of error have been quantified, minimized, and, where possible, eliminated.

Spectroradiometric measurements exhibit many of the above sources of error to some extent. However, the accuracy of spectroradiometric results is generally much better since:

- Spectra contain many data points. The calculation of the integrals as described in Sec. 8.6 effectively reduces the contribution of errors associated with each data point in proportion to the $V(\lambda)$ function and integrates noise in much the same way as multiple sampling does.
- The systematic error of matching the photometer response to the $V(\lambda)$ function can often represent the greatest limit on the accuracy of broadband measurements. For spectroradiometric systems, the actual $V(\lambda)$ function is used in calculations, so it may be considered to represent an ''ideal'' photometer.
- Differences between the spectral distributions of the calibration and test source are not relevant since these are determined.
- Autoranging can occur at each wavelength of measurement, rather than for the entire measurement as a whole, giving a better representation of the source, minimizing errors and extending the dynamic range.

REFERENCES

1. Grum, F., and Becherer R. J., *Optical Radiation Measurements* (Academic, New York, 1979), Vol. 1, Chap. 2, pp. 12–13.
2. Schneider, W. E., Lasers and Optronics **12** (1), p. 29–32 (1993).
3. Schneider, W. E., and Miller, K. A., *The Photonics Design and Application Handbook*, 41st ed. (Laurin, Pittsfield, MA, 1995), pp. 36–44.
4. Eckerle, K. L., Hsia, J. J., Mielenz, K. D., and Weidner, V. R., *Regular Spectral Transmittance*, NBS Special Publication No. 250-6 (1987).
5. Hecht, E., and Zajac, A., *Optics* (Addison-Wesley, Reading, MA, 1974), Chap. 4, pp. 62–63.
6. Boyd, R. W., *Radiometry and the Detection of Optical Radiation* (Wiley, New York, 1983), Chap. 6, pp. 95–97.
7. Smith, W. J., *Modern Optical Engineering*, 2nd ed. (McGraw-Hill, New York, 1990), Chap. 8, p. 217.
8. Kirchoff, G., Philos. Mag. J. Sci. 4th Ser. **20**, No. 130 (1860).
9. Wolfe, W. L., and Zissis, C. J., *The Infrared Handbook*, revised ed. (Office of Naval Research, Dept. Of The Navy, Washington D.C., 1989), Chap. 1, p. 29.
10. DeVos, J. C., Physica **20**, p. 669 (1954).
11. Gouffé, A., Rev. d'Opt. **24**, p. 1 (1945).
12. Stair, R., Johnston, R. G., and Halbach, E. W., NBS J. Res. **64A**, p. 291–296 (1960).
13. Kostkowski, H. J., Erminy, D. E., and Hattenburg, A. T., Adv. Geophys. **14**, p. 111–127 (1970).
14. Walker, J. H., Saunders, R. D., and Hattenburg, A. T., *Spectral Radiance Calibrations*, NBS Special Publication No. 250-1 (1987).

15. Mielenz, K. D., Saunders, R. D., Parr, A. C., and Hsia, J. J., J. Res. Nat. Inst. Stand. Technol. **95**, p. 621 (1990).
16. Stair, R., Schneider, W. E., and Jackson, J. K., Appl. Opt. **2**, p. 1151–1154 (1963).
17. Saunders, R. D., and Shumaker, J. B., *The 1973 NBS Scale of Spectral Irradiance*, NBS Technology Note No. 594-13 (1977).
18. Walker, J. H., Saunders, R. D., Jackson, J. K., and McSparren, D. A., *Spectral Irradiance Calibrations*, NBS Special Publication No. 250-20 (1987).
19. Schneider, W. E., and Goebel, D. G., Int. J. Opt. Eng. Proc. **262**, p. 74–83 (1981).
20. Schneider, W. E., and Goebel, D. G., Laser Focus/Electro-Opt. **20**, No. 9, p. 82–96 (1984).
21. Schneider, W. E., ''Spectral Ultraviolet Measurements and Utilization of Standards,'' presented at the Critical Issues in Air Ultraviolet Metrology Workshop at NIST, Gaithersburg, MD, May, 1994.
22. Zalewski, E. F., *The NBS Photodetector Spectral Response Calibrations Transfer Program*, NBS Special Publication No. 250-17 (1988).
23. Eppeldauer, G., and Hardis, J. E., Appl. Opt. **30**, p. 3091–3099 (1991).
24. Zalewski, E. F., and Duda, C. R., Appl. Opt. **22**, p. 2867–2873 (1983).
25. Cromer, C. L., ''A New Spectral Response Calibration Method using a Silicon Photodiode Trap Detector,'' J. Res. Nat. Inst. Stand. Technol. (to be published).
26. Schneider, W. E., Test Measure. World **6**, p. 159–169 (1985).
27. Young, R., and Schneider, W. E., Laser Focus World **31**, No. 5, p. 215–220 (1995).
28. Zappa, F., Lacaita, A. L., Cova, S. D., and Lovati, P., Opt. Eng. **35**, No. 4, p. 938–945 (1996).
29. Robertson, A. R., J. Opt. Soc. Am. **58**, p. 1528 (1968).
30. Grum, F., and Bartleson, C. J., *Optical Radiation Measurements* Volume 2 (Academic, New York, 1980), Vol. 2, Chap. 3, pp. 134–142.
31. *Method of Measuring and Specifying Colour Rendering of Sources*, CIE Publication No. 13.2 (1974).
32. *Method of Measuring and Specifying Colour Rendering of Sources*, 2nd ed. (corrected), CIE Publication No. 13.2 (1988).
33. *Method of Measuring and Specifying Colour Rendering Properties of Light Sources*, CIE Publication No. 13.3 (1995).

9

Retroreflection

Justin J. Rennilson

Advanced Retro Technology, a part of Gamma Scientific, Inc., San Diego, California 92123

9.1 INTRODUCTION

9.1.1 Examples

Natural

The phenomena of retroreflection is easily observed from an aircraft flying over the desert terrain. One needs to position the aircraft such that the shadow of the aircraft is discerned on the ground. If the brightness of the ground in the immediate vicinity of the shadow is carefully observed it can be seen that the ground is brighter closer to the shadow than farther away. If the aircraft flies at a height sufficiently far where the sun's rays from each edge of the solar disk join after passing the edge of the aircraft (Fig. 9.1), the ground below will reach a peak brightness. The rays of the sun are now reflecting directly back to the observer. We call this directional reflection *retro*—from the Latin, meaning returned to the source.

All natural surfaces exhibit this *retroreflection*, even those that look on the surface to be uniform and devoid of texture. For example, the surface of barium sulfate, an often used reference standard for reflectance, will increase its absolute reflectance over 100% (Fig. 9.2) when the illumination and viewing geometry is less than a few degrees [1]. This is the reason why such reference standards are not suitable as standards for the measurement of retroreflection.

One of the most obvious natural retroreflectors is our nearest neighbor, the moon. For many centuries astronomers noted that the moon greatly increased in brightness when it was full [2] (Fig. 9.3). Here the angle (known as the phase angle in astronomy and is equivalent to the observation angle) subtended at the moon between the sun and the earth was at a minimum. The edge of the moon also appeared sharp and of almost equal brightness, excluding the varitions due to the ''maria'' and ''highlands'' areas. If the moon were a uniform perfect diffuser, the edges would fall off in brightness to zero at the very edge. Figure 9.4 is a photograph from one of the Apollo 14 astronauts looking downsun. The halo appearance is very obvious. Since the advent of the Apollo lunar landings and the collection of the lunar soil, another contributor to the peak retroreflection was the presence of very small glass spheres or beads [3] (Fig. 9.5). These beads resulted from the tremendous transfer of energy from impacting bodies, generating temperatures high enough to cause the rock and soil to become molten. In their trajectory from the force of impact in an airless atmosphere, they became spherical. The glass sphere thus became the original optical element enhancing the phenomena of "retroreflection." We will examine how this occurs later in this chapter.

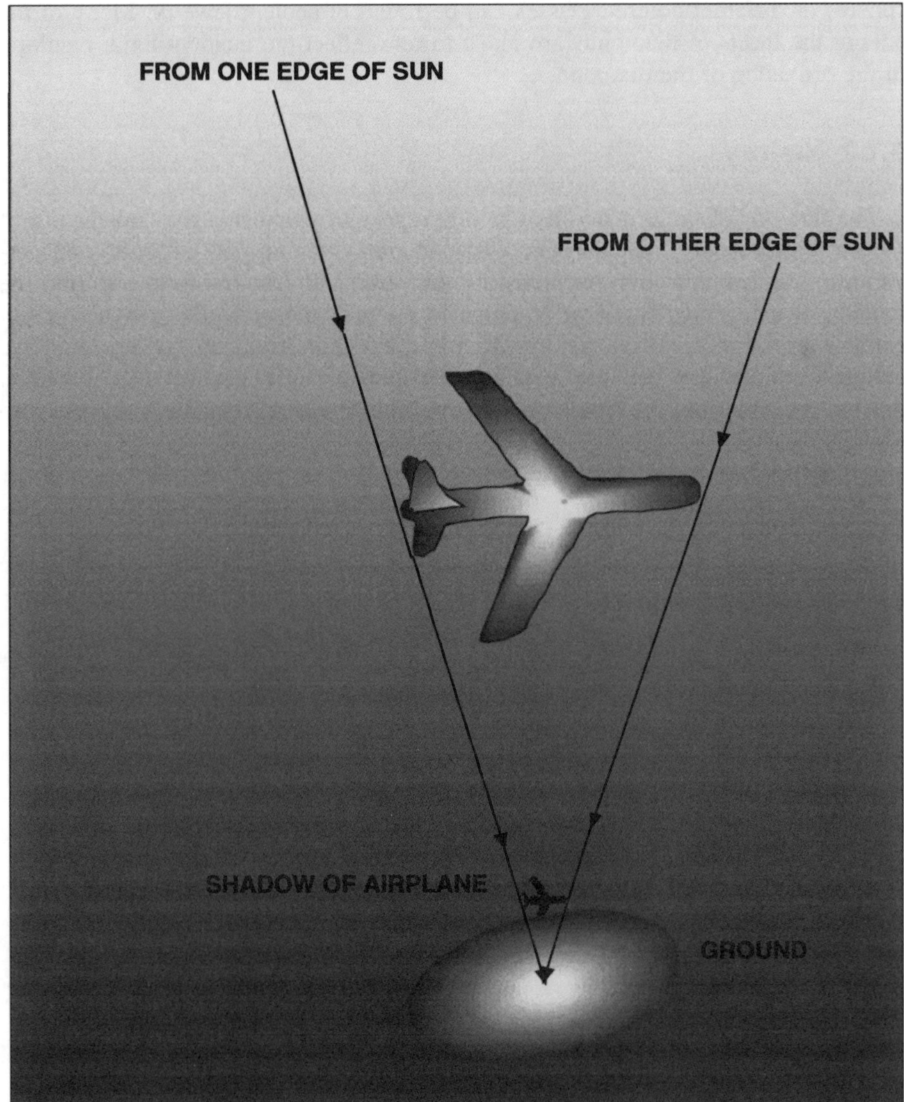

NOTE: ANGLES ARE EXAGGERATED FOR EFFECT

FIGURE 9.1. Illustration of the retroreflection of sunlight from the ground terrain as seen from a low-flying aircraft from a height where the plane's shadow does not touch the ground.

Man-Made

The earliest recorded formation of glass into spherical shapes dates back to early Chinese history. Observation of light reflected back toward the light source from these spheres was used to heighten the brightness or brilliance of art objects. Glass spheres of different colored glasses can be found in ancient jewelry. Many of the cuts of the facets of diamonds are made to retroreflect the incident light, resulting in the brilliance of the diamond.

9.1.2 History

The first use of these materials for the purpose of enhancing the conspicuity of objects was probably created in Great Britain around the turn of the century. Observation of light reflected from animals eyes, especially those of cats, started the thought of using this technique to enhance the brightness of objects. The current explanation of the glowing nature of cat's eyes stems from the highly reflective nature of their retinas. Because these animals were primarily nocturnal and relied on making the most use of the available low light levels, the retina is coated with

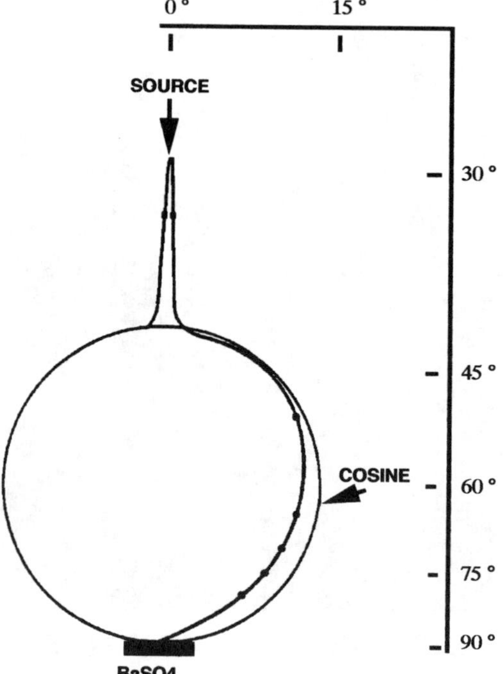

FIGURE 9.2. The increased reflectance greater than 100% of barium sulfate at very small angles of incidence.

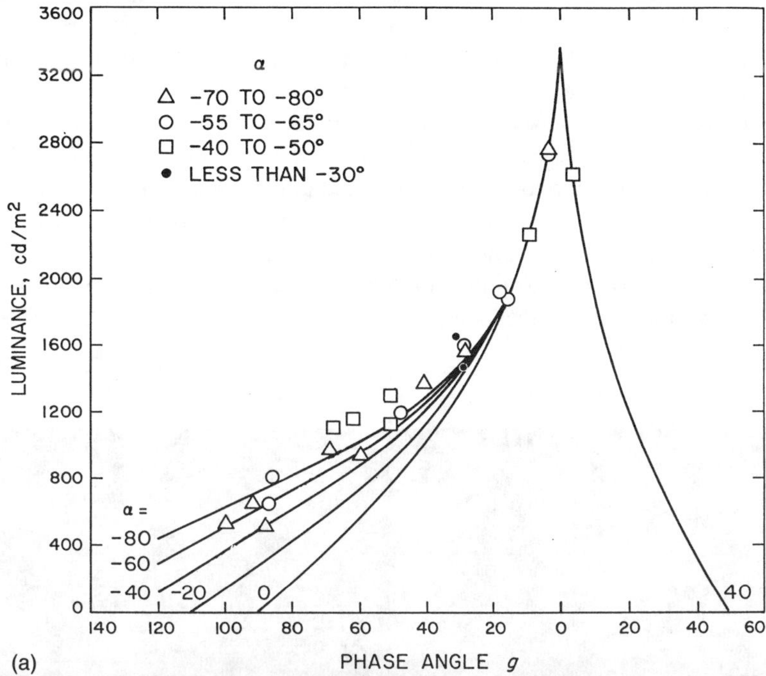

(a)

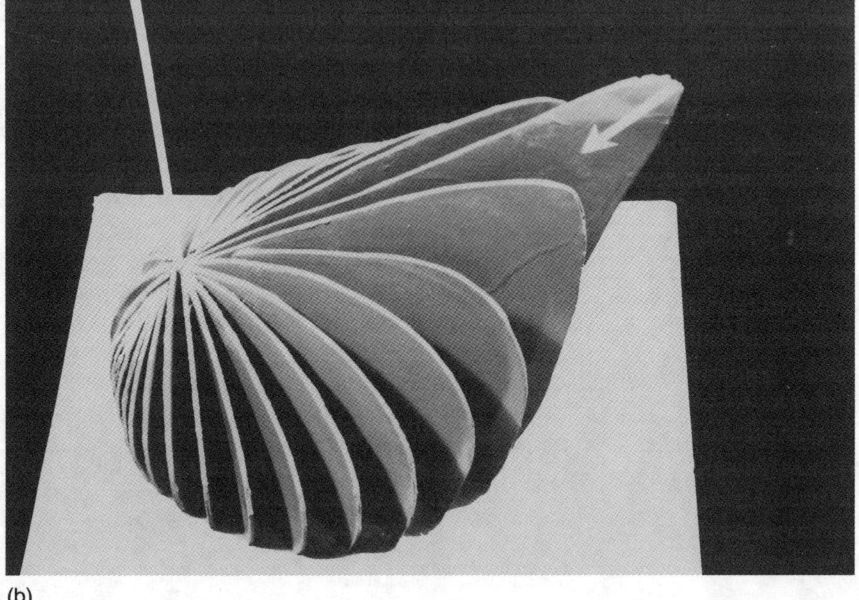

(b)

FIGURE 9.3. The luminance of the moon as a function of it's phase (a) and the BRDF (b) at a solar angle of 45°.

material which reflects over 80% of the incident light, and traps this light in the eye to increase the detectivity of objects. An often told story of the first application of glass spheres toward the safety of drivers in the area of transportation illustrates how circumstances dictate a direction. A driver of a horse-drawn carriage was returning to his house one very dark night and fell asleep. As the carriage was about to leave the road and plunge over a cliff, a startled cat appeared, spooked the horse, and awoke the driver in time to see the retroreflection in the cat's eyes from the dim light of the carriage lantern, thereby avoiding an accident. This shock awakened an idea in this man and his investigations of glass spheres to aid in warning such drivers with the first use of retroreflection for traffic control. Thus the popular name of ''cat's eyes'' for such devices was born.

FIGURE 9.4. Photograph of Apollo 14 astronaut looking downsun, illustrating the retroreflective nature of the lunar surface.

9.1.3 Uses

The most important use of retroreflectors is in enhancing the light return from vehicle headlights to enable the drivers to detect warnings, impart information and aid the driver through the necessary maneuvers required. Other uses include the retroreflection of laser light for distance measurement and location. An array of retroreflectors was left on the lunar surface by the crew of Apollo 11 to enable the distance from the earth to the moon to be determined with an accuracy of a few centimeters [4]. These retroreflectors are of a precision which is not suitable for transportation applications because of the geometry involved. The tetrahedral prisms employed in the above experiments have faces mutually perpendicular to a few arc seconds, as the spread of the returned light over the many kilometers is still significant. For traffic situations where distances are in meters a completely different retroreflector must be used.

9.2 TERMINOLOGY

The International Commission on Illumination (CIE), the international standardizing body responsible for light and color measurement, defines retroreflection as

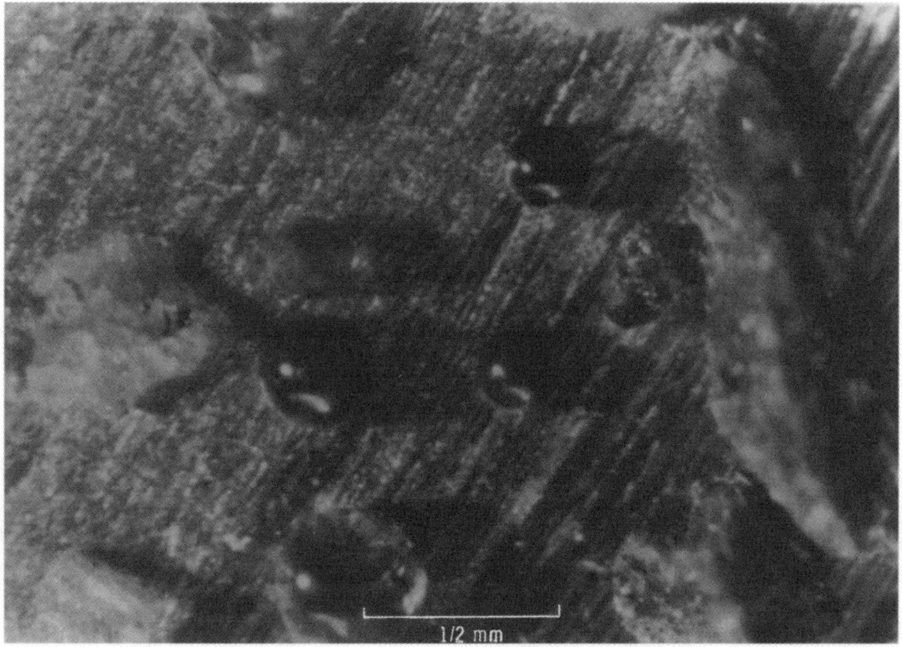

FIGURE 9.5. NASA figure (5-7) of glass spherules of various colors found in the fine-grained material. The lunar fines contain many glass beads which increase the retroreflectance of the lunar surface. (From Ref. [3].)

follows: "Reflection in which the reflected rays are preferentially returned in directions close to the opposite of the direction of the incident rays, this property being maintained over wide variations of the direction of the incident rays" [5]. Figure 9.6 is a diagram of the CIE angular system for specifying and measuring retroreflection.

9.2.1 Geometric Definitions

Illumination Axis

This is a line segment from the effective center of the light source aperture to the retroreflector center.

Observation Axis

This axis is a line segment from the effective center of the receiver aperture to the retroreflector center.

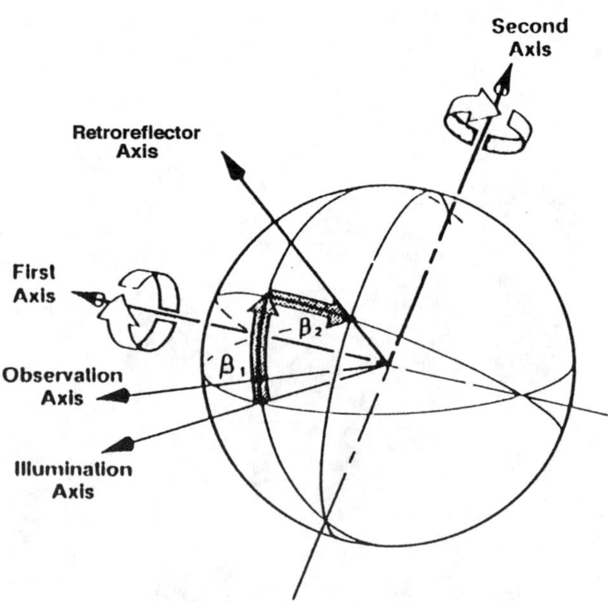

FIGURE 9.6. CIE angular system for specifying and measuring retroreflectors. The first axis is perpendicular to the plane containing the observation and illumination axes. The second axis is perpendicular both to the first and retroreflector axes. All axes, angles, and direction of rotation are shown positive.

Observation Angle (α)

This angle is between the illumination axis and the observation axis. The observation angle is always positive and in the context of retroreflection restricted to small acute angles.

Observation Half-Plane

This is the half-plane which originates on the illumination axis and which contains the observation axis.

Retroreflector Axis

This designated line segment originates on the retroreflector center which is used to describe the angular position of the retroreflector. For a flat retroreflector, the retroreflector axis is normally perpendicular to the flat surface of the retroreflector.

Entrance Angle (Illumination Angle) (β)

This is the angle from the illumination axis to the retroreflector axis. The angle is usually not larger than 90°. In order to specify the orientation in full, this angle is characterized by two components: β_1 and β_2.

Axes of Importance

First Axis

This axis is through the retroreflector center and perpendicular to the observation half-plane.

Second Axis

This axis is through the retroreflector center and perpendicular to both the first axis and the retroreflector axis.

Components of the Entrance Angle

The β_1 component is the angle from the illumination axis to the plane containing the retroreflector axis and the first axis. The β_2 component is the angle from the plane containing the observation half-plane to the retroreflector axis.

When testing samples of retroreflective material or devices other angles and axes are often specified. Figure 9.7 illustrates these angles.

Datum Mark and Axis

A mark on the retroreflector is used to indicate the orientation of the retroreflector with respect to rotation about the retroreflector axis. The datum axis is the axis that passes through the retroreflector center and the datum mark. The datum mark must not lie on the retroreflector axis.

Rotation Angle (ϵ)

This is the angle from the datum axis to the second axis.

Presentation Angle (γ)

This is the angle from the observation half-plane and the half-plane containing the illumination and retroreflector axes.

Orientation Angle (ω)

The angle from the half-plane contains the retroreflector and datum axes to the half-plane contain the illumination and the retroreflector axes.

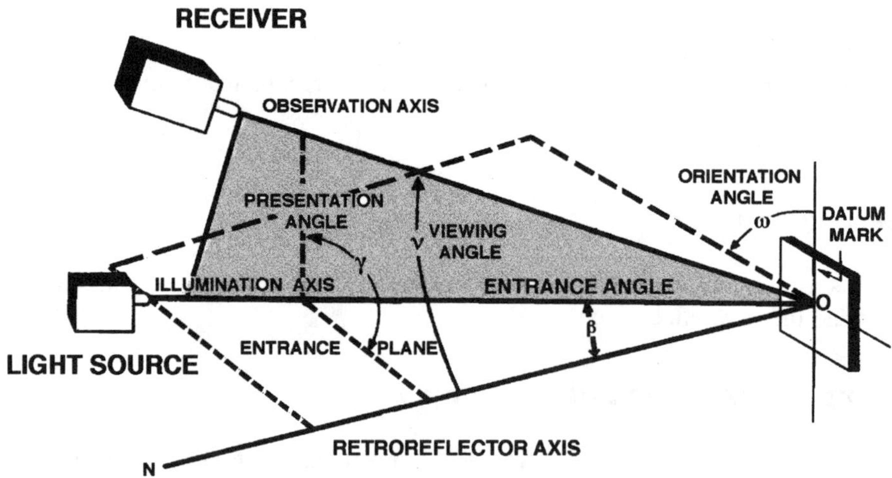

FIGURE 9.7. Other angles illustrated in a different configuration of CIE geometry.

Viewing Angle (ν)

The angle between the retroreflector axis and the observation axis is the viewing angle.

9.3 UNITS OF MEASUREMENT

The photometric properties of retroreflectors are dependent on the luminous intensity of the illuminating source of light. This is similar to the luminance of standard reflecting materials that depend on the amount of incident luminous flux. The efficiency of such materials is measured as a ratio with respect to a perfectly uniform diffuser. Thus if such a diffuser is used as a retroreflector its efficiency is equal to $\rho/\pi=0.328$. Since history has used retroreflectors at night as markers for traffic control, the device appears as a source of light and thus the first unit of measurement was based on the same units as used in lamp intensity, i.e., *candelas*. The absolute photometric amount depends on the amount of illumination incident on the retroreflector; to increase the amount of returned flux the illumination is simply increased. The unit of retroreflection may then be thought of as simply a ratio of the amount of returned flux to that of the incident flux. Illuminance incident on the retroreflector is measured perpendicular to the light in candelas per square meter (lux).

9.3.1 Coefficient of Luminous Intensity

This is the quotient of the luminous intensity (I) of the retroreflector in the direction of observation by the illuminance ($E_\perp$) at the retroreflector on a plane perpendicular to the direction of the incident light:

$$R_I=I/E_\perp \ \text{cd/lx}. \tag{9.1}$$

9.3.2 Coefficient of Retroreflection

This is the quotient of the coefficient of luminous intensity (R_A) of a plane retroreflective surface by its area (A):

$$R_A=R_I/A=I/(E_\perp A) \ \text{cd/(lx m}^2\text{)}. \tag{9.2}$$

9.3.3 Coefficient of Retroreflected Luminance

This is the quotient of the luminance of an extended surface to the illuminance at the retroreflector perpendicular to the direction of the incident light:

$$R_L=L/E_\perp=I/(E_\perp A\cos\nu) \ \text{(cd/m}^2\text{)/lx}. \tag{9.3}$$

9.4 TYPES OF RETROREFLECTIVE DEVICES

9.4.1 Cat's Eyes

This was the first type of retroreflective device used in the transportation industry. It can take a variety of shapes, as Fig. 9.8 shows. Here the incident light from a distant source is focused on the rear of an elongated glass sphere. On the rear surface is a metalized coating (silver on the first elements) which reflects the incident light rays returning them in the approximate direction of the source. These elements can be arranged side by side to form a larger retroreflective unit.

9.4.2 Spherical Glass Elements

The optics of a single glass sphere has the property of focusing the light rays, the focal point being determined by the index of refraction of the material:

$$f = n' r / 2(n' - n), \tag{9.4}$$

where f is the focal length is measured from the sphere's center, n' is the refractive index of the glass, n is the refractive index of the surrounding medium, and r is the radius of the sphere.

When the index of refraction of the glass surrounded by air is equal to 2.0 the paraxial rays from an infinitely distant object will come to a focus on the rear surface of the glass sphere.

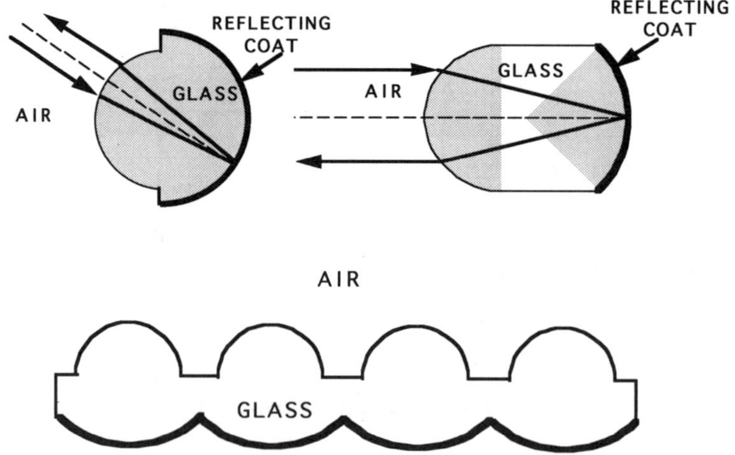

FIGURE 9.8. Three different arrangements of cat's eyes retroreflectors. The lower one is molded from a single piece of glass and metalized. This type was very common in the 1930s and 1940s.

The spherical aberration causes the edge rays to come to a focus shorter than the paraxial rays. The optimum position for the metalized coating is on the rear surface, and the index of refraction should equal 1.913 so the minimum cross section of all the incident light rays is located on the rear surface (Fig. 9.9). The spherical aberration will cause the rays to spread somewhat and provide an advantage for the traffic control application. For glasses with indices less than 1.913 the greatest efficiency of return lies outside the rear surface of the sphere. For the greatest retroreflection to occur the rear surface needs to be coated with aluminum for spheres with an index of 1.913 and for glass spheres with an index less than 1.913 the metalized surface needs to placed at separation of $d=(n'-1.913)/3.65(n'-1)$ from the rear surface. If the glass sphere is embedded in an optical medium different than air ($n=1.0$) the separation will change. The common term for these glass spheres is beads. Spherical lens retroreflectors can be divided into two categories: (1) large spherical lens retroreflectors (cat's eyes), and (2) microsphere retroreflecting sheetings.

Three distinct types of spherical glass elements are used in retroreflective materials. They are designated as exposed glass beads, enclosed glass beads, and encapsulated glass beads. The easiest optical elements to apply are the exposed glass beads, which can be simply dropped onto the coating material. Minute glass spheres of various diameters between 0.05 and 0.9 mm and refractive indices from 1.5 to 1.9 mixed together are commonly used for pavement markings. Almost all pavement marking materials use this type of bead, which is applied by air pressure spraying over the pigment material or remixed in the paint or liquid plastic or both. Such beads are also used for the retroreflectorization of adhesive plastic strips that are stuck to the pavement. The optics of this retroreflective type is shown in Fig. 9.10. Here the pigment of the binder in which the glass bead is embedded acts as a diffusing element, and only a small portion of the incident light is returned to the source using the glass sphere as a refocusing lens. The optical properties of the embedment medium here play a very significant role in the amount of retroreflected flux. The most efficient use of exposed glass beads is in the preparation of retrore-

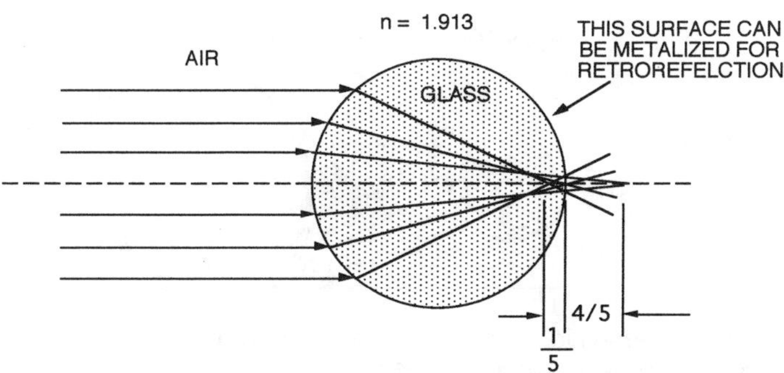

FIGURE 9.9. The simplified optics of a perfect glass sphere.

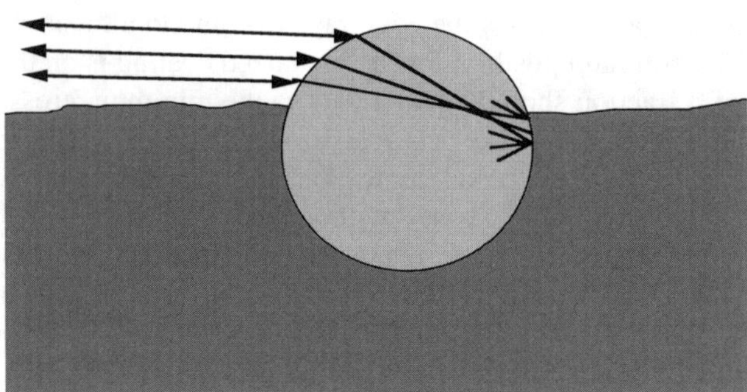

FIGURE 9.10. The optics of the exposed glass beads embedded in traffic pavement marking materials. The bead acts as a lens focusing the incident light onto the highly reflecting diffuse pigments. Only a small portion of this diffusely reflected light is collected by the same bead and returned in the direction from which the light came.

flective striping and garment apparel. Special proprietary techniques allow these glass spheres to be bonded to fabric and to retain their highly efficient retroreflective properties. This material has seen increasing use for pedestrian and worker safety at night.

Exposed glass beads were used for traffic signs in the early 1920s and 1930s but have a major disadvantage when used where moisture is present. The index of refraction of the water changes the position of the focal surface and greatly decreases the amount of light returned. The first major improvement in traffic sign technology began when enclosed lens material was introduced. As Fig. 9.11 shows, a built-in separation was used between the rear of the glass surface and the metallic coating. Water incident on the flat exterior of the material had almost no deleterious effect on the efficiency.

The improvement in the late 1960s used high-index glass spheres with metalized coatings in contact with the rear surface of the glass. The glass spheres had to be protected so that an air-glass interface existed and a clear transparent plastic acting

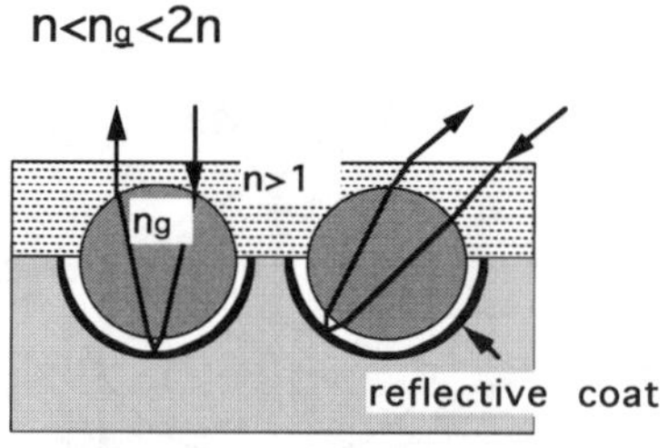

FIGURE 9.11. Glass beads in the enclosed type of retroreflective materials. Here the index of refraction lies somewhere between 1.5 and 2.0. The metalized reflective coating cannot lie on the rear surface of the bead, but must be separated from the rear surface by a distance *d*.

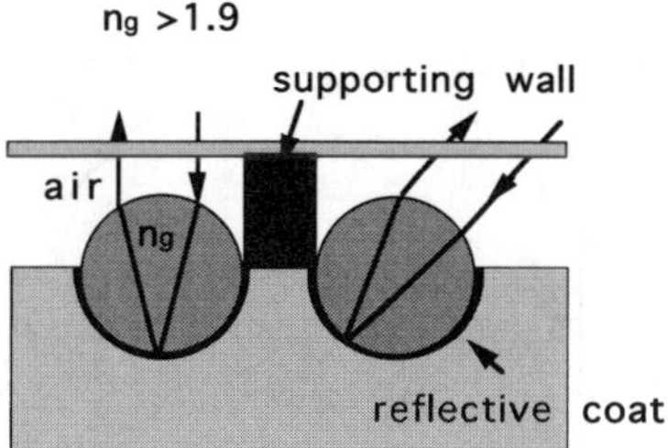

FIGURE 9.12. By constructing a plastic window over the glass beads an air-glass interface can be maintained maximizing the optical efficiency. The metalizing coating may now be placed in contact with the glass when the index of refraction is high enough.

as a window was mounted on a supporting wall. Thus rain has a minimum effect on the optical performance. This type of material is designated an *encapsulated lens* (Fig. 9.12). The efficiency of this type is about three times that of the enclosed lens type.

9.4.3 Cube-Corner, Prismatic, and Tetrahedral Prisms, Both Macro and Micro

The other popular type of retroreflective material involves the use of the reflection of light from three mutually perpendicular reflecting surfaces. Figure 9.13 illustrates the path of light through such a device.

As mentioned before, if the surfaces are glass and are precision ground and polished, the light is returned after three total internal reflections exactly in the direction of the incident light source. If the surfaces are at not exactly 90°, then the

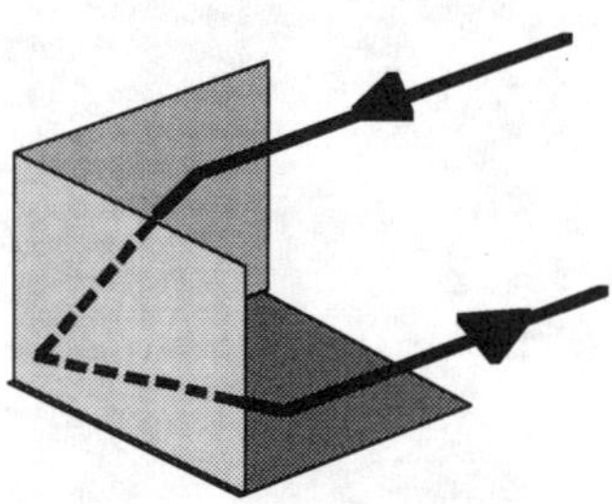

FIGURE 9.13. Ray path through a cube-corner or prismatic retroreflector. The rays are reflected three times in the approximate direction from which they came. The three sides of the cube must be mutually perpendicular for the rays to return to the source exactly. For traffic applications, the returned light rays should spread out, requiring that the sides not be exactly at 90°.

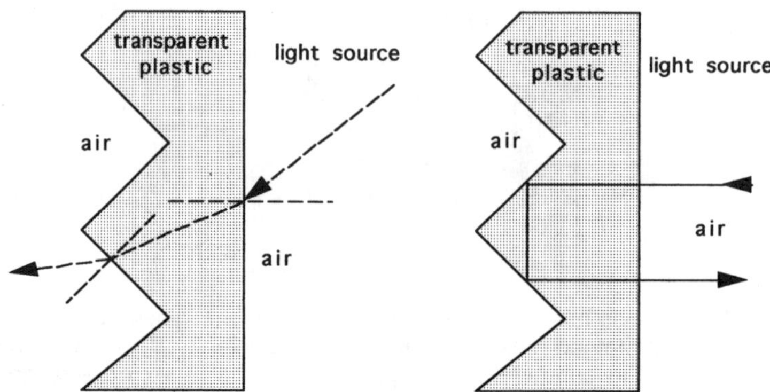

FIGURE 9.14. Ray path through a cube-corner retroreflector. The diagram on the left shows what happens when the angle of incidence on the rear surface is less than 42°.

returned light is spread into six beams (two sides are reflected by the other remaining one). With solid cube-corner retroreflection, two component principles are involved, refraction and total internal reflection (TIR). For TIR to occur the angle of incidence on the surfaces of the cube must be less than the critical angle. This angle for some transparent plastic materials in air is about 42°. Thus, for a plastic corner cube all the rays striking the plastic–air interface at an angle larger than 42° are reflected specularly inside the material. Those rays striking the interface at less than 42° partially go through the plastic boundary surface into the air (Fig. 9.14). Figure 9.15 shows the spread of light from such a cube corner.

In order to distribute light in a more even manner the sides of the adjacent cube corners are slightly rotated with respect to each other. Even in the best of cube-

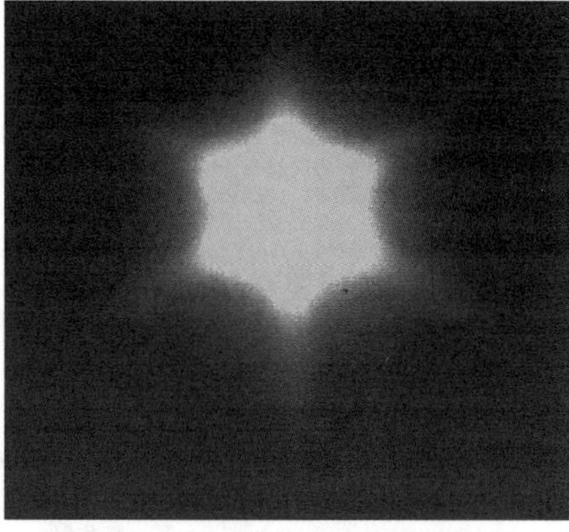

FIGURE 9.15. A photograph of the light return from a micro-cube-corner retroreflector. The angular spread of the light is about 4°.

corner or prismatic material, as it is sometimes called, rotational symmetry is difficult to achieve. Cube corners are made in plastic material in large size (1 mm on a side) and in microsizes (about 0.03 mm). The latter can be formed by using retroreflective elements bonded to vinyl so that the material is flexible. Such materials are used to provide conspicuity at night for worker's garments. These cube corners can be assembled side by side to form a composite retroreflector of any shape (square, circular, triangular, rectangular) and can be used for letters, figures, and symbols.

9.5 GEOMETRICAL CHARACTERISTICS OF THESE DEVICES

9.5.1 Spherical Glass Elements (Glass Beads)

As mentioned earlier, the geometrical characteristics of these elements can be illustrated by three-dimensional diagrams showing how the light retroreflected from these elements is spatially distributed. For spherical glass elements this distribution is symmetrical around the retroreflector axis. The different types of material using glass beads generate different distributions largely caused by the size, index of refraction, and the medium in which they are placed. For the very small glass beads diffraction enters into the distribution. The wavelength dependency of diffraction also causes the spectral distribution to change as a function of the geometry (see Sec. 9.6). The spatial distributions for the various material configurations are shown in Figs. 9.16 and 9.17. The major differences between these configurations is increased coefficient of retroreflection and the light return at larger entrance angles. These are important for traffic signs and work zone warnings utilized in urban areas where the distances from the vehicle to the sign is short.

9.5.2 Exposed Glass Beads Used in Pavement Markings

This use of glass beads is commonplace throughout the world to enhance the conspicuity of pavement markings. The geometrical distribution of the different types of marking materials are shown in Figs. 9.18–9.20. The differences are caused by the sizes of the glass beads and the indices of refraction as well as the diffuse characteristics of the pigment or binder. The geometry of illumination and viewing plays a very important part in these materials. Some testing geometries will provide values of the coefficient of retroreflected luminance which are high with respect to the geometry available to the passenger car driver. These are unrealistic and the present direction of fabrication is toward increased performance at the driver's geometry.

9.5.3 Cube Corners

This type of retroreflector has the greatest variation of geometrical characteristics. Not only the size of the cube corners affects the distribution but the orientation

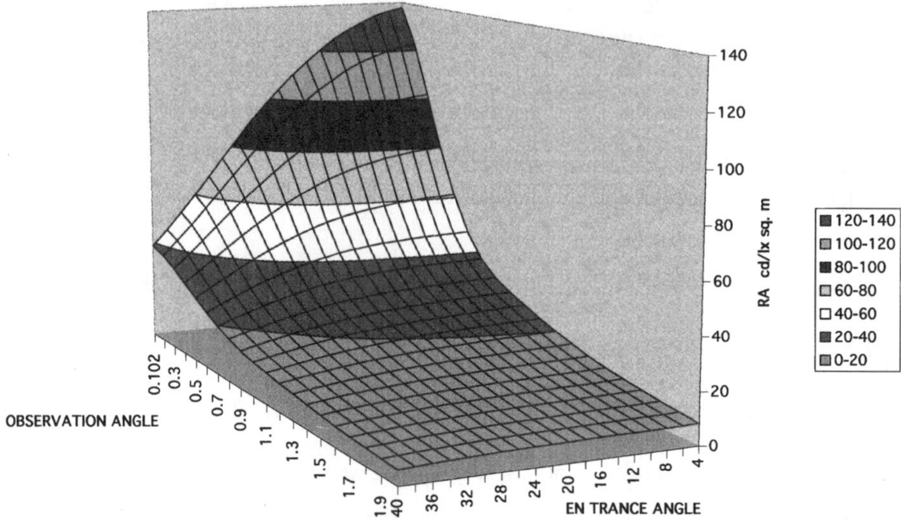

FIGURE 9.16. A three-dimensional plot of the spatial distribution of the retroreflected light from an enclosed lens type glass bead sheeting material. The observation and entrance angles are shown on the x and y axes and the coefficient of retroreflection on the z axis. The entrance angle is in the observation half-plane ($\beta_2, \epsilon=0$). The light pattern is generally symmetrical, and only one component (β_1) is needed to describe the light return from such materials.

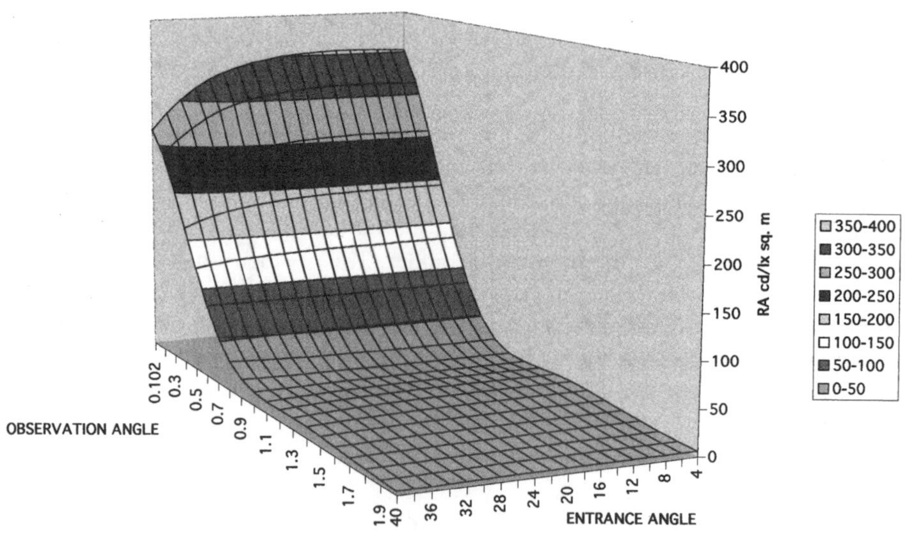

FIGURE 9.17. A three-dimensional plot of the spatial distribution of retroreflected light from an encapsulated lens glass bead sheeting material.

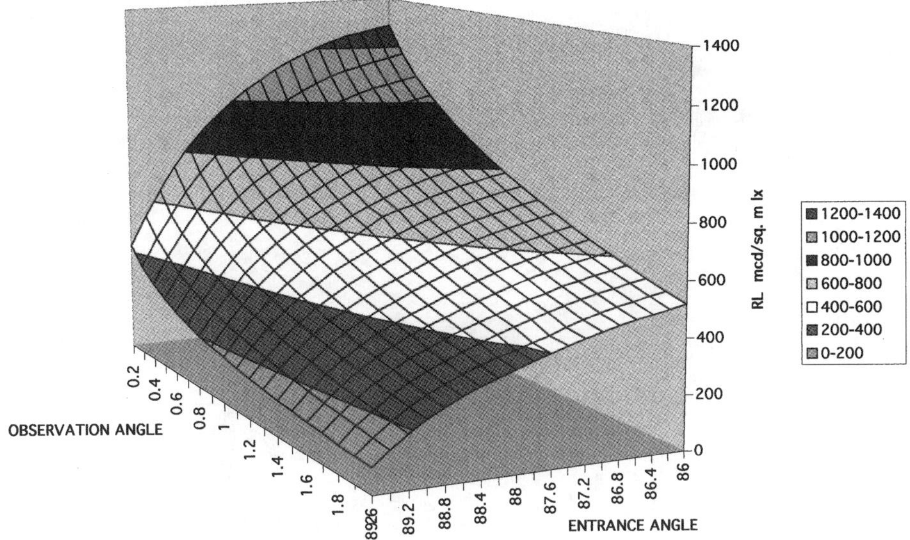

FIGURE 9.18. A three-dimensional plot of exposed glass beads in a preformed plastic tape for pavement markings.

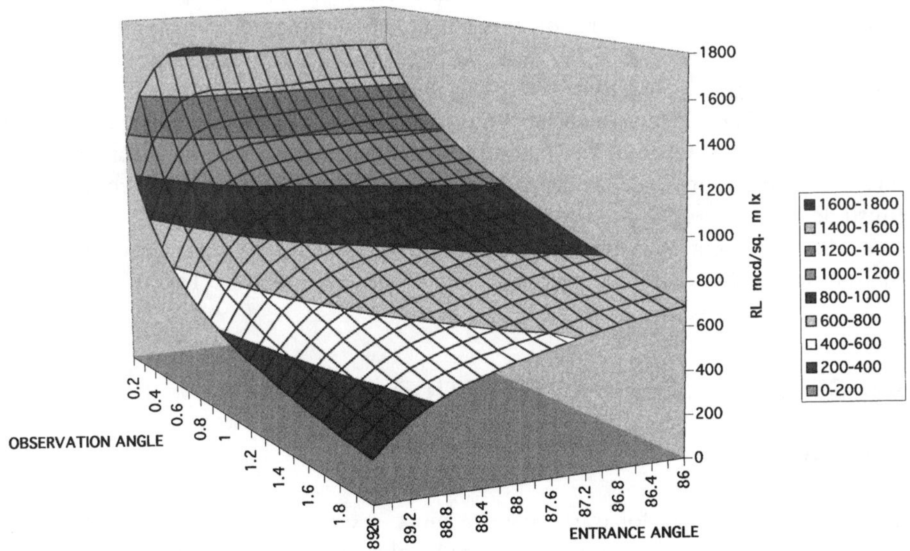

FIGURE 9.19. A three-dimensional plot of exposed glass beads in a profiled tape. The profiling allows the retroreflectance of the marking to continue under wet conditions.

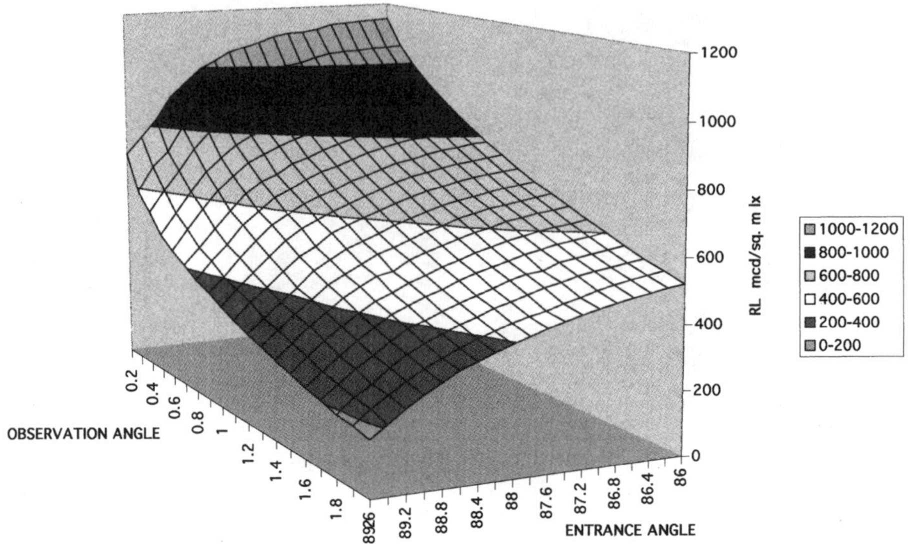

FIGURE 9.20. A three-dimensional plot of large glass beads embedded in thermoplastic material.

of the principle axis of the cube corner, the relative sizes of the three perpendicular planes, the surface smoothness of the planes, the index of refraction of the material, and whether the rear surfaces have a metal coating or not. The spatial distribution of this type of retroreflector is not symmetrical about the principle axis. Some small examples of this type can be seen in Figs. 9.21–9.26. This type of material possesses certain advantages over the glass beads. The degrees of freedom involved in redirecting the light incident on the retroreflector is much greater than for glass spheres. The distribution of the light return can therefore be optimized for any given observing condition. Indeed the pattern of light return can be quite complex, as illustrated in Figs. 9.22–9.25.

Raised pavement markers are an example where the geometrical distribution of the returned light is oriented primarily in the vertical direction. This distribution is illustrated in Fig. 9.25.

9.6 SPECTRAL CHARACTERISTICS

9.6.1 Use

Retroreflectors used for traffic control must also possess color properties which signal certain driving behaviors. The red color of a sign, for example, signals extreme maneuvers, i.e., STOP, DO NOT ENTER, WRONG WAY, etc. Yellow signals are warning signs: CAUTION, CURVES, etc. Orange implies work zone areas, and green and blue imply guidance and information. In certain recreational

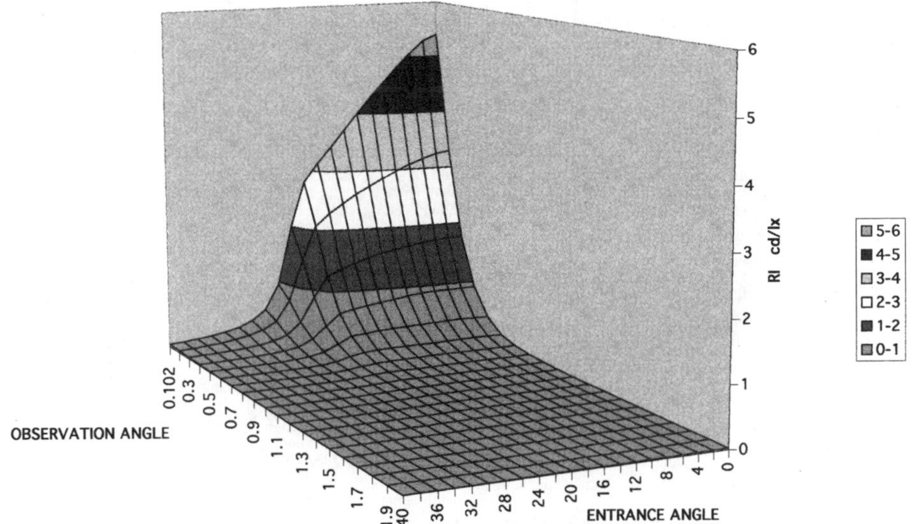

FIGURE 9.21. A three-dimensional plot of macro-cube-corner retroreflectors used as road delineators. All the cube-corner plots which follow in this chapter are distributions only in the observation half-plane where the rotation and β_2 components are set equal to zero. The shape of the distribution of light return when these other angles are considered will change significantly.

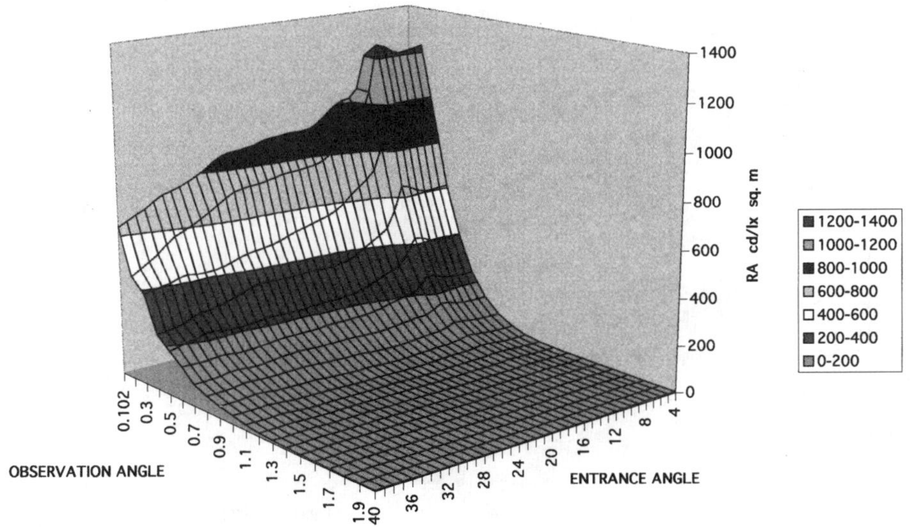

FIGURE 9.22. A three-dimensional plot of one type of micro-cube-corner or prismatic retroreflecting material used for signing.

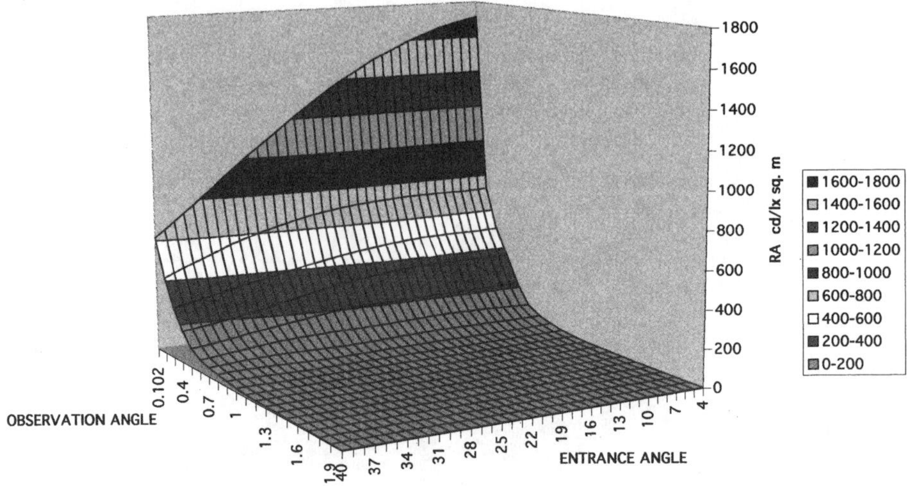

FIGURE 9.23. A three-dimensional plot of a different type of micro-cube-corner or prismatic retroreflecting material.

areas the color brown is used. The colors of the materials may be applied in two ways. One way is to screen-print transparent inks over a white background. This is common for many small jurisdictions. Since light passes twice through this transparent ink its thickness must be carefully controlled; otherwise the amount of

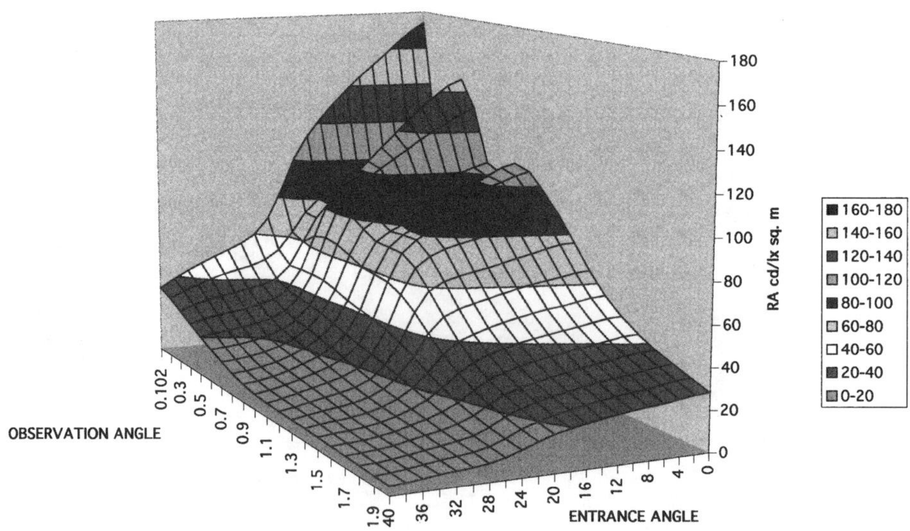

FIGURE 9.24. A three-dimensional plot of micro-cube-corner flexible material. The shape of the distribution is heavily influenced by the nonflatness of the material.

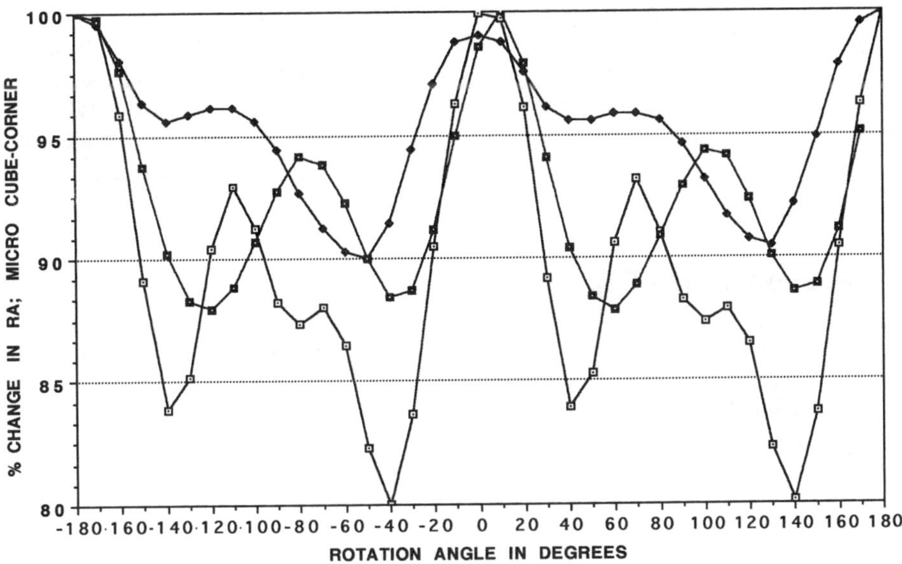

FIGURE 9.25. The rotational variation in the cube-corner materials.

retroreflection will be reduced. The more common way is to die the clear transparent film over the encapsulated lens, some cube corners, or the upper layer of enclosed lens material. This then acts as a standard filter and can be carefully controlled in the manufacturing process to assure maximum retroreflective perfor-

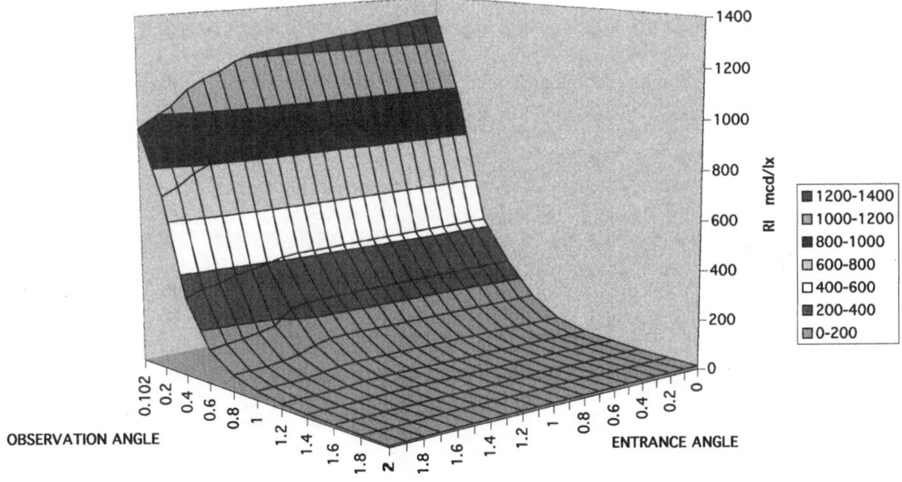

FIGURE 9.26. A three-dimensional plot of the spatial distribution of the light return from a macro-cube-corner raised pavement marker.

mance. The color of the material is specified by its CIE chromaticity coordinates (x,y) for the 2° standard observer. The color gamuts are determined from the International Standards Organization (ISO) and CIE publications for surface colors used in signaling [6]. Because the retroreflective material is used under daylight as well as nighttime, illumination consideration is made to the boundary of the color gamut in CIE color space to yield a similar color appearance both day and night (Fig. 9.27).

9.6.2 Geometrical Dependence

As is the case in using filters at other than normal incidence, the transmission changes as a function of the entrance angle. In addition the diffraction of the light in the very small glass beads and micro-cube-corner materials has a shifting effect

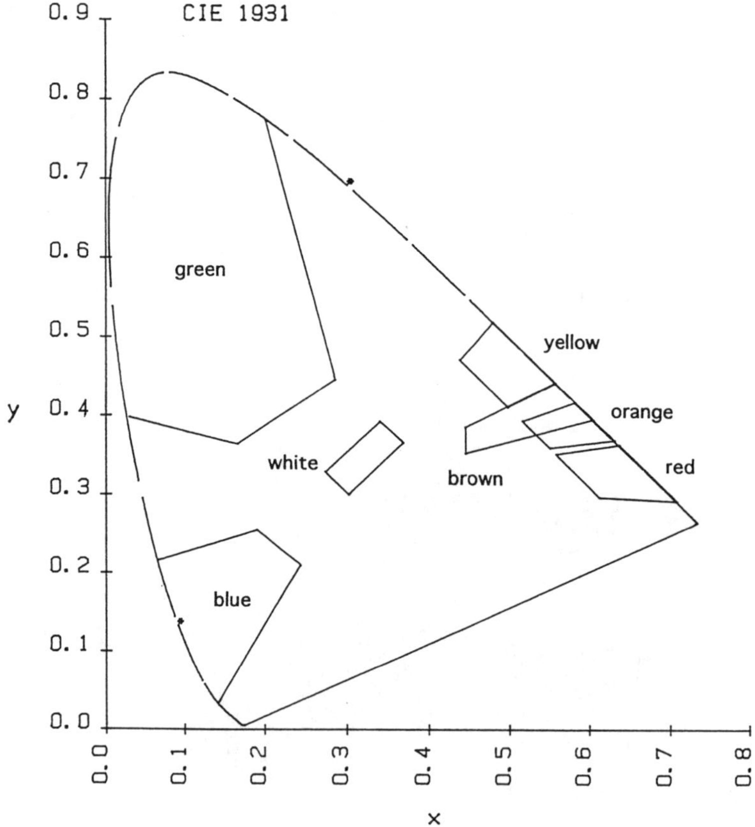

FIGURE 9.27. The CIE chromaticity diagram for surface colors under CIE D65 illumination. Initial gamuts for new material are much smaller but are not shown here because they differ in various countries and have yet to be adopted by the CIE.

on the spectral reflectance of this type of material [7]. Figures 9.28 and 9.29 illustrates what happens to the spectral characteristics of encapsulated lens material as a function of the observation and the entrance angles. The chromaticity coordinates can shift as the driver approaches a traffic sign, and this may be observed in shifts of a bluish sign becoming more greenish and the white sign becoming more yellowish.

These conditions are in general not significant except where the original color lies close to the boundary of the color gamut. In this case the geometrical change in illumination and viewing angles as the driver approaches a sign or device may shift the color outside of the gamut (see Fig. 9.30).

9.7 MEASUREMENT

9.7.1 Test Methods

The measurement of retroreflection may be divided into two parts. The first and most common method determines the photometric characteristics of the retroreflector when illuminated by a standard CIE Illuminant such as A or for daylight D65. The resultant light return is evaluated by a detector equipped with a filter

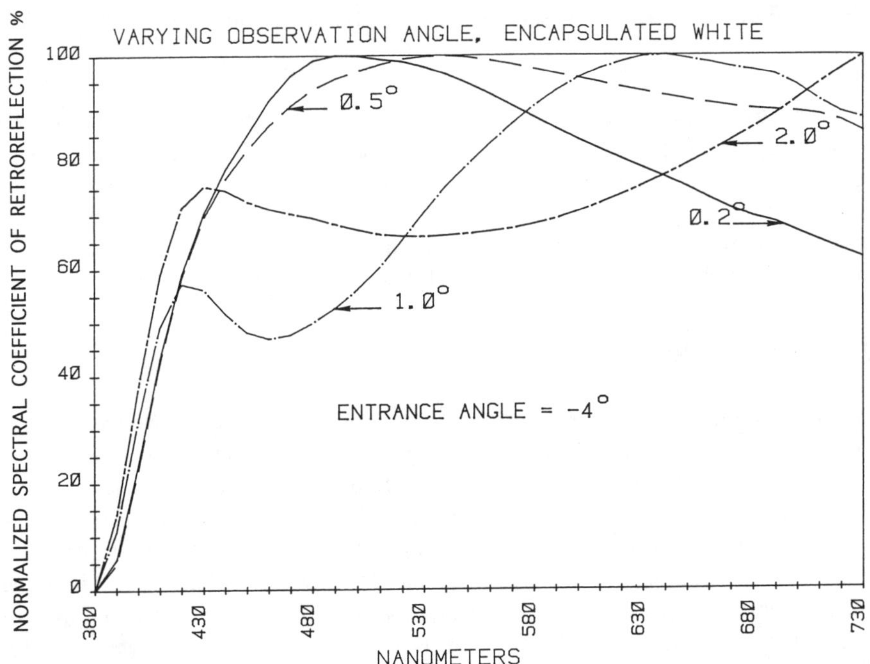

FIGURE 9.28. A plot of the spectral coefficient of retroreflection for a white encapsulated sheeting material. The geometry of measurement is indicated.

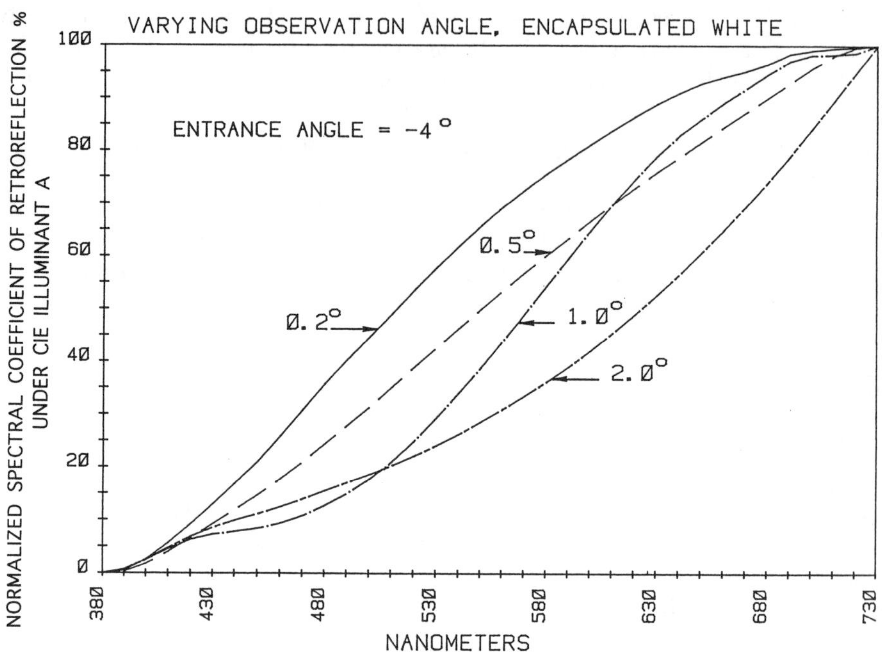

FIGURE 9.29. A plot of the same spectral distribution under CIE Illuminant A.

whose spectral responsivity closely matches the CIE 2° standard photopic observer. The second method measures the spectral characteristics of the retroreflector using a spectroradiometer. This latter method has the advantage that illuminating sources different than CIE Illuminant A may be used and that the light return may be evaluated by mathematical means using the tabulated values for the CIE 2° or 10° standard observer. These methods are briefly described in the following paragraphs.

Photometric Methods

Two common methods are in use for measuring the absolute performance characteristics of retroreflective devices. In CIE Publication No. 54 [5] these are defined as the *relative* and the *direct* methods. The relative method uses the same detector [known as the receiver in the American Society for Test and Measurement (ASTM) terminology] to measure the amount of illumination incident on the retroreflective device and the amount retroreflected. This has the advantage that the receiver need not be calibrated in absolute photometric terms. A photopic filter and exact linearity is all that is required. The direct method uses photometrically calibrated detectors (illuminometers and luminance meters) and calibrated reference lamps of known luminous intensity. Today most organizations, such as ASTM, DIN, and Convention National European (CEN), use the former method with the exception of the

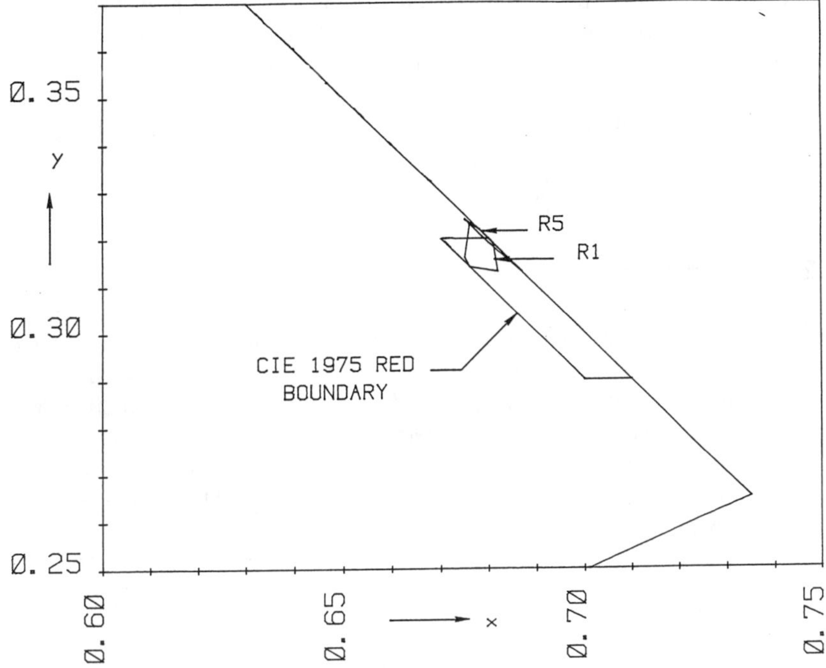

FIGURE 9.30. The CIE chromaticity diagram for nighttime conditions showing the shift of chromaticity with geometry for a red material.

Society for Automotive Engineers (SAE). Because of the SAE use of photometric measurements for headlights and other self-luminous sources, that organization prefers the direct method. It is also important to recognize that all the measurements in these methods are at specific geometries and all results must include the geometrical parameters of the measurement.

Substitution Method

This method is a variation of the *relative* method. It uses a previously measured retroreflective device as a reference standard. The device under test is interchanged with the calibrated device, and the ratio of the readings multiplied by the calibrated value will then yield the absolute value of the test sample. This technique is particularly applicable to portable instruments used in the field or on the production line.

Spectral Methods

This method may be considered the best technique for the measurement of colored retroreflectors. It also is the most accurate from the standpoint that the exact match of the detector to the CIE $V(\lambda)$ function and the spectral distribution of the light source to CIE Illuminant A is not necessary. The only disadvantage is the level of radiance available which will affect the signal-to-noise ratio. The measurement technique is the same as in the two preceding sections except that the normal photometer is replaced by a spectroradiometer. This method is most important for the measurement of the daytime color where the CIE Illuminant D65 is specified as well as the geometry (45°,0°).

9.7.2 Instrumentation

Two types of instrumentation are used in the measurement of retroreflection. These are the laboratory system and the portable or field systems. The most accurate one is the laboratory where the separation between source and retroreflector varies from 10 to 30 m. This type uses the latest in electronic instrumentation and in computer control. The objective is to achieve the greatest accuracy. Generally four components make up such a laboratory photometric range. The components of such a system are listed below.

Light Source

This is generally of the projector type with control over the area illuminated and a careful match to the spectral power distribution of CIE Illuminant A (2856 K). The light source should be stable using either a well-regulated power supply or a feedback system of monitoring the light output. The aperture angle of the exit optics is also an important parameter which must be controlled. This is specified in all the test methods which have been standardized.

Photometer Head or Receiver

This instrument should be a photopically corrected detector equipped with an optical system which can be focused on the test sample. It should have a variable field of view to limit the area of light return collected to minimize the contribution from stray light. A viewing system is also advisable to use to enable the sample to be positioned accurately within the collection solid angle.

Goniometer

This component is the mounting device for the retroreflective device. It should be a three-axis assembly to allow complete freedom in characterizing the sample under test. The movement of the axes should allow the entrance angle components (β_1 and β_2) to be set to 0°–90° with a resolution of at least 6 min of arc. The third axis is the rotation axis and it should be capable of 0°–360° in movement. These axes can be either manually set by dial indicators or computer driven using stepper motors.

Observation Angle Positioner (OAP)

This component is the mounting device for the light source and the photometer. It should have the capability of moving the photometer either vertically or horizontally with respect to the light source corresponding to a change in observation angle. The movement may be manual with settings determined by a scale or computer operated by a lead screw stepper motor drive. Figure 9.31 illustrates a typical goniometer and observation angle positioner.

9.7.3 Portable Instruments

For the convenience of quality control on the production line and field measurements, hand-held instruments are common. Some of these closely follow the geometry of measurements prescribed for the laboratory equipment. It is important that the measurements obtained with these instruments yield similar results with those obtained in the laboratory. If careful observance of the optical requirements for measurement of retroreflective devices is followed, the results obtained by these instruments agree within 5–10 % with laboratory methods. Some of the optical parameters which must be observed in the design of such instruments are numbered 1–8 below.

1. *Collimation:* The measurement in the laboratory is designed to approximate the driving conditions, i.e., long distances. These can be achieved by using collimation optics such that the light source and detector are at optical infinity or

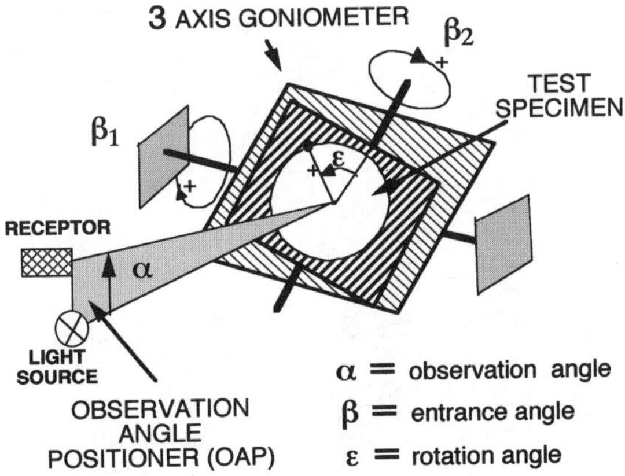

FIGURE 9.31. A diagram of a typical observation angle positioner (OAP) and a three-axis goniometer. The appropriate four angles defining the characteristics of a retroreflector are indicated. The first axis (fixed) is horizontal and the observation angle is vertical. By using combinations of these four angles, auxillary angles such as the presentation and orientation angles can be set. Under computer-controlled laboratory systems these combinations are easily achieved.

equivalent distances to that in the laboratory. When cube-corner retroreflective material is measured the distance of the light source and detector from the sample under test becomes increasingly important for accurate results.

2. *Aperture angles:* This is an element of the instrument that is often neglected. To measure retroreflective devices in the laboratory the exit and entrance apertures of the light source and receptor, respectively, are kept small (3–10 min of arc depending on the material). The same criterion must be adhered to when designing portable instruments; otherwise the values will differ from those obtained in the laboratory photometric range (Fig. 9.32).

3. *Measurement area:* As is usual in the laboratory, the size of the area measured is an important parameter. For portable instruments the reference standard area should be the same as the measurement area of the instrument. The area should be many times the size of the retroreflective elements that make up the material. A good rule to follow is a size which is 10 times the element.

4. *Precision of measurement geometry:* The observation, entrance, and rotation

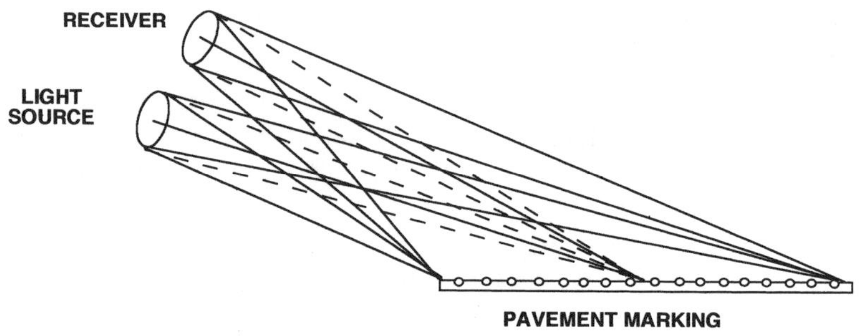

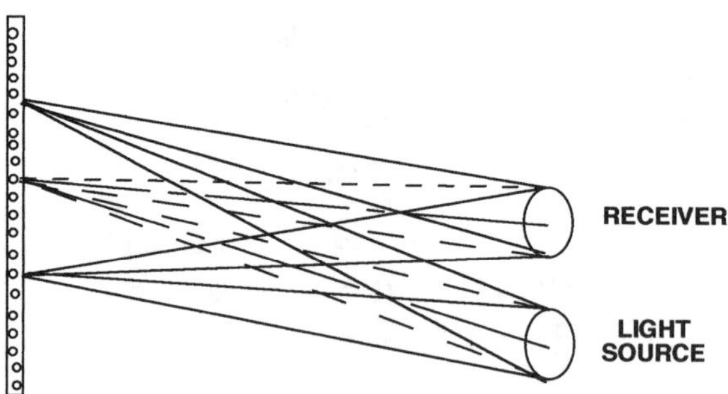

ENTRANCE AND OBSERVATION ANGLE TOLERANCES DEPEND ON: 1. SETTINGS OF ENTRANCE AND OBSERVATION ANGLES; 2. APERTURE ANGLES

FIGURE 9.32. The importance of the geometry of measurement is illustrated.

angles should possess the desired precision necessary to achieve good agreement with the laboratory photometric system.

5. *Light source:* The spectral power distribution should approximate CIE Illuminant A (2856±50 K).

6. *Receiver:* This should be matched to the CIE standard photopic observer 2° field. The light source and receiver emittance and responsivity may be designed to match the combined spectral distribution of CIE Illuminant A and the $V(\lambda)$ spectral luminousity function.

7. *Avoidance of regular reflection:* This is the reflection from the front surface of sheeting material that can affect the accuracy of measurement.

8. *Avoidance of stray light:* Proper baffling and light traps should receive a high priority in the design, as this effect can generate significant errors which may often be buried in the results.

9.8 STANDARDS

9.8.1 International

The necessity for worldwide standardization in the transportation area has become very important. The traffic control device requires recognition by all types of operators in all types of vehicles—aircraft, surface vehicles, and marine vehicles. The conspicuity of these devices requires that the retroreflective elements remain in good condition. The start for standardization began in Vienna in 1968 with the Convention of Road Signs and Signals. This was followed by a Convention in Geneva in 1973 where the European standards were agreed upon (CEN). The CIE (International Commission on Illumination) is now the international organization for the standards in light and color. Many technical reports dealing with the recommendations for signals, retroreflection, and roadway lighting are the basis for international agreements. Some of these are listed in the reference section [8–11]. The geometry of measurement differs somewhat among various countries; however, today's international trend is to specify the characteristics of retroreflectors in terms of the driving or transportation scenario.

9.8.2 National

The need for standardizing the traffic control devices first arose in the United States in the 1930s. Organizations such as the SAE and the Illuminating Engineering Society (IES) adopted specifications for these devices. A significant step toward harmonization of the specifications occurred with the publication in the Journal of the Optical Society of America of the three-part paper by Dr. George Van Lear in 1940 [12]. The federal government started the standardization with the first *Manual for Uniform Traffic Control Devices* in 1935 and has continued it ever since

[13]. In all subsequent editions of this manual, however, no quantitative values of the retroreflection were required, either for new or in-place traffic control devices. This is expected to change with the publication of the new manual expected in 1998 or 1999. For the first time in the United States, guidelines for new and minimum performances of retroreflective devices will be formalized. Much of the details for measurement and procurement specification was left up to other government agencies [14] and consensus organizations such as ASTM. The reference section lists the important documents dealing with retroreflection written by ASTM [15–26].

9.8.3 Deterioration Causes

The retroreflective performance of the traffic control device deteriorates with time in a nonuniform manner. The deterioration rate or service life of these devices depends on many factors and their influence is type dependent [27]. Some of these causes are listed below for different materials.

Cat's eyes. Most of these types of retroreflectors depend on total internal reflection; when dirt and other gases come in contact with the optical faces the TIR is partially destroyed and the efficiency drops. This was recognized early in the application of these devices, and metalized coatings and sealing of the rear of the retroreflector minimized this deterioration to a large degree.

Plastic cube-corner (prismatic) materials. The color may change due to pigment fading and loss of transparency caused by UV exposure. The material for rigid cube corners may crack allowing moisture to penetrate and destroy the TIR, and of course abrasion of the outer surface is very common. For raised pavement markers the most common cause is due to the impact from vehicles. In addition to the former, microprismatic materials that are in flexible materials will deform, changing the orthogonality of the faces and spreading the returned light, thus lowering the value at the same geometry.

Glass bead materials, exposed, enclosed, and encapsulated. The exposed glass beads either on the pavement marking or on fabric will loose their retroreflection by bead removal and dirt. Enclosed glass beads will lose their efficiency by the change in distance from the rear bead surface to the metalized coating, resulting in a focal surface displacement and a spreading of the returned light. Encapsulated glass beads are less subject to this latter effect, as the metalized coating is usually on the rear of the bead. All other degradation effects—color fading, abrasion, or inappropriate solvents—are always present.

9.9 EXAMPLES OF MEASUREMENT

9.9.1 Step by Step

The following measurement procedure will be that recommended by the ASTM. In a large part the methods described by the ASTM are very similar to those adopted in many other countries belonging to the CIE.

Procedure A

This procedure follows the relative method where measurement of the retroreflective material at specified geometries does not require any standards or reference materials. The steps are written in detail in ASTM E-809 [16] and E-810 [17] and will only be generally described here. The receiver is adjusted to accept the radiant flux from an area slightly in excess of the sample under test. The photometer is then positioned in place of the test sample and used to measure the radiant flux from the projection light source. The receiver does not have to measure this radiant flux absolutely. The linearity of the photometer and its close match to the CIE $V(\lambda)$ function must be adhered to. Without altering the gain or the solid angle of the optical collection of the receiver, it is returned to the OAP and set at the appropriate observation angle. The sample is now returned to the goniometer and the appropriate entrance and rotation angles set. The radiant flux from the sample under test is now recorded. The measurement equations are as follows:

The *coefficient of luminous intensity* is

$$R_1 = m_1' d^2 / m_2. \tag{9.5}$$

The *coefficient of retroreflection* is

$$R_A = m_1' d^2 / m_2 A. \tag{9.6}$$

The *coefficient of retroreflected luminance* is

$$R_L = m_1' d^2 / m_2 A \cos \nu, \tag{9.7}$$

where d is the observation distance (in m), A is the area of the test specimen (in m^2), ν is the viewing angle, α is the observation angle, β is the entrance angle, m_1' is the receiver reading minus the stray light of the retroreflected luminous flux from the specimen at the observation angle positioner (OAP) (in relative units), and m_2 is the receiver reading of the retroreflected luminous flux from the projection light source at the specimen location in relative units.

The following equations are also used to specify the photometric characteristics of retroreflective materials.

For the *viewing angle*:

$$\cos \nu = \sin \alpha \sin \beta_1 \cos \beta_2 + \cos \alpha \cos \beta_1 \cos \beta_2. \tag{9.8}$$

For horizontal coating materials (pavement markings) the co-viewing $(90-\nu)$ angle is useful.

For the *presentation angle*:

$$\cos \gamma = (\cos \beta_2 \sin \beta_1)/\sin \beta, \quad \sin \gamma = \sin \beta_2 / \sin \beta, \tag{9.9}$$

where

$$\sin \beta = (\sin^2 \beta_2 + \cos^2 \beta_2 \sin^2 \beta_1)^{1/2}. \tag{9.10}$$

Orientation angle:

$$\cos\omega=(\sin\epsilon\cos\beta_1\sin\beta_2-\cos\epsilon\sin\beta_1)/\sin\beta, \tag{9.11}$$

$$\sin\omega=(\cos\epsilon\cos\beta_1\sin\beta_2+\sin\epsilon\sin\beta_1)/\sin\beta. \tag{9.12}$$

Procedure B

This procedure is commonly used for routine measurements in the laboratory and field. It relies on the results of the samples measured in procedure A and thus are designated reference or working standards. The receiver is not moved and only the reference standard and the unknown specimen are placed on the goniometer. The measurement equations are as follows.

The *coefficient of luminous intensity* is

$$R_{IT}=(m'_{1T}/m'_{1S})R_{IS}. \tag{9.13}$$

The *coefficient of retroreflection* is

$$R_{AT}=(A_s m'_{1T}/A_T m'_{1S})R_{AS}. \tag{9.14}$$

The *coefficient of retroreflected luminance* is

$$R_{LT}=(A_s m'_{1T}/A_T m'_{1S})R_{LS}, \tag{9.15}$$

where the subscript T represents the test specimen, and S represents the reference or working standard.

Other useful transfer equations are

$$R_L=(L/E_\perp)=(R_I/A\cos\nu)=(I/E_\perp A\cos\nu)=(R_A/\cos\nu). \tag{9.16}$$

Direct method. This technique uses a standard lamp to calibrate the photometer. A illuminometer is used to measure the illuminance falling on the sample. The same illuminometer is then used to measure the returned light from the sample. For horizontal coatings a luminance meter is used to measure the luminance. If a standard lamp is placed in the sample position the illuminance at the OAP can be measured, and then the lamp is replaced by the sample under test and the illuminance measured. The ratio of the two illuminances divided by the illuminance of the projection light source illuminating the sample at the sample position then determines the coefficient of retroflected luminance.

Measurement of the spectral coefficient of retroreflection [28]. This method requires the use of a spectroradiometer. The coupling optics should use the same type of collection system as in the photometric methods. Especial care must be exercised to avoid nonuniform illumination of the entrance slit of the monochrometer by either coupling with a fiber-optics bundle or/and an integration device (sphere, etc.) of some type. The equations for this method are as follows:

$$R_A = \frac{d^2 \int_{380\ \mathrm{nm}}^{780\ \mathrm{nm}} S_p(\perp, \lambda) R_A(\lambda) V(\lambda) d\lambda}{\int_{380\ \mathrm{nm}}^{780\ \mathrm{nm}} S_p(\perp, \lambda) V(\lambda) d\lambda}, \tag{9.17}$$

where R_A is the coefficient of retroreflection, d is the source to specimen distance, $S_p(\perp,\lambda)$ is the spectral power distribution of the light source perpendicular to source, $R_A(\lambda)$ is the spectral coefficient of retroreflection, and $V(\lambda)$ is the spectral luminous efficiency for photopic vision.

Once the spectral coefficient of retroreflection for that material at the desired geometry has been obtained, the photometric values may be computed using the tabulated values for CIE Illuminant A and the 2° standard observer $V(\lambda)$. The resultant distribution may be used to compute the CIE tristimulus values and the chromaticity coordinates for the measured geometry. If a well-corrected tristimulus colorimeter is used, the chromaticity coordinates may be measured directly. Care must be exercised to ensure that the aperture angles are maintained. The major disadvantage of a tristimulus colorimeter is that other light sources and responsivity functions cannot be mathematically evaluated because the power distribution is not available.

The daytime measurement of the retroreflector can be measured in the usual manner with spectrophotometers, but the values may be slightly different than those obtained using a spectroradiometer at 45° and the projector light source illuminating the specimen at 0°. This is especially apparent in the colorimetric measurement of micro-cube-corner materials. The probable cause of these differences is due to the collection geometry available in standard commercial spectrophotometers.

9.9.2 Characterization

The previous methods of retroreflection measurement, if properly employed, will yield a complete characterization of the behavior of the light return from the retroreflector. For complete characterization where the spatial resolution must be high, a computer-controlled photometric range system is the most efficient manner to collect the data. In the development of new devices and materials the full characterization is essential. The time to characterize a retroreflector such as that shown in Fig. 9.22 would consume a full day's work using manual techniques and can be accomplished in 15 min under computer control. For testing of retroreflectors for specification compliance only a few geometries are necessary and manual methods will suffice.

9.9.3 Output Formats

In today's computer age the format for data transfer is an important factor. Two types of compilation of retroreflective data are useful. One uses the vector approach where the number of data points is limited and located at widely differing geometries such as occurs in the specifications for the retroreflectors. The other approach is the complete characterization of the retroreflector over a range of entrance and observation angles. For this approach the matrix compilation is best. If the data are collected in ASCII format software is available to illustrate the characterization and enable the designer to formulate their products efficiently.

9.10 SOURCES OF COMMON ERRORS

There are many sources of error in the measurement of retroreflection. Excellent papers exist on the mathematical treatment of these errors and the reference section should be used for more detail in these papers [29–31]. It will suffice to mention the common errors here.

Geometrical errors of placement. These are errors in the settings of the four angles necessary to characterize the retroreflector. For a 1% error in setting the observation angle of 0.2° a 1.3% error can occur. A 1-mm uncertainty in the distance between the source and the specimen at 30 m results in a 1% error. The centroid of the radiance pattern eminating from the source and the centroid of responsivity of the receiver must be determined to accurately set the observation angle. The variation of the coefficient of luminous intensity with entrance angle is particularly significant for cube-corner materials. The setting of the entrance angle of horizontal coatings (pavement markings) is a difficult task and the R_L values will reflect errors when the entrance angle β assumed is in actuality a different β. Rotational effects of cube-corner materials are very susceptible to changes in the rotation angle, and it is necessary to set these angles to a precision of 0.1°.

Geometrical errors of extent. These are errors due to the finite sizes of the specimen, the source, and receiver apertures. The magnitude of these errors depend on the shape of the characteristic function of the retroreflector. For cube-corner materials a 1.5% error is equivalent to a 3-min aperture, whereas pavement marking materials for the same error is equivalent to 16 min of arc.

Field nonuniformity. The greatest error is the nonuniformity of the illumination beam over the specimen area and the corresponding nonuniform spatial responsivity of the receiver.

Inverse-square error. The error in measurement of the separation between source and specimen is squared. Thus a 1% error in separation measurement yields a 2% error in measured retroreflectance.

Chromaticity errors. These are the usual errors associated with spectroradiometry such incorrect wavelength settings and spectral mismatches of tristimulus colorimeters to the CIE color matching functions.

Polarization errors. The state of polarization must be considered as the retroreflector may alter the incident light in a manner depending on the material. Therefore the source and receptor should be insensitive to polarization.

9.10.1 Assessment

Many of the foregoing errors may be evaluated using the methods denoted in the respective ASTM documents and the reader is referred to these.

REFERENCES

1. Egan, W. G. and Hilgeman, T., ''Retroreflectance Measurements of Photometric Standards and Coatings,'' Appl. Opt. **15**, 1845–1849 (1976).
2. Rennilson, J. J., Holt, H. E. and Morris, E. C., ''In Situ Measurements of the Photometric Properties of an Area on the Lunar Surface,'' J. Opt. Soc. Am. **58**, 747–755 (1968).
3. Preliminary Science Report, Apollo 11, Preliminary Examination of Lunar Samples, Part 5, NASA Special Publication No. SP-214 (1969).
4. Alley, C. O. *et al.*, Laser Ranging Retroreflector, Preliminary Science Report, Apollo 11, Part 7, NASA Special Publication No. SP-214 (1969).
5. *Retroreflection, Definition and Measurement*, CIE Publication No. 54 (1982).
6. *Recommendations for Surface Colours for Visual Signalling*, 2nd ed., CIE Publication No. 39.2 (1983).
7. Rennilson, J. J., ''Chromaticity Measurements of Retroreflective Material under Nighttime Geometry,'' Appl. Opt. **19**, 1260–1267 (1980).
8. *Guide to the Properties and Uses of Retroreflectors at Night*, CIE Publication No. 70 (1985).
9. *Visual Aspects of Road Markings*, CIE/PIARC Publication No. 73 (1988).
10. *Roadsigns*, CIE Publication No. 74 (1989).
11. *Maintained Night-Time Visibility of Retroreflected Road Signs*, CIE Publication No. 113 (1995).
12. Van Lear, Jr., G. A., ''Reflectors Used in Highway Signs and Warning Signals, Parts I, II and III,'' J. Opt. Soc. Am. **30**, 462–487 (1940).
13. *Manual on Uniform Traffic Control Devices*, Federal Highway Administration, Govn. Supt. of Document (1988).
14. *Retroreflectance MAP Service of Coefficient of Luminous Intensity*, NBS Special Publication No. 671 (1984).
15. *Standard Practice for Describing Retroreflection*, ASTM Publication No. E-808 (1994).
16. *Standard Practice for Measuring Photometric Characteristics of Retroreflectors*, ASTM Publication No. E-809 (1994).
17. *Standard Test Method for Coefficient of Retroreflection of Retroreflective Sheeting*, ASTM Publication No. E-810 (1994).
18. *Standard Practice for Measuring Colorimetric Characteristics of Retroreflectors under Nighttime Conditions*, ASTM Publication No. E-811 (1995).
19. *Standard Test Method for Retroreflectance of Horizontal Coatings*, ASTM Publication No. D-4061 (1994).
20. *Standard Test Method for Measurement of Retroreflective Pavement Marking Materials with CEN-Prescribed Geometry Using a Portable Retroreflectometer*, ASTM Publication No. E-1710 (1995).
21. *Standard Test Method for Field Measurement of Raised Retroreflective Pavement Markers Using a Portable Retroreflectometer*, ASTM Publication No. E-1696 (1995).
22. *Standard Test Method for Measurement of Retroreflective Signs Using a Portable Retroreflectometer*, ASTM Publication No. E-1709 (1995).
23. *Standard Specification for Preformed Plastic Pavement Marking Tape for Extended Service Life*, ASTM Publication No. D-4505 (1992).
24. *Standard Specification for Extended Life Type, Nonplowable, Prismatic, Raised, Retroreflective Pavement Markers*, ASTM Publication No. D-4280 (1995).
25. *Standard Specification for Plowable, Raised Retroreflective Pavement Markers*, ASTM Publication No. D-4383 (1995).
26. *Standard Specification for Retroreflective Sheeting for Traffic Control*, ASTM Publication No. D-4956 (1995).
27. *Service Life of Retroreflective Traffic Signs*, FHWA Publication No. FHWA-RD-90-101 (1991).
28. Rennilson, J. J., ''Definition, Measurement, and Use of the Spectral Coefficient of Retroreflection,'' Color Res. Appl. **7**, No. 4, 315–318 (1982).
29. Venable, Jr., W. H., and Johnson, N. L., ''Unified Coordinate System for Retroreflectance Measurements,'' Appl. Opt. **19**, 1236–1241 (1980).
30. Venable, Jr., W. H., Stephenson, H. F., and Terstiege, H., ''Factors Affecting the Metrology of Retroreflecting Materials,'' Appl. Opt. **19**, 1242–1246 (1980).
31. Johnson, N. L., and Stephenson, H. F., ''Influence of Aperture Size on the Photometry of Retroreflectors,'' Appl. Opt. **19**, 1247–1252 (1980).

10

Colorimetry

János D. Schanda

Commission International de l'Eclairage, Central Bureau, Kegelgasse 27, A-1030 Vienna, Austria and Department for Image Processing and Neural Computing, University Veszprém, Hungary

10.1 INTRODUCTION

The term *color* is used with a number of different meanings in different technologies. Lamp manufacturers understand the physical stimulus produced by their lamps. In everyday life often the pigment is termed *color*; in psychophysics the human sensation is called *color*. Thus it is wise to distinguish between color sensation or color perception, the very fundamental aspects of color that are formed by our visual system and our brain, and the color stimulus, the physical quantity producing the color perception. Materials used for coloring objects can be pigments, inks, paints, etc., but should not be called color.

Colorimetry is the science and technology used to quantify and describe by the help of mathematical models the human color perceptions.

As color perception is a psychophysical phenomenon, to understand colorimetric methods it is proper to give first an overview of the physiological basis of color perception. Then some psychophysical experiments that lead to a quantified description of color matching will be discussed. Based on these findings a standardized method for describing tristimulus color matches, the Commission Internationale de l'Eclairage (CIE) system of colorimetry will be presented, followed by the introduction to different color spaces and color appearance models.

From the point of view of photometry one of the most important aspects of colorimetry is the quantification of the color-rendering properties of light sources. This will be dealt with together with the introduction of the concept of correlated color temperature.

Color measurement is performed by spectrophotometric, spectroradiometric, and tristimulus measurements. The technique to conduct such measurements will be discussed together with the description of the fundamentals of the instruments.

10.2 VISUAL BASIS OF COLORIMETRY

10.2.1 Human Visual System

Figure 10.1 shows a probable model of the human visual system. Part 1 of the visual system is the optics of the eye, a living camera, that focuses a picture of the visual field onto the retina (part 2 in the figure). The optical nerves of the two eyes are combined and then split in the optic chiasma (part 3), the switching board of the

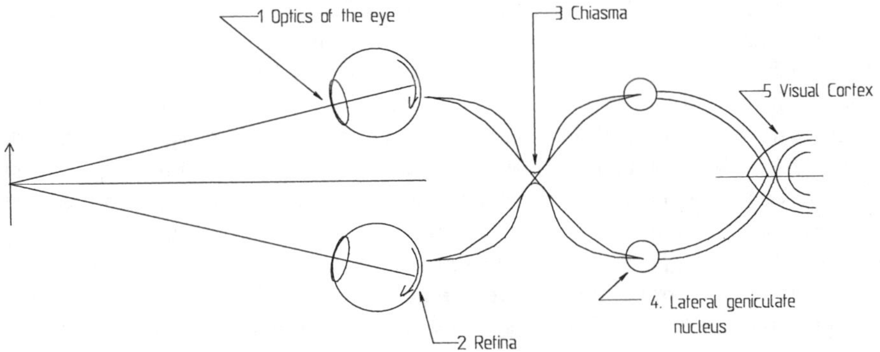

FIGURE 10.1. Model of the human visual system.

optical signals. Within the lateral geniculate nucleus (part 4) the optical information is subdivided into parts relating to color, motion, form, and depth information, which are first processed separately by the brain and then reunited into a mental picture (part 5). Our knowledge on the functioning of the different parts of the optical system gets fainter and fainter as we move from part 1 to part 5.

Figure 10.2 shows the cross section of the eyeball in more detail. Most of the refractive power of the eye comes from the cornea. The ciliary muscle can change the shape of the eye lens, so that the joint refractive power of the cornea and the lens can focus objects at different distances onto the retina. When we look at an object, we try to focus its picture onto the fovea, the part of the retina with the highest

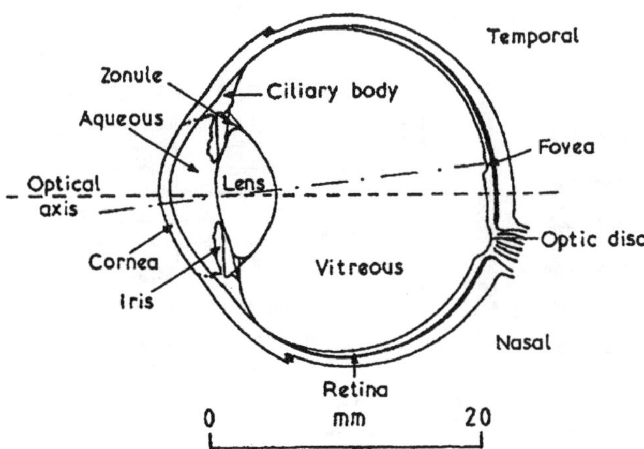

FIGURE 10.2. Cross section of the human eye. [From *Handbook of Optics*, edited by M. Bass (McGraw–Hill, New York, 1995). Reprinted by permission from McGraw–Hill Companies.]

visual acuity. (Observe that the visual and optical axes of the eye are approximately 4° apart from each other.) The iris is the colored annular shaped part of the eye that can be seen from the outside. Its shape changes with light intensity. The ''black hole'' in the middle of the eye is the pupil; its diameter depends on the shape of the iris and can change from about 2 mm diameter in bright light to about 8 mm in dim light. This is the system to adjust the light entering the eye. From fundamental optics a 1- to 16-fold intensity change can be calculated as the pupil diameter changes from 2 to 8 mm. In reality, rays that pass the peripheral regions of the pupil are less effective in producing a light sensation (Stiles–Crawford effect of the first kind). The dynamic range of our eye is much larger; further sensitivity adjustment is of physiological nature. To explain this, we have to investigate the human retina in more detail.

10.2.2 Human Retina

Figure 10.3 is a schematic drawing of a part of the retina. Light is reaching the retina from the pupil through the vitreous body. First it has to penetrate the layer of the optical nerve, layers of different cells (ganglion cells, amacrine cells, bipolar cells) before it reaches the layer of the rods and cones responsible for the transformation of optical radiation into light perception. The back of the eye is covered by an almost black layer (the pigment epithelium) that absorbs the light not being absorbed effectively in the rods and cones, and thus decreases the amount of scattered light produced in the eye.

In the human retina there are about 6.8 million cones responsible for vision at daytime light levels and evoking color vision, and about 115 million rods, functioning mainly at nighttime and providing peripheral vision. The distribution of rods and cones is not uniform throughout the retina. The retina has a small pit called the

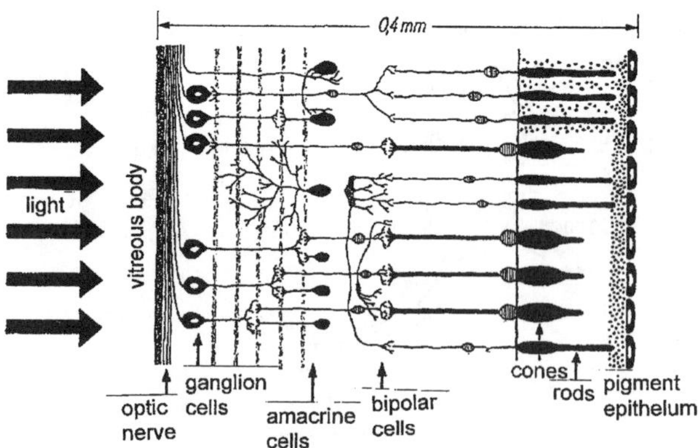

FIGURE 10.3. Schematic drawing of a part of the retina. [Reprinted by permission from M. Richter, *Einführung in die Farbmetrik*, 2. Auflage (Walter De Gruyter, Inc., Berlin, New York, 1981).]

fovea. In the fovea we have the highest visual acuity. The approximate diameter of the fovea is—in angular measurement—1.5°. Here the rod population drops considerably. In the central part of the fovea (approximate diameter 1°), called the foveola, no rods are present. (This is the reason, e.g., that if we look at a faint star, it can be observed by looking at it indirectly; when trying to look at it directly, it seems to disappear.) Figure 10.4 shows the distribution of rods and cones within the retina. Here we see also that in the direction of the blind spot, where the optical nerve leaves the eye, the eye is insensitive. It is interesting that we do not observe this, not even in monocular vision: the brain fills in the missing information. This phenomenon becomes visible only if one specifically searches for it.

There are several other features of the interior of the eye that have not been dealt with here. We would like to draw attention to only one single entity not shown in the drawing of the eyeball: The *macula lutea* or *yellow spot*: a pigmented area of approximately 4° visual angle covering the fovea and its neighboring regions. Owing to individual changes in the pigmentation of this layer and to the discoloration of other transmissive parts of the eye, the spectral sensitivity of individual observers varies, and it also changes with age [4,5].

As already mentioned, the rods are our receptors at very low light levels, at the so called scotopic luminance levels (below 10^{-3} cd m^{-2}, as, e.g., at moonlight). The rods contain a single photopigment that absorbs the light and produces neural signals. These signals are preprocessed in the layers of the different bipolar, amacrine, ganglion, etc., cells and fed via the optical nerves to the brain. Owing to the fact that the rods contain only a single type of photopigment—called

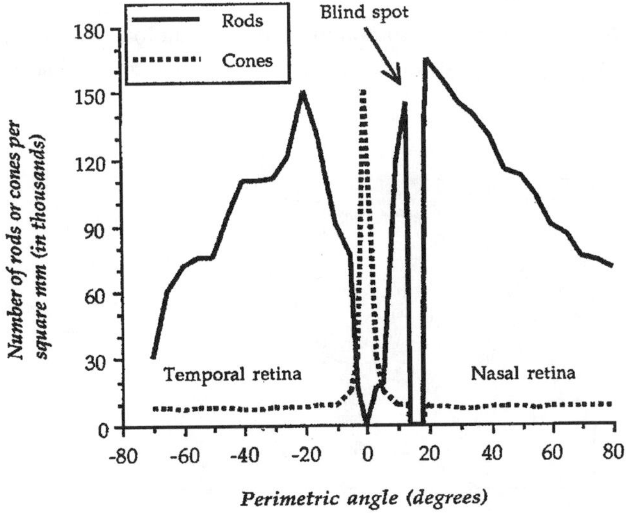

FIGURE 10.4. Distribution of rods and cones within the retina. [Reprinted by permission from D. Travis, *Effective Colour Displays* (Academic Press Ltd., London, England, 1991).]

rhodopsin—the rod vision is color-blind: at low light levels we cannot distinguish among different colors, only light-intensity differences are sensed. The rod absorption spectrum and the psychophysically determined $V'(\lambda)$ function (see Chap. 2) show reasonably similar spectral dependence, so that one can be pretty sure that the psychophysically determined curve describes the physiological visual mechanism. The $V'(\lambda)$ function has been reproduced in Fig. 2.1 (Sec. 2.3).

To determine the primary physiological color vision mechanism, i.e., the cone photopigment absorption spectra is not that easy. Microspectrophotometric measurements of human retinal tissue have shown that there are three distinctly different photopigments in the human cones. The measurements show reasonable agreement with so called suction electrode photoelectric measurements [6] obtained on monkey cone cells where the minute electric currents produced by the illuminated cone sucked into a micropipette were investigated. Thus from physiological measurements one can assume that the cones of the human retina contain three different photopigments, where two have rather similar spectral sensitivity and the third one is distinctly different. It is usual to call the three cone sensitivities long-wavelength sensitive (or L), medium-wavelength sensitive (or M), and short-wavelength sensitive (or S) cone sensitivity (see e.g., [7]). [In older literature we find also the symbols R (red), G (green) and B (blue), or to distinguish them from real light the symbols ρ, γ, and β, to identify these visual sensitivities.] The precision of direct physiological investigations is not sufficient to develop from these measurements cone-sensitivity functions with an accuracy high enough to be used to model the visual mechanism. They serve merely for comparison with cone-sensitivity functions derived from psychophysical experiments (e.g., heterochromatic brightness matching experiments). Some psychophysically derived cone excitation functions are seen in Fig. 10.5.

The L, M, and S cones are distributed more or less randomly in the retinal mosaic of receptors, but their concentration in the retina is not equal. The relative abun-

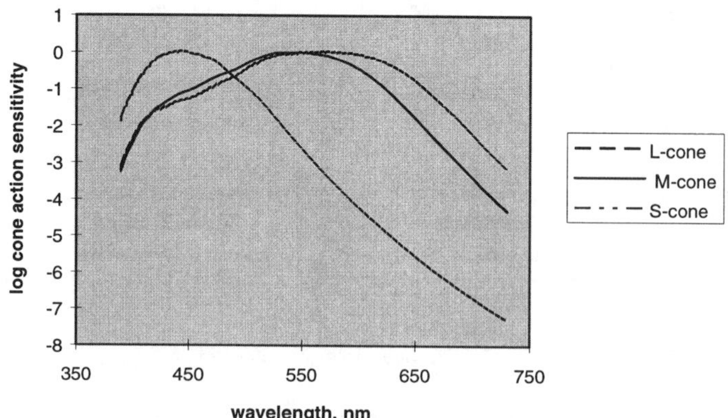

FIGURE 10.5. Relative logarithm of the cone action spectra.

dance of the L to M to S cones is approximately 40:20:1 [8]. This ratio holds for a large part of the retina. The central part of the fovea, the foveola, however, lacks S cones; thus it is blue color blind, or tritanop. (In everyday life we do not realize this because our eyes permanently scan the visual field, and via the so-called *micronystagmus* and *saccades* the brain receives more than one packet of information on the field just viewed.)

One can assume that the primary visual physicochemical process leading to color sensation is the absorption of photons by the cone photopigments, producing electrical signals. We will not go into detail on the production of these signals (see, e.g., [6]), we would only like to mention that there is physiological evidence that the primary signals are processed by the layers of the bipolar, amacrine, and ganglion cells, leading to a signal (pair) that is coded to light intensity and two signals (pairs) that are color coded. The primary photochemical process is most probably a linear one: the primary photoelectric signal produced by a cone is linear, depending on the number of photons absorbed by the cone; the subsequent coding processes are, however, highly nonlinear. This makes the investigation of the visual processes by psychophysical techniques (experimental techniques where a psychical response is registered, produced by a physical stimulus) most difficult.

As a working model the retinal color vision can be modeled by the scheme as seen in Fig. 10.6.

The nerve pulses can be followed by physiological techniques, and it can be shown—with some simplifications—that three kinds of signals are carried along the visual pathway. All signals are ''frequency coded,'' i.e., in an unexcited state the nerve cells fire at a given rate, and if excited, the firing rate will increase or decrease. The three kinds of signal carriers are:

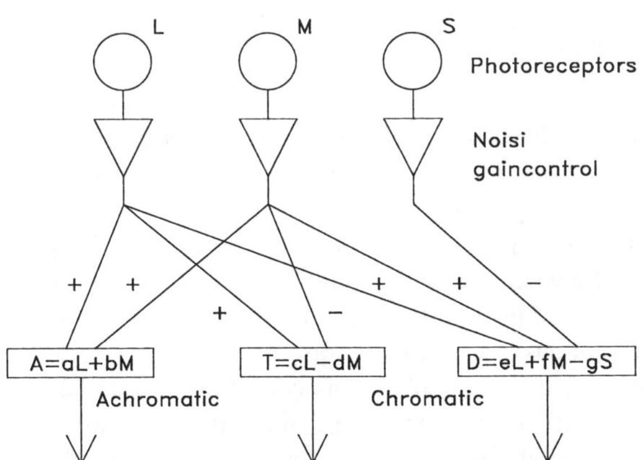

FIGURE 10.6. Schematic model of initial stages of human color vision: The L, M, and S cone signals are processed in a noisy gain control system and fed into an achromatic channel and two opponent chromatic channels. *a–g* give percentage contributions of the three cone processes in the establishment of the achromatic and chromatic signals (nerve pulse chains).

- Nerves that respond to optical excitation differences of neighboring sites (contrast-sensitive cells); at present it is assumed that only the L and M cone signals contribute to this achromatic, contrast signal.
- Nerves that respond if the excitation becomes overwhelmingly red or green; this is an opposing L versus M excitation.
- Nerves that respond if excitation becomes overwhelmingly yellow or blue; produced by antagonistic excitations coming from the added L plus M cone excitation versus the S cone excitation.

Also differences in response speed have been observed, where the contrast channel proved to be the quickest and the blue channel inherently slower than the other two.

Physiological experiments have shown that, with some simplification, beyond the retina the achromatic (contrast) information is carried by the magnocellural pathway and the chromatic information by the parvocellular pathway (see, e.g., [9,10]).

10.2.3 Color Deficiency

Human color vision varies from one observer to another. The majority of human observers belongs to one class, having very similar cone spectral sensitivity functions. As they are the majority, their color vision is called *normal*. What has been discussed up to now is valid for the so-called normal observer. About 8% of men and less then 0.5% of women show congenitally different cone spectral sensitivities. They are called color-deficient observers, as most of them can differentiate less precisely among colors than the normal observer. It is usual to classify the color-deficient observers into the following main classes:

- Dichromats;
- Anomalous trichromats;
- Monochromats.

Dichromats cannot distinguish among all the colors that normal observers (often called also normal trichromats) can. They can produce color matches using only two—not three—properly chosen primary colors.

The most common type of dichromatism is *protanopia*. Protanopes find it difficult to distinguish between red and green hues. The red sensitivity of protanopes is much lower than that of normal trichromats (approximately 1% of men and 0.02% of women are protanopes). A probable cause of protanopia could be that the L cones are missing (and probably their place has been filled up by M cones).

Deuteranopia is a dichromatism, where again the discrimination ability between red and green hues is impaired, but no color appears appreciably dimmer than that for normal trichromats. It is unclear whether in the case of deuteranopia the M cones are missing or whether it is caused by a second-stage abnormality, an impairment of the red–green channel. About 1.1% of men and 0.01% of women are deuteranopes.

Tritanopia is a very rare form of dichromatism. It manifests itself in a severely reduced discrimination of the bluish and yellowish content of colors. A probable cause could be missing S cones. As S cones presumably do not contribute to the luminance channel (see Fig. 10.6), there is no loss in luminance perception. We have seen that there are no S cones in the foveola and therefore the foveola is tritanopic. Only about 0.002% of men are tritanopes; in women this type of dichromatism is extremely rare.

Monochromatism can have two forms: *Rod monochromats* have no color discrimination ability and their brightness sensation resembles that in scotopic vision. Most probably their cone system does not function. Its occurrence frequency is 0.003% in men and 0.002% in women.

Cone monochromats have no color discrimination ability whatsoever, but approximately normal brightness sensation. This deficiency might be caused by either missing of both the M and S cones or problems with the color-difference channels.

Anomalous trichromats can discriminate among every color, similar to normal trichromats but their ability to do this is different from that of normal trichromats.

Protanomaly manifests itself in some reduction in the discrimination of the reddish and greenish hues, with reddish colors appearing dimmer than normal. It might be caused by a somewhat shifted absorption spectrum of the L cones. Protanomalous observers make up about 1% of the male and 0.02% of the female population.

Deuteranomalous observers have some reduction in the discrimination of the reddish and greenish content of colors, which might be caused by an abnormal M-cone spectrum, having its maximum shifted towards that of the L cones. Deuteranomaly is the most common form of anomalous trichromacy. Almost 4.9% of the male population and 0.38% of the female population show this deviation from normal trichromacy.

Recent molecular genetic investigations (see, e.g., [11,12]) showed correlation between different kinds of anomalous forms of trichromacy and a single amino acid substitution at a given position of the M- and L-cone pigment.

10.2.4 Color-Pseudostereopsis

Before we leave the realm of physiology, we have to deal with one further subject: *color pseudostereopsis*, a phenomenon in which red and blue letters printed on a dark background look as if they were not in the same plane. Figure 10.7 shows the phenomenon. For most observers the red letters are standing out in front of the plane and the blue letters are lying behind it. Some observers might not see the phenomenon or even might find an opposite effect. An explanation of the effect, after Hunt [13], is the following: The effect is caused by the fact that the pupils of the eyes are not always centered about the optical axis. The eye is unable to focus both the red light and the blue light simultaneously onto the retina. As known, blue light has always a larger refraction than red light. Thus if the pupils of the observer are displayed outwards relative to the optical axis, as seen in Fig. 10.7(a), the rays

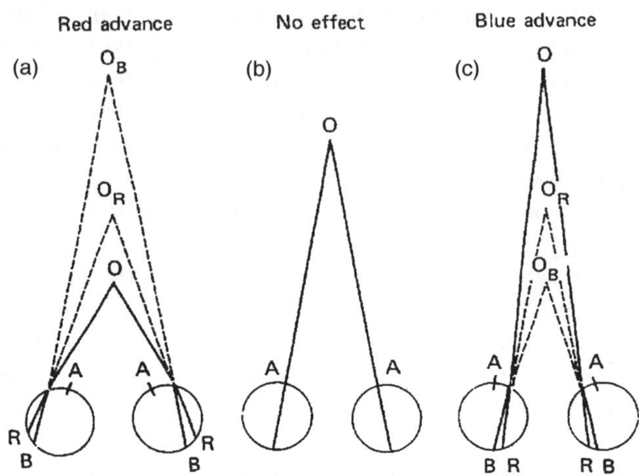

FIGURE 10.7. An explanation for pseudostereopsis, after Hunt. [From R. W. G. Hunt, *Measuring Colour* (Ellis Horwood Ltd., Chichester, U.K., 1987).]

from an object at O are dispersed by refraction: the images of blue (B) light in the two eyes will be closer to one another than in the case of red (R) light. The red light therefore appears to emanate from an object that is closer than an object from which blue light appears to emanate. Figure 10.7(b) shows the situation for an observer who sees no color pseudostereopsis effect, and Fig. 10.7(c) shows the situation for an observer whose pupils are displaced inwards relative to the optical axis and therefore for whom the effect is reversed. Color pseudostereopsis has to be considered, e.g., in computer display color design, as the unconscious readjustment of the focusing of the eyes is fatiguing.

10.2.5 Color Vision Model

Based on the considerations above, a color vision model can be constructed. Such a model is always an oversimplification of the visual mechanism, but it helps to memorize the functioning of the system and to explain the fundamental experimental findings.

First Level: Cone Signals

Both the physiological data and the psychophysical finding that colors can be matched with the mixture of three independent colors (see Sec. 10.3) suggest that the primary visual mechanism is trivalent. We have seen that there is enough evidence to assume three cone types in the retina (L, M, and S cones) as shown in Fig. 10.5. In that retinal area where we have good color and form recognition capability [a central region of approximately 10°–15° diameter, but not in the very central region of less than 1° diameter (the foveola)] the relative abundance of the three cone classes is, as mentioned, L:M:S=40:20:1. The L and M cones behave rather similarly, their spectral sensitivity is not very much different, their speed of response is similar, etc. The S cones behave differently: they respond much slower than the L and M cones, and their spectral sensitivity is also markedly different; there are also much fewer S cones in the retina than L or M cones.

Second Level: Antagonistic Colorsignals

Anthropologists speculate that the L and M cones differentiated from each other only at a late stage of evolution and still show many common features. One of these could be that the L- and M-cone signals combine to form the *luminance channel* of vision, i.e., one of the visual pathways from the eyes to the brain contains signals, where both L and M cones feed into spatially antagonistic structures, leading to contrast information signals. It is possible to build from the spectral sensitivity of the L and M cones a spectral sensitivity curve that practically duplicates the $V(\lambda)$ function. In Fig. 10.6 the M- and L-cone signal represents this schematically. A signal in which L and M cones work antagonistically against each other propagates in a color-difference channel, carrying red versus green information. A second color difference channel could combine the M- and L-cone response antagonistically with the S-cone response, producing a yellow versus blue signal. Neurophysiological investigations have shown the existence of such pathways. All three signals have such structures that there is a quiescent firing rate when there is no excitation in the pathway (i.e., there is no light stimulation difference between a small central area and a larger surrounding area for the luminance channel, and it is achromatic for the two color-difference channels). As soon as differences occur in the center and edges, and/or red–green, and/or yellow–blue primary excitation, the firing rate in the respective channels will change (increase or decrease) and the information on luminance and hue is mediated in going to the brain.

In this simplified model other couplings between the single levels of neural pathway have been omitted. Such couplings function as feedback to control the overall sensitivity and relative sensitivity in the different pathways, and make the system nonlinear. Also the influence of the rods has been left out of this picture. Their signal most probably combines with the L and M signals. Some authors also consider inputs from the S cones into the M–S and L–M channels, but most recent publications assume the model as depicted in Fig. 10.6.

The above-noted model of color vision corresponds well with the psychophysical observation that in human perception a mental picture exists of pure red, yellow, green, and blue, and we can well think of a yellowish green or red but never of a yellowish blue. Similarly red and green are exclusive, no reddish green exists (or vice versa).

Third Level: Mental Processing

The first two levels of color perception are—at least at this elementary level—reasonably well understood. Far less is known about the final stages of color perception function. There are a few peculiarities of color vision, most probably stage-three phenomena, which are important for understanding the colorimetric issues of photometry.

Color constancy is one phenomenon whose interpretation can be given only partly on the basis of first- and second-stage mechanisms: If daylight and incandescent light are seen side by side it is apparent that the incandescent light is much more yellow than daylight. If, however, we look at a white paper, illuminated by daylight or incandescent light only, our visual mechanism adjusts itself to the illuminant and in both cases we will say that paper is white. Part of this adjustment—chromatic adaptation—is probably taking place at retinal levels: It can be due to the bleaching of one of the cone pigments and to selective gain control between the different channels. The phenomenon is, however, partly mental: for well-known objects the color of illumination can be greatly changed and the observer will still seem to see the object in its original color.

Also, a further ''transformation'' of color information takes place in the brain: From the luminance and two opposing color-difference signals, brightness, hue, and saturation perception is formed: luminance, determined by flicker-photometric experiments (see Chap. 2), corresponds to brightness only if the color is almost achromatic. Highly chromatic colors show higher brightness at the same luminance level as the less chromatic ones (so-called Helmholtz–Kohlrausch effect). In the concept of a model one assumes that the brain adds information from the chromatic (i.e., color-difference) channels to the luminance information in creating the perception of brightness. (As we will see in Sec. 10.5.2, some model builders try to construct a brightness descriptor using vector addition analogy.)

The concept of *hue*—as used in the elementary description of colors based on the hue circle—is also a mental concept, and we do not find its physiological analogy in level-one or level-two color perception.

Further features, where higher mental processing seems to come into play, are *lightness* and *saturation*. The color stimulus should (transformed to the mental picture) produce signals of brightness, hue, and *chromaticness* (called also *colorfulness*). As mentioned, brightness can be thought to be composed of the achromatic signal and the chromatic signals. Hue is somehow related to the chromatic signals, and for an engineering approach it is easy to visualize it as the angle between a starting point in a three-dimensional diagram, as seen in Fig. 10.8, and the particular hue described by the vectorial sum of the red–green and yellow–blue signals. Also the ''strength'' of the color, the chromaticness, can be visualized in such a diagram. The human visual system has, however, the peculiar property that looking at a real scene it relates the brightness and the chromaticness to that of a mental reference to white: to the brightness of a surface in the picture that it assumes to be white. Thus, if in a scene the brightness of a particular object is decreased, e.g., due to the fact that it is moved from direct sunlight into a shaded area, the perceived *lightness* of

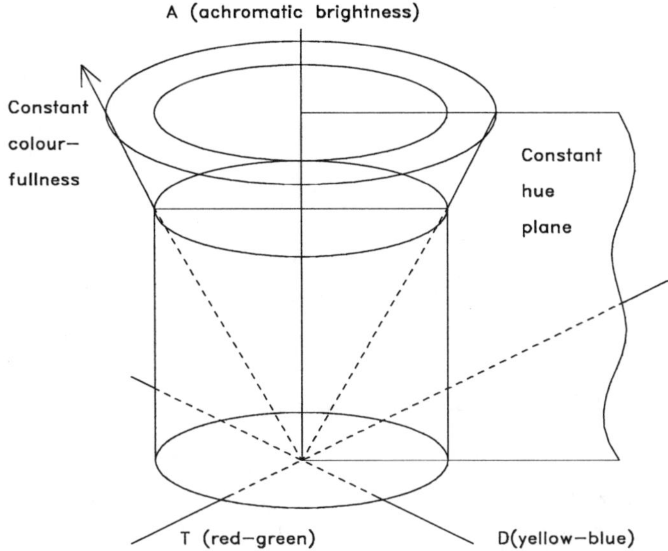

FIGURE 10.8. A three-dimensional representation of the mental representation of a color perception.

this object will stay unchanged. We are able to distinguish a white surface that is in the shadow from a grey one in sunlight, even if the stimulus reaching the eye from the white surface (i.e., its luminance) is lower than that of the grey one.

Similarly, for colors of objects (so-called related colors) the absolute length of the colorfulness arrow is not the most important mental characteristic of what is the result of the color channel signals, but a quantity somehow related to the apparent brightness. This characteristic is called *saturation*, and is something that seems again to stay constant when the colored object is moved from a place with high illumination to one where the illumination level is lower.

In the third level of the visual mechanism, different features, such as form, movement, or color, of the real world seen are preprocessed at different locations of the brain and then are integrated into the final mental picture. Although little is known in this field, the area of physiological and psychophysical research is developing fast, and it is a continuous endeavor in engineering research to apply the newest findings of vision research. For further reading on the topography of the human visual cortex see the excellent review of Schwartz [14].

10.3 PSYCHOPHYSICAL EXPERIMENTS TO QUANTIFY COLOR MATCHES

10.3.1 Additive-Color-Mixture Experiments

The fundamental experiment in colorimetry is depicted in Fig. 10.9. A test color stimulus C is selected. This color stimulus can be an arbitrary light, colored or

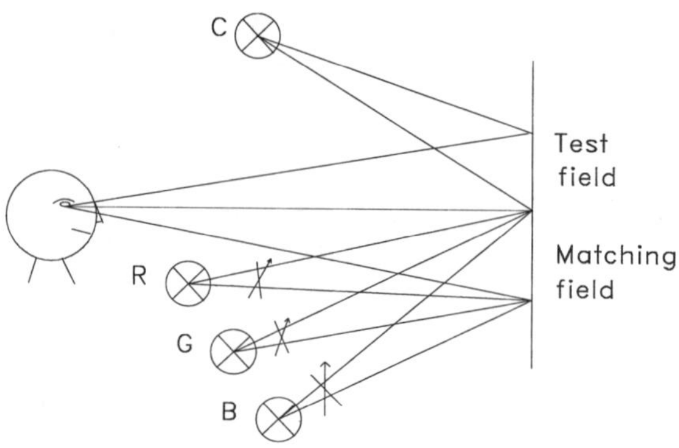

FIGURE 10.9. Basic color-matching experiment.

white, but its intensity should be so high that when the light reflected from the white screen is viewed, as depicted in the figure, the eye should be stimulated in the photopic region, preferably to a level as in normal daylight (i.e., an illumination of the white screen of approximately 1000 lx should be set). This color stimulus illuminates one half of a bipartite field (the test field) in the masking screen and is observed by the observer's eye. The other half of the bipartite field (the matching field) is illuminated by the additive mixture of three lights. *Additive mixture* means here that the three lights irradiate the common white surface in such a form that by viewing this surface the three sources cannot be seen as independent light sources any more, only the mixed light reflected from the screen can be perceived. There is just one constraint on the selection of the color of the three sources: the three colors have to be independent, i.e., it should not be possible to produce the color of one of the lights via mixing given proportions of the other two. It is convenient to select saturated red (R), green (G), and blue (B) lights as constituents of the additive mixture. Means to adjust the intensity of the three sources are shown as well. These are essential parts of the experiment. The intensity of the three lights has to be chosen so that they are strong enough to produce matches with the test stimulus. It is convenient to choose the units of the three lights (i.e., unity setting of the light-adjusting means) so that they produce a white mixture.

Let us first choose a color stimulus C_1 (e.g., the light of an incandescent lamp filtered with an orange filter). One can adjust the intensity of the R, G, and B lights to match this light. These quantities shall be given as R_1, G_1, and B_1 (we will call them the *tristimulus values*). In mathematical form one describes the color match as

$$C_1 \equiv R_1 + G_1 + B_1, \tag{10.1}$$

where the relation symbol $\equiv$ is read as "matches."

Let us now choose a second color stimulus C_2 (e.g., the light of the incandescent lamp filtered with a green filter). This will be matched by the quantities R_2, G_2, and B_2:

$$C_2 \equiv R_2 + G_2 + B_2. \tag{10.2}$$

Experimenting with different C color stimuli, one will see that to every color a unique setting of the R, G, and B values is needed. In an experiment with saturated test colors, we may observe that we are unable to establish a color match. By the mixture of two matching stimuli, we reach an approximate match, and adding the third one makes the match worse. Slightly modifying the experimental setup the color match can be established: the third light whose intensity was set to zero to establish the best approximation is redirected to fall onto the test half screen, i.e., one establishes a two–two-color match. Let us assume for simplicity that the red matching stimulus is the one that behaves peculiarly; naturally with other test stimuli it could be the green or the blue just as well:

$$C_3 + R_3 \equiv G_3 + B_3. \tag{10.3}$$

As we would like to express the test stimulus by the three matching stimuli we rearrange the equation to read:

$$C_3 \equiv G_3 + B_3 - R_3. \tag{10.4}$$

The adding of one of the matching stimuli to the test stimulus is mathematically expressed as subtracting the stimulus on the matching side (naturally there is nothing like a negative color stimulus).

10.3.2 Grassmann's Laws

As a next step in our experiments let us irradiate the test field side once with C_1 and C_2 separately and then with the sum of C_1 and C_2 (with the orange and green lights of the previous paragraph it will produce a yellow light patch). We will observe that color match is obtained if the matching field is illuminated with intensities R_1+R_2, G_1+G_2, B_1+B_2, i.e., color matching is additive.

Mathematically one can formulate the laws of additive color matching in the following form [15]:

1. *Symmetry law:* If color stimulus A matches color stimulus B, then color stimulus B matches color stimulus A.
2. *Transitivity law:* If A matches B and B matches C, then A matches C.
3. *Proportionality law:* If A matches B, then αA matches αB, where α is any positive factor.
4. *Additivity law:* If A,B,C,D are any four color stimuli, then if any two of the following three conceivable color matches hold:

$$A \equiv B, \quad C \equiv D, \quad A + C \equiv B + D, \tag{10.5}$$

then so does the remaining match

$$(A+D)\equiv(B+C), \tag{10.6}$$

where $(A+C)$, $(B+D)$, $(A+D)$, and $(B+C)$ denote, additive mixtures of A and C, B and D, A and D, and B and C, respectively.

These four laws are often called the stronger form of trichromatic generalization of color matching. It is a mathematical description of what was stated by Grassmann in 1853–1854 [16]. A modern verbal formulation of Grassmann's laws is given by Hunt [13] in the following form:

1. To specify a color match, three independent variables are necessary and sufficient.
2. For an additive mixture of color stimuli, only their tristimulus values are relevant, not their spectral composition.
3. In additive mixtures of color stimuli, if one or more components of the mixture are gradually changed, the resulting tristimulus values also change gradually.

There are some constraints that have to be considered if Grassmann's laws are applied.

1. Matches have to be always made under similar observational situation. If an observer sets a match where the fields have different spectral compositions, the match will not necessarily hold if the observer's eyes move, the observer does not look centrally at the fields, or the sizes of the fields are changed.
2. The light exposure of the eyes will effect the state of *adaptation*. Although the persistence of color matches is rather high, with very high irradiation the spectral sensitivity of the eye can be influenced.
3. If a larger area of the retina (e.g., a 10° field) is used for color match, a failure of the proportionality law is found.

10.3.3 Towards a System of Colorimetry

Grassmann's laws will be followed as long as above constraints are obeyed, and a colorimetric system based on additivity and proportionality can be used.

As will be seen in Sec. 10.4, such a system can be built on the analogy of the photometric system. Since color is three dimensional (we need three independent variables to describe color, or to obtain a color match), three integrals of this form (see Sec. 2.3) will be needed to describe color.

$$Ti=k\int_{380\ \mathrm{nm}}^{780\ \mathrm{nm}} \Phi_{e\lambda}\overline{t_i}(\lambda)\,d\lambda, \quad i=1,2,3 \tag{10.7}$$

where $\overline{t_i}(\lambda)$ stand for three weighting functions of the stimulus, called color-matching functions, and T_i describes the color stimulus in the trichromatic system.

The principles to reach an expression as Eq. (10.7) are the following: First, the units of the matching stimuli have to be set. This can be done, e.g., by setting them so that the mixture of the unit amounts provide a color match with a particular white stimulus, such as that produced by an equienergetic spectrum (having equal amounts of energy per wavelength interval throughout the visible spectrum). Using the symbols [R], [G], [B] for these unit amounts of matching stimuli, Eq. (10.1) can be written in the form:

$$C_1 \equiv r_1[\mathrm{R}] + g_1[\mathrm{G}] + b_1[\mathrm{B}], \tag{10.8}$$

where r_1, g_1, b_1 represent the amount of light taken from the [R], [G], [B] matching stimuli (called also tristimulus values). C_1 has a spectral distribution $[C_1(\lambda)]$ and one could write an equation like (10.8) for every wavelength band of the spectrum and then add these to obtain C_1. The additivity and multiplicativity of colorimetry permits this and via this technique we can reach an equation like (10.7), where $\overline{t_i}(\lambda)$ provide the monochromatic constituents of the tristimulus values (they were actually called spectral tristimulus values; at present the term *color-matching functions* is preferred, since by the help of them color matching is performed).

As will be seen from the next section, this principle can be used in praxis; the CIE system of colorimetry is built around these fundamental experimental findings.

10.4 CIE COLORIMETRY

10.4.1 Color Equations

We have seen, that despite the fact that color is a perception and underlies individual variations, some fundamental rules can be elaborated. These permit predictions to outline, how colors change, when the stimuli are changed, and how color matches can be obtained. Based on them, a color measuring system enabling the specification of colors can be worked out, at least to that extent that colors with the same numerical specification will look similar to the vast majority of observers. CIE (from its French name, Commission Internationale de l'Éclairage, International Commission on Illumination) is the international organization that undertook the task of developing these specifications. As seen in the previous section, colorimetry needs conventions. Such conventions are necessary because two color stimuli that look similar under one viewing condition might look different when seen under other conditions (field of view, adaptation, direction of viewing, etc.; see Sec. 10.3.2).

Condition 1 of CIE colorimetry is that the color matches on which the system is based have to be performed with an approximately 2° bipartite visual field, central fixation, and dark surround.

The schematic drawing of such a trichromatic match experiment is shown in Fig. 10.10. In one of the basic investigations, performed by Wright [17] the monochromatic wavelengths 700, 546.1, and 435.8 nm were used as red (R), green (G) and blue (B) stimuli. (The red stimulus can be selected using an interference filter, and the green and blue stimuli are emission lines of a mercury discharge). In the trichromator means have to be available to set the intensity of the three matching stimuli and to measure these intensities. To construct the color-matching functions to be used in the integrals, as discussed in Sec. 10.3.3, monochromatic radiation has to be used as a test stimulus.

The three matching stimuli are mixed in a photometer sphere (often called Ulbricht sphere) and viewed via the exit port of the sphere. A mirror blends into one half of the 2° viewing field the light coming from the test stimulus. A black light trap absorbs the light of the test lamp not hitting the white reflecting plate. Two light baffles in the sphere (not shown in the figure) ensure that no direct light from the sources or the mixture field reaches the observer

Condition 2 of the CIE trichromatic match concerns the magnitude of the selected 700-, 546.1-, and 435.8-nm matching stimuli (called also reference color stimuli, primary stimuli, or primaries): The units of the three primaries have been set to such values that their mixture should provide a color match with an equienergetic white test stimulus, i.e., a stimulus where at every wavelength in the visible spectrum the same amount of power is propagating $[dS(\lambda)/d\lambda = S_\lambda = \text{const}]$. Taking 1 cd/m^2 of red light as 1 red unit, to achieve such a match 4.5907 cd/m^2 of green and 0.0601 cd/m^2 of blue light had to be taken. The mixture of these matched the equienergetic white stimulus of 5.6508 cd/m^2. The luminance of the three new units chosen is given in Table 10.1.

If now an arbitrary color is matched with the so selected R, G, B matching stimuli, the amounts of the three matching stimuli, expressed in the above new units are called (R,G,B) tristimulus values, and the color match is expressed as:

$$C \equiv R[\mathrm{R}] + G[\mathrm{G}] + B[\mathrm{B}], \tag{10.9}$$

where the relation symbol $\equiv$ is read as "matches."

If monochromatic test stimuli are used one observes that, e.g., a 520-nm green stimulus cannot be matched by any combination of the three matching stimuli. One

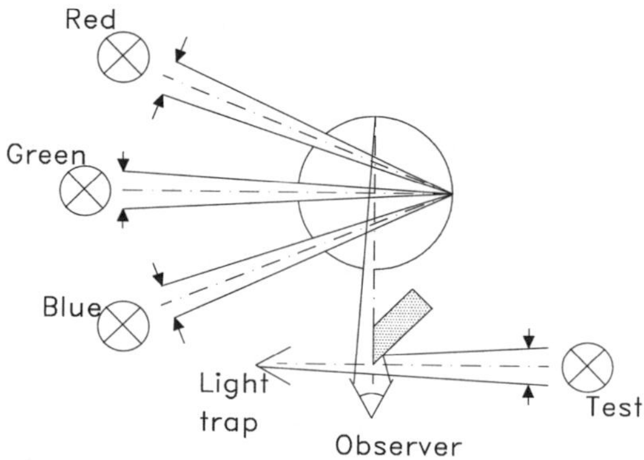

FIGURE 10.10. Schematic drawing of the principles behind a visual colorimeter.

TABLE 10.1. *Luminance of the selected R, G, B matching stimuli.*

Color	Luminance (cd/m^2)	
Red	1.000	1 new R unit
Green	4.5907	1 new G unit
Blue	0.0601	1 new B unit

comes nearest to the match if only blue and green matching stimuli are used, but an exact color match can only be obtained if some red light is mixed to the test stimulus:

$$C(520\ \text{nm})+R[\text{R}]\equiv G[\text{G}]+B[\text{B}], \tag{10.10}$$

i.e., the color mixture of given amounts of the green and blue matching stimuli will match the mixture of the test and red stimuli. This is shown schematically in Fig. 10.11. Here the pivoting mirror Mp in the red channel directs the light onto the side of the test stimulus to obtain a color match. Mathematically the fact that from one of the primaries light is mixed with the light of the test stimulus is written with a negative sign for that stimulus:

$$C\equiv -R[\text{R}]+G[\text{G}]+B[\text{B}]. \tag{10.11}$$

It should be, however, quite clear that this does not mean ''negative'' light!

Tristimulus Values and Color-Matching Functions

If between 380 and 780 nm all the monochromatic radiation of equal power within a small wavelength band are matched with the combination of the R,G,B primaries, the so-called color-matching functions (CMF) designated by the symbols $\bar{r}(\lambda)$, $\bar{g}(\lambda)$, and $\bar{b}(\lambda)$ are obtained. The CIE standardized such CMF as average

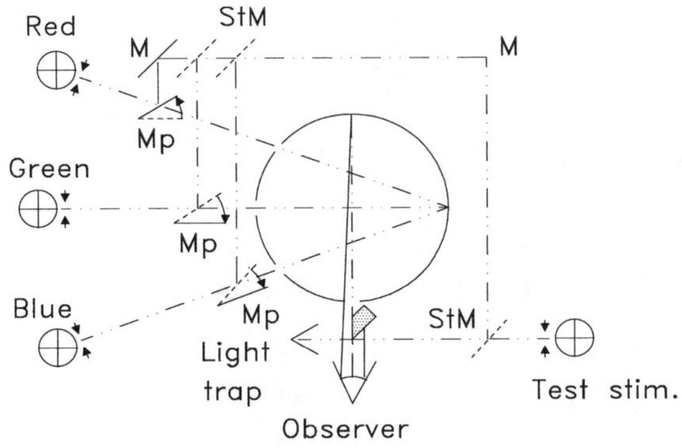

FIGURE 10.11. Schematic view of a visual colorimeter.

curves in 1931. They are called the *color-matching functions* of the *CIE 1931 Standard Colorimetric Observer* and are shown in Fig. 10.12.

As color matches are additive, if 1 unit of power of light with wavelength λ_1 $[C(\lambda_1)]$ is matched with the R,G,B primaries,

$$C(\lambda_1)\equiv R_1[\mathrm{R}]+G_1[\mathrm{G}]+B_1[\mathrm{B}], \tag{10.12}$$

and 1 unit of power of light with wavelength λ_2 $[C(\lambda_2)]$ is matched with the R,G,B primaries,

$$C(\lambda_2)\equiv R_2[\mathrm{R}]+G_2[\mathrm{G}]+B_2[\mathrm{B}], \tag{10.13}$$

then the additive mixture of the two monochromatic lights $C(\lambda_1)+C(\lambda_2)$ will be matched with the additive mixture of the two amounts of the primaries:

$$C(\lambda_1)+C(\lambda_2)\equiv(R_1+R_2)[\mathrm{R}]+(G_1+G_2)[\mathrm{G}]+(B_1+B_2)[\mathrm{B}]. \tag{10.14}$$

This reasoning can be extended to three, four, etc., monochromatic components, so that if the spectral power distribution of a stimulus is known its tristimulus values can be determined by adding the R_i, G_i, B_i weighted spectral constituents of the stimulus. The R,G,B tristimulus values of a stimulus with $P(\lambda)$ spectral power distribution are

$$R=k\sum_{380\ \mathrm{nm}}^{780\ \mathrm{nm}} P(\lambda)\bar{r}(\lambda)\Delta\lambda,$$

$$G=k\sum_{380\ \mathrm{nm}}^{780\ \mathrm{nm}} P(\lambda)\bar{g}(\lambda)\Delta\lambda, \tag{10.15}$$

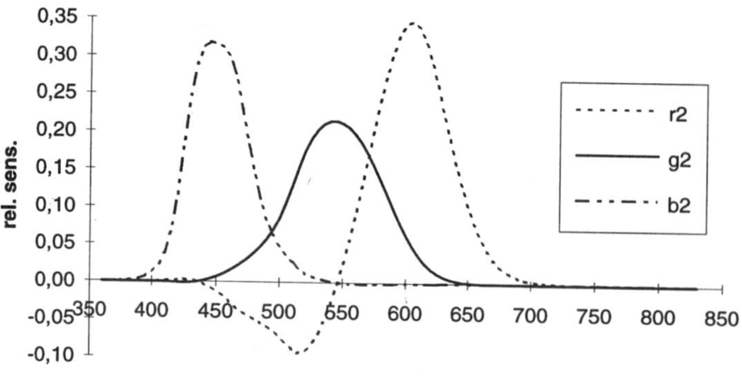

FIGURE 10.12. $\bar{r}(\lambda), \bar{g}(\lambda), \bar{b}(\lambda)$ color-matching functions of the *CIE* 1931 Standard Colorimetric Observer.

$$B=k\sum_{380\ \mathrm{nm}}^{780\ \mathrm{nm}} P(\lambda)\bar{b}(\lambda)\Delta\lambda,$$

or using the integral form

$$R=k\int_{380\ \mathrm{nm}}^{780\ \mathrm{nm}} P(\lambda)\bar{r}(\lambda)d\lambda,$$

$$G=k\int_{380\ \mathrm{nm}}^{780\ \mathrm{nm}} P(\lambda)\bar{g}(\lambda)d\lambda, \qquad (10.16)$$

$$B=k\int_{380\ \mathrm{nm}}^{780\ \mathrm{nm}} P(\lambda)\bar{b}(\lambda)d\lambda.$$

Here k is a constant that can be used to couple the colorimetric (photometric) quantities to the radiometric ones: As seen from Table 10.1 the luminance of a color matched by the amounts R red units, G green units and B blue units will be

$$L=1.0000R+4.5907G+0.601B. \qquad (10.17)$$

This means that the constant k can be chosen in such a way that if $P(\lambda)$ is a radiometric quantity, the L value will be a corresponding photometric quantity, e.g., if $P(\lambda)$ is measured in W/(sr m^2) then L will be provided in cd/m^2.

G, R, B Color Space

To visualize a trichromatic color system the **R, G, B** primaries can be thought of as the basis vectors of a three-dimensional vector space. Figure 10.13 shows such a representation. Here a plane has been drawn where the **R**, **G**, and **B** vectors have unity values. Also the cone of monochromatic color stimuli is shown. The **R**, **G**, and **B** vectors are located on the surface of this cone, as they are monochromatic lights

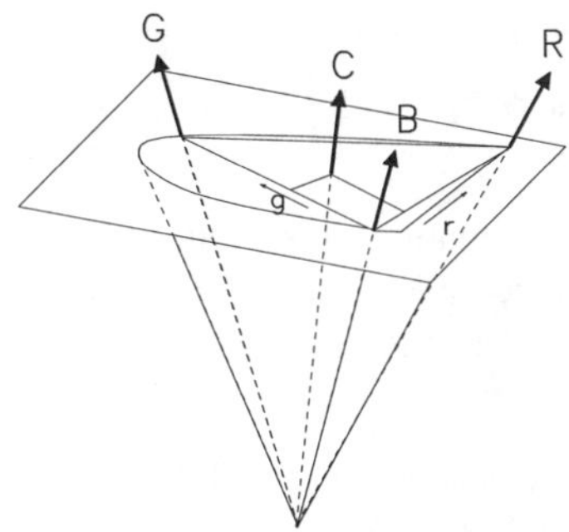

FIGURE 10.13. Schematic representation of the **R-G-B** color space.

(i.e., if the monochromatic test stimulus has the value of 700, 546.1, or 435.8 nm, only one of the R,G,B tristimulus values will be different from zero). The two extremes of the monochromatic spectrum locus (380 and 780 nm) have been connected by a straight line, and we will see that purple colors are located here. From this representation it is also easy to see that only color stimuli that lie within the tetrahedron traced by the three basic vectors can be realized with a simple additive mixture. For all other stimuli, thus also most of the monochromatic ones, a negative amount from one of the basic vectors has to be taken, i.e., in the color mixture one of the primaries has to be mixed with the test stimulus to achieve a color match. This is equivalent with the finding in Fig. 10.12 that the $\bar{r}(\lambda),\bar{g}(\lambda),\bar{b}(\lambda)$ functions show negative lobes and that at the wavelengths of the matching stimuli two of the functions are always zero and a single one shows a positive value.

10.4.2 (*X*, *Y*, *Z*) Color Space

CIE 1931 Standard Colorimetric Observer

For those who have a good concept of vector calculus, the construction of the **R**, **G**, **B** color space can be easily conceived in the following form: Let us assume an n-dimensional spectrum space, where a color stimulus is represented by an n-dimensional vector. The coordinates of this vector are the power (or radiance, etc.) values at the n wavelengths (e.g., an 81-dimensional space if the color vector has been defined by its values at 81 wavelength coordinates [(780−380)/5+1, if measurements have been performed at 5-nm wavelength increments between 380 and 780 nm]. The three-dimensional color space is a subspace of this spectrum space, where any three independent vectors can be used as basis vectors of the color space. Thus we can choose, e.g., basis vectors such that the total volume of the color cone is within the positive quadrant of the space. This has been done by the CIE at the time the CIE trichromatic system was built [18,19], as it was felt that the manipulation with negative values in the color-matching functions may lead to calculation errors.

The basis-vector transformation was done in such a form that

1. The tristimulus values of the color stimulus of the equienergetic spectrum should again be equal;
2. All the photometric information (luminance, if the stimulus is measured in radiance units) should be in a single value, i.e., one of the color-matching functions should be equal to the $V(\lambda)$ function;
3. The tristimulus values of all real colors should be positive and the volume of the tetrahedron should be as small as possible, i.e., the borders of the color cone should touch the tetrahedron at as many as possible places.

(For further details see Ref. [20].)

With the information above in mind the transformation adopted by the CIE was

$$\begin{vmatrix} X \\ Y \\ Z \end{vmatrix} = \begin{vmatrix} 2.768\,88 & 1.751\,75 & 1.130\,16 \\ 1.000\,00 & 4.590\,70 & 0.060\,10 \\ 0.000\,00 & 0.056\,51 & 5.594\,27 \end{vmatrix} \cdot \begin{vmatrix} R \\ G \\ B \end{vmatrix}, \tag{10.18}$$

or in an expanded form,

$$\begin{aligned} X &= 2.768\,88R + 1.751\,75G + 1.130\,16B, \\ Y &= 1.000\,00R + 4.590\,70G + 0.060\,10B, \\ Z &= 0.000\,00R + 0.056\,51G + 5.594\,27B. \end{aligned} \tag{10.19}$$

In these equations we see that the Y component is identical with Eq. (10.17) and thus contains the luminance information. If the matrix elements are multiplied with 0.176 967 [the coefficient of $Y(R)$ to get the unit Y value], the matrix gets a simple form,

$$\begin{vmatrix} 0.49 & 0.31 & 0.2 \\ 0.176\,967 & 0.812\,397 & 0.010\,636 \\ 0 & 0.01 & 0.99 \end{vmatrix}, \tag{10.20}$$

where only the Y component is carrying six precisely chosen significant digits, all other matrix elements having only two nonzero decimal digits. From this it is also easily seen that the X value is composed mainly from the R value, but the G and B values also have significant weight. The Z value is almost entirely composed by the B value.

The inverse transformation from the XYZ system into the GRB one can be done by using the inverse matrix of (10.18):

$$\begin{vmatrix} 0.418\,46 & -0.158\,66 & -0.082\,83 \\ -0.091\,17 & 0.252\,43 & 0.015\,71 \\ 0.000\,92 & -0.002\,55 & 0.178\,60 \end{vmatrix}. \tag{10.21}$$

One can see that all three matrices (10.18), (10.20), and (10.21) fulfill the requirement that an equienergetic stimulus has equal tristimulus values in the system.

Color-matching functions can be built in this system on the analogy of the $\bar{r}(\lambda)$, $\bar{g}(\lambda)$, and $\bar{b}(\lambda)$ color-matching functions as introduced in "Tristimulus Values and Color-Matching Functions" in Sec. 10.4.1 and shown in Fig. 10.12. The values of the $\bar{x}(\lambda)$, $\bar{y}(\lambda)$, and $\bar{z}(\lambda)$ color-matching functions are the tristimulus values of the monochromatic stimuli. Thus Eq. (10.22) holds for their transformation:

$$\begin{vmatrix} \bar{x}(\lambda) \\ \bar{y}(\lambda) \\ \bar{z}(\lambda) \end{vmatrix} = \begin{vmatrix} 2.768\,88 & 1.751\,75 & 1.130\,16 \\ 1.000\,00 & 4.590\,70 & 0.060\,10 \\ 0.000\,00 & 0.056\,51 & 5.594\,27 \end{vmatrix} \cdot \begin{vmatrix} \bar{r}(\lambda) \\ \bar{g}(\lambda) \\ \bar{b}(\lambda) \end{vmatrix}. \tag{10.22}$$

These functions (see Appendix A) form the *CIE 1931 Standard Colorimetric Observer*. They are tabulated in an ISO/CIE Standard [21] at 1-nm intervals to seven significant digits between 360 and 830 nm. A graphical representation is given in Fig. 10.14. As seen all the values of these functions are non-negative, and the $\bar{y}(\lambda)$ function has exactly the shape as the $V(\lambda)$ function. One peculiarity of this transformation is that the $\bar{x}(\lambda)$ function turned out to have two maximums.

Tristimulus Values and Chromaticity Coordinates

Using the $\bar{x}(\lambda)$, $\bar{y}(\lambda)$, and $\bar{z}(\lambda)$ functions the X, Y, and Z tristimulus values of a color stimulus $[S(\lambda)]$ can be calculated exactly in the same way as described in connection with Eq. (10.16):

$$X=k\int_{380\ \mathrm{nm}}^{780\ \mathrm{nm}} S_\lambda(\lambda)\bar{x}(\lambda)d\lambda,$$

$$Y=k\int_{380\ \mathrm{nm}}^{780\ \mathrm{nm}} S_\lambda(\lambda)\bar{y}(\lambda)d\lambda, \tag{10.23}$$

$$Z=k\int_{380\ \mathrm{nm}}^{780\ \mathrm{nm}} S_\lambda(\lambda)\bar{z}(\lambda)d\lambda.$$

If $S(\lambda)$ is a radiometric quantity (e.g., spectral radiance) then with $k=683$ lm/W the corresponding photometric quantity (in our case luminance) is received for Y. This is the proper use of these equations for *self-luminous colors*. (For surface colors see Sec. 10.4.4.)

Unfortunately the X, Y, and Z tristimulus values are not very easy to interpret, and it is not very easy to "see" the color they specify. As the luminance measure

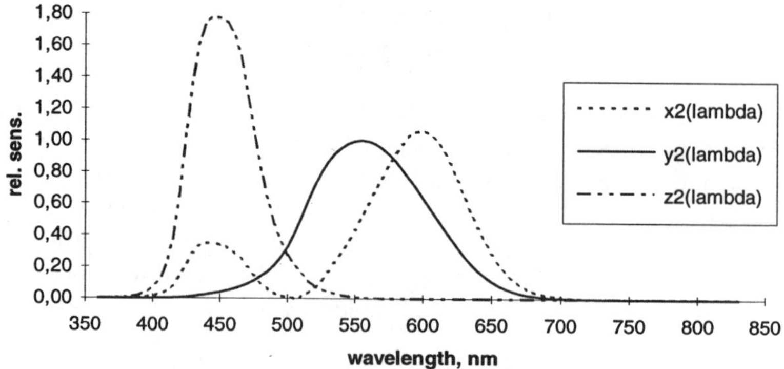

FIGURE 10.14. $\bar{x}(\lambda)$, $\bar{y}(\lambda)$, and $\bar{z}(\lambda)$ color-matching functions of the CIE 1931 Standard Colorimetric Observer.

has been condensed into the Y tristimulus value it seemed to be reasonable to transform from the (X,Y,Z) space into another space where Y is one of the coordinates and the other two describe chromaticity. To do this the CIE introduced the chromaticity coordinates (x,y,z) and defined them in the following form:

$$x=\frac{X}{X+Y+Z}, \quad y=\frac{Y}{X+Y+Z}, \quad z=\frac{Z}{X+Y+Z}, \tag{10.24}$$

where $x+y+z=1$, thus it is enough to use two of the chromaticity coordinates to describe the chromaticity of the stimulus. It is usual to use x and y, and to plot the chromaticities in a *chromaticity diagram.* Figure 10.15 shows the CIE (x,y) *chromaticity diagram* with the spectrum locus, the equienergetic stimulus, and the CIE red, green, and blue matching stimuli drawn in.

In the chromaticity diagram the chromaticity point of two additive mixed colors is located on the line joining the chromaticity points of the two constituent colors. The mixture of red and blue produces purple colors, and thus the two ends of the spectrum locus has been joined by a line, the purple boundary of the chromaticity chart. As seen the triangle traced by the **R**, **G**, and **B** primaries does cover only a part of the locus of the possible chromaticities, the area located within the boundaries of the spectrum locus and the purple boundary. Chromaticities outside of the triangle can be matched with the **R**, **G**, and **B** primaries only if one of the primaries is mixed with the test stimulus.

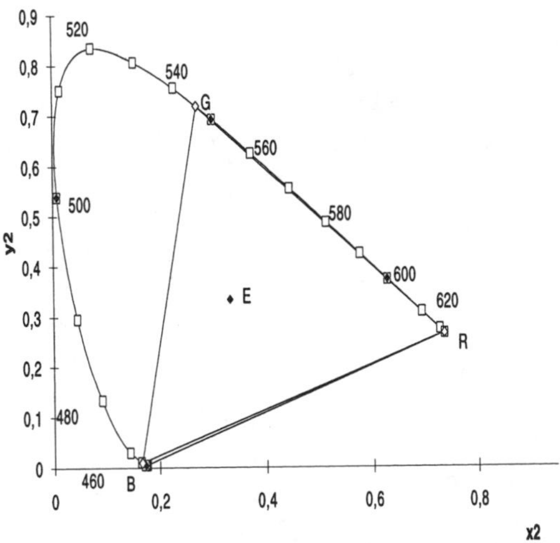

FIGURE 10.15. (x,y) chromatically diagram with the spectrum locus, the equienergetic stimulus (E) and the CIE 1931 **R,G,B** matching stimuli.

Alternative Description of Chromaticity

It is relatively easy to learn the approximate hue of a monochromatic light described by its wavelength. Thus it is convenient to use as one of the descriptors of a color a wavelength-related quantity. This is done in the chromaticity diagram in the following way: One assumes as reference an achromatic (*N* for neutral) chromaticity (called often the ''white point''). Such a reference chromaticity could be the chromaticity of the equienergetic radiation, shown in Fig. 10.15 by the point *E*. We will see that in practice many other reference points are possible, e.g., the chromaticity of average daylight illumination. In Fig. 10.16 *N* is such a reference point.

Let us first consider the chromaticity by point *C* and draw from point *N* a straight line through the chromaticity point of the test color (point *C*) to the spectrum locus. The intersection of this line and the spectrum locus is shown in the figure by DW. The wavelength of the monochromatic radiation corresponding to this spectrum chromaticity is called dominant wavelength (λ_D).

If the test chromaticity is within the triangle set by the reference point *N* and the two end points of the spectrum locus, as shown by *C'*, then one uses the concept of the complementary wavelength, as shown by the point CW in the diagram. The concept of dominant wavelength is a first approximation for a correlate with hue. We will see later that constant dominant (or complementary) wavelength does not mean exactly constant hue.

As a second descriptor of chromaticity the concept of excitation purity is used. The distance between the neutral point and the test color ($\overline{NC}$) divided by the distance between the neutral point and the chromaticity point of the corresponding

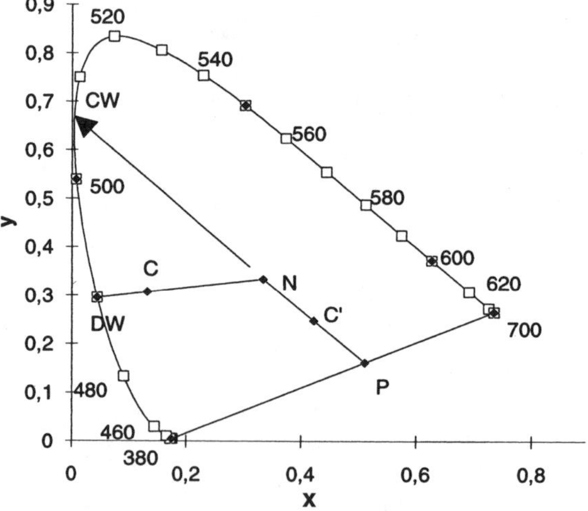

FIGURE 10.16. Concept of dominant wavelength, complementary wavelength, and excitation purity in the (x, y) diagram.

dominant wavelength ($\overline{N\ \mathrm{DW}}$). Point DW describes the chromaticity of maximum purity for the test color C. In case of purple colors the chromaticity point on the purple line where the line defining the complementary wavelength intersects the purple line is used as the chromaticity of maximum purity.

For chromaticity point C,

$$p_e = (y_C - y_N)/(y_{\mathrm{DW}} - y_N)$$

or

$$p_e = (x_C - x_N)/(x_{\mathrm{DW}} - x_N), \tag{10.25}$$

and for chromaticity point C',

$$p_e = (y_{C'} - y_N)/(y_P - y_N)$$

or

$$p_e = (x_{C'} - x_N)/(x_P - x_N). \tag{10.26}$$

Excitation purity is the first approximation to a correlate of saturation. Again, just as in case of dominant wavelength, this is only a very crude approximation, but helps to visualize some phenomena.

To describe a color stimulus, three quantities are always needed; alternative possibilities are to use:

- X, Y, and Z tristimulus values;
- a Y tristimulus value (for a self-luminous object a photometric quantity, e.g., luminance L) and the (x,y) chromaticity coordinates;
- a Y tristimulus value (i.e., luminance L), dominant (or complementary) wavelength λ_D, and excitation purity p_e.

One should never forget that three quantities are needed to describe a color; thus λ_D and p_e alone are not enough—a brightness correlate (e.g., luminance) is needed to get a full description!

The Chromaticity of the Additive Mixture of Two Stimuli

If two color stimuli are given, e.g., the red and green phosphor primaries of a cathode-ray-tube (CRT) display, then the chromaticity of the additive mixture of the two stimuli can be calculated in the following way:

Let the tristimulus values of the red phosphor at maximum intensity setting be X_{R}, Y_{R}, and Z_{R}, and those of the green phosphor X_{G}, Y_{G}, and Z_{G}. The corresponding chromaticity coordinates are $(x_{\mathrm{R}}, y_{\mathrm{R}})$ and $(x_{\mathrm{G}}, y_{\mathrm{G}})$. In the mixture an amount a_{R} of red and an amount a_{G} of green light will be used. Thus the tristimulus values of the additive mixture will be

$$X = a_{\mathrm{R}} X_{\mathrm{R}} + a_{\mathrm{G}} X_{\mathrm{G}},$$

$$Y = a_R Y_R + a_G Y_G, \tag{10.27}$$

$$Z = a_R Z_R + a_G Z_G,$$

and the corresponding chromaticity coordinates are

$$x = \frac{a_R Y_R + a_G X_G}{a_R(X_R + Y_R + Z_R) + a_G(X_G + Y_G + Z_G)},$$

$$y = \frac{a_R Y_R + a_G Y_G}{a_R(X_R + Y_R + Z_R) + a_G(X_G + Y_G + Z_G)}. \tag{10.28}$$

Expressing X_R, Y_R, and Z_R and X_G, Y_G, and Z_G with the corresponding chromaticity coordinates and luminance values (Y_R and Y_G), the following equations are reached:

$$x = \frac{\dfrac{a_R Y_R x_R}{y_R} + \dfrac{a_G Y_G x_G}{y_G}}{\dfrac{a_R Y_R}{y_R} + \dfrac{a_G Y_G}{y_G}},$$

$$y = \frac{a_R Y_R + a_G Y_G}{\dfrac{a_R Y_R}{y_R} + \dfrac{a_G Y_G}{y_G}}. \tag{10.29}$$

Here $a_R Y_R$ is the luminance of the amount of the red stimulus taken, and $a_G Y_G$ is the amount of the green stimulus used in the mixture. As seen the chromaticity coordinates of the mixture are the luminance-weighted averages of the two constituents. This is often called the center of gravity law of color mixture.

CIE 1964 Supplementary Standard Colorimetric Observer

As discussed in Sec. 10.2.2, the central area of the retina is covered by the macula lutea or yellow spot. This yellow pigmentation functions as a color filter and masks the real absorption spectra of the cone pigments. Due to this the color-matching functions differ if considerably larger then 2° fields are used for the color matches.

In 1964 the CIE standardized another set of color-matching functions to be used ''whenever more accurate correlation with visual colour matching of fields of larger angular subtense (more then 4° at the eye of the observer is desired'' [22,23]) based on the extensive work of Stiles and Burch [24,25] as well as of Speranskaya [26], where the color matches were performed in a 10° field.

It turned out that luminance is nonadditive in this 10° field, because in this larger field of view rod vision also participates in the perceived brightness (termed *rod intrusion*), and this participation depends on the intensity of the stimulus (at lower luminance levels rod intrusion becomes more important [27,28]). Due to this fact this colorimetric system was not coupled to a photometric one. Otherwise the colorimetric system for the 10° observer was created in the same form as that for the 2° observer. It is called the CIE 1964 Supplementary Standard Colorimetric Observer (sometimes termed also the 10° observer). The color-matching functions of the CIE 1964 Supplementary Standard Colorimetric Observer are distinguished from those of the CIE 1932 Standard Colorimetric Observer by a subscript 10. This subscript is used also for the tristimulus values and chromaticity coordinates calculated by using the 10° observer. Figure 10.17 shows the 10° color-matching functions; for comparison also the 2° color matching functions are shown in this figure.

Thus in case of the 10° observer equations similar to Eq. (10.23) are used:

$$X_{10}=k\int_{380\text{ nm}}^{780\text{ nm}} S_\lambda(\lambda)\bar{x}_{10}(\lambda)d\lambda,$$

$$Y_{10}=k\int_{380\text{ nm}}^{780\text{ nm}} S_\lambda(\lambda)\bar{y}_{10}(\lambda)d\lambda, \qquad (10.30)$$

$$Z_{10}=k\int_{380\text{ nm}}^{780\text{ nm}} S_\lambda(\lambda)\bar{z}_{10}(\lambda)d\lambda,$$

where, however, the use of $k=683$ lm/W is not recommended. In this case, for self-luminous stimuli $k=Y_{10}$ is used.

Chromaticity coordinates are obtained in the same way as in the 2° system:

$$x_{10}=\frac{X_{10}}{X_{10}+Y_{10}+Z_{10}}, \quad y_{10}=\frac{Y_{10}}{X_{10}+Y_{10}+Z_{10}}, \quad z_{10}=\frac{Z_{10}}{X_{10}+Y_{10}+Z_{10}}. \qquad (10.31)$$

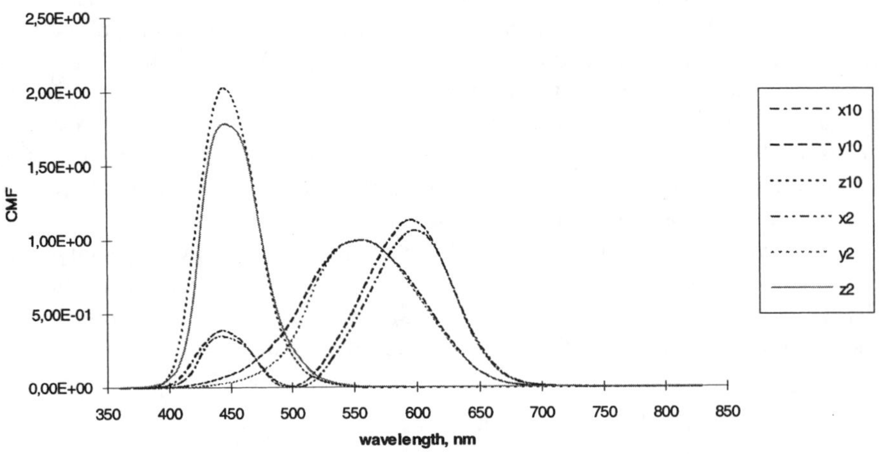

FIGURE 10.17. CIE 1932 2° and CIE 1964 10° color-matching functions.

By using the (x_{10},y_{10}) chromaticity coordinates a (x_{10},y_{10}) chromaticity diagram can be drawn, the properties of which are similar to those of the 2° chromaticity diagram.

The CIE 1964 Supplementary Standard Colorimetric Observer is used if the color patches to be matched are larger than 4° in diameter. It is important to realize that in a field of view as large as 5°–10° the influence of rod vision (termed *rod intrusion*) cannot be neglected. Thus the 10° observer should be used only if the light intensity is high enough so that rod intrusion can be neglected [CIE is at present (1997) working on a recommendation on this subject].

Small-Field Colorimetry

It is well known that at visual angles below 1° the color-matching properties of the eye change again—the eye becomes more and more tritanopic. Although such small field color matches are common in many modern applications of colorimetry, e.g., in video display unit (VDU) color graphics, there are no standards to describe the color matches for field angles smaller than 2°. Photometric investigation have shown that the $V_M(\lambda)$ function [29] describes the brightness of small fields well [30], but no extension to colorimetry has been officially agreed upon (CIE TC 1-36 is working on a recommendation that could be used down to a 1° visual angle).

10.4.3 Uniform Chromaticity Scales and Their Use in Defining Correlated Color Temperature

Although the CIE (x,y) and (x_{10},y_{10}) chromaticity diagrams can be used to describe properties of color stimuli, they suffer from the drawback that equal perceptual color differences are at widely different distances in these diagrams. This drawback was recognized shortly after the introduction of the CIE 1931 colorimetric system. Distances corresponding to 10 times the just perceptible color differences in the case of constant luminance is seen in Fig. 10.18 [31]. In a later paper Brown and MacAdam [32] extended these investigations to colors with different luminances. It has also been realized that color space is not Euclidean, i.e., a surface of constant luminance in this space will not be a two-dimensional surface, and no projection exists that will transform this into a flat surface showing equal chromaticity differences as equal distances.

Over the years many transformations have been tried to obtain reasonably equidistant chromaticity scales. At present we are using the CIE 1976 uniform chromaticity diagram, also called the (u',v') diagram. The (u',v') coordinates are obtained from the (x,y) coordinates by the following equations:

$$u'=4X/(X+15Y+3Z)=4x/(-2x+12y+3),$$
$$v'=9Y/(X+15Y+3Z)=9y/(-2x+12y+3). \tag{10.32}$$

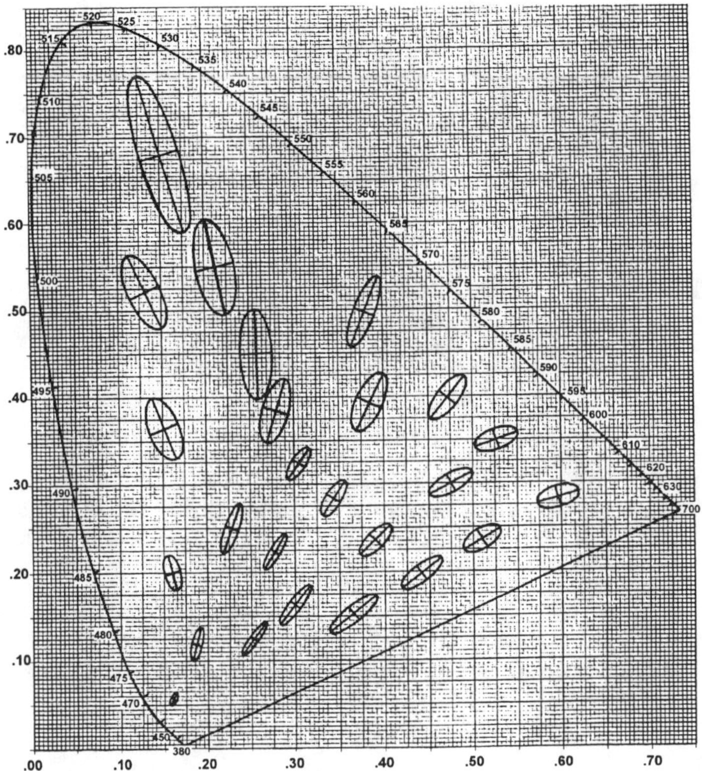

FIGURE 10.18. The CIE (x,y) diagram with ellipses representing small color differences. [Reprinted by permission from D. L. MacAdam, Visual sensitivities to color differences in daylight. *J. Opt. Soc. Am.* **32**, 271 (1942).]

In some applications still an older uniform chromaticity scale diagram [(u,v) diagram] is used, the coordinates of which can be calculated from the (u',v') coordinates:

$$u=u', \quad v=(2/3)v'. \tag{10.33}$$

Figure 10.19 shows the (u',v') diagram with the location of the equienergetic stimulus S_n. The (u',v') diagram can be used to define some correlates of color perception. Let C be an arbitrary stimulus, a correlate of its hue in this diagram is the CIE 1976 (u,v) hue angle,

$$h_{uv}=\arctan[(v'-v_n')/(u'-u_n')]=\arctan(v^*/u^*). \tag{10.34}$$

Here h_{uv} lies between 0° and 90° if both $u'-u_n'$ and $v'-v_n'$ are positive, between 90° and 180° if $v'-v_n'$ is positive and $u'-u_n'$ is negative, between 180° and 270° if both $v'-v_n'$ and $u'-u_n'$ are negative, and between 270° and 360° if $v'-v_n'$ is

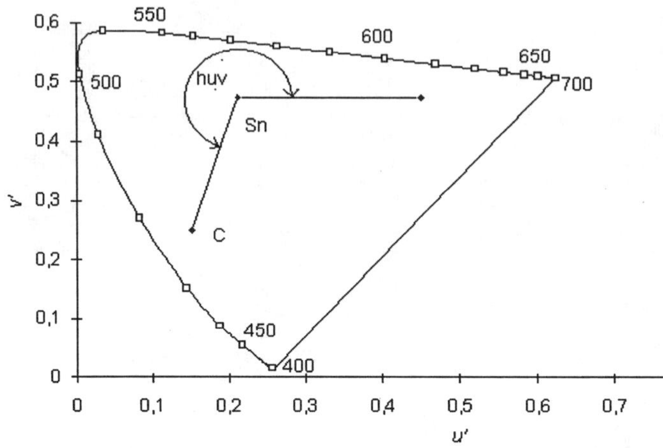

FIGURE 10.19. CIE 1976 (u', v') diagram with the location of the equienergetic stimulus S_n, a color stimulus C; shown also are the CIE 1976 (u, v) hue angle h_{uv} and the location of some dominant wavelengths along the spectrum locus.

negative and $u' - u'_n$ is positive (for u^* and v^* see below).

The CIE 1976 u,v saturation is a correlate of the color perception saturation:

$$s_{uv} = 13[(u' - u'_n)^2 + (v' - v'_n)^2]^{1/2}. \tag{10.35}$$

In Fig. 10.19 S_n is shown as the location of the equienergetic stimulus, but it can be any other suitable neutral stimulus.

Color Temperature, Correlated Color Temperature, and Distribution Temperature

We have seen that in a chromaticity diagram, for the purpose of defining a colorimetric system, the locus of the spectral stimuli and of the equienergetic stimulus have a particular importance. A further locus of special significance is the locus of the black body radiators.

It is well known that blackbodies, or cavity, Planckian, or full radiators, are radiators where the spectral power distribution is defined by the absolute temperature (measured in K) of the cavity. Human eye experiences light reflected from an aselective surface (whose spectral radiance factor is near to unity and wavelength independent) as ''white'' if irradiated by a full radiator, the temperature of which is between approximately 2700 and 20 000 K if no other clue of ''white'' light is in the field of vision. This wide range of ''white'' light sensation gives special significance to the chromaticity locus of full radiators.

The spectral power distribution of a full radiator can be calculated using Planck's formula:

$$M_e = c_1\lambda^{-5}[\exp(c_2/\lambda T - 1]^{-1}, \tag{10.36}$$

where as long as we deal only with chromaticities, i.e., with relative spectral power distributions, only the constant c_2 is of importance (see Sec. 10.4.4):

$$c_2 = 1.4388 \times 10^{-2} \text{ mK}. \tag{10.37}$$

Figure 10.20 shows the loci of Planckian radiators in the (x,y) chromaticity diagram. It is convenient to describe the chromaticity of a full radiator by simply stating its temperature.

As we have seen in discussing color matches two color stimuli are indistinguishable for the human eye if their tristimulus values are equal and they are seen under the same observational conditions. Thus lights of non-Planckian distribution but with a chromaticity equal to that of a full radiator will evoke the same visual color effect then a full radiator. (Colors of surfaces irradiated with the full radiator and the radiator with non-Planckian spectral distribution might have different color; this phenomenon will be described in "Color Rendering"; in Sec. 10.4.5) Due to this fact it is convenient to characterize these nonfull radiators by stating the absolute temperature of the full radiator with the same chromaticity. This "temperature" is called *color temperature.*

Sometimes the chromaticity of a radiator does not equal exactly with the chromaticity of a full radiator but lies very near to it. Also in this case one would like to use the term color temperature. To distinguish between the cases of an exact color match with a full radiator and an approximate one, the term *correlated color temperature* is used to describe the nonperfect match. At present the correlated color temperature is defined as the temperature of that full radiator whose chroma-

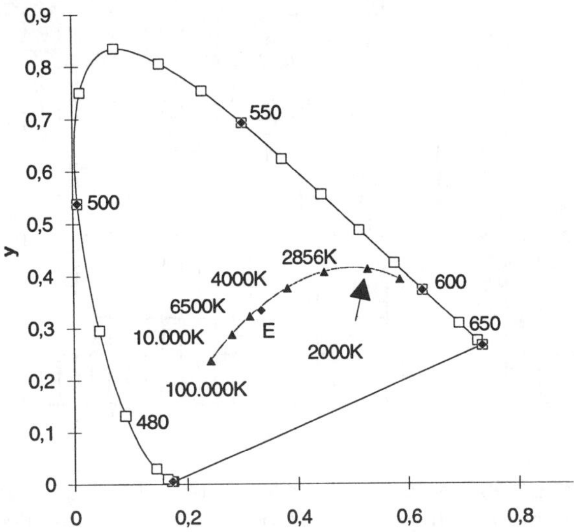

FIGURE 10.20. (x,y) diagram with locus of Planckian radiators.

ticity approximates most closely the chromaticity of the test radiator in the CIE 1960 (u,v) chromaticity diagram (i.e. the two chromaticities have to lie on the line that is perpendicular to the tangent of the Planckian locus at the given point, on a so-called *isotemperature line*). Figure 10.21 shoes a section of the (x,y) diagram with some isotemperature lines (see [33]); for the possibility to describe correlated color temperature in the CIE LUV and CIE LAB spaces see Refs. [34,35], and for the calculation of correlated color temperature, see Refs. [36–38].

Further terms used sometimes in colorimetry and photometry are the distribution temperature and ratio temperature.

Distribution temperature is the "temperature of the Planckian radiator whose relative spectral distribution $S_b(\lambda)$ is the same or nearly the same as that of the radiator considered $[S_t(\lambda)]$ in the spectral range of interest" [39]. Mathematically this means that the distribution temperature of a source in a given wavelength range, $\lambda_1 - \lambda_2$, is the temperature T_D of the Planckian radiator for which the following integral is minimized by adjusting of a and T:

$$\int_{\lambda_1}^{\lambda_2} [1 - S_t(\lambda)/aS_b(\lambda,T)]^2 d\lambda. \tag{10.38}$$

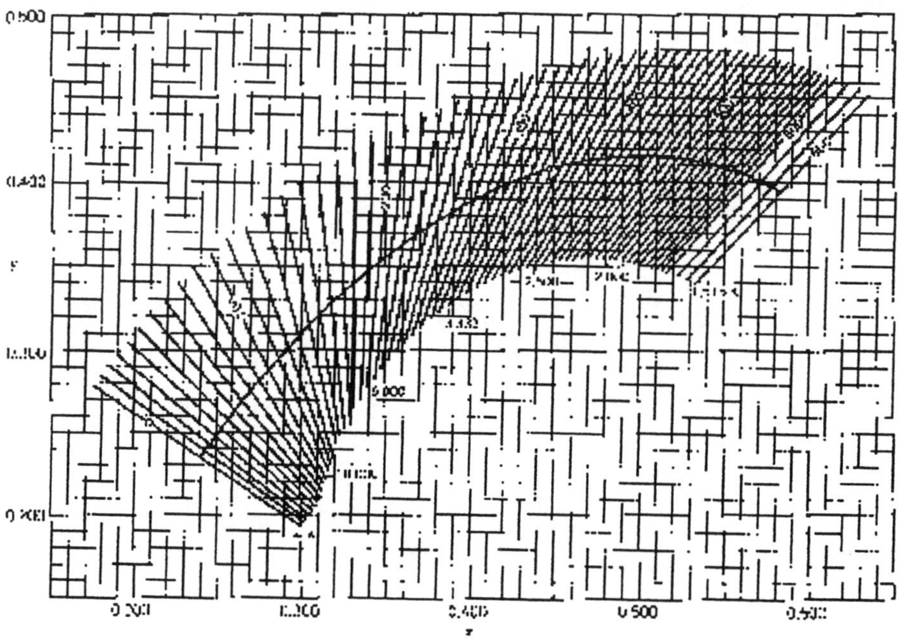

FIGURE 10.21. CIE 1931 (x,y) chromaticity diagram showing isotemperature lines. [Reprinted by permission from K. L. Kelly, Lines of constant correlated color temperature based on MacAdam's (u,v) uniform chromaticity transformation of the CIE diagram. *J. Opt. Soc. Am.* **53**, 1001 (1963).]

For the expression of $S_b(\lambda)$ see Eqs. (10.36) and (10.37); in colorimetry and photometry it is usual to select λ_1=400 nm and λ_2=750 nm.

The concept of T_D should be used only for an extended wavelength range as stated above and for radiation whose spectral distribution is a continuous function of wavelength in that range.

If the value of the distribution temperature has to be transferred from one incandescent lamp to another of the same type, the concept of *ratio temperature* might be used. Two irradiation measurements are made using narrow-band filters near 460 and 660 nm on the calibrated source and the ratio of the two readings is taken. Then the current and tension of the second lamp is adjusted until the same ratio reading is obtained. In this case the two lamps have the same ratio temperature and it is assumed that also their distribution temperature will be the same (for further details see Refs. [39,40]).

10.4.4 Colorimetry of Surface Colors

Colorimetric Properties of Materials

The stimulus reaching the eye is usually radiation reflected by a surface or radiation penetrating through a layer of material. These substances may modify the spectral power distribution of the radiation by reflection [specular and diffuse (see Chap. 9)] or transmission. Let us assume an opaque surface. As discussed in Chap. 9, such a surface can be characterized by its spectral radiance factor $\beta(\lambda)$. If this material surface is irradiated by a light source the color stimulus will be the radiation reflected back into the eye. Thus in Eqs. (10.23) and (10.30) $S_\lambda\beta(\lambda)$ has to be written instead of S_λ:

$$X=k\int_{380\text{ nm}}^{780\text{ nm}} S(\lambda)\beta(\lambda)\bar{x}(\lambda)d\lambda,$$

$$Y=k\int_{380\text{ nm}}^{780\text{ nm}} S(\lambda)\beta(\lambda)\bar{y}(\lambda)d\lambda, \tag{10.39}$$

$$Z=k\int_{380\text{ nm}}^{780\text{ nm}} S(\lambda)\beta(\lambda)\bar{z}(\lambda)d\lambda.$$

Here we used the symbol $S(\lambda)$ and not $S_\lambda(\lambda)$ to denote that the absolute spectral power distribution is not needed for the calculations. For characterizing the colorimetric properties of the surface, we need only the relative spectral power distribution function. As discussed in "Third Level: Mental Processing" in Sec. 10.2.5 (see discussion on saturation) our mental picture of a colored surface does not change with the level of illumination (within the limits of normal photopic vision, say, 300 and 30 000 lx illumination). This is reflected in CIE colorimetry by the fact that in calculating the tristimulus values of surface colors only the relative spectral power distribution of the illuminant is used, and the normalizing constant k is calculated as:

$$k=\frac{1}{\int S(\lambda)\bar{y}(\lambda)d\lambda}, \tag{10.40}$$

i.e., the tristimulus values are normalized to the Y value of the illumination. Chromaticity coordinates (x,y) and (u',v') both for the 2° observer and the 10° observer are calculated exactly in the same way, as discussed in Secs. 10.4.2 and 10.4.3 [see Eqs. (10.24), (10.31), and (10.32)]:

$$x=\frac{X}{X+Y+Z},\quad y=\frac{Y}{X+Y+Z},\quad z=\frac{Z}{X+Y+Z}, \tag{10.41}$$

$$x_{10}=\frac{X_{10}}{X_{10}+Y_{10}+Z_{10}},\quad y_{10}=\frac{Y_{10}}{X_{10}+Y_{10}+Z_{10}},\quad z_{10}=\frac{Z_{10}}{X_{10}+Y_{10}+Z_{10}}, \tag{10.42}$$

$$\begin{aligned} u'&=4X/(X+15Y+3Z)=4x/(-2x+12y+3),\\ v'&=9Y/(X+15Y+3Z)=9y/(-2x+12y+3), \end{aligned} \tag{10.43}$$

and similarly for the 10° observer.

CIE Standard Illuminants and Sources

Standard Illuminants

Equations (10.39)–(10.43) can be calculated by using any illuminant spectral distribution. The attained tristimulus values and chromaticity coordinates depend both on the reflectance characteristics of the material and the emission spectrum of the illuminant. Therefore, to be able to compare colorimetric characteristics of materials, it was necessary to agree on a smaller number of representative spectral distributions. This has been done by defining some spectral power distributions, called *illuminants*. These are described only with the tabulated values of the spectral power distribution and are used to calculate the tristimulus values and chromaticity coordinates. In order to perform visual observations, realizable sources are needed. These are called *standard sources*.

In 1931 the CIE standardized three illuminants and standard sources (in these definitions we give the temperature values as correct in the International Practical Temperature Scale of 1968):

CIE Standard Illuminant A: An illuminant having the same relative spectral power distribution as a Planckian radiator at a temperature of 2856 K.

CIE Standard Illuminant B: An illuminant having the relative spectral power distribution near to that of direct sunlight. (This illuminant is now obsolete.)

CIE Standard Illuminant C: An illuminant representing average daylight with a correlated color temperature of about 6800 K. (For the definition of correlated color temperature see Sec. 10.4.3. This illuminant is now obsolete.)

A series of daylight illuminants were suggested in 1963 and standardized in 1967 [19,41]. The relative spectral power distribution of these illuminants can be calculated along the following lines.

First the correlated color temperature T_c of the daylight (D) has to be chosen. With this value of T_c one calculates the 1931 x_D chromaticity of the daylight:

(1) For correlated color temperatures from approximately 4000 to 7000 K:

$$x_D=-4.6070\,\frac{10^9}{T_c^3}+2.9678\,\frac{10^6}{T_c^2}+0.099\,11\,\frac{10^3}{T_c}+0.244\,063. \tag{10.44}$$

(2) For correlated color temperatures from 7000 to approximately 25 000 K:

$$x_D=-2.0064\,\frac{10^9}{T_c^3}+1.9018\,\frac{10^6}{T_c^2}+0.247\,48\,\frac{10^3}{T_c}+0.237\,040. \tag{10.45}$$

With a known x_D value the y_D value can be obtained, and by the help of both, the spectral power distribution of the daylight illuminant can be calculated:

$$y_D=-3.000x_D^2+2.870x_D-0.275 \tag{10.46}$$

and

$$S(\lambda)=S_0(\lambda)+M_1S_1(\lambda)+M_2S_2(\lambda) \tag{10.47}$$

where $S_0(\lambda)$, $S_1(\lambda)$, and $S_2(\lambda)$ are functions of wavelength λ given in Appendix 10.3, and M_1 and M_2 are factors whose values are related to the chromaticity coordinates x_D and y_D as follows:

$$M_1=\frac{-1.3515-1.7703x_D+5.9114y_D}{0.0241+0.2562x_D-0.7341y_D},$$

$$M_2=\frac{0.0300-31.4424x_D+30.0717y_D}{0.0241+0.2562x_D-0.7341y_D}. \tag{10.48}$$

Although based on this calculation procedure the relative spectral power distribution of any daylight illumination between 4000 and 25 000 K can be calculated, for practical reasons it has been recommended to restrict the daylight illuminants used to a few illuminants. Besides CIE Standard Illuminant A, CIE Standard Illuminant D65 is the other primary standard illuminant of colorimetry: CIE Standard Illuminant D65 is an illuminant representing a phase of daylight with a correlated color temperature of approximately 6504 K.

Note: At the time of defining CIE Standard Illuminant A and preparing the definition how to calculate the daylight illuminant spectral power distribution, older versions of the International Practical Temperature Scale were valid. To keep the spectral power distributions unaltered, the assigned color temperatures have been changed. In 1931 the temperature of that blackbody whose spectrum equals that of Standard Illuminant A was 2848 K. [The c_2 constant in Planck's equation, see Eq. (10.36), was 1.4350×10^{-2} m K; now it is 1.4388×10^{-2} m K; see Eq. (10.37).] This brings the temperature of the Standard Illuminant A on the present-day International Practical Temperature Scale to 2856 K. Similar changes were necessary for the daylight illuminants (D65 originally had a correlated color temperature of 6500 K, due to the change in the 1948 International Practical Temperature Scale as amended in 1960, with c_2 changing from 1.4380×10^{-2} m K to the 1968 value of $c_2=1.4388\times10^{-2}$ m K). The correlated color temperature of illuminant D65 increased by the factor of 1.4388/1.4380, and due to this fact the originally defined spectral power distribution is now reached at $T_c=6504$ K.

The CIE standardized the Illuminants A and D_{65} (often written as D65) as its primary colorimetric illuminants and one of them should be used whenever possible. As secondary colorimetric standards the illuminants D50, D55, and D75 are recommended, having correlated color temperatures of approximately 5000, 5500, and 7500 K. Figure 10.22 shows the relative spectral power distribution of CIE Standard Illuminants A and D65. Tabulated values are reproduced in Appendix 10.2.

Standard Sources

For practical colorimetry one needs sources with a spectral power distribution corresponding to those of the standard illuminants. Illuminant A is to be realized by a gas-filled tungsten filament lamp operating at a correlated color temperature of 2856 K. In the original definition a source corresponding to Illuminant C was also defined, where source A had to be filtered by two filter solutions [19]. There is no recommendation for an artificial source to realize Illuminant D65 or other D illuminants. Thus, in case of daylight sources we can speak only about *daylight simulators*.

Daylight Simulators

CIE elaborated a method to assess the quality of D55, D65, and D75 daylight simulators [42], where the evaluation is done separately in the visible and in the

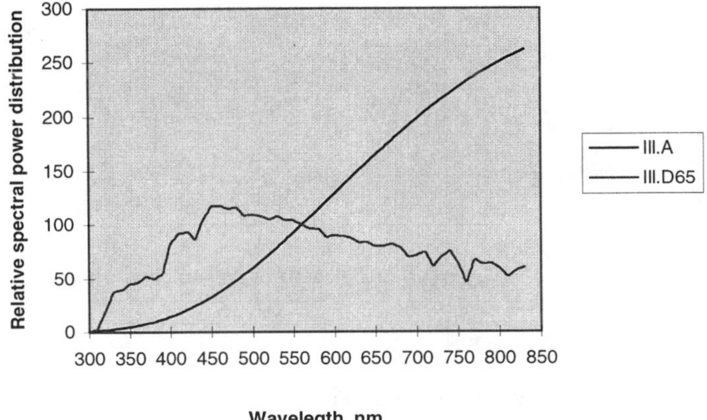

FIGURE 10.22. Relative spectral power distribution of CIE Standard Illuminants A and D65.

ultraviolet range of the spectrum: For each of these standard illuminants, spectral radiance factor data are supplied for five nonfluorescent metameric sample pairs (see ''Whiteness,'' Sec. 10.4.4, p. 375). The colorimetric differences of the five pairs are computed for the test illuminant; the average of these differences is taken as the visible range metamerism index and is used as a measure of the quality of the test illuminant as a simulator for nonfluorescent samples. For fluorescent samples (see ''Metamerism,'' Sec. 10.4.4, p. 373, and Sec. 10.6.3), the quality is assessed in terms of an ultraviolet range metamerism index, defined as the average of the colorimetric differences computed with the test illuminant for three further pairs of samples, each pair consisting of a fluorescent and a nonfluorescent sample that are metameric under the standard illuminant. Test sources are then categorized by two letters, depending on the calculated metamerism indices for the visible and UV ranges. Category letter A means excellent simulation and B is still very good, and for C to E category sources the simulation becomes worse and worse. The CIE published also a computer program to calculate the visible and UV category classes [43].

During the years many attempts were made to build good daylight simulators (for a review of the subject see Ref. [44]). Practical needs call for two types of sources: those used in instruments (spectral or tristimulus) and those used in viewing booths, where test and reference samples are illuminated side by side and viewed visually. The best laboratory instrument source uses a high-pressure xenon arc lamp as source, and a threefold light path with individual correction filters in each path [45,46]. Other commercial systems might use filtered xenon arc lamps or halogen incandescent lamps. It is usual to have a fine tuning possibility to adjust UV to visible radiation proportions. Liu and co-workers described an optimization algorithm for designing D65 simulators [47]. For viewing booth application, special fluorescent lamps are often used. Figure 10.23 shows the spectral power distribution

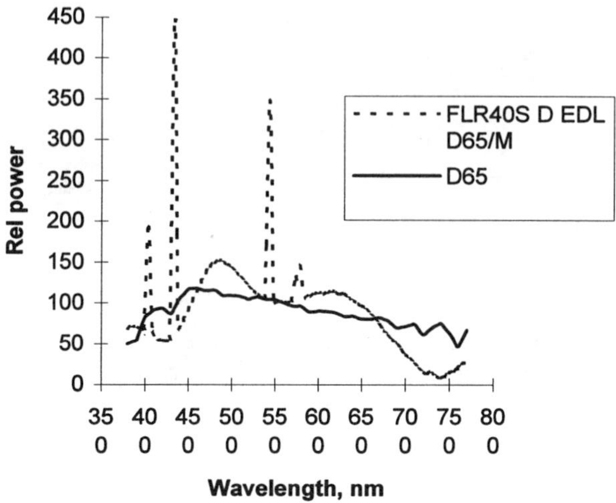

FIGURE 10.23. Relative spectral power distribution of Illuminant D65 and of a fluorescent lamp built to simulate this illuminant.

of such a fluorescent lamp compared with Illuminant D65. As seen, the spectral match is poor, mainly due to the mercury emission lines.

Secondary Standard Illuminants and Sources

Many practical lighting situations differ considerably from the standard incandescent light (Illuminant A) and daylight (D65) situations. Problems are encountered also if visual estimates are needed and no practical source is available. There is a considerable move among colorimetrists to standardize a realizable daylight spectral distribution, suggestions were made to accept filtered tungsten lamp light as indoor daylight (excluding the UV part of the emission) [48], and to standardize filtered Xe-lamp or fluorescent-lamp light as approximations of D65 and D50 [49].

The spectra of 12 representative fluorescent lamps have been published by the CIE [19] for the use when colors of surfaces have to be judged under different indoor lighting situations. Three of them should take priority whenever such problems occur. Figure 10.24 shows the relative spectral power distribution of these lamps. F2 is a standard fluorescent lamp of medium color temperature, F7 is one of daylight color temperature and broad-band spectrum, while F11 comes nearest to modern so-called ''tree band'' fluorescent lamps (such phosphors are now widely used in the modern small diameter compact fluorescent lamps).

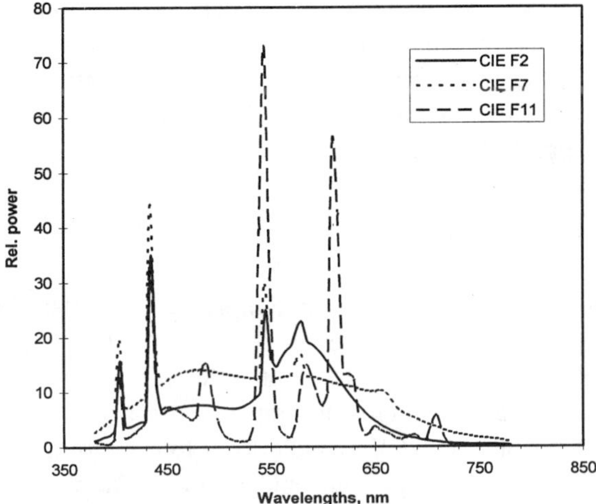

FIGURE 10.24. Relative spectral power distribution of three selected fluorescent lamps of 4230 K (F2), 6500 K (F7), and 4000 K (F11) correlated color temperature.

Recommendations on Surface Color Measurement

Standard of Reflectance Factor

The perfect reflecting diffuser is recommended as the reference standard for surface color measurement [19]. It is defined as the ideal isotropic diffuser with a reflectance equal to unity.

For practical colorimetry we need material standards. The classical secondary reference reflectance factor standard is the pressed barium sulfate plate. It is prepared in using a special $BaSO_4$ press and high-grade $BaSO_4$ powder. Pressed barium sulfate tablets can be made with reasonable reproducibility by using colorimetric grade powder. They show good diffuse reflectance characteristics (are Lambertian reflectors). The $BaSO_4$ plates are, however, highly fragile and have to be prepared fresh before use (maximum recommended shelf life 2–3 days).

In recent years the $BaSO_4$ secondary reflectance factor standards have been superceded by the so called ''halon'' white standards: This material is a polyfluorinated resin. The raw material has to be pressed into a sample holder. A thickness of 5 mm is recommended, to assure perfect opacity. An ideal packing density is about 2.5 g cm^{-3} [50,51]. Such fluorocarbon samples, now commercially available under the trade mark Spectralon, are robust and can be used for a considerable time before replacement. Recently also colored Spectralon samples have been produced.

Another class of secondary color standards are the white and colored ceramic and glazed tiles. These are very robust and can be cleaned relatively easily, but are usually glossy. They can be used only for checking the calibration stability of a colorimeter or as a transfer standard if the two instruments have the same measuring geometry. Problems can also occur if the laboratory temperature changes, as many of the ceramic tiles are thermochromic, i.e., change their reflectance spectrum with temperature (see, e.g., Refs. [52–54]).

Standard Measuring Geometry

As the reflection properties of average samples might differ considerably from those of an ideal standard, i.e., they are neither totally diffuse (Lambertian) nor are they reflecting only regularly (mirrorlike), at different locations and with different instruments reproducible measurements can only be achieved if the measuring geometry used is the same. The CIE standardized four measuring geometries [19]: 45°-normal, normal-45° diffuse-normal, and normal-diffuse. Here the first angle description refers to irradiation, the second to observation.

(a) *45°-normal (symbol: 45/0):* The specimen is irradiated by one or more beams whose effective axes are at an angle of 45°±2° from the normal to the specimen surface. The angle between the direction of viewing and the normal to the specimen should not exceed 10°. The angle between the axis and any ray of an irradiating beam should not exceed 8°. The same restriction should be observed in the viewing beam.

(b) *Normal-45° (symbol: 0/45):* The specimen is irradiated by a beam whose effective axis is at an angle not exceeding 10° from the normal to the specimen surface. The specimen is viewed at an angle of 45°±2° from the normal to the specimen surface. The angle between the axis and any ray of the irradiating beam should not exceed 8°. The same restriction should be observed in the viewing beam.

(c) *Diffuse-normal* (*symbol d/n*): The specimen is irradiated diffusely by an integrating sphere. The angle between the normal to the specimen surface and the axis of the viewing beam should not exceed 10°. The integrating sphere may be of any diameter provided the total area of the ports does not exceed 10% of the internal reflecting sphere area. The angle between the axis and any ray of the viewing beam should not exceed 5°.

(d) *Normal-diffuse* (*symbol: 0/d*): The specimen is irradiated by a beam whose axis is at an angle not exceeding 10° from the normal to the specimen surface. The reflected flux is collected by means of an integrating sphere. The angle between the axis and any ray of the irradiating beam should not exceed 5°. The integrating sphere may be of any diameter provided the total area of the ports does not exceed 10% of the internal reflecting sphere area.

For the conditions "diffuse-normal" and "normal-diffuse" the regularly reflected component of specimens with mixed reflection may be excluded by the use of a gloss trap. If a gloss trap is used, details of its size, shape, and position should be given.

In the "normal-diffuse" condition the sample should not be measured with a strictly normal axis of irradiation if it is required to include the regular component of reflection. Similarly, in the "diffuse-normal" condition the sample should not be measured with a strictly normal axis of view if it is required to include the regular component of reflection.

Cases (a), (b), and (c) give values of the reflectance factor $R(\lambda)$. For directional viewing with a sufficiently small angular spread, these reflectance factors become identical to radiance factors. For case (d), for viewing with an integrating sphere, in ideal case the reflectance is measured. Thus, in the limit, the 45/0 condition gives the radiance factor $\beta_{45/0}$, the 0/45 condition gives the radiance factor $\beta_{0/45}$, the $d/0$ condition gives the radiance factor $\beta_{d/0}$, and the $0/d$ condition gives the reflectance ρ.

For the use of integrating spheres see Chap. 5 and 8.

Color Space and Color Difference

One could build a color space by plotting, e.g., the X, Y, Z tristimulus values or the (x,y) chromaticity coordinates and the Y tristimulus value in a rectangular coordinate system. As seen in Sec. 10.4.3, in such a representation the color differences in different parts of the color space would be very unequal. During the past decades several attempts were made to transform the (Y,x,y) system into a color space that is more even. Owing to the fact that there was no single new color spaces that were really much superior to all others, in 1976 the CIE standardized two alternative spaces as an interim solution to reach international agreement of usage.

CIE Color Spaces [19]

CIE 1976 (L^*, u^*, v^*)—CIE LUV color space. This space builds on the uniform chromaticity scale described in Sec. 10.4.3. It is a three-dimensional, approximately uniform, color space produced by plotting in rectangular coordinates (L^*, u^*, v^*) quantities defined by

$$\begin{aligned} L^* &= 116(Y/Y_n)^{1/3} - 16 \quad \text{for } Y/Y_n > 0.008\,856, \\ u^* &= 13L^*(u' - u'_n), \\ v^* &= 13L^*(v' - v'_n), \end{aligned} \tag{10.49}$$

where Y, u', and v' describe the color stimulus considered and Y_n, u'_n, and v'_n describe the specified white object color stimulus, i.e., the stimulus that would be produced if the illuminant illuminated the ideal perfect reflecting diffuser (in practice, this is the stimulus produced by the illuminating source).

In "Third Level: Mental Processing" in Sec. 10.2.5, we have mentioned that our mental picture of a color can be described by the dimensions of lightness, saturation, chroma, and hue. In the CIE LUV system the following approximate correlates have been defined:

CIE 1976 lightness:

$$L^*=116(Y/Y_n)^{1/3}-16 \quad \text{for } Y/Y_n>0.008\,856. \tag{10.50}$$

CIE 1976, u,v saturation:

$$s_{uv}=13[(u'-u'_n)^2+(v'-v'_n)^2]^{1/2}. \tag{10.51}$$

CIE 1976 u,v chroma:

$$C^*_{uv}=(u^{*2}+v^{*2})^{1/2}=L^*s_{uv}. \tag{10.52}$$

CIE 1976 u,v hue angle:

$$h_{uv}=\arctan[(v'-v'_n)/(u'-u'_n)]=\arctan(v^*/u^*). \tag{10.53}$$

CIE 1976 (L^*,a^*,b^*)—CIE LAB color space. This is another approximately uniform color space produced by plotting in rectangular coordinates (L^*,a^*,b^*) quantities defined by the equations:

$$\begin{aligned} L^*&=116(Y/Y_n)^{1/3}-16,\\ a^*&=500[(X/X_n)^{1/3}-(Y/Y_n)^{1/3}],\\ b^*&=200[(Y/Y_n)^{1/3}-(Z/Z_n)^{1/3}], \end{aligned} \tag{10.54}$$

where

$$X/X_n>0.008\,856, \quad Y/Y_n>0.008\,856, \quad Z/Z_n>0.008\,856, \tag{10.55}$$

X, Y, and Z describe the color stimulus considered, and X_n, Y_n, and Z_n describe the specified white object color stimulus.

Similarly to the CIE LUV system, attributes corresponding to the perceived lightness, chroma, and hue have been defined (no correlate of saturation is possible in this system as it has no chromaticity diagram):

CIE 1976 lightness (as in the CIE LUV system):

$$L^*=116(Y/Y_n)^{1/3}-16 \quad \text{for } Y/Y_n>0.008\,856. \tag{10.56}$$

CIE 1976 a,b chroma:

$$C^*_{a,b}=(a^{*2}+b^{*2})^{1/2} \tag{10.57}$$

CIE 1976 a,b hue angle:

$$h_{a,b}=\arctan(b^*/a^*). \tag{10.58}$$

Note: Both in the equations for L^* and in those for a^* and b^* a modified formula can be used if $X/X_n<0.008\,856$, $Y/Y_n<0.008\,856$, or $Z/Z_n<0.008\,856$:

$$L^*=903.3(Y/Y_n) \quad \text{for } Y/Y_n \leqslant 0.008\,856 \tag{10.59}$$

and

$$\begin{aligned} a^* &= 500[f(X/X_n)-f(Y/Y_n)], \\ b^* &= 200[f(Y/Y_n)-f(Z/Z_n)], \end{aligned} \tag{10.60}$$

where

$$\begin{aligned} f(X/X_n) &= 7.787(X/X_n)+\tfrac{16}{116} \quad \text{for } X/X_n \leqslant 0.008\,856, \\ f(Y/Y_n) &= 7.787(Y/Y_n)+\tfrac{16}{116} \quad \text{for } Y/Y_n \leqslant 0.008\,856, \\ f(Z/Z_n) &= 7.787(Z/Z_n)+\tfrac{16}{116} \quad \text{for } Z/Z_n \leqslant 0.008\,856. \end{aligned} \tag{10.61}$$

For further specification details see the original CIE publication [19].

CIE Color-Difference Specifications

Both in CIE LUV and in CIE LAB spaces the color difference between two stimuli can be calculated as the vector difference between the (L^*,u^*,v^*) or the (L^*,a^*,b^*) coordinates:

$$\Delta E^*_{uv}=[(\Delta L^*)^2+(\Delta u^*)^2+(\Delta v^*)^2]^{1/2} \tag{10.62}$$

and

$$\Delta E^*_{ab}=[(\Delta L^*)^2+(\Delta a^*)^2+(\Delta b^*)^2]^{1/2}. \tag{10.63}$$

In both cases also a lightness $(\Delta L^*=L^*_1-L^*_2)$, a chroma $(\Delta C^*=C^*_1-C^*_2)$, and a hue difference can be defined. This latter is done as the color difference minus the sum of the lightness and chroma differences.

CIE 1976 u,v hue difference:

$$\Delta H^*_{uv}=[(\Delta E^*_{uv})^2-(\Delta L^*)^2-(\Delta C^*_{uv})^2]^{1/2}. \tag{10.64}$$

CIE 1976 a,b hue difference:

$$\Delta H^*_{ab}=[(\Delta E^*_{ab})^2-(\Delta L^*)^2-(\Delta C^*_{ab})^2]^{1/2}. \tag{10.65}$$

As neither the CIE LUV nor the CIE LAB color spaces are perfectly equidistant, the calculated color difference will not correspond to the perceived one well enough. Research continued after 1976 to reach to a better color-difference formula, both for threshold data (i.e., just perceptible color differences) and acceptability data (industrial permissible color differences between standard and test batch).

Among the many suggested modifications of the CIE formulas the most successful was the so-called CMC formula, adopted among others by the ISO Textile Committee to evaluate industrial color differences [55]. It has been also established that perceived color differences depend on the individual color vision of the observer, the viewing situation (color and lightness of background, size of samples, distance between the samples compared, etc.), texture of the sample, luminance of the sample, etc. [56].

CIE 1994 ($\Delta L^*, \Delta C^*_{ab}, \Delta H^*_{ab}$) *color-difference model (CIE 94).* Based on above consideration, the CIE recommended in 1995 a new color-difference formula, called the CIE 1994 color difference, ΔE^*_{94} [57] as an extension of the CIE LAB formula:

$$\Delta E^*_{94} = \left[\left(\frac{\Delta L^*}{k_L S_L} \right)^2 + \left(\frac{\Delta C^*_{ab}}{k_C S_C} \right)^2 \left(\frac{\Delta H^*_{ab}}{k_H S_H} \right)^2 \right]^{1/2}, \tag{10.66}$$

where ΔL^*, ΔC^*_{ab}, and ΔH^*_{ab} are the lightness, chroma, and hue differences [see Eq. (10.65)].

The S_L, S_C, and S_H *weighting functions* adjust the total color difference to account for variation in perceived color difference with variation in the color standard location in CIE LAB color space. Experiments have shown that increasing chroma of the standard results in decreased chroma and hue difference [58,59]. Current best estimates are

$$S_L = 1, \quad S_C = 1 + 0.045 C^*_{ab}, \quad S_H = 1 + 0.015 C^*_{ab}, \tag{10.67}$$

where C^*_{ab} is the chroma of the standard. A note in the CIE report states that when neither member of the color-difference pair can logically be assigned as standard, their mean chroma $(C^*_{ab,1} C^*_{ab,2})^{1/2}$ might be used in the calculation.

The k_L, k_C, and k_H are *parametric factors* and are thought to correct for variation in experimental conditions from the defined reference conditions, for which

$$k_L = k_C = k_H = 1. \tag{10.68}$$

In the textile industry it is common practice to use

$$k_L = 2, \quad k_C = k_H = 1. \tag{10.69}$$

A test of the new formula is described in Ref. [60].

The *reference conditions* are summarized in the CIE report as follows:

- Illumination: source simulating the spectral relative irradiance of CIE Standard Illuminant D65.
- Illuminance: 1000 lux.
- Observer: normal color vision.
- Background field: uniform, neutral grey with $L^* = 50$.
- Viewing mode: object.
- Sample size: larger than 4° subtended visual angle.

- Sample separation: minimum sample separation achieved by placing the sample pair in direct edge contact.
- Sample color-difference magnitude: equal to 5 CIE LAB units.
- Sample structure: homogeneous color without a visually apparent pattern or nonuniformity.

This enumeration presents at the same time the most important factors where for a deviation from the reference condition a parametric factor could be introduced. Up to now no such factors have been recommended, except the mentioned practice of textile industry (where at present the CMC formula is still prevailing).

Further Recommendations on Surface Color Measurement

Metamerism

Two stimuli could have a different spectral distribution with identical tristimulus values. Such stimuli are called to be *metameric* (in contrast to a metameric match, if the two stimuli have identical spectral distribution, they are called *nonmetameric* matches). Figure 10.25 shows the reflectance spectra (spectral radiance factors) of two samples that are metameric if illuminated with CIE Standard Illuminant D65 and observed with the CIE 1964 Supplementary Standard Observer (they are two samples used in calculating the visible goodness of fit of a D65 simulator; see "Daylight Simulators," [42] Sec. 10.4.4, p. 364).

The tristimulus values are defined by Eqs. 10.39. Two material samples having spectral radiance factors $\beta_1(\lambda)$ and $\beta_2(\lambda)$ are metameric under illuminant m and for observer n if for

$$T_{i,1}=k\int_{380\ \mathrm{nm}}^{780\ \mathrm{nm}} S_{\lambda,m}\beta_1(\lambda)\bar{t}_{i,n}(\lambda),\quad T_{i,2}=k\int_{380\ \mathrm{nm}}^{780\ \mathrm{nm}} S_{\lambda,m}\beta_2(\lambda)\bar{t}_{i,n}(\lambda),\tag{10.70}$$

it holds that

$$T_{i,1}=T_{i,2},$$

where $T_{1,1}=X_1$, $T_{2,1}=Y_1,\ldots,\bar{t}_{3,2}=\bar{z}_2$.

If another illuminant is used, or an observer looks at the samples whose spectral sensitivities deviate from the standard curves, the two stimuli will not match. We describe the mismatch under the test source or the deviate observer by the help of the observed color difference.

In practice we distinguish two special metamerism indices: change in illuminant and change in observer.

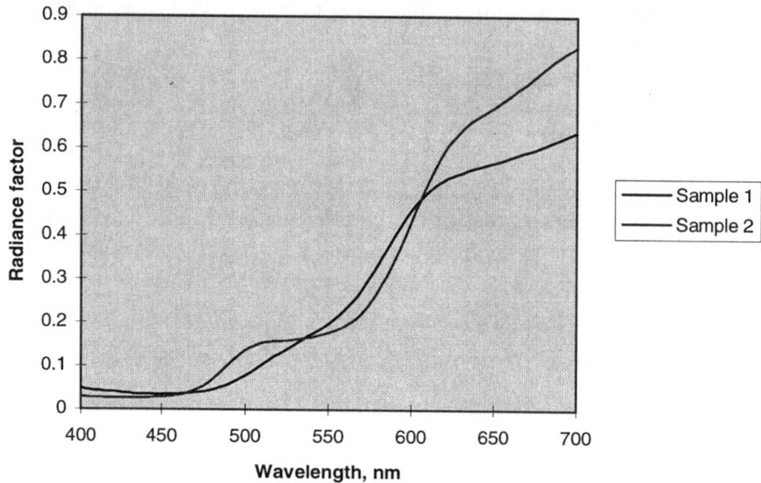

FIGURE 10.25. Spectral radiance factor of two metameric reflecting samples (for D65 and CIE 1964 Observer).

The *special metamerism index: change in illuminant* [19] is obtained for two specimens [$\beta_1(\lambda)$ and $\beta_2(\lambda)$] whose tristimulus values are identical for one illuminant (usually D65 is used as reference illuminant) but differ for a test illuminant. Suitable test illuminants are CIE Standard Illuminant A or one of the F illuminants discussed in "Secondary Standard Illuminants and Sources," Sec. 10.4.4, p. 366. (Practical application has to show which test illuminant is most suitable: if the question is whether the two samples look very different or not both under daylight and tungsten light, then the D65 reference and A test illuminant will be suitable; if comparison has to be made under industrial or commercial lighting, then probably one of the F illuminants will be more suitable.)

The measure of the metamerism index is the ΔE^* color difference calculated using either CIE LAB or CIE LUV color-difference equation [Eq. (10.62) or (10.63)] and either the 2° or the 10° standard observer (see "CIE 1931 Standard Calorimetric Observer" and "CIE 1964 Supplementary Standard Colorimetric Observer" in Sec. 10.4.2), depending on the practical situation.

The *special metamerism index, change in observer*, is based on similar calculations [61]: Four deviate observers have been defined. First one calculates the color difference using the Standard Observer (CIE 1931 or 1964) and the Deviate Observer in Eq. (10.70). This supplies a metamerism index. All four deviate observer spectra are used to determine a range of mismatches from which an ellipse showing the range of mismatches is calculated.

Recent investigations have shown [62–64] that while the illuminant metamerism index is well established and widely used in industry, the observer metamerism index needs further reformulation.

Whiteness

Whiteness is a psychometric quantity. Whiteness perception is experienced in a rather large part of the chromaticity diagram, and one is able to scale the three-dimensional color space—with reasonable accuracy—into a one-dimensional whiteness scale. The presently used whiteness scale was developed (see Refs. [65–68]) and standardized by the CIE [19] in the mid seventies.

The CIE whiteness descriptor consists of a whiteness scale and a tint formula describing greenishness or reddishness:

$$W = Y + 800(x_n - x) + 1700(y_n - y), \tag{10.71}$$

where W is the whiteness descriptor, Y is the Y tristimulus value of the sample, (x,y) are the chromaticity coordinates of the sample, (x_n, y_n) are the chromaticity coordinates of the perfect diffuser. A higher W value means greater whiteness. The whiteness formula can be used both with the CIE 1931 and CIE 1964 Standard Colorimetric Observer.

The tint formula is different for the 2° and the 10° observers:

$$T_W = 1000(x_n - x) - 650(y_n - y) \tag{10.72}$$

and

$$T_{W,10} = 900(x_{n,10} - x_{10}) - 650(y_{n,10} - y_{10}). \tag{10.73}$$

A more positive T_W or $T_{W,10}$ value indicates greater greenishness; a more negative value indicates more reddishness.

The use of the whiteness and tint formulas is restricted to the range

$$40 < T_W < 5Y - 280, \quad -3 < T_W < +3 \tag{10.74}$$

both for the 2° and the 10° observers. The above scales are nonlinear; thus equal whiteness- or tint-scale differences do not mean perceptually equal whiteness or tint differences. Ganz and Pauli [69] have recently described a method how the whiteness and tint scale values can be visualized in the CIE LAB color space.

10.4.5 Colorimetry of Sources

Chromatic Adaptation

In Sec. 10.3.3 we have already discussed that the human visual organ has the peculiar capability to adjust itself to the chromaticity of the illumination; thus, e.g., a white paper is perceived as white when seen both under incandescent light and daylight. This faculty of the visual system is called *chromatic adaptation*. For its mathematical description many attempts were made during the past century.

The theory and method to describe chromatic adaptation postulated by von Kries is one of the best-known and widely applied methods, although it has many flaws and can be regarded only as a first approximate description of the phenomenon.

The von Kries hypothesis assumes that the L, M, and S cone sensitivity functions (see Sec. 10.2.2) show in a nonexcited state given sensitivities, so that one can assign to their relative sensitivity unity coefficients. The stimulation of the retina will produce a new equilibrium between the nonbleached and bleached amount of the three photopigments, leading to a new state of the relative sensitivity of the three cone types. According to the von Kries hypothesis, to reach the new equilibrium, the sensitivities of those cone types that were more strongly excited have to be multiplied by a coefficient smaller than one. In a photographic film the three film sensitivities are in a fixed relationship. If the film has been designed for daylight exposures and one uses it in tungsten light, the abundance of long-wavelength radiation in the incandescent light produces a reddish-yellowish haze. The human visual system counteracts the development of this haze by turning the long- and medium-wavelength sensitive cones to a lower sensitivity state.

For complete chromatic adaptation one assumes that a reference white surface will be seen white both under the test and reference light. Let us denote the cone responses for the white surface for the test illuminant by $L_{w,t}, M_{w,t}, S_{w,t}$ and for the reference illuminant by $L_{w,r}, M_{w,r}, S_{w,r}$. The cone responses for a chromatic stimulus will be L_t, M_t, S_t, and L_r, M_r, S_r, respectively. The visual signal under the test light source will depend on the $L_t/L_{w,t}, M_t/M_{w,t}, S_t/S_{w,t}$ ratios and similarly for the reference conditions.

We are seeking the cone signals of that stimulus that produces the same color perception under the reference illuminant as the stimulus with the L_t, M_t, S_t cone responses produced under the test source; it is necessary that for both states of adaptation the white reduced visual signals should be equal:

$$L_t/L_{w,t}=L_r/L_{w,r}, \quad M_t/M_{w,t}=M_r/M_{w,r}, \quad S_t/S_{w,t}=S_r/S_{w,r}. \qquad (10.75)$$

Thus the *corresponding color stimulus* under the reference conditions will be

$$L_r=L_t(L_{w,r}/L_{w,t}), \quad M_r=M_t(M_{w,r}/M_{w,t}), \quad S_r=S_t(S_{w,r}/S_{w,t}). \qquad (10.76)$$

Here $L_{w,r}/L_{w,t}, M_{w,r}/M_{w,t}, S_{w,r}/S_{w,t}$ are the von Kries coefficients that scale the cone sensitivities in such a form that irrespective of the chromaticity of the illuminant the white point stays constant. This is only approximately the case in practice.

To answer the question of what the tristimulus values of a color stimulus will be under the reference light source (X_r, Y_r, Z_r), if its tristimulus values under the test light source are X_t, Y_t, Z_t, first these and the corresponding achromatic stimulus (white-point) tristimulus values have to be transformed into the LMS space. This is a similar transformation to that described in "CIE 1931 Standard Colorimetric in Observer" in Sec. 10.4.2, when we transformed from the *RGB* space into the *XYZ* one [see Eq. (10.18)]. The coefficients of the matrix transformation depend on the chosen L,M,S primaries. The chromatic adaptation transformation procedure recommended by the CIE [70] uses the transformation proposed by

Hunt and Pointer [71]:

$$\begin{vmatrix} L \\ M \\ S \end{vmatrix} = \begin{vmatrix} 0.400\,24 & 0.707\,60 & -0.080\,81 \\ -0.226\,30 & 1.165\,32 & 0.045\,70 \\ 0 & 0 & 0.918\,22 \end{vmatrix} \cdot \begin{vmatrix} X \\ Y \\ Z \end{vmatrix}. \tag{10.77}$$

Earlier transformations (e.g. Ref. [72]) were based on other assumptions regarding the most probable form of the *LMS* primaries, which lead to somewhat different numerical results (see the CIE method of color-rendering calculation [73]).

To be able to determine the $L_{w,r}/L_{w,t}, M_{w,r}/M_{w,t}, S_{w,r}/S_{w,t}$ coefficients, this transformation has to be performed first for the achromatic stimuli under both illuminants and for the test stimulus $(X_t, Y_t, Z_t,)$. Then the cone excitations for the corresponding color can be determined by using Eq. (10.76). To reach the X,Y,Z tristimulus values of the corresponding color the matrix transformation uses the inverse matrix of Eq. (10.77):

$$\begin{vmatrix} X \\ Y \\ Z \end{vmatrix} = \begin{vmatrix} 1.859\,95 & -1.129\,39 & 0.219\,90 \\ 0.361\,19 & 0.638\,81 & 0 \\ 0 & 0 & 1.089\,06 \end{vmatrix} \cdot \begin{vmatrix} L \\ M \\ S \end{vmatrix}. \tag{10.78}$$

The present CIE recommendation [70] applies a nonlinear formula instead of the simple von Kries coefficient law (for details see the original publication and the papers by Nayatani and co-workers that led to this recommendation [74,75]).

Color Rendering

Light-source color is described by stating the chromaticity of the emitted light or the corresponding correlated color temperature (CCT), see "Color Temperature, Correlated Color Temperature, and Distribution Temperature" in Sec. 10.4.3. As general purpose light sources are used to illuminate objects, beside the light-source color (or lamp color) also the color appearance of the illuminated objects is of importance. This is done by comparing the appearance of the object colors with that as seen when illuminated by a reference illuminant.

The basic idea behind the currently used color-rendering index calculation method [73] is that we assume that colors are seen "natural" if illuminated by daylight or by blackbody radiation. The color of the objects illuminated by the test lamp and by the selected reference illuminant is then compared and the color differences calculated. From these color differences the color-rendering indices for each test color sample (selected object color) are calculated. The average of some of these special color-rendering indices provides the general color rendering index.

Figure 10.26 shows the flow chart of calculating the color-rendering indices. First the correlated color temperature of the test lamp has to be determined (see "Color Temperature, Correlated Color Temperature, and Distribution Temperature" in Sec. 10.4.3), and based on this result a reference illuminant is chosen. For a CCT below 5000 K the blackbody with equal CCT has to be selected; above 5000 K a

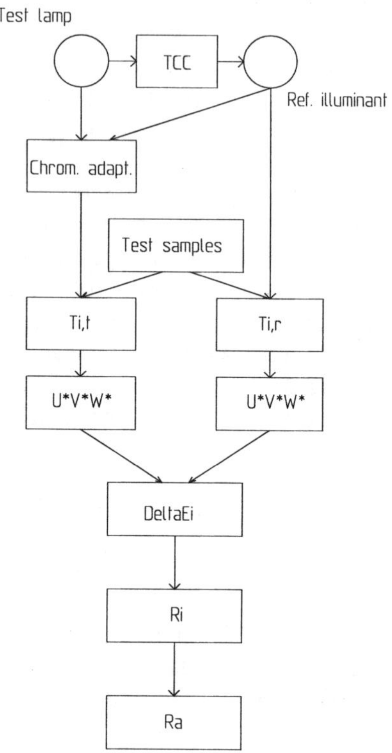

FIGURE 10.26. Flow chart for calculating color-rendering indices.

daylight phase with equal CCT is taken. The small remaining chromaticity difference between the test lamp and reference illuminant chromaticity is corrected by a von Kries–type chromatic adaptation transformation (see ''Chromatic Adaptation'' in Sec. 10.4.5).

The next step is to select representative test color samples. The CIE test method uses eight less saturated Munsell (see Sec. 10.5.1) samples of medium lightness spanning the color gamut of general-purpose object colors, and six further test samples, partly highly saturated, partly representative of often encountered object colors, as e.g., human complexion and leaf green. Then the tristimulus values for all 14 test samples are then calculated both for the test lamp and the reference illuminant [$T_{i,t}$ and $T_{i,r}$, using the terminology of Eq. (10.70), where i refers to the test sample].

At the time the CIE test method for color-rendering calculation was elaborated, an older transformation into an approximately equidistant color space [the so-called (U^*, V^*, W^*) space] was used. This is still used to perform these calculations. In this space color differences are calculated in the same form as shown in Eq. (10.62), only the $\Delta U^*, \Delta V^*, \Delta W^*$ quantities are used.

The ΔE color difference will be zero if the colors under the test lamp are rendered exactly as under the reference illuminant. It was thought that in this case the color-rendering index should be equal to 100. The scale was devised so that one of the most common fluorescent lamps of those days, the so-called warm white lamp, should have a general color-rendering index (see below) of approximately 50. Based on these principles the special color-rendering index formula is as follows:

$$R_i = 100 - 4.6\Delta E_i . \tag{10.79}$$

The general color-rendering index is calculated from the first eight R_i values as their arithmetic mean:

$$R_a = \tfrac{1}{8}\sum_{i=1}^{8} R_i . \tag{10.80}$$

At present the CIE is working on updating the color-rendering-index calculation method. Most probably first an update to the now accepted chromatic adaptation transformation [70] and color-difference calculation [57] will take place; the final goal is to use an accepted color-appearance model (see Sec. 10.5.2) to describe color rendering.

10.5 FURTHER QUESTIONS OF COLORIMETRY

In the preceding section we summarized the most important colorimetric rules and procedures where international agreement exists. There are two major items where there is yet no international consensus, but where the concepts are used widely and where very active research is going on. These items are the color-order systems (coupled to them is the question of color atlases), and the color-appearance models. In this section a short overview of these problems of colorimetry will be presented.

10.5.1 Color-Order Systems

During the centuries many artists and scientists tried to systematize colors. One of the first color solid devised along the lines of color perception (see "Third Level: Mental Processing" in Sec. 10.2.5) is the Forsius color solid [76]. Owing to limited space we will not be able to discuss items such as Newton's or Goethe's color circle, the many trials to devise from pigment mixtures color-order systems, and related color atlases. We will concentrate on the two systems used most often at present, the Munsell and the natural color system (NCS) system, and will mention shortly a few other systems that are also in use.

Munsell System

Munsell, an artist, started with the development of a color notation system in 1905 (see [77]), which was followed by the first collection of samples, the *Atlas of the Munsell Colour System* in 1915. The system was developed in a number of steps leading up to the Optical Society of America Munsell Renotation System in 1943 [78].

The Munsell system is built along the three perceptual quantities: hue, chroma (describing saturation), and value (describing lightness). Figure 10.27 shows the schematic structure of the Munsell system.

The hue circle is divided into five major steps, yellow (Y), red (R), purple (P), blue (B), green (G), and its intermediate steps, YR, RP, PB, BG, and GY. Every such step is further subdivided into 10 substeps; see Fig. 10.28. Figure 10.29 shows a page of constant Munsell hue, with the white–black axis.

The Munsell color notation first specifies the hue, then states the value and chroma in the form value/chroma, e.g., 5 YR 8/4 for a light yellowish pink (this is the sample used in color-rendering calculations to represent human complexion color). *The Munsell Book of Color* is the manifestation of the system.

Unfortunately there is no exact transformation between the CIE system and the Munsell system. Only the Y tristimulus value has an exact definition:

$$Y = 1.2219V - 0.231\,11V^2 + 0.239\,51V^3 - 0.021\,009V^4 + 0.000\,840\,4V^5, \tag{10.81}$$

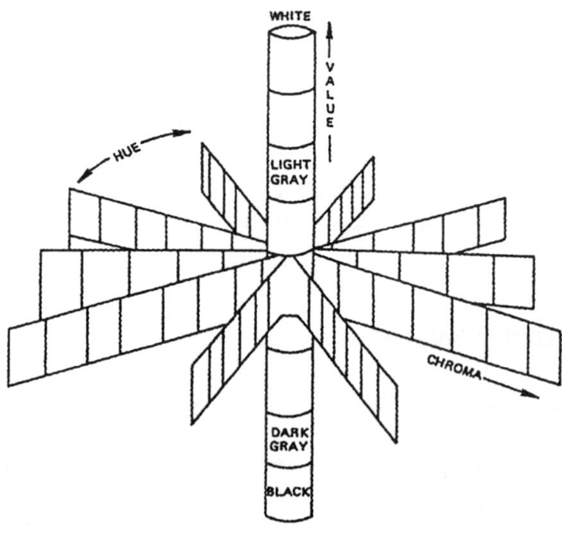

FIGURE 10.27. Schematic view of the Munsell system. (From R. S. Hunter, *The Measurement of Appearance*. Copyright © 1975 John Wiley & Sons. Reprinted by permission of John Wiley & Sons, Inc.)

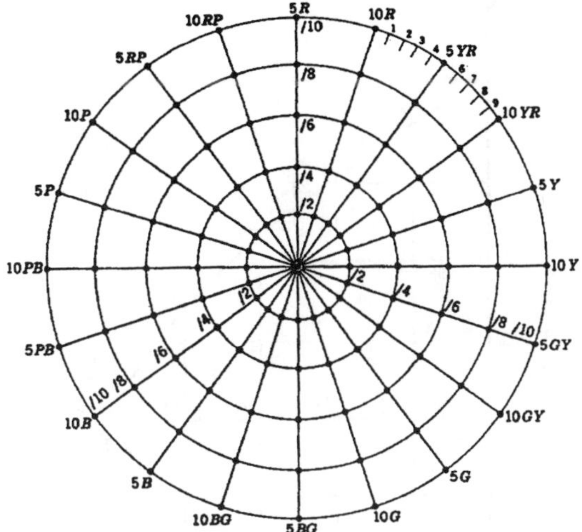

FIGURE 10.28. Organization of the Munsell hue circle. [From W. S. Stiles and G. Wyszecki, *Color Science, Concepts and Methods, Quantitative Data and Formulae*, Second Edition. © 1982 John Wiley & Sons. Reprinted by permission of John Wiley & Sons, Inc.)

where V is the Munsell value describing the lightness of the sample. This equation is, however, noninvertible. Approximations for the value scale [81] have been proposed and lookup table programs exist also for value and chroma calculations [82].

Natural Color System (NCS) System

The Swedish NCS system is built on the Hering color theory [83] that states that "there are six elementary colors, each of which shows no resemblance to any of the others"; and that "all perceived colors can be described by their resemblance to the elementary colors only." The six elementary colors are the two achromatic ones, white and black, and the four chromatic ones, yellow, red, blue and green. Tonnquist states that "for each elementary color there is an elementary attribute (whiteness w, blackness s, yellowness y, redness r, blueness b, and greenness g, giving the resemblance of a given color to the elementary color on a scale of 0 to 100)."

Yellow–blue and red–green are mutually exclusive color perceptions; therefore the NCS system places these perpendicularly to each other, with 90° between the adjacent colors, as seen in Fig. 10.30.

Figure 10.31 shows the NCS hue circle and Fig. 10.32 shows one hue leaf of the NCS system with equal blackness and chromaticness (similarity with pure color, i.e., with a color without blackness or whiteness content) lines.

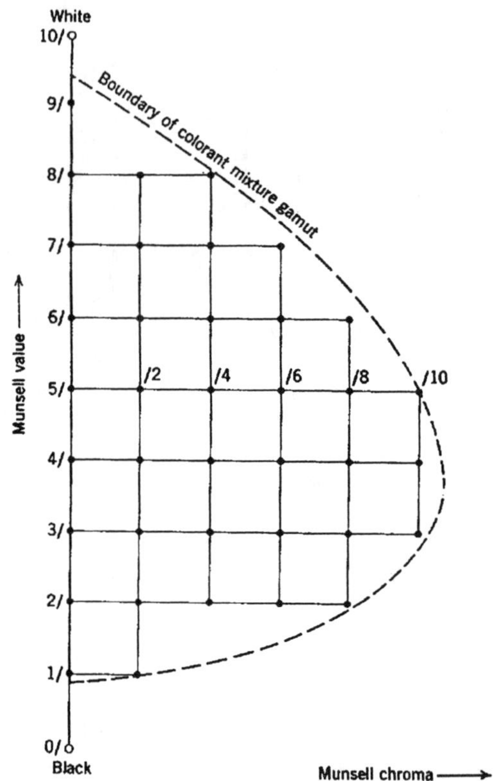

FIGURE 10.29. Organization of colors of constant Munsell hue. (From W. S. Stiles and G. Wyszecki, *Color Science, Concepts and Methods, Quantitative Data and Formulae*, Second Edition. © 1982 John Wiley & Sons. Reprinted by permission of John Wiley & Sons, Inc.)

The NCS system is backed by an NCS atlas, which is used, however, only to exemplify the system and is not the actual manifestation. These samples of the NCS atlas have been measured and the CIE tristimulus values for each color sample are available. Derefeldt and Sahlin have shown that there is no simple relationship between the NCS and CIE LAB systems [84].

Further Color-Order Systems

At this point we would like to briefly mention some of the other color-order systems that find application in different application areas.

The Optical Society of America–Uniform Color Scales (OSA–UCS) system [77,85,86] has been built by placing the samples on a cuboctahedron lattice. A sample located in the center of the cuboctahedron has 12 nearest neighbors at the 12

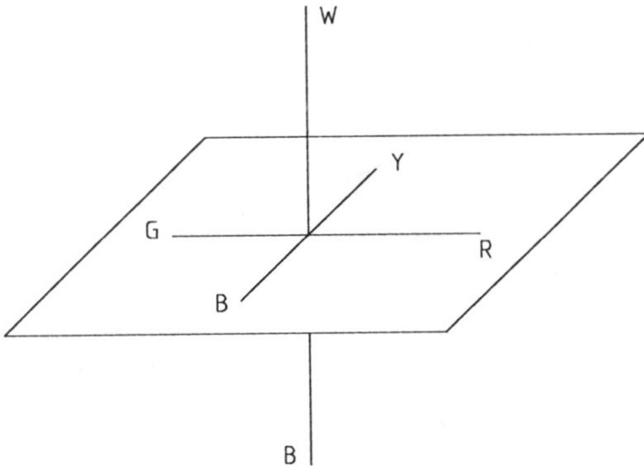

FIGURE 10.30. Antagonistic red–green, yellow–blue, and white–black axes in the NCS system.

corner points of the solid. The lattice has three axes with colorimetric meaning, an L axis representing lightness and two opponent axes: j and g. The system has a transformation into the CIE XYZ system.

The DIN system [87] attempts to show equal color distances in defined color series of hue, saturation, and darkness degrees. It has been planned to be an aid for technical purposes. The hue and saturation scales of the system do not differ essentially from those of other color-order systems, but the brightness description is considerably different. For different colors not the equal luminous reflectance but

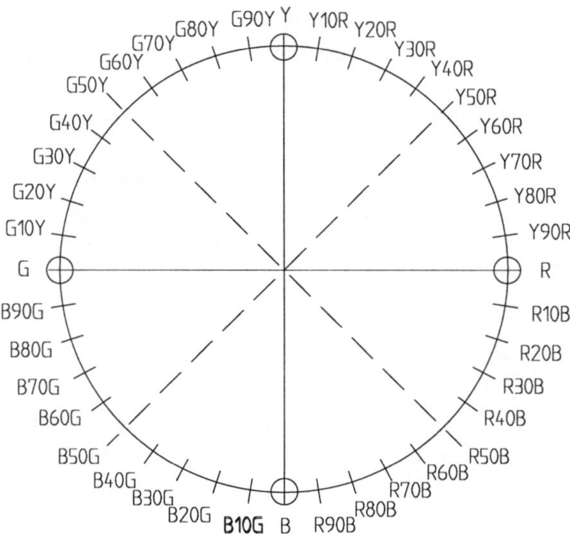

FIGURE 10.31. NCS hue circle with the four unichromatic hues.

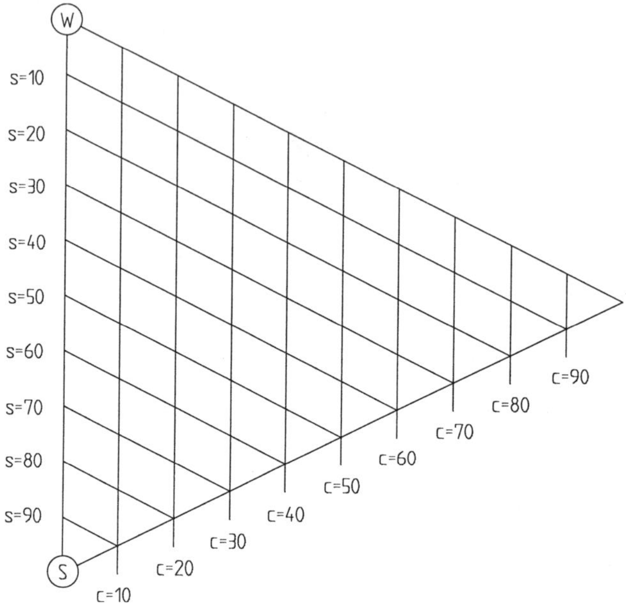

FIGURE 10.32. NCS constant-hue triangle with lines of equal chromaticness and equal blackness.

the relative luminance factors are kept equal. The relative luminance factor is the quotient of the luminous reflectance Y of the color and of the optimal color of equal chromaticity Y_0. Optimal colors are those whose reflectance curve consists of only zero or unity values and have not more than two transitions between unity and zero values.

The DIN system uses many ideas derived from the Ostwald system [88,89] of ordering colors.

The Coloroid color system [90,91] has been developed to show color harmony series and is intended for the use of architects and interior designers.

In a series of articles Smith and co-workers have described the interrelationship between different color-order systems [92–96].

10.5.2 Color-Appearance Models

The CIE system of colorimetry has been designed to predict whether two stimuli will match or not. Color-order systems are capable of systematizing color perceptions but are not able to predict color sensations yet as they will be perceived under different adaptation conditions. Color-appearance models have been designed to predict the human perception of colors. They should model color vision and present measurement tools for quantities such as brightness, colorfulness or unique hue.

There is no international agreement on a color-appearance model; therefore we will describe here—very briefly—four models that compete for acceptance. None of them will be described in detail as they are usually quite complex and need much explanation. The reader has to be referred back to the original literature.

One of the common features of all the color-appearance models is that neither of them is in a final form. From time to time model builders modify their calculation method, introducing some fine-tuning to the model. Therefore the reader is advised to search for the most recent literature to find the latest version of a given model (although every effort was taken to describe the models in their most up-to-date form at the time of writing this chapter, the author would like to apologize had he missed a facet of development in one case or the other).

ATD Color Vision Model

The ATD model developed by Guth [97–100] is a multizonal color vision model. It contains a nonlinear receptor gain control and two opponent color processing stages together with a compression stage of the signals. The first stage mediates apparent brightness and visual discrimination, while the second stage mediates apparent hues and saturation. Brightness is described in this model as the vectorial sum of the ATD signals derived from the LSM cone signals after adding noise and performing a gain control.

RLAB Color-Appearance Model

Based on Fairchild's chromatic adaptation model [101,102] Berns and Fairchild described an extension of the CIE LAB color space [103,104] that could be used as a color-appearance space, especially for the reproduction industry. The original tristimulus values are transformed to a D65 1000-lux illumination reference condition by the above chromatic adaptation transformation, then a somewhat modified CIE LAB transformation is performed, depending on the surround conditions. If the corresponding colors under another illuminant have to be determined, inverse transformations can be performed.

Nayatani and Co-Workers' Model of Color Appearance

A color-appearance model based on the chromatic adaptation model described in ''Chromatic Adaptation'' in Sec. 10.4.5 [70,74,75] was developed by Nayatani and co-workers [105]. This model has been tested and extended several times during the past years; some of the more important phases are described in Refs. [106–108]. The model predicts brightness and lightness, chroma (saturation), colorfulness, and hue angle of the test sample.

In the calculations the adapting luminance is taken into consideration, and transformations into the Esteves–Hunt–Pointer primary system are performed, followed by nonlinear transformations. From these data lightness–brightness, redness–greenness, and yellowness–blueness coordinates are calculated. These are then used to calculate the hue angle, the chroma, and the colorfulness correlates.

Hunt Color-Appearance Model

The other elaborate color-appearance model is that developed by Hunt [109–113]. The model takes into consideration the adapting field and the proximal field luminance level, can scope with the brightness and chromatic surround induction effect, and predicts hue, colorfulness, saturation lightness, and brightness of the test sample. The model can also be reversed to calculate the CIE colorimetric data of the corresponding color for another visual situation.

The model first transforms CIE tristimulus data into a LMS cone sensitivity space, performs there a nonlinear transformation, and considers a number of influencing factors (surround, adapting field, induction, etc.). Then signals that could correspond to the antagonistic yellow–blue, red–green second-stage visual signals are calculated. From these signals hue, colorfulness, brightness, lightness chroma, and whiteness–blackness quantities are calculated.

The system counts as one of the most advanced representations of quantitative visual modeling, and a number of experimental studies dealt with its verification and determining the limits of its applicability (see e.g., Refs. [114–119]). Recently also a CIE guide has been published [120] dealing with the question of testing color-appearance models.

10.6 COLORIMETRIC PRACTICE

In this section some general issues of colorimetric practice will be discussed. Within the limited space it is certainly not possible to cover all the practical aspects of colorimetry. We will be selective in choosing items of interest regarding photometry. First an overview on the available instrumentation will be given, followed by some selected items.

10.6.1 Colorimetric Instrumentation

Tristimulus Colorimetry

A tristimulus colorimeter consists of a measuring head containing some input optics (cosine corrector for ''illuminance'' type of measurements or lens optics for ''luminance'' type of measurement) and three or four filter packages to correct the spectral sensitivity of the detector(s) to the color-matching functions (see Fig. 10.14 in Sec. 10.4.2). At present most ''illuminance'' measuring instruments use Si photovoltaic cells as detectors, while the ''luminance'' measuring instruments might contain photoelectric multipliers.

The detector(s) are usually followed by operational amplifiers to convert the current of the detector into a voltage signal, which is then converted from analog to digital form and digitally processed. Figure 10.33 shows the scheme of a tristimulus colorimeter.

The most critical part of a tristimulus colorimeter is the filter correction. Usually optical colored glass filters are used and several such filter glasses have to be ground and polished to the proper thickness and glued together to achieve a good filter correction.

To achieve better spectral correction in some instruments chips of filter glasses are glued side by side, called partial filtering (see Fig. 10.34).

There is no internationally accepted method for characterizing the goodness of fit of the spectral match of the colorimetric detector heads. At present many manufacturers use the method suggested by the CIE for radiometers and photometers [122]. This method is based on the so called f_1' error expression of spectral mismatch:

$$f_1' = \frac{\int_0^\infty |s^*(\lambda)_{\mathrm{rel}} - \bar{t}_i(\lambda)|\, d\lambda}{\int_0^\infty \bar{t}_i(\lambda)\, d\lambda} \times 100\%, \tag{10.82}$$

where $\bar{t}_i(\lambda)$ is one of the color-matching functions [see Eqs. (10.39) and (10.70)], $s^*(\lambda)_{\mathrm{rel}}$ is the normalized relative spectral responsivity,

$$s^*(\lambda)_{\mathrm{rel}} = \frac{\int_0^\infty S(\lambda)_m \bar{t}_i(\lambda)\, d\lambda}{\int_0^\infty S(\lambda)_m s(\lambda)_{\mathrm{rel}} \mathrm{d}\lambda}\, s(\lambda)_{\mathrm{rel}}, \tag{10.83}$$

and $S(\lambda)_m$ is one of the standard illuminants, in photometry usually Standard Illuminant A, in colorimetry often Standard Illuminant D65. $s(\lambda)_{\mathrm{rel}}$ is the spectral responsivity of the detector–filter combination. f_1' values of the $\bar{y}(\lambda)$-function approximation can be as good as 1–1.5 %, and for the $\bar{x}(\lambda)$ and $\bar{z}(\lambda)$ functions the approximations are usually poorer, depending to a large extent on the permissible

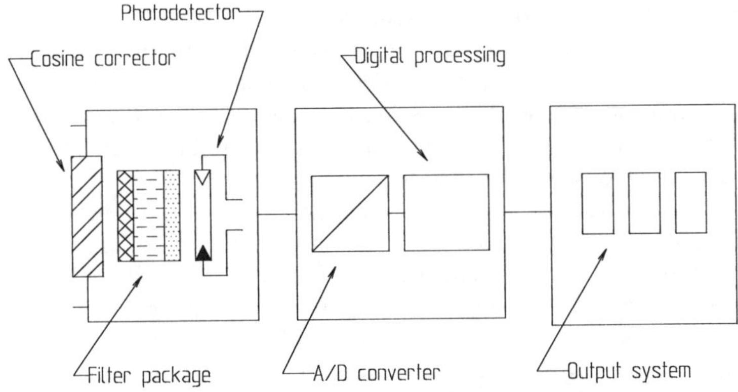

FIGURE 10.33. Schematic view of a tristimulus colorimeter (only one channel is shown).

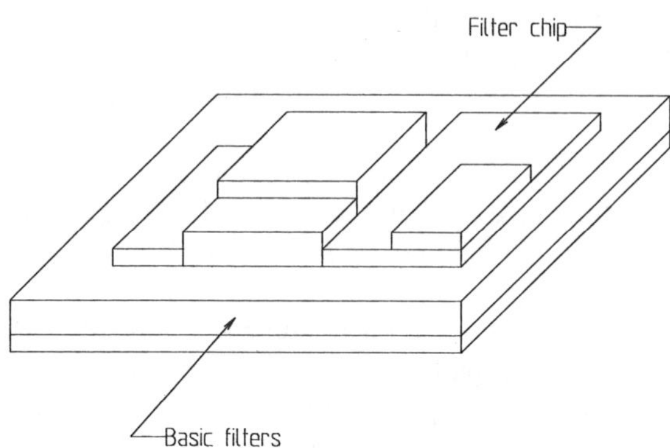

FIGURE 10.34. Spectral correcting filter using both full filtering and partial filtering (courtesy of LMT, Berlin [121]).

total absorption. (The CIE is working on a method to quantify the spectral mismatch of a colorimeter head, but this technique has not received general acceptance yet [123].) Figure 10.35 shows state-of-the art instrumental spectral responsivity color-matching functions. Figure 10.35(a) shows spectral matches achieved by using full-filtering techniques, while Fig. 10.35(b) shows similar data for a partially filtered color measuring head.

Depending on the application there are other characteristics of a tristimulus colorimeter, and of a radiometer in general, that have to be observed. The second most important item is the angular responsivity: For a cosine-corrected detector head the irradiance responsivity has to change with the cosine of the angle of incidence; for a radiance type of detector head it should stay constant for a small angular subtense and then drop to zero within a very small range.

Should the colorimeter be used with sources having considerable amount of radiation outside the visible spectrum range, the UV and IR sensitivity is also of importance, as it might lead to stray radiation interference.

If the source to be measured has a nonconstant temporal emission, as do all ac powered gas discharge lamps and all video display units, then the synchronization of the measurement cycle to the emission frequency is of outstanding importance, as otherwise extremely long measurement times have to be selected to average out the temporal variations of the signal to be measured.

There are measurement situations when the color has to be determined at different parts of a pictorial scene, as e.g., on a computer picture displayed on a cathode ray tube (CRT) screen. For such purposes tristimulus filtered CCD cameras can be used. Figure 10.36 shows instrumental color-matching filters developed for such an instrument [125]. The state-of-the-art filter match is in case of the CCD cameras not as good as for Si photovoltaic cells, as in this case large surface full filtering has to made, covering all the pixels of the camera. An alternative way would be to filter the pixels of the camera individually as done in the home video

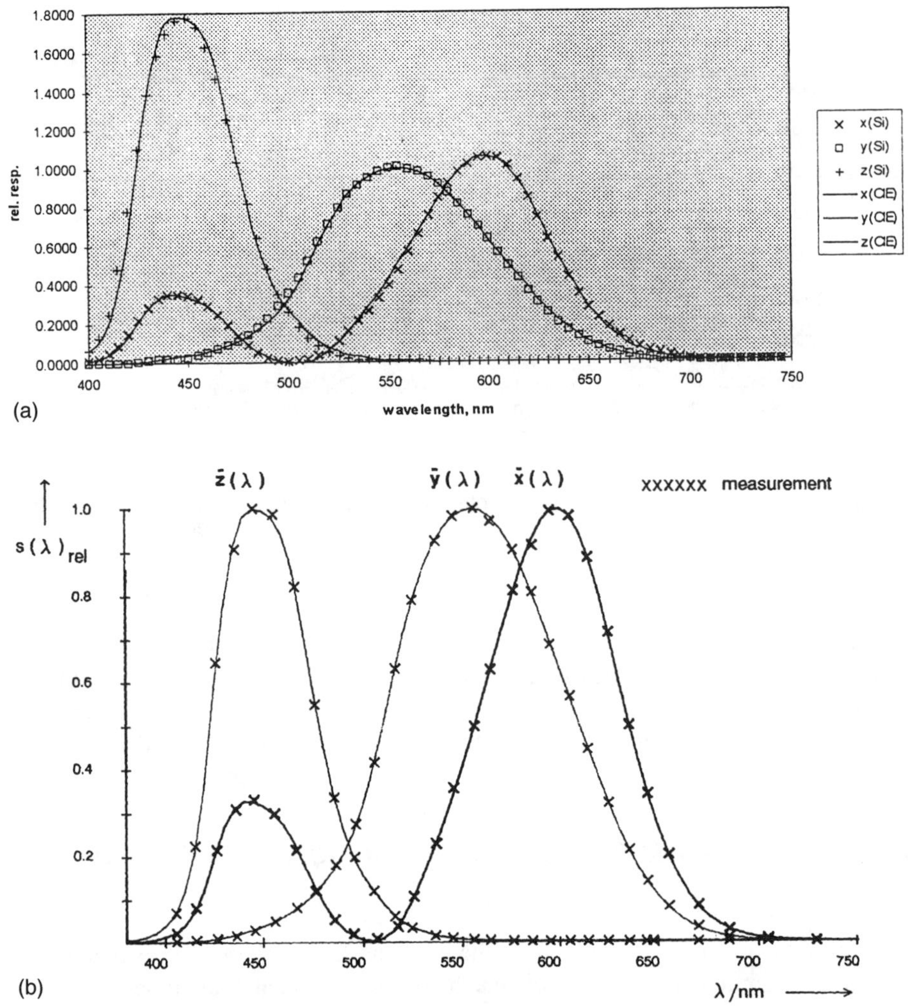

FIGURE 10.35. Spectral sensitivity functions of a tristimulus colorimeter with Si-cell and (a) full-filter correction [$f_1(x)=2.6$, $f_1(y)=2.2$, $f_1(z)=4.3$, courtesy of InPhoRa Corp. [124]], and (b) with partial-filter correction (courtesy of LMT [121]).

recorders. At present the pixel-filtered CCD cameras colorimetric mismatch is still too high to permit meaningful colorimetric measurements. However, if the light to be measured is composed of three and only three spectral emissions, as is the case with a CRT display, matrix transformation techniques can be applied to correct for the given CRT–CCD combination [126].

A further problem with all types of tristimulus colorimeters is that both the color-matching filters and the detectors show a spectral and absolute sensitivity

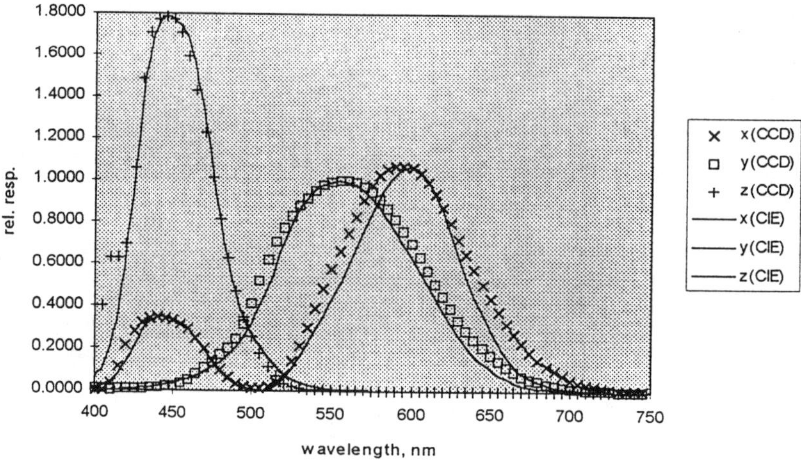

FIGURE 10.36. Spectral sensitivity functions of a tristimulus colorimeter built with a CCD camera [$f_1(x)=7.5$, $f_1(y)=3.6$, $f_1(z)=3.9$, courtesy InPhoRa Corp. [125]].

change with temperature. As the measuring head can be temperature controlled, the main concern occurs in the case of measuring relatively strong radiation with heat absorption in the filters, as a 1 °C temperature raise can produce a 0.12% decrease in absolute sensitivity and a marked shift in the spectral responsivity (depending on the glass types used) [127].

Spectroradiometric Techniques: Scanning and CCD-Type Instruments

Tristimulus instruments suffer from the inherent drawback that the available type of filters is limited; thus a perfect match to the color-matching functions is not possible. The way out is only to determine the emission spectrum and calculate the tristimulus values. Traditionally scanning spectrometers are used for this purpose. There are many good commercial instruments available for the measurement of transmission or reflection spectra. Owing to differences in optics and partly in the built-in algorithms to calculate color coordinates, the repeatability of these instruments is usually much better than their absolute accuracy [128–130], which should be checked carefully by using standards calibrated in national laboratories [131].

Most of the high-end spectroradiometers for colorimetric purposes are custom-built instruments. Some items that have to be checked with such systems are summarized in Refs. [132,133]. Most important is that the bandwidth of the instrument and the wavelength increment by which measurements are made have to be carefully set equal. If one is larger than the other, the instrument will oversample or undersample the spectrum, i.e., will measure a spectral band twice or will leave it out completely, leading to measurement errors, especially if the radiation consists of superposed spectral lines and broad spectral bands.

Recently another type of spectrometer has become available, where the detector is a CCD linear array (or some other variant of that detector type where a high number of single detector elements are sitting side by side). This system permits one to use a spectrograph arrangement, i.e., to focus the entire spectrum onto the measuring sensor and thus collect information in parallel over a broad wavelength range. By this technique spectrometers without moving parts can be built.

10.6.2 Light-Source Colorimetry

General-Purpose Sources

The color of incandescent lamps is often determined from distribution temperature measurements (see Sec. 10.4.3). For gas-discharge lamps either tristimulus colorimeters or spectroradiometric techniques are used. As the color of metal halide lamps depends on the angle of view (whether light is collected, e.g., in a horizontal plane or vertically upwards or downwards) [134], it is usual to measure the color, together with the photometric characteristics in a photometer sphere. If a tristimulus colorimeter is used, the spectral match of the filters has to include also the spectral selectivity of the wall paint. As this is difficult to achieve, and shifts of spectral reflection with time cannot be excluded, it is more appropriate to use spectroradiometric methods, where the calibration with an incandescent lamp of known spectral power distribution can correct for selective absorption effects.

It is general practice to measure general-purpose gas-discharge lamp spectra at 5-nm intervals with 5-nm bandwidth as measurements with higher resolution often only decrease the available light intensity to levels where noise contribution becomes unacceptably high. The attained colorimetric errors produced by the 5-nm abridged measurement compared with a more accurate 1-nm measurement are usually negligible.

If gas-discharge lamps are used with magnetic ballast (i.e., not high-frequency electronic ballast are used), it is also important to synchronize the measuring cycle to the main frequency, so that at every wavelength radiation is collected for the same amount of time.

Proper stabilization time of the source is important both in case of low- and high-pressure gas discharge lamps.

Light-Emitting Diodes (LEDs)

Light-emitting diodes are selective emitters with bandwidths of only a few tens of nanometers. That is why for their colorimetric measurement tristimulus instruments have to be carefully calibrated in that particular wavelength range where the LED emits. Instrument manufacturers frequently offer special LED measuring tristimulus filter sets tailored for one class of LEDs (i.e., red, yellow, or green).

Owing to the narrow emission spectrum of LEDs, spectroradiometric measurements have to be performed at 1-nm steps (with 1-nm bandwidth) in order to achieve the required accuracy.

The spatial light distribution of LEDs can be very different, from narrow beam emitters to almost Lambertian radiators. The spectral characteristics vary somewhat also with angular distribution, but this is generally of smaller importance. Therefore it has become general practice to measure the emitted radiant flux of a LED in a cone of several degrees opening angle, although techniques with integrating spheres have also been described [135].

Video Display Units (VDUs)

While the colorimetry of general-purpose light sources and of LEDs is more of a technical character, in case of VDUs there are a number of fundamental unsolved problems. In this section we will deal only with cathode-ray-tube (CRT) monitor colorimetry, as the more modern flat-panel display monitors are not advanced well enough to permit real colorimetric work. Flat-panel displays suffer from viewing-angle-dependent color, temporal instability, etc. Thus at present they cannot be used in applications where colorimetric fidelity is of importance.

The white point on the VDU, the chromaticity produced when all three electron guns are fully energized, is usually set by the manufacturer, although with high-end VDUs nowadays there is the possibility of user-defined white-point selection. Computer manufacturers often prefer a high-temperature (bluish) white point (with a correlated color temperature of about 9300 K), as this permits higher luminance also in the blue channel. The international standards call for a D65 white point, while graphic arts has settled for an even lower temperature white point (D50) to compromise between daylight and electrical light situations [136,137]. Table 10.2 shows the chromaticities of the most often used white points.

The white-point chromaticity and luminance are important not only as they provide the fundamental setting of the monitor, but also as they define the anchor point of any transformation from the CIE XYZ system into a more equidistant and appearance conforming system (e.g., CIE LUV or CIE LAB; see ''Color Space and Color Difference'' in Sec. 10.4.4).

The calibration of the CRT-monitor, i.e., the determination of the relationship between the digital-to-analog converter (DAC) values set by the software and the

TABLE 10.2. *Most often used CRT white-point settings.*

	CCT	x_n	y_n
D50	5000	0.346	0.359
D65	6504	0.313	0.329
D93	9300	0.280	0.312

produced chromaticity coordinates and luminance of the emitted color, is based usually on the assumption that the luminance versus DAC value relationship can be described by an exponential function and that the luminance produced by one of the electronic channels of one of the pixels is independent of the setting of the other channels and for that of the other pixels. This is usually termed spatial and channel independence [138–141]. Depending on the quality of the monitor, neither channel nor spatial independence is fulfilled and more elaborate calibration techniques might be necessary [142,143].

With a well calibrated VDU it becomes possible to produce photorealistic pictures. However, their color appearance will be still different from those produced on hardcopy media. A CIE committee is working on the evaluation of color-appearance models for hard copy and soft-copy comparison [144].

One of the problems in VDU colorimetry is that the colored patches are seen under other visual angles as for most reflecting objects. Thus, e.g., a colored line has a much smaller angular subtense than prescribed in CIE colorimetry (2° observation angle), and the human color-matching functions differ under these circumstances from those determined under standard conditions [145], see also "Small-Field Colorimetry" in Sec. 10.4.2. Visual ergonomic colorimetric standards are laid down in an ISO standard [146].

Brightness Description

In "Third Level: Mental Processing" in Sec. 10.2.5 we already mentioned that chromatic colors seem to be brighter than achromatic ones with the same luminance. There is no standardized metric to describe this so called "brightness–luminance discrepancy," but a CIE Research Note [147] gave a tentative recommendation to quantify this phenomenon. The different color-appearance models (see Sec. 10.5.2) all deal with the problem, but usually give a less satisfactory result than the CIE empirical approach [148]. This model states that for two stimuli that stimulus will seem to be brighter for which the following expression is larger:

$$L^{**}=\ln(L)+C, \tag{10.84}$$

where L is the luminance of the stimulus and C is a chromaticity dependent quantity:

$$C=0.256-0.184y-2.527xy+4.656x^3y+4.657xy^4. \tag{10.85}$$

If the two stimuli have equal $\log_{10}(L)+C$ values they will be seen equally bright. Figure 10.37 shows contour lines of equally bright lights of equal luminance. (For further details on the brightness–luminance relationship see Ref. [149].)

Signal Colors

Signal lights have their own photometric and colorimetric problems. They have to be visible, and their color unambiguously recognizable even from very long

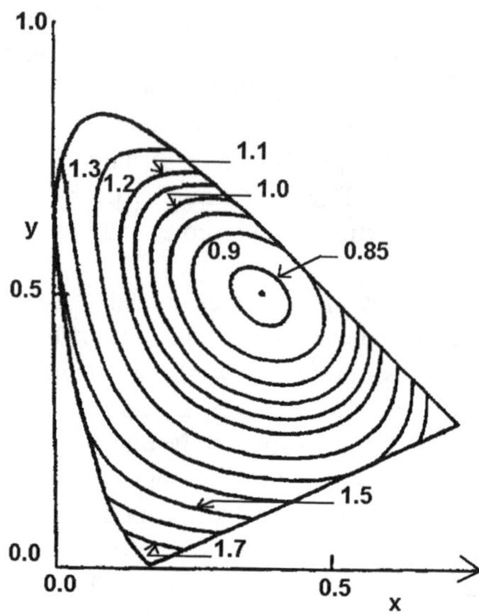

FIGURE 10.37. Contour lines of equiluminous lights of equal brightness.

distances; this means one has to be able to see them under small angular subtense and low intensity. They should be designed in such a form that even color-deficient people have no difficulty in identifying them.

The first CIE international signal color recommendations date back to the 1950s. Studies performed in different countries led to new recommendations, restricting some of the signal light chromaticity boundaries [150]. Figure 10.38 shows in the CIE chromaticity diagram the old and revised signal light color boundaries of red, yellow, green, blue, and white signal colors. It is expected that in the not too distant future these new signal color boundaries will become an international standard.

The measurement of signal lights provides similar technical problems as those of LEDs, only the strength of the light is usually much higher. If tristimulus instruments are used the goodness of the colorimetric fit is important. Often from each color to be measured a laboratory standard is prepared by high-precision spectroradiometric measurements and the tristimulus instrument is used as a color-difference meter comparing the color of the test light to that of the standard light. Again, if the fit of the colorimeter is not really good, the spectral distribution of the test should not deviate too much from that of the standard.

Recently signal lights with dichroic filters have been introduced on the market. The spectral transmittance of dichroic filters—as that of all systems working on the principle of interference—shows an angular dependence. To describe completely the colorimetric characteristics of such lights, it is necessary to measure the angle dependence both of the luminance and of the chromaticity. This is especially important if the signal light can be seen under different angles. One has to ascertain that the chromaticity stays for all the relevant angles within the permitted tolerances. If

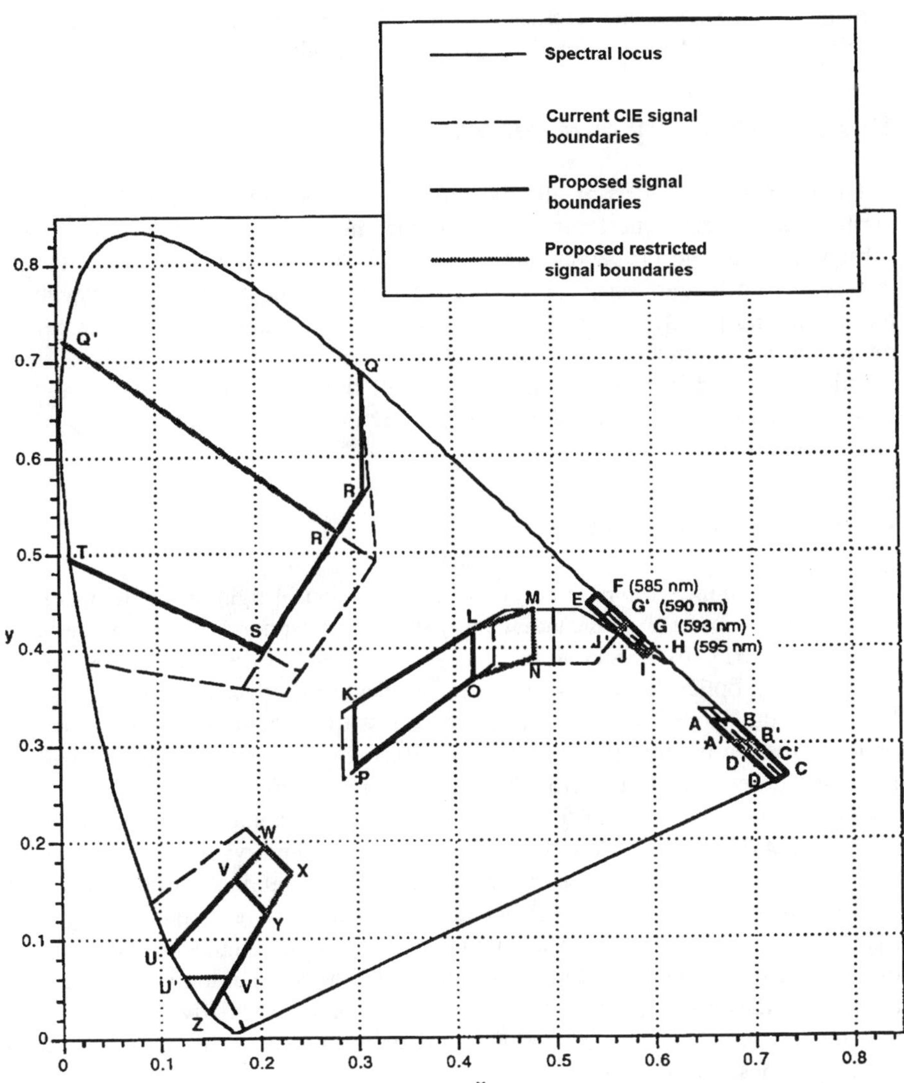

FIGURE 10.38. Original and revised boundaries for colors of signal lights.

it is a signal light with an incandescent lamp, this has to be checked for the lowest and highest main voltage value, as the color temperature of the incandescent lamp is highly voltage dependent. In the case of gas-discharge lamps, the light-source chromaticity tolerances have to be observed. In the case of glass-filter signal lights, it is permitted to shine the light into an integrating sphere and measure the average chromaticity of the integrating-sphere wall. For dichroic filter signal lights this is not permitted, as the chromaticity of dichroic filtered lights is changing with viewing angle. Such a measurement is allowed only if it is ensured that the goniometric variations of the chromaticity are small enough.

10.6.3 Colorimetry of Materials

The overwhelming part of all colorimetric work is done on materials: textiles, plastics, paper, etc. Questions related to the measurement technology of these substances would go far beyond the limits of this book, and we would like to refer in this respect to the many specialized monographs and books dealing with the theoretical and practical aspects of measuring material sample color. (See general references at the end of the bibliographic section.) At this point only a single question will be discussed, the measurement of fluorescing materials, as the fluorescence measurement is actually a photometric one and consequently falls within the scope of the present book.

Fluorescing Materials

The general concept *luminescence* is ''the emission of light at low temperatures by any process other than incandescence, such as phosphorescence or chemiluminescence'' [151]. Its special form *fluorescence* refers to emission produced by excitation by optical radiation or particles (electrons, ions, etc.) and where the lifetime of the excited atoms or molecules is less than about 10^{-8} s. Excitation by optical radiation is often termed *photofluorescence*, but in this section we will use the term fluorescence to describe this latter phenomenon.

Fluorescing materials transform shorter-wavelength radiation into longer-wavelength radiation, usually UV radiation or short-wavelength visible (blue) light into green, orange, or red light. Such materials are, e.g., the luminophores (or phosphors) used in gas-discharge lamps (fluorescent tubes, high-pressure mercury lamps, etc.). Paper, textiles, and other materials often contain optical brighteners, a special class of fluorescent material, used to increase the ''whiteness'' sensation. A further class are some fluorescent pigments or inks (used in marker pens, advertisements, etc.) that produce a strong colored radiation when irradiated by ultraviolet or short-wavelength visible radiation.

Figure 10.39 shows an example of the four characteristic spectra of a fluorescent specimen from an orange fluorescent material [152]. β_T is the total spectral radiance factor measured when the sample is properly irradiated by a D65 source; β_S is the reflected radiance factor one would obtain if no fluorescence occurred. β_L is the fluorescent part of the radiance factor, i.e., the emitted radiation depending on the irradiation at shorter wavelength, and X is the excitation spectrum showing the relative efficiency of different wavelength radiation to excite the fluorescence. Problems are encountered if the excitation spectrum and the emission spectrum overlap, as shown on this example. (The emission occurring in wavelength regions where the fluorescence can be still excited is called anti-Stokes emission.)

To be able to measure the color of such surfaces reproducibly, not only the

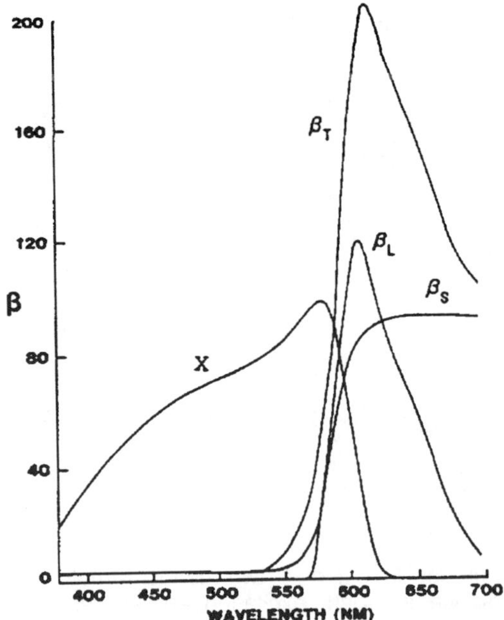

FIGURE 10.39. Total spectral radiance factor β_T and its components for an orange fluorescent sample. (From F. W. Billmeyer, Metrology, documentary standards, and color specifications for fluorescent materials, *COLOR Res. & Appl.* **19**, 413–425. © 1994 John Wiley & Sons. Reprinted by permission of John Wiley & Sons, Inc.)

reflected light has to be evaluated but also the emitted one, and the emission has to be excited by a well-known amount of radiation of known spectral composition. The recommended illuminant for such measurements is the CIE Standard Illuminant D65. Unfortunately, the practical realization of this illuminant (the production of a D65 source) is not easy, and there are no standardized methods how to proceed [42].

Material samples are often measured in photometer spheres where either the irradiation is hemispherical, or the collection of the radiation is done in the upper hemisphere. Measuring fluorescent samples in such measuring geometry is further complicated by the fact that the absorption by the test sample influences the spectral distribution of the radiation falling onto the sample. Ways to correct for this effect are described in Ref. [153].

It is better practice to use a 0°-45° measuring geometry as in this case the spectral distribution of the excitation is not influenced by the absorption of the specimen. Such a measurement is described for the very important class of fluorescent retroreflective materials in Ref. [154].

The most exact technique to measure fluorescent materials is to use the so-called two-monochromator method and to scan both the excitation spectrum and the emission spectrum. The schematic layout of such a system is seen in Fig. 10.40.

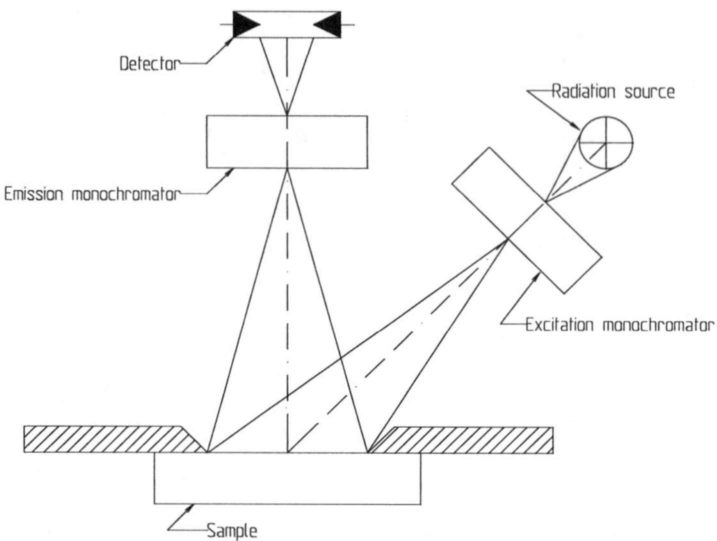

FIGURE 10.40. Schematic layout of a two-monochromator spectroradiometer to measure the total radiance factor, the excitation, and emission spectra of fluorescent samples.

The two-monochromator method was first introduced by Donaldson [155], a modern version of the method together with the proper colorimetric evaluation of the measurement results was published in Ref. [156]. When the two monochromators are scanned in tandem, one gets the true reflection spectrum of the sample. Scanning the excitation monochromator for the preselected emission monochromator setting provides the excitation spectrum. Usually a complete matrix of intensities determined at the different settings of the excitation monochromator and of the emission monochromator has to be measured. Proper calibration of the responsivity of the two systems, band-pass corrections, etc., restricts such measurements to well-equipped laboratories even in the days of fully computerized spectroradiometers.

10.7 SUMMARY AND CONCLUSIONS

The present chapter intended to provide an overview of one of the most intriguing issues of optical radiation measurement, searching for objective correlates of our visual system. Color is, on the one hand, an everyday experience, and most of us with more or less good trichromatic vision take it for granted that we see colors and that physics should be able to describe these colors objectively.

Even from the short presentation of the physiological factors of our color vision system, it became obvious that there are a number of phenomena that cannot be described by simple objective equations. Colorimetry has to make use of simplification, only this makes it possible to obtain quantifiable results.

It is remarkable that CIE colorimetry, despite the great number of simplifications and the fact that it has been designed to predict only color perception equality, performs so well in a number of industries, especially those for which it was not originally intended, as, e.g., the textile coloring industry. Applications that were in mind when the system was founded have the most problems. The colorimetry of light sources, color-rendering descriptions, e.g., are yet not completely adequate. The lighting industry, and in more modern times the color graphics and color computer industry, need a color-appearance description, which is outside the classical limits of CIE colorimetry.

In this chapter first we tried to summarize the techniques where international consensus exists. We covered other items where only tentative recommendations are available, or where measurement methods are only available as suggested by individual authors in subsequent sections.

For further reading and a detailed description of the single colorimetric methods we had to refer to the international literature. For North American use further details can be found in the ASTM Standards. A compilation of relevant standards was published in 1994 [157].

At the end of the References section a ''Further Reading'' list has been included, giving the bibliographic references of some books presenting an overview on colorimetry or treating color as a tutorial subject.

APPENDIX 10.1

CIE Standard Illuminant A and D65 relative spectral power distribution. 300–830 nm at 5-nm intervals; extract from Table 1.1, CIE Publication No. 15.2-1986.

λ	Illuminant A	Illuminant D65
300	0.930483	0.03410
305	1.128210	1.66430
310	1.357690	3.29450
315	1.622190	11.76520
320	1.925080	20.23600
325	2.269800	28.64470
330	2.659810	37.05350
335	3.098610	38.50110
340	3.589680	39.94880
345	4.136480	42.43020
350	4.742380	44.91170
355	5.410700	45.77500
360	6.144620	46.63830

λ	Illuminant A	Illuminant D65
365	6.947200	49.36370
370	7.821350	52.08910
375	8.769800	51.03230
380	9.795100	49.97550
385	10.899600	52.31180
390	12.085300	54.64820
395	13.354300	68.70150
400	14.708000	82.75490
405	16.148000	87.12040
410	17.675300	91.48600
415	19.290700	92.45890
420	20.995000	93.43180
425	22.788300	90.05700
430	24.670900	86.68230
435	26.642500	95.77360
440	28.702700	104.86500
445	30.850800	110.93600
450	33.085900	117.00800
455	35.406800	117.41000
460	37.812100	117.81200
465	40.300200	116.33600
470	42.869300	114.86100
475	45.517400	115.39200
480	48.242300	115.92300
485	51.041800	112.36700
490	53.913200	108.81100
495	56.853900	109.08200
500	59.861100	109.35400
505	62.932000	108.57800
510	66.063500	107.80200
515	69.252500	106.29600
520	72.495900	104.79000
525	75.790300	106.23900
530	79.132600	107.68900
535	82.519300	106.04700
540	85.947000	104.40500
545	89.412400	104.22500
550	92.912000	104.04600
555	96.442300	102.02300
560	100.000000	100.00000
565	103.582000	98.16710
570	107.184000	96.33420
575	110.803000	96.06110
580	114.436000	95.78800
585	118.080000	92.23680
590	121.731000	88.68560
595	125.386000	89.34590

λ	Illuminant A	Illuminant D65
600	129.043000	90.00620
605	132.697000	89.80260
610	136.346000	89.59910
615	139.988000	88.64890
620	143.618000	87.69870
625	147.235000	85.49360
630	150.836000	83.28860
635	154.418000	83.49390
640	157.979000	83.69920
645	161.516000	81.86300
650	165.028000	80.02680
655	168.510000	80.12070
660	171.963000	80.21460
665	175.383000	81.24620
670	178.769000	82.27780
675	182.118000	80.28100
680	185.429000	78.28420
685	188.701000	74.00270
690	191.931000	69.72130
695	195.118000	70.66520
700	198.261000	71.60910
705	201.359000	72.97900
710	204.409000	74.34900
715	207.411000	67.97650
720	210.365000	61.60400
725	213.268000	65.74480
730	216.120000	69.88560
735	218.920000	72.48630
740	221.667000	75.08700
745	224.361000	69.33980
750	227.000000	63.59270
755	229.585000	55.00540
760	232.115000	46.41820
765	234.589000	56.61180
770	237.008000	66.80540
775	239.370000	65.09410
780	241.675000	63.38280
785	243.924000	63.84340
790	246.116000	64.30400
795	248.251000	61.87790
800	250.329000	59.45190
805	252.350000	55.70540
810	254.314000	51.95900
815	256.221000	54.69980
820	258.071000	57.44060
825	259.865000	58.87650
830	261.602000	60.31250

APPENDIX 10.2

CIE 1931 Standard Colorimetric Observer and CIE 1964 Supplementary Standard Colorimetric Observer color matching functions, 360–830 nm at 5-nm intervals; extract from tables taken from CIE Publication No. 15.2-1986.

λ	$\bar{x}_2(\lambda)$	$\bar{y}_2(\lambda)$	$\bar{z}_2(\lambda)$	$\bar{x}_{10}(\lambda)$	$\bar{y}_{10}(\lambda)$	$\bar{z}_{10}(\lambda)$
360	0.0001	0.0000	0.0006	0.0000	0.0000	0.0000
365	0.0002	0.0000	0.0011	0.0000	0.0000	0.0000
370	0.0004	0.0000	0.0019	0.0000	0.0000	0.0000
375	0.0007	0.0000	0.0035	0.0000	0.0000	0.0001
380	0.0014	0.0000	0.0065	0.0002	0.0000	0.0007
385	0.0022	0.0001	0.0105	0.0007	0.0001	0.0029
390	0.0042	0.0001	0.0201	0.0024	0.0003	0.0105
395	0.0077	0.0002	0.0362	0.0072	0.0008	0.0323
400	0.0143	0.0004	0.0679	0.0191	0.0020	0.0860
405	0.0232	0.0006	0.1102	0.0434	0.0045	0.1971
410	0.0435	0.0012	0.2074	0.0847	0.0088	0.3894
415	0.0776	0.0022	0.3713	0.1406	0.0145	0.6568
420	0.1344	0.0040	0.6456	0.2045	0.0214	0.9725
425	0.2148	0.0073	1.0391	0.2647	0.0295	1.2825
430	0.2839	0.0116	1.3856	0.3147	0.0387	1.5535
435	0.3285	0.0168	1.6230	0.3577	0.0496	1.7985
440	0.3483	0.0230	1.7471	0.3837	0.0621	1.9673
445	0.3481	0.0298	1.7826	0.3867	0.0747	2.0273
450	0.3362	0.0380	1.7721	0.3707	0.0895	1.9948
455	0.3187	0.0480	1.7441	0.3430	0.1063	1.9007
460	0.2908	0.0600	1.6692	0.3023	0.1282	1.7454
465	0.2511	0.0739	1.5281	0.2541	0.1528	1.5549
470	0.1954	0.0910	1.2876	0.1956	0.1852	1.3176
475	0.1421	0.1126	1.0419	0.1323	0.2199	1.0302
480	0.0956	0.1390	0.8130	0.0805	0.2536	0.7721
485	0.0580	0.1693	0.6162	0.0411	0.2977	0.5701
490	0.0320	0.2080	0.4652	0.0162	0.3391	0.4153
495	0.0147	0.2586	0.3533	0.0051	0.3954	0.3024
500	0.0049	0.3230	0.2720	0.0038	0.4608	0.2185
505	0.0024	0.4073	0.2123	0.0154	0.5314	0.1592
510	0.0093	0.5030	0.1582	0.0375	0.6067	0.1120
515	0.0291	0.6082	0.1117	0.0714	0.6857	0.0822
520	0.0633	0.7100	0.0782	0.1177	0.7618	0.0607
525	0.1096	0.7932	0.0573	0.1730	0.8233	0.0431
530	0.1655	0.8620	0.0422	0.2365	0.8752	0.0305
535	0.2257	0.9149	0.0298	0.3042	0.9238	0.0206
540	0.2904	0.9540	0.0203	0.3768	0.9620	0.0137
545	0.3597	0.9803	0.0134	0.4516	0.9822	0.0079
550	0.4334	0.9950	0.0087	0.5298	0.9918	0.0040
555	0.5121	1.0000	0.0057	0.6161	0.9991	0.0011

λ	$\bar{x}_2(\lambda)$	$\bar{y}_2(\lambda)$	$\bar{z}_2(\lambda)$	$\bar{x}_{10}(\lambda)$	$\bar{y}_{10}(\lambda)$	$\bar{z}_{10}(\lambda)$
560	0.5945	0.9950	0.0039	0.7052	0.9973	0.0000
565	0.6784	0.9786	0.0027	0.7938	0.9824	0.0000
570	0.7621	0.9520	0.0021	0.8787	0.9556	0.0000
575	0.8425	0.9154	0.0018	0.9512	0.9152	0.0000
580	0.9163	0.8700	0.0017	1.0142	0.8689	0.0000
585	0.9786	0.8163	0.0014	1.0743	0.8256	0.0000
590	1.0263	0.7570	0.0011	1.1185	0.7774	0.0000
595	1.0567	0.6949	0.0010	1.1343	0.7204	0.0000
600	1.0622	0.6310	0.0008	1.1240	0.6583	0.0000
605	1.0456	0.5668	0.0006	1.0891	0.5939	0.0000
610	1.0026	0.5030	0.0003	1.0305	0.5280	0.0000
615	0.9384	0.4412	0.0002	0.9507	0.4618	0.0000
620	0.8544	0.3810	0.0002	0.8563	0.3981	0.0000
625	0.7514	0.3210	0.0001	0.7549	0.3396	0.0000
630	0.6424	0.2650	0.0000	0.6475	0.2835	0.0000
635	0.5419	0.2170	0.0000	0.5351	0.2283	0.0000
640	0.4479	0.1750	0.0000	0.4316	0.1798	0.0000
645	0.3608	0.1382	0.0000	0.3437	0.1402	0.0000
650	0.2835	0.1070	0.0000	0.2683	0.1076	0.0000
655	0.2187	0.0816	0.0000	0.2043	0.0812	0.0000
660	0.1649	0.0610	0.0000	0.1526	0.0603	0.0000
665	0.1212	0.0446	0.0000	0.1122	0.0441	0.0000
670	0.0874	0.0320	0.0000	0.0813	0.0318	0.0000
675	0.0636	0.0232	0.0000	0.0579	0.0226	0.0000
680	0.0468	0.0170	0.0000	0.0409	0.0159	0.0000
685	0.0329	0.0119	0.0000	0.0286	0.0111	0.0000
690	0.0227	0.0082	0.0000	0.0199	0.0077	0.0000
695	0.0158	0.0057	0.0000	0.0138	0.0054	0.0000
700	0.0114	0.0041	0.0000	0.0096	0.0037	0.0000
705	0.0081	0.0029	0.0000	0.0066	0.0026	0.0000
710	0.0058	0.0021	0.0000	0.0046	0.0018	0.0000
715	0.0041	0.0015	0.0000	0.0031	0.0012	0.0000
720	0.0029	0.0010	0.0000	0.0022	0.0008	0.0000
725	0.0020	0.0007	0.0000	0.0015	0.0006	0.0000
730	0.0014	0.0005	0.0000	0.0010	0.0004	0.0000
735	0.0010	0.0004	0.0000	0.0007	0.0003	0.0000
740	0.0007	0.0002	0.0000	0.0005	0.0002	0.0000
745	0.0005	0.0002	0.0000	0.0004	0.0001	0.0000
750	0.0003	0.0001	0.0000	0.0003	0.0001	0.0000
755	0.0002	0.0001	0.0000	0.0002	0.0001	0.0000
760	0.0002	0.0001	0.0000	0.0001	0.0000	0.0000
765	0.0001	0.0000	0.0000	0.0001	0.0000	0.0000
770	0.0001	0.0000	0.0000	0.0001	0.0000	0.0000
775	0.0001	0.0000	0.0000	0.0000	0.0000	0.0000
780	0.0000	0.0000	0.0000	0.0000	0.0000	0.0000
785	0.0000	0.0000	0.0000	0.0000	0.0000	0.0000
790	0.0000	0.0000	0.0000	0.0000	0.0000	0.0000

λ	$\bar{x}_2(\lambda)$	$\bar{y}_2(\lambda)$	$\bar{z}_2(\lambda)$	$\bar{x}_{10}(\lambda)$	$\bar{y}_{10}(\lambda)$	$\bar{z}_{10}(\lambda)$
795	0.0000	0.0000	0.0000	0.0000	0.0000	0.0000
800	0.0000	0.0000	0.0000	0.0000	0.0000	0.0000
805	0.0000	0.0000	0.0000	0.0000	0.0000	0.0000
810	0.0000	0.0000	0.0000	0.0000	0.0000	0.0000
815	0.0000	0.0000	0.0000	0.0000	0.0000	0.0000
820	0.0000	0.0000	0.0000	0.0000	0.0000	0.0000
825	0.0000	0.0000	0.0000	0.0000	0.0000	0.0000
830	0.0000	0.0000	0.0000	0.0000	0.0000	0.0000

APPENDIX 10.3

Components $S_0(\lambda)$, $S_1(\lambda)$, and $S_2(\lambda)$ of daylight used in the calculation of relative spectral power distribution of daylight illuminants of different correlated color temperatures, for wavelengths 300–830 nm at 5-nm intervals (see CIE Publication 15.2 Table 1.2 [19]).

λ (nm)	S_0	S_1	S_2
300	0.04	0.02	0.00
305	3.02	2.26	1.00
310	6.00	4.50	2.00
315	17.80	13.45	3.00
320	29.60	22.40	4.00
325	42.45	32.20	6.25
330	55.30	42.00	8.50
335	56.30	41.30	8.15
340	57.30	40.60	7.80
345	59.55	41.10	7.25
350	61.80	41.60	6.70
355	61.65	39.80	6.00
360	61.50	38.00	5.30
365	65.15	40.20	5.70
370	68.80	42.40	6.10
375	66.10	40.45	4.55
380	63.40	38.50	3.00
385	64.60	36.75	2.10
390	65.80	35.00	1.20
395	80.30	39.20	0.05
400	94.80	43.40	−1.10
405	99.80	44.85	−0.80
410	104.80	46.30	−0.50
415	105.35	45.10	−0.60
420	105.90	43.90	−0.70
425	101.35	40.50	−0.95
430	96.80	37.10	−1.20

λ (nm)	S_0	S_1	S_2
435	105.35	36.90	−1.90
440	113.90	36.70	−2.60
445	119.75	36.30	−2.75
450	125.60	35.90	−2.90
455	125.55	34.25	−2.85
460	125.50	32.60	−2.80
465	123.40	30.25	−2.70
470	121.30	27.90	−2.60
475	121.30	26.10	−2.60
480	121.30	24.30	−2.60
485	117.40	22.20	−2.20
490	113.50	20.10	−1.80
495	113.30	18.15	−1.65
500	113.10	16.20	−1.50
505	111.95	14.70	−1.40
510	110.80	13.20	−1.30
515	108.65	10.90	−1.25
520	106.50	8.60	−1.20
525	107.65	7.35	−1.10
530	108.80	6.10	−1.00
535	107.05	5.15	−0.75
540	105.30	4.20	−0.50
545	104.85	3.05	−0.40
550	104.40	1.90	−0.30
555	102.20	0.95	−0.15
560	100.00	0.00	0.00
565	98.00	−0.80	0.10
570	96.00	−1.60	0.20
575	95.55	−2.55	0.35
580	95.10	−3.50	0.50
585	92.10	−3.50	1.30
590	89.10	−3.50	2.10
595	89.80	−4.65	2.65
600	90.50	−5.80	3.20
605	90.40	−6.50	3.65
610	90.30	−7.20	4.10
615	89.35	−7.90	4.40
620	88.40	−8.60	4.70
625	86.20	−9.05	4.90
630	84.00	−9.50	5.10
635	84.55	−10.20	5.90
640	85.10	−10.90	6.70
645	83.50	−10.80	7.00
650	81.90	−10.70	7.30
655	82.25	−11.35	7.95
660	82.60	−12.00	8.60
665	83.75	−13.00	9.20

λ (nm)	S_0	S_1	S_2
670	84.90	−14.00	9.80
675	83.10	−13.80	10.00
680	81.30	−13.60	10.20
685	76.60	−12.80	9.25
690	71.90	−12.00	8.30
695	73.10	−12.65	8.95
700	74.30	−13.30	9.60
705	75.35	−13.10	9.05
710	76.40	−12.90	8.50
715	69.85	−11.75	7.75
720	63.30	−10.60	7.00
725	67.50	−11.10	7.30
730	71.70	−11.60	7.60
735	74.35	−11.90	7.80
740	77.00	−12.20	8.00
745	71.10	−11.20	7.35
750	65.20	−10.20	6.70
755	56.45	−9.00	5.95
760	47.70	−7.80	5.20
765	58.15	−9.50	6.30
770	68.60	−11.20	7.40
775	66.80	−10.80	7.10
780	65.00	−10.40	6.80
785	65.50	−10.50	6.90
790	66.00	−10.60	7.00
795	63.50	−10.15	6.70
800	61.00	−9.70	6.40
805	57.15	−9.00	5.95
810	53.30	−8.30	5.50
815	56.10	−8.80	5.80
820	58.90	−9.30	6.10
825	60.40	−9.55	6.30
830	61.90	−9.80	6.50

REFERENCES

1. *Handbook of Optics*, edited by M. Bass (McGraw-Hill, New York, 1995), Chap. 24.2, p. 24.5.
2. Richter, M., *Einführung in die Farbmetrik* (de Gruyter, Berlin, 1981), Chap. 4, p. 34.
3. Travis, D. *Effective Colour Displays* (Academic, London, 1991), Chap. 2, p. 37.
4. Pokorny, J. Smith, V. C., and Lutze, M. Appl. Opt. **26**, 1437–1440 (1987).
5. van Norren, D. V., and Voss, J. J., Vision Res. **14**, 1237–1244 (1974).
6. Schnapf, J. L., and Baylor, D. A., Sci. Am. **256**, 32–39 (1987).
7. Stockman, A., MacLeod, D. I. A., Johnson, N. E., J. Opt. Soc. Am. **10A**, 2491–2521 (1993).
8. Walraven, P. L., and Bouman, M. A., Vision Res. **6**, 567 (1966).
9. Pokorny, J., and Smith, V. C., ''Scientific Basis of Visual Performance,'' in *Proceedings of the CIE Symposium on Advances in Photometry, Vienna, 1994.*

10. Martinez-Uriegas, E. ''Chromatic-Achromatic Multiplexing in Human Color Vision,'' in *Visual Science and Engineering, Models and Applications*, edited by D. H. Kelly (Dekker, New York, 1994).
11. *OSA Advances in Color Vision*, 1992 Technical Digest Series, Vol. 4, (1992).
12. Piantanida, T. P., ''The Molecular Genetics of Human Color Vision: From Nucleotides to Nanometers,'' in *Visual Science and Engineering, Models and Applications*, Ref. 10.
13. Hunt, R. W. G., *Measuring Colour* (Ellis Horwood, Chichester, 1987).
14. Schwartz, E. L., ''Topographic Mapping in Primate Visual Cortex: History, Anatomy, and Computation,'' in *Visual Science and Engineering, Models and Applications*, Ref. 10.
15. Wyszecki, G., and Stiles, W. S., *Color Science, Concepts and Methods, Quantitative Data and Formulae*, 2nd ed. (Wiley, New York, 1982).
16. Grassmann, H. G., ''Theory of Compound Colours,'' Philos. Mag. **4** (7), 254–264 (1854), see also in *Selected Papers on Colorimetry—Fundamentals*, edited by D. L. MacAdam and B. J. Thomson, SPIE Milestone Series, MS 77, (SPIE, Bellingham, WA, 1993).
17. Wright, W. D., ''The Historical and Experimental Background to the 1931 CIE System of Colorimetry,'' in *Proceedings of the Symposium on Golden Jubilee of Colour in the CIE*, edited by K. McLaren (Soc. Dyers and Colourists, Bradford, 1981), pp. 3–18.
18. *CIE Proceedings, 8th Session, Cambridge 1931* (Cambridge University Press, Cambridge, 1932), pp. 19–29.
19. *Colorimetry*, CIE Publication No. 15.2-1986, 2nd ed. (CIE Central Bureau, Vienna, 1986).
20. Morren, L., ''A Few Suggestions Regarding the CIE Publications on Colorimetry,'' in *Proceedings of the Symposium on Advanced Colorimetry, Vienna*, edited by J. Schanda and C. Hermann (CIE Central Bureau, Vienna, 1993), pp. 50–54.
21. CIE Standard Colorimetric Observer, ISO/CIE Standard 10527-1991.
22. *Compte Rendu Quatorzième Session, Bruxelles, 1959*, CIE Publication No. 4-1960, Vol. A, (Publisher, City, 1960), pp. 91–109.
23. *Compte Rendu Quinzième Session, Vienne, 1963*, CIE Publication No. 11-1964, Vol. A (Publisher, City, 1964), p. 35.
24. Stiles, W. S., and Burch, J. M., ''Interim Report to the Commission Internationale de l'Eclairage, Zurich, 1955, on the National Physical Laboratory's Investigation of Colour-Matching,'' Opt. Acta **2**, 168 (1955).
25. Stiles, W. S. and Burch, J. M., ''N. P. L. Colour-Matching Investigation: Final Report (1958),'' Opt. Acta **6**, 1–26 (1959).
26. Speranskaya, N. I., ''Determination of Spectrum Colour Co-ordinates for Twenty-Seven Normal Observers,'' Opt. Spectrosc. **7**, 424–428 (1959).
27. Stiles, W. S., ''The Average Colour-Matching Functions for a Large Matching Field,'' in *Visual Problems of Color* (Chemical Publ., New York, 1961), pp. 213–249.
28. Shapiro, A. G., Pokorny, J., and Smith, C. V., ''Rod Contribution to Large-Field Color Matching,'' Color Res. Appl. **19**/4, 236–245 (1994).
29. *CIE 1988 2° Spectral Luminous Efficiency Function for Photopic Vision*, CIE Publication No. 86-1990 (CIE Central Bureau, Vienna, 1990).
30. *Spectral Luminous Efficiency Functions Based upon Brightness Matching for Monochromatic Point Sources, 2° and 10° Fields*, CIE Publication No. 75-1988 (CIE Central Bureau, Vienna, 1988).
31. MacAdam, D. L., ''Visual Sensitivities to Color Differences in Daylight,'' J. Opt. Soc. Am. **32**, 247–274 (1942).
32. Brown, W. R. J., and MacAdam, D. L., ''Visual Sensitivities to Combined Chromaticity and Luminance Differences,'' J. Opt. Soc. Am. **39**, 808–834 (1949).
33. Kelly, K. L., ''Lines of Constant Correlated Color Temperature Based on MacAdam's (u,v) Uniform Chromaticity Transformation of the CIE diagram,'' J. Opt. Soc. Am. **53**, 999–1002 (1963).
34. Schanda, J., and Dányi, M., ''Correlated Color-Temperature Calculations in the CIE 1976 Chromaticity Diagram,'' Color Res. Appl. **2** (4), 161–163 (1977).
35. Schanda, J., ''Correlated Colour Temperature and the ΔE^*_{ab} Colour-Difference Formula,'' in *Proceedings of the AIC COLOR 77*, edited by F. W. Billmeyer and G. Wyszecki (Hilger, Bristol, 1978), pp. 292–295.
36. Robertson, A. R., ''Computation of Correlated Color Temperature and Distribution Temperature,'' J. Opt. Soc. Am. **58**, 1528–1535 (1968).
37. Schanda, J., Mészáros, M., and Czibula, G., ''Calculating Correlated Color Temperature with a Desktop Programmable Calculator,'' Color Res. Appl. **3**, 56–69 (1978).

38. McCamy, C. S., ''Correlated Color Temperature as an Explicit Function of Chromaticity Coordinates,'' Color Res. Appl. **17**, 142–144 (1992).
39. *Distribution Temperature and Ratio Temperature*, CIE Publication No. 114/4-1994 (CIE Central Bureau, Vienna, 1994).
40. Mori, L., Sugiyama, H., and Kambe, N., ''An Absolute Method of Color Temperature Measurement,'' Acta Chrom. **1**, 93–102 (1964).
41. CIE Recommendations on Standard Illuminants for Colorimetry, CIE Publication No. 14A-1968 *Compte Rendu Seizième Session, Washington, 1967*, pp. 95–97.
42. *A Method for Assessing the Quality of Daylight Simulators for Colorimetry*, CIE Publication No. 51-1981 (CIE Central Bureau, Vienna, 1981).
43. CIE D005-1994, A Method for Assessing the Quality of Daylight Simulators for Colorimetry, CIE Publication No. D005-1994 (based on CIE 51.181) (CIE Central Bureau, Vienna, 1994).
44. McCamy, C. S., ''Simulation of Daylight for Viewing and Measuring Color,'' Color Res. Appl. **19**(6), 437–445 (1994).
45. Gundlach D., ''Annäherung von Normlichtart D_{65} für Zwecke der Farbmessung,'' in *AIC Color 77*, (Hilger, Bristol, 1978), pp. 218–221.
46. Terstiege, H., ''Artificial Daylight for Measurement of Optical Properties of Materials,'' Color Res. Appl. **14** (3), 131–138 (1989).
47. Liu, Y., Berns, R. S., and Shu, Y., ''Optimization Algorithm for Designing Colored Glass Filters to Simulate CIE Illuminant D65,'' Color Res. Appl. **16** (2), 89–96 (1991).
48. Clarke, F. J. J., ''Practical Standard Illuminant Representative of Interior Daylight,'' in *Proceedings of the 19th Session of the CIE, Kyoto 1979*, CIE Publication No. 50-1980, pp. 73–78.
49. Hunt, R. W. G., ''Standard Sources to Represent Daylight,'' Color Res. Appl. **17** (4), 293–294 (1992).
50. Grum, F., and Saltzmann, M, ''A New White Standard of Reflectance,'' in *Proceedings of the 18th Session of CIE, London*, CIE Publication No. 36-1975.
51. *A Review of Publications on Properties and Reflection Values of Material Reflection Standards*, CIE Publication No. 46-1979 (CIE Central Bureau, Vienna, 1979).
52. Malkin, F., and Verrill, J. F., ''The New Series of Ceramic Colour Standards,'' in *Proceedings of the 20th Session of the CIE, Amsterdam, 1983*, E37/1-2, CIE Publication No. 56-1983.
53. Fillinger, L., Lukacs, Gy, and Andor, Gy, ''Színetalonok Termokromimusa'' (''Thermochromism of Color Standards''), Mérés Aut. **24**, 342–347 (1976).
54. Fairchild, M. D., and Grum, F., ''Thermochromism of Ceramic Reference Tiles,'' Appl. Opt. **24**, 3432 (1985).
55. ISO Draft International Standard, *Textiles—Test for Colour Fastness—Part J01: Calculation of Small Colour Differences*, ISO DIS Publication No. 105-J01/1992.
56. ''*Parametric Effects in Colour-Difference Evaluation*,'' CIE Technical Report CIE 101-1993 (CIE Central Bureau, Vienna, 1993).
57. *Industrial Colour-Difference Evaluation*, CIE Technical Report CIE 116-1995 (CIE Central Bureau, Vienna, 1995).
58. Berns, R. S., Alman, D. H., Reniff, L., Snyder, G. D., and Balonon-Rosen, M. R., ''Visual Determination of Suprathreshold Colour-Difference Tolerances Using Probit Analysis,'' Color Res. Appl. **16**, 297–316 (1991).
59. Luo, M. R., and Rigg, B., ''Chromaticity-Discrimination Ellipses for Surface Colours,'' Color Res. Appl. **11**, 25–42 (1986).
60. Witt, K., ''Modified CIELAB Formula Tested Using a Textile Pass/Fail Data Set,'' Color Res. Appl. **19**, 273–276 (1994).
61. *Special Metamerism Index: Change in Observer*, CIE Technical Report, CIE 80-1989 (CIE Central Bureau, Vienna, 1989).
62. North, A., and Fairchild, M., ''Measuring Color Matching Functions, Part 1,'' Color Res. Appl. **18**, 155–162 (1993).
63. North, A., and Fairchild, M., ''Measuring Color Matching Functions, Part 2,'' Color Res. Appl. **18**, 163–170 (1993).
64. Rich, D. C., and Jalijali, J., ''Effects of Observer Metamerism in the Determination of Human Color-Matching Functions,'' Color Res. Appl. **20**, 29–35 (1995).
65. Berger-Schun, A., ''Description of Samples Used and Their Colorimetric Measurement,'' Die Farbe **26**, 7–16 (1977).
66. Ganz, E., ''Whiteness: Photometric Specification and Colorimetric Evaluation,'' Appl. Opt. **15**, 2039–2058 (1976).
67. Ganz, E., ''Whiteness Formulas: A Selection,'' Appl. Opt. **18**, 1073–1078 (1979).

68. Brockes, A., "The Evaluation of Whiteness," CIE J. **1**, 38–39 (1982).
69. Ganz, E., and Pauli, H. K. A., "Whiteness and Tint Formulas of the Commission Internationale de l'Eclairage: Approximations in the $L^*a^*b^*$ Color Space," Appl. Opt. **34**, 2998–2999 (1995).
70. A Method of Predicting Corresponding Colours under Different Chromatic and Illuminance Adaptations, CIE Technical Report CIE 109-1994 (CIE Central Bureau, Vienna, 1994).
71. Hunt, R. W. G., and Pointer, M. R., A Color-Appearance Transform for the CIE 1931 Standard Colorimetric Observer," Color Res. Appl. **10**, 165–179 (1985).
72. Judd, D. B., "Standard Response Functions for Protanopic and Deuteranopic Vision," J. Opt. Soc. Am. **35**, 199–220 (1945).
73. *Method of Measuring and Specifying Colour Rendering Properties of Light Sources*, CIE Technical Report, CIE 13.3-1995 (CIE Central Bureau, Vienna, 1995).
74. Takahama, K., Sobagaki, H., and Nayatani Y., "Formulation of a Nonlinear Model of Chromatic Adaptation for a Light-Gray Background," Color Res. Appl. **9**, 106–115 (1984).
75. Nayatani, Y., Takahama, K., and Sobagaki, H., "Field Trials on Colors Appearance of Chromatic Colors Under Various Light Sources," Color Res. Appl. **13**, 307–317 (1988).
76. Forsius, S. A., "Physica," manuscript (1611), published in Acta Bibliothecae Stockholmiensis, 315–321 (1971).
77. Billmeyer, F. W., Jr., "Survey of Color Order Systems," Color Res. Appl. **12**, 173–186 (1987).
78. Newhall, S. M., Nickerson, D., and Judd, D. B., "Final Report of the OSA Subcommittee on the Spacing of the Munsell Colors," J. Opt. Soc. Am. **33**, 385–418 (1943).
79. Hunter, R. S., *The Measurement of Appearance* (Wiley, New York, 1975).
80. Wyszecki, G., and Stiles, W. S., *Color Science, Concepts and Methods, Quantitative Data and Formulae*, 2nd ed. (Wiley, New York, 1982), pp. 508–509.
81. McCamy, C. S., "Munsell Value as Explicit Functions of CIE Luminance Factor," Color Res. Appl. **17**, 205–207 (1992).
82. Rheinboldt, W. C., and Menard, J. P., "Mechanized Conversion of Colorimetric Data to Munsell Renotations," J. Opt. Soc. Am. **50**, 802–807 (1960).
83. Tonnquist, G., "Philosophy of Perceptive Color Order Systems," Color Res. Appl. **11**, 51–55 (1986).
84. Derefeldt, G., and Sahlin, C., "Transformation of NCS Data into CIELAB Colour Space," Color Res. Appl. **11**, 146–152 (1986).
85. MacAdam, D. L., "Colorimetric Data for Samples of OSA Uniform Color Scales," J. Opt. Soc. Am. **68**, 121–130 (1978).
86. Nickerson, D., "History of the OSA Committee on Uniform Color Scales," Opt. News **3**, 8–17 (1977).
87. Richter, M., and Witt, K., "The Story of the DIN Color System," Color Res. Appl. **11**, 138–145 (1986).
88. Ostwald, W., *Die Farbenlehre* (Unesma, Leipzig, 1915).
89. Ostwald, W., *The Color Primer*, translated and edited by F. Birren (Van Nostrand Reinhold, New York, 1969).
90. Nemcsics, A., "Color Space of the Coloroid Color System," Color Res. Appl. **12**, 135–146 (1987).
91. Nemcsics, A., "Spacing in the Munsell Color System Relative to the Coloroid Color System," Color Res. Appl. **19**, 122–125 (1994).
92. Smith, N. S., Whitfield, T. W. A., and Wiltshire, T. J., "Research Note on the Accuracy of the NCS, DIN and OSA-UCS Colour Atlases," Color Res. Appl. **15**, 197–299 (1990).
93. Smith, N. S., Whitfield, T. W. A., and Wiltshire, T. J., "Comparison of the Munsell, NCS, DIN and Coloroid Colour Order Systems Using the OSA-UCS Model," Color Res. Appl. **15**, 327–337 (1990).
94. Smith, N. S., Whitfield, T. W. A., and Wiltshire, T. J., "Accuracy of the NCS Atlas Samples," Color Res. Appl. **16**, 108–113 (1991).
95. Smith, N. S., Whitfield, T. W. A., and Wiltshire, T. J., "NCS, and DIN Notations for the OSA-UCS Atlas Samples," Color Res. Appl. **17**, 273–283 (1992).
96. Smith, N. S., and Billmeyer, F. W., Jr., "Comparison of the Colorcurve and SCA-2541 Colour Order Systems using the OSA-UCS Model," Color Res. Appl. **19**, 363–374 (1994).
97. Guth, S. L., "Model for Color Vision and Light Adaptation," J. Opt. Soc. Am. A **8**, 976–993 (1991); **9**, 344 (E) (1992).
98. Guth, S. L., "Unified Model for Human Color Perception and Visual Adaptation II," Proc. SPIE **1913**, 440–448 (1993).
99. Guth, S. L., "ATD Model for Color Vision I: Background; II: Applications," Proc. SPIE **2170**, 149–161 (1994).

100. Guth, S. L., ''Further Applications of the ATD Model for Color Vision,'' Proc. SPIE Conf. on Device-Independent Color Imaging II, **2414**, 12–26 (1995).
101. Fairchild, M. D., ''Formulation and Testing of an Incomplete-Chromatic Adaptation Model,'' Color Res. Appl. **16**, 243–250 (1991).
102. Fairchild, M. D., ''A Model of Incomplete Chromatic Adaptation,'' in *Proceedings of the 22nd Session of the CIE*, CIE 91-1991 (CIE Central Bureau, Vienna, 1991), pp. 33–34.
103. Fairchild, M. D., and Berns, R. S., ''Image Color-Appearance Specification through Extension of CIELAB,'' Color Res. Appl. **18**, 178–190 (1993); **18**, 437 (E) (1993).
104. Fairchild, M. D., ''Visual Evaluation and Evolution of the RLAB Color Space,'' in *IS&T and SID's 2nd Color Imaging Conference: Color Science Systems and Applications* (IS&T, Springfield, Va., 1994), pp. 9–13.
105. Nayatani, Y., Hashimoto, K., Takahama, K., and Sobagaki, H., ''A Nonlinear Color-Appearance Model using Estevez–Hunt–Pointer Primaries,'' Color Res. Appl. **12**, 231–242 (1987).
106. Nayatani, Y., Takahama, K., Sobagaki, H., and Hashimoto, K., ''Color-Appearance Model and Chromatic-Adaptation Transform,'' Color Res. Appl. **15**, 210–221 (1990).
107. Nayatani, Y., ''Revision of the Chroma and Hue Scales of a Nonlinear Color-Appearance Model,'' Color Res. Appl. **20**, 143–155 (1995).
108. Nayatani, Y., Sobagaki, H., Hashimoto, K., and Yano, T., ''Lightness Dependency of Chroma Scales of a Nonlinear Color-Appearance Model and its Latest Formulation,'' Color Res. Appl. **20**, 156–167 (1995).
109. Hunt, R. W. G., ''A Model of Colour Vision for Predicting Colour Appearance in Various Viewing Conditions,'' Color Res. Appl. **12**, 297–314 (1987).
110. Hunt, R. W. G., ''Errata: A Model of Colour Vision for Predicting Colour Appearance in Various Viewing Conditions,'' Color Res. Appl. **13**, 132 (1988).
111. Hunt, R. W. G., ''Revised Colour-Appearance Model for Related and Unrelated Colours,'' Color Res. Appl. **16**, 146–165 (1991).
112. Hunt, R. W. G., ''Erratum: Revised Colour-Appearance Model for Related and Unrelated Colours,'' Color Res. Appl. **16**, 351 (1991).
113. Hunt, R. W. G., ''An Improved Predictor of Colourfulness in a Model of Colour Vision,'' Color Res. Appl. **19**, 23–26 (1994).
114. Luo, M. R., Clarke, A. A., Rhodes, P. A., Schappo, A., Scrivener, S. A. R., and Tait, C. J., ''Quantifying Colour Appearance, Part I. LUTCHI Colour Appearance Data,'' Color Res. Appl. **16**, 166–180 (1991).
115. Luo, M. R., Clarke, A. A., Rhodes, P. A., Schappo, A., Scrivener, S. A. R., and Tait, C. J., ''Quantifying Colour Appearance, Part II. Testing Colour Models Performance using LUTCHI Colour Appearance Data,'' Color Res. Appl. **16**, 181–197 (1991).
116. Luo, M. R., Gao, X. W., Rhodes, P. A., Xin, H. J., Clarke, A. A., and Scrivener, S. A. R., ''Quantifying Colour Appearance, Part III. Supplementary LUTCHI Colour Appearance Data,'' Color Res. Appl. **18**, 98–113 (1993).
117. Luo, M. R., Gao, X. W., Rhodes, P. A., Xin, H. J., Clarke, A. A., and Scrivener, S. A. R., ''Quantifying Colour Appearance, Part IV. Transmissive Media,'' Color Res. Appl. **18**, 191–209 (1993).
118. Hunt, R. W. G., and Luo, M. R., ''Evaluation of a Model of Colour Vision by Magnitude Scalings: Discussion of Collected Results,'' Color Res. Appl. **19**, 27–33 (1994).
119. Luo, M. R., Gao, X. W., and Scrivener, S. A. R., ''Quantifying Colour Appearance, Part V. Simultaneous Contrast,'' Color Res. Appl. **20**, 18–28 (1995).
120. Fairchild, M. D., ''Testing Colour-Appearance Models: Guidelines for Coordinated Research,'' Color Res. Appl. **20**, 262–267 (1995).
121. Technical leaflet: Colorimeter head, LMT Lichmesstechnik, Berlin, Germany.
122. *Methods of Characterizing Illuminance Meters and Luminance Meters*, CIE Technical Report, CIE 69-1987 (CIE Central Bureau, Vienna, 1987).
123. Terstiege, H., and Gundlach, D., ''Characterizing the Quality of Colorimeters,'' SID 91 Digest 641–644 (1991).
124. Technical leaflet: Colorimeter head, InPhoRa California.
125. Technical leaflet: CCD-colorimeter, InPhoRa California.
126. Schanda, J., and Lux, G., ''On the Electronic Correction of Errors in a Tristimulus Colorimeter,'' in *Colour 73, Proceedings of the AIC Congress, York, 1973* (Hilger, London, 1973), pp. 466–469.
127. Andor, Gy, ''Temperature Dependence of High Accuracy Photometer Heads,'' Appl. Opt. **28**, 4733–4734 (1989).

128. Raggi., A., and Barbiroli G., ''Colour-Difference Measurement: The Sensitivity of Various Instruments Compared, Color Res. Appl. **18**, 11–27 (1993).
129. Rodgers, J., Wolf, K., Willis, N., Hamilton, D., Ledbetter, R., and Stewart, C., ''A Comparative Study of Color Measurement Instrumentation,'' Color Res. Appl. **19**, 322–327 (1994).
130. Hirschler, R., ''Colour Standardisation—Are We There Yet?,'' in Colour Communication UMIST 1995 Conference.
131. A Review of Publications on Properties and Reflection Values of Material Reflection Standards, CIE Technical Report, CIE 46-1979 (CIE Central Bureau, Vienna, 1979).
132. *The Spectroradiometric Measurement of Light Sources*, CIE Technical Report, CIE 63-1984 (CIE Central Bureau, Vienna, 1984).
133. *Spectroradiometry of Pulsed Optical Radiation Sources*, CIE Technical Report, CIE 105-1993 (CIE Central Bureau, Vienna, 1993).
134. Makai, J., Czibula, G., Vida, D., and Schanda, J., Spatial Distribution of Colorimetric Characteristics of Metal Halide Lamps,'' in *Proceedings of the CIE Symposium on Light and Colour Measurement '81*, 1981, pp. 51–56.
135. Martin, G., Muray, K., Réti, I., Diós, J., and Schanda, J., ''Miniature Integrating Sphere—Silicon Detector Combination for LED Total Power Measurement,'' Measurement **8**, 84–89 (1990).
136. Engeldrum, P. G., and Ingraham, J. L., ''Analysis of White Point and Phosphor Set Differences of CRT Display,'' Color Res. Appl. **15**, 151–155 (1990).
137. Brill, M. H., and Derefeldt, G., ''Comparison of Reference-White Standards for Video Display Units,'' Color Res. Appl. **16**, 26–30 (1991).
138. Cowan, W. B., and Rowell, N., ''On the Gun Independence and Phosphor Constancy of Colour Video Monitors,'' Color Res. Appl. **11**, S34–S38 (1986).
139. Brainard, D. H., ''Calibration of a Computer Controlled Color Monitor,'' Color Res. Appl. **14**, 23–34 (1989).
140. Berns, R. S., Motta, R. J., and Gorzynski, M. E., ''CRT colorimetry, Part I: Theory and Praxis,'' Color Res. Appl. **18**, 229–314 (1993).
141. Berns, R. S., Gorzynski, M. E., and Motta, R. J., ''CRT Colorimetry, Part I: Metrology,'' Color Res. Appl. **18**, 315–325 (1993).
142. Bodrogi, P., and Schanda, J., ''Testing the Calibration Model of Colour CRT Monitors,'' Displays **16**, 123 (1995).
143. Bodrogi, P., Schanda, J., Muray, K., and Kránicz, B., ''Accurate Colorimetric Calibration of CRT Monitors,'' SID Conference, Orlando, 1995.
144. Alessi, P. J., ''CIE Guidelines for Coordinated Research on Evaluation of Colour Appearance Models for Reflection Print and Self-Luminous Display Image Comparisons,'' Color Res. Appl. **19**, 48–58 (1994).
145. Schanda, J., ''Colour and the Visual Display Unit,'' in Proc. AIC '93 Congress, Budapest, 1993, pp. I11-1–12.
146. ISO Draft Standard, *Visual Display Terminals (VDTs) Used for Office Tasks—Ergonomic Requirements, Part 8: Requirements for Displayed Colours*, IS/DIS Publication No. 9241-8 1994.
147. *Models of Heterochromatic Brightness Matching*, CIE Research Note, CIE 118/2-1995 (CIE Central Bureau, Vienna, 1995).
148. Schanda, J., ''Brightness Description using Different Metrics,'' UMIST Internat. Conf. on Colour Communication, 1995.
149. *Brightness–Luminance Relations: Classified Bibliography*, CIE Technical Report, CIE 78-1988 (CIE Central Bureau, Vienna, 1988).
150. *Review of the Official Recommendations of the CIE for the Colours of Signal Lights*, CIE Technical Report, CIE 107-1994 (CIE Central Bureau, Vienna, 1994).
151. *Collins Dictionary of the English Language*, edited by P. Hans, T. H. Long, and L. Urdang (Collins, London, 1983).
152. Billmeyer, E. W., Jr., ''Metrology, Documentary Standards, and Color Specifications for Fluorescent Materials,'' Color Res. Appl. **19**, 413–425 (1994).
153. *Intercomparison on Measurement of (Total) Spectral Radiance Factor of Luminescent Specimens*, CIE Technical Report, CIE 77-1988 (CIE Central Bureau, Vienna, 1988).
154. Burns, D. M., Johnson, N. L., and Pavelka, L. A., ''Colorimetry of Durable Fluorescent Retroreflective Materials,'' Color Res. Appl. **20**, 93–107 (1995).
155. Donaldson, R., ''Spectrophotometry of Fluorescent Pigments,'' Brit. J. Appl. Phys. **5**, 210–224 (1954).
156. Gundlach, D., and Terstiege, H., ''Problems in Measurement of Fluorescent Materials,'' Color Res. Appl. **19**, 427–436 (1994).

157. ASTM Standards on Color and Appearance Measurement, sponsored by ASTM Committee E-12, ASTM, 1994.

Further Reading

ASTM Standards on Color and Appearance Measurement, sponsored by ASTM Committee E-12, ASTM, 1994.

Color in Business, Science and Industry by D. Judd and G. Wyszecki (Wiley, New York, 1975).

Color Science; Concepts and Methods, Quantitative Data and Formulae, 2nd ed., by G. Wyszecki and W. S. Stiles (Wiley, New York, 1982).

Computer Generated Color, A Practical Guide to Presentation and Display by R. Jackson, L. MacDonald, and K. Freeman (Wiley, Chichester, 1994).

Effective Color Display, Theory and Practice by D. Travis (Academic, London, 1991).

Einführung in die Farbmetrik, 2nd ed., by M. Richter (de Gruyter, Berlin, 1981).

IESNA Lighting Handbook, Reference & Application, 8th ed., edited by M. S. Rea, Chapter 3, Vision and Perception, Chapter 4, Color (Illuminating Engineering Society of North America, New York, 1993).

Measuring Colour, 2nd ed., by R. W. G. Hunt (Ellis Horwood, New York, 1991).

OSA Handbook of Optics, edited by M. Bass, D. R. Williams, and W. L. Wolfe (McGraw-Hill, New York, 1995).

Praktische Farbmessung by Anni Berger-Schunn (Muster-Schmidt, Göttingen, 1991).

The Measurement of Appearance, by R. S. Hunter (Wiley, New York, 1975).

The Reproduction of Colour, 5th ed. (Fountain, Kingston-upon-Thames, England, 1995).

11

Future Trends in Photometry

János D. Schanda

Commission International de L'Eclairage, Kegelgasse 27, A-1030 Vienna, Austria

11.1 INTRODUCTION

Photometry has a unique position in physics. It is radiometry evaluating optical radiation in a form that should reflect visual perception. Thus, on the one hand, it is influenced by vision science; on the other hand, it is a branch of optical radiometry.

Some 70 years ago, when the fundamentals of modern photometry were laid down, a number of simplifications were introduced, shaping photometry into a linear additive system of radiometry.

During the past decades a number of new discoveries were made, both in visual science and in the technology of optical radiometry, which will influence the development of photometry during the next decades. It is difficult to foresee how quickly the new findings will penetrate practical photometry, especially as in everyday life giant industries rely on photometric measurements, and to change their measuring and evaluating system certainly needs time.

If we try to analyze the foreseeable trends in photometry we have to approach the question from different aspects:

- Fundamental visual science is working towards a better understanding of the functioning of the visual mechanism, thereby providing information on the visual-impression–optical-stimulus relationship. The task of photometry is to find the mathematical description of this relationship under different visual situations (lighting level, accommodation, speed of response, color compositions, etc.) and to develop meaningful quantitative evaluations.
- Metrologists would like to develop radiometry and photometry into an exact science where the accuracy of the measurement results are comparable to those obtained in other branches of physics. Unfortunately, precision and accuracy attainable in photometry lag far behind those of other areas of physics, e.g., of length and time measurement.
- New branches of technology, e.g., information technology, formulate new demands for photometry. In image technology and in long-distance transmission of such images a proper description of pixel brightness and perceived contrast is needed. Photometry is faced with the request to develop new brightness magnitude descriptors for imaging technologies.
- As new technologies of transforming optical radiation into electrical signals become available, photometry applies them, leading partly to new devices the photometrist can apply, but also to new evaluation methods and techniques.

In the following sections the items noted above will be discussed, and we will try to reach conclusions on the directions photometry may take in the coming years.

11.2 LUMINANCE AND BRIGHTNESS DESCRIPTION

In photometry we have the fundamental problem that light perception is a human cognitive process, and in the framework of a photometric system corresponding quantitative physical measures have to be found. In 1924 the *photometric system* and the concept of *luminance* were coined to provide a psychophysical correlate of brightness. One had to realize, however, that luminance is not an accurate descriptor of brightness. First of all, the relationship between luminance and brightness is nonlinear, and fundamental difficulties occur when the intensity drops to mesopic levels or when the light stimulus becomes chromatic (nonadditivity, etc.).

Vision research has provided some information on these phenomena. We know now that within the human retina we have two types of photoreceptors: the cones, used at normal daylight light levels and with them we perceive also colors, and the rods, which function at low light levels and are color blind. At mesopic luminance levels both the rods and cones function, and this fact perturbs the simple relationship between luminance and brightness considerably [1]. Within the human visual pathway more than one channel transmits light sensation; one of them produces sensations corresponding more or less to the luminance excitation [2]. Brightness sensation is composed of the luminance and chromatic constituents of the light stimulus.

Luminance, measured by the Commission International de L'Eclairage (CIE) $V(\lambda)$ function, is a measure of minimal flicker, but is a good descriptor of visual acuity as well; thus it can be used to evaluate the degree of work performance. Brightness is another dimension of the light stimulus and one should not use luminance as its correlate. The term *equivalent luminance* has been coined to describe a psychophysical quantity that correlates with brightness sensation. At a 1994 symposium the CIE formed a group to investigate the issue of equivalent luminance in more detail [3].

The official definition of *equivalent luminance* so far is the following [3a]:

> Luminance of a comparison field in which the radiation has the same relative spectral distribution as that of a Planckian radiator at the temperature of freezing platinum and which has the same brightness as the field considered under the specified photometric conditions of measurement; the comparison field must have a specified size and shape which may be different from that of the field considered.
>
> Note: A comparison field may also be used in which the radiation has a relative spectral distribution different from that of a Planckian radiator at the temperature of freezing platinum (T=2042 K), if the equivalent luminance of this field is known under the same conditions of measurement.

There are several problems with this definition. First of all the present photometric system is based on a radiation of 555 nm wavelength (see Chap. 5) and not a blackbody radiator. Second, the definition is too vague to be applied unambiguously. It presents only guidelines on how to create a system to describe brightness. Third, the experts of the CIE Symposium 95 [3] were of the opinion that the principle of the definition should be used also for other phenomena, not only brightness (e.g., ease of recognition of objects). At present a CIE Technical Committee is working on a new, more exact definition of equivalent luminance with the task "to write a Technical Report describing the fundamental concepts of equivalent luminance and to provide guidelines on how to apply those concepts" [4].

Equivalent luminance is based on equal brightness between the test and reference stimulus; it should only be used to describe whether the test stimulus is smaller than, equal to, or larger than the reference stimulus, i.e., it reflects an ordinal scale [5]. No magnitude scale of brightness has been accepted yet. In the case of achromatic stimuli a model was put forward by Bodmann and co-workers some 20 years ago [6]. In this model brightness depended on the luminance of the target L_{T}, and on the luminance of the surround, L_{u}, as well as the angular subtence of the target ϕ:

$$B = C_{\mathrm{T}}(\phi) L_t^n - B_0(L_{\mathrm{u}}, \phi),$$

where

$$B_0(L_{\mathrm{u}}, \phi) = C_{\mathrm{T}}(\phi)[S_0(\phi) + S_1(\phi) L_{\mathrm{u}}^n]$$

and where C_T, S_0, and S_1 depend on ϕ. Authors found good correlation between observed brightness-to-luminance ratio and prediction based on above equation in the luminance range between 0.1 and 10^4 cd m^{-2}. At the lower limit of applicability rod intrusion becomes important; in the photopic range deviations are caused by the chromatic contribution to brightness.

Two recent models that try to extend the model in both directions are those put forward by Kokoschka [7] and by Sagawa and Takeichi [8]. Both systems can be used to construct a measuring instrument to determine an equivalent brightness measure. Color appearance models (see, e.g., [9,10] and Chap. 10, Sec. 10.5.2) provide a brightness axis, but correlation with brightness perception is still not too good [11,12].

Other experimental findings have also interpreted rod contribution in setting optimal visual performance [13] by influencing pupil diameter and, through this effect, visual acuity. Although this theory is still debated [14], proposals for photometric instruments using both a $V(\lambda)$, and $V'(\lambda)$-corrected detector to achieve a "pupil lumen" measure of some sort of visual performance describing equivalent luminance have been put forward [15].

Many industries need a brightness magnitude description. This should work both at photopic and mesopic luminance levels and should describe the magnitude of not only achromatic stimuli but also for chromatic stimuli for different surround and background luminance levels. It is most probable that such a formula will be constructed using both $V(\lambda)$- and $V'(\lambda)$-type spectral sensitivity functions together with $\bar{x}(\lambda)$- and $\bar{z}(\lambda)$-type functions. The final aim in this research is to develop a color appearance model with a brightness axis. At the recent CIE Symposium Dr. Robert Hunt formulated the requirements a generally accepted color appearance model has to fulfill in the following form [16]. The model:

- should be based on static two-dimensional images rather than dynamic or three-dimensional images,
- should provide limits for very dark or bright images,
- should work in light ranges stretching from starlight to daylight (though rod-related measurements should be optional),
- should allow for at least white, grey, and black backgrounds with average, dim, and dark surrounds (but not colored backgrounds at this stage),
- should be mappable to XYZ or $X_{10}Y_{10}Z_{10}$,
- should allow for all degrees of chromatic adaptation and include options for incorporating cognitive effects and the Helson–Judd effect,
- should incorporate parameters for measuring hue angle and quadrature, brightness, lightness, colorfullness, chroma, and saturation,
- should be reversible,
- should be as simple as possible,
- should allow the development of simplified versions that are subsets of the main model,
- should provide good performance,
- should provide for unrelated colors (e.g., by using the Hunt equienergy stimulus model).

Visual science sheds also some light on the cause and extent of the individual variations in human brightness and color perception. Photopigments in the human eye show individual variations and these can now be traced back to molecular genetics [17].

11.3 PHOTOMETRY AND FUNDAMENTAL METROLOGY

11.3.1 Detector- versus Source-Based Photometry

As mentioned in the Introduction the metrological situation of photometry is by no means satisfactory. The accuracy and reproducibility of photometric measurements is by far not as good as those in other branches of physics. The fundamental equation coupling photometry to radiometry is the following (see Eq. (2.7), p. 38)

$$\Phi = K_m \int_{380\ \text{nm}}^{780\ \text{nm}} \Phi_e(\lambda) V(\lambda) d\lambda. \tag{11.1}$$

Here K_m is the maximum value of the luminous efficacy of radiation. Its value is set to 683 lm/W. $\Phi_e(\lambda)$ is the spectral radiant power. A similar equation holds for scotopic luminous flux with $V'(\lambda)$ and $K'_m = 1700$ lm/W. By these equations photometry is coupled to radiometry, the only physiologically based functions are the $V(\lambda)$ and the $V'(\lambda)$ functions.

For radiometric measurements of light sources it has been the traditional technique to compare light source to light source and to maintain light-source primary standards. Thus in photometry we have had oil-lamp standards, the ''Heffner lamp,'' incandescent-lamp standards and finally the blackbody standard, where the spectral power distribution could be calculated by using Planck's law. The main drawback of this standard was that the realization was quite complicated, and only major national laboratories were in a position to afford it. The temperature of a reasonably stable standard had to be quite low: the freezing point of platinum ($T = 2042$ K) was chosen as the reference temperature.

That is the reason why some 30 years ago metrologists started to work on detector-based standards. This work culminated in the international adoption of a new definition of the candela based on the power of monochromatic radiation [18]. By this change photometry could be based on detector standards [19–21] (see Chap. 3) instead of a light-source standard. It is now possible to achieve a 0.021% relative standard uncertainty by using a high-accuracy cryogenic radiometer [22].

At first thermal detectors were used as detector standards, their cryogenic versions providing the highest radiometric accuracy attainable. Later also semiconductor photodiodes were tried as it became feasible to predict the spectral responsivity of Si photodiodes from their material characteristics and structure [23,24].

But in photometry higher precision can be obtained if light sources are compared with light sources and detectors with detectors; thus researchers were eager to test new possibilities to develop source standards. The development of new high-temperature pyrographite materials enabled the building of blackbodies in the 3200–

3500-K range [25] and the reconsideration of the use of blackbody-based photometry. It can be expected that in the near future further investigations on the comparison of detector- and source-based radiometry and photometry will take place and, hopefully, will result in the increase of attainable measurements accuracy.

11.3.2 Derivation of Luminous-Flux Standards from Luminous-Intensity Measurement

Equation (11.1) couples luminous flux to radiant flux and provides the bridge between photometry and radiometry. In photometry it is often necessary to determine the total luminous flux emitted by a light source. This is done by determining either the spatial integral of the luminous intensity, or the integrated illuminance on a (spherical) surface around the light source (see Chap. 3):

$$\Phi = \int_{4\pi} I_v(\Omega) d\Omega = \int_{(A)} E_v dA, \tag{11.2}$$

where I_v is the luminous intensity, E_v the illuminance, and A the surface where the illuminance is measured.

The traditional technique to determine the spatial integral is by using goniophotometry [26]. In the case of gas-discharge lamps goniophotometry has to be done in such a way that the lamp does not change its position in the gravitational field of the earth and that the vibrations and acceleration of the lamp in the goniophotometer are negligible. Also the air movement around the lamp has to be low to avoid cooling of the lamp. The practical consequence is that the goniophotometer has to be relatively large and robust, the lamp can be turned only around a vertical axis, and the movement of the goniometer arm(s) has to be slow, which leads to long measuring times. Recently NIST took a new look at an alternative technique: the use of a photometer sphere as transfer instrument. Up to now photometer spheres were used mainly to compare the luminous flux of sources with similar geometric construction and light distribution. Based on a detailed study of the sphere performance and analyzing all possible error sources, it became possible to create a luminous-flux scale with a relative expanded uncertainty ($k=2$) of 0.5% [27,28]. This development certainly shows new directions for the photometric scale realization and to measure the total luminous flux of sensitive lamps.

11.4 PRACTICAL PHOTOMETRY

In practical photometry we expect three major breakthroughs: increased accuracy and reproducibility in laboratory measurements, implementation of new theoretical findings and new techniques in field-photometry.

11.4.1 Possibilities of Increasing Accuracy and Reproducibility

The metrological chain from the primary standard to the field measuring instrument is relatively long: national standards institutions [e.g., NIST in the United States, National Research Council (NRC) in Canada, National Physical Laboratory (NPL) in the United Kingdom, Physikalisch Technische Bundesanstalt (PTB) in Germany] establish the national photometric scales [usually luminous intensity and (spherical) total luminous flux] and provide them for the applied laboratories in the form of standard lamps and detectors as well as calibrated instruments.

These standards are used in a calibration laboratory to calibrate secondary standard lamps and detectors as well as instruments to be used in the practical photometric laboratory. As photometric measuring precision is only in the 0.1–1 % range, the determined photometric values degrade rather rapidly. Substantially better results can be achieved if the precision of the calibration is improved.

A careful analysis of the error budget of photometric measurements shows that the uncertainty in the determination of mechanical parameters is one of the most important causes of precision degradation.

In the development of the luminance scale from the luminous-intensity scale the apertures of light stops have to be measured with high accuracy. Modern semiconductor technology and metrology provide highly accurate circular light stops, where also the influence of diffraction effects can be better evaluated.

The reproducibility of photometric measurements is influenced by the alignment precision of the different components. Here laser alignment tools may be of significant advantage. A 2–3 times increase in reproducibility has already been achieved [29] and there are further possibilities in these techniques.

11.4.2 Implementation of New Theoretical Findings

As discussed in Sec. 11.2, scientists are working on brightness descriptors that should provide correct measures both in the case of chromatic lights and at lower than photopic luminance levels. A possible advanced luminance meter may be built in the form as seen in Fig. 11.1.

The input radiation is focused on a diaphragm that sets the measuring field size. Behind this is a beam-splitter device that directs the light onto four photodetectors: three channels correspond to the $\bar{x}(\lambda)$, $\bar{y}(\lambda)$, and $\bar{z}(\lambda)$ functions, and the fourth has a spectral sensitivity similar to the $V'(\lambda)$ function. After proper amplification the signals are combined and nonlinearly processed to produce L_{equ} signals for photopic and scotopic vision. In the next two blocks these two signals are combined and further processed, with a possible feedback. Finally the signal is output to a display unit. The building blocks of such a system are available, only the algorithms have to be internationally agreed upon and standardized before such a photometer can be built.

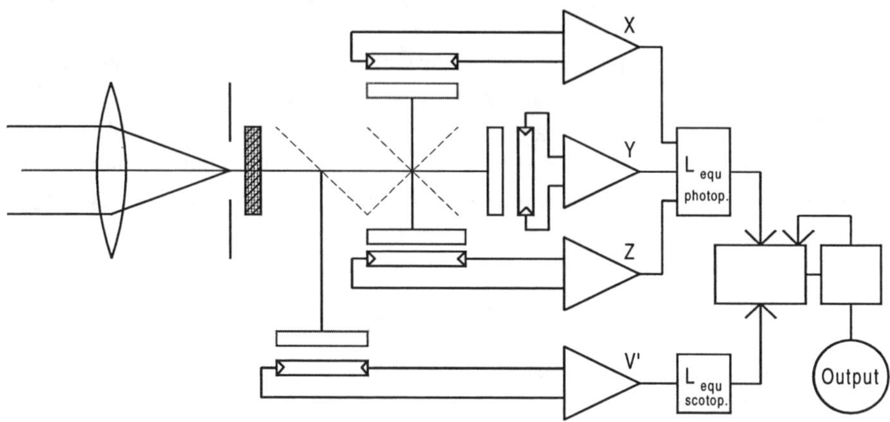

FIGURE 11.1. Block diagram of a possible equivalent luminance meter.

11.4.3 Spatial Luminance Distribution

The equivalent luminance meter discussed in the above section could provide a better measure of brightness of unrelated light patches. As soon as the light is seen in context, however, not only the brightness of the target alone influences our visual mechanism, but also the brightness of the surround. Thus the contrast of the scene has to be evaluated. Modern image capture and processing can provide a possibility to accomplish this.

Two-dimensional charge-coupled device (CCD) cameras have already high enough dynamic range and low enough cross-talk to be used also for photometric purposes [30,31]. Due to non-uniform spatial sensitivity and local differences in the spectral sensitivity the filter correction of such a photometer is still somewhat poorer than that of a traditional illuminance or luminance meter.

CCD cameras produce an overwhelming amount of information, and new ways to analyze these data are needed before wider use can be expected. But it is certainly the direction of progress, as luminance distribution mapping is needed in a number of applications, both for interior lighting applications [32] and for road lighting measurements [33]. The high information content of an image taken with a CCD camera enables also to evaluate further illuminating engineering tasks that were up to now hard to study [34].

11.5 CONCLUSIONS

The present short enumeration of some of the items photometry is dealing with at the moment could not give a complete overview of all the subjects where rapid progress can be expected. The use of computer techniques to design illumination opens up the possibilities to include also effects that could not be considered by traditional design methods. This will certainly focus the attention of vision scientists on new subjects and will influence the development of new photometric concepts.

The automation of the measurements permits many readings keeping the statistical errors low. The use of image-processing techniques will enable the evaluation of complex scenes, hopefully leading not only to increased measuring accuracy but also to the design of better lighting and thus to the improvement of quality of human life.

REFERENCES

1. *Mesopic Photometry: History, Special Problems and Practical Solutions*, CIE Publication No. 81-1989 (CIE Central Bureau, Vienna, 1989).
2. Lennie, P., Pokorny, J., and Smith, V. C., ''Luminance,'' J. Opt. Soc. Am. **10**, 1283–1293 (1993).
3. Bodmann, H. W., ''General Conclusions,'' in *Proceedings of the CIE Symposium on Advances in Photometry*, CIE Publication No. X009 (1995).

3a. *International Lighting Vocabulary*, CIE Pub. 17.4-1987.

4. *Concept and Application of Equivalent Luminance*, CIE TC 1-46.
5. Bodmann, H. W., ''The Scientific Basis of Photometry,'' in *Proceedings of the CIE Symposium on Advances in Photometry* (CIE, Vienna, 1995), p. 7–16.
6. Bodmann, H. W., Haubner, P., and Marsden, A. M., ''A Unified Relationship Between Brightness and Luminance,'' in *Proceedings of CIE 19th Session Kyoto, 1979* (CIE, Vienna, 1980), pp. 99–102.
7. Kokoschka, S., ''Advantage of Four Component System and its Application to Photometer,'' in *Proceedings of the 1994 Annual Conference and '94 International Symposium of the Illuminated Engineering Institute of Japan* (Illuminated Engineering Institute of Japan, 1994), pp. 55–62.
8. Sagawa, K., and Takeichi, K., ''System of Mesopic Photometry for Evaluating Lights in Terms of Comparative Brightness Relationships,'' J. Opt. Soc. Am. **9**, 1240–1246 (1992).
9. Nayatani, Y., Sobagaki, H., Hashimoto, K., and Yano, T., ''Lightness Dependency of Chroma Scales of a Nonlinear Color-Appearance Model and its Latest Formulation,'' Color Res. Appl. **20**, 156–167 (1995).
10. Hunt, R. W. G., ''Revised Colour-Appearance Model for Related and Unrelated Colours,'' Color Res. Appl. **16**, 146–165 (1991).
11. Bodmann, H. W., ''Zur Metrik der visuellen Strahlungsbewertung,'' Die Farbe **32/33**, 235–249 (1985/86).
12. Schandà, J., ''Brightness Description Using Different Metrics,'' in Proceedings of the UMIST International Conference on Colour Communication, 1995.
13. Berman, S., ''Visual Performance and Light Spectrum: The Inadequacy of Conventional Photometry,'' in *Proceedings of the CIE Symposium on Advances in Photometry*, CIE Publication No. X009 (1995).
14. McGowan, T., and Rea, M., ''Visibility and Spectral Composition; Another Look in the Mesopic,'' in *Proceedings of the CIE Symposium on Advances in Photometry*, CIE Publication No. X009 (1995).
15. Berman, S., Jewett, D. J., Fein, G., Saika, G., and Ashford, F., ''Photopic Luminance Does Not Always Predict Perceived Room Brightness,'' Lighting Res. Technol. **22**, 37–41 (1990).
16. *Proceedings of the CIE Expert Symposium '96 on Colour Standards for Image Technology*, CIE Publication No. x010 (1996).
17. Piantanida, T. P., in ''The Molecular Genetics of Human Color Vision: From Nucleotides to Nanometers,'' in *Visual Science and Engineering*, edited by D. H. Kelly (Dekker, New York, 1994).
18. *Proceedings of the 16th General Conference of Weights and Measures, 1979* (BIPM, Sevre, 1979).
19. Biovin, L. P., Gaertner, A. A., and Gignac, D. S., ''Realisation of the new candela (1979) at NRC,'' Metrologia **24**, 139–152 (1987).
20. Goodman T. M., and Key, P. J., ''The NPL radiometric realisation of the candela,'' Metrologia **25**, 29–40 (1988).
21. Cromer, C. L., Eppeldauer, G., Hardis, J. E., Larason, T. C., and Parr, A. C., ''National Institute of Standards and Technology Detector-Based Photometric Scale,'' Appl. Opt. **32**, 2936–2948 (1993).

22. Gentile, T. R., Houston, J. M., Hardis, J. E., Cromer, C. L., and Parr, A. C., ''National Institute of Standards and Technology High-Accuracy Cryogenic Radiometer,'' Appl. Opt. **35**, 1056–1068 (1996).
23. Zalewski, E. F., and Hoyt, C., ''Comparison between Cryogenic Radiometry and the Predicted Quantum Efficiency of *pn* Silicon Photodiode Light Traps,'' Metrologia **28**, 203–206 (1991).
24. Ohno, Y., ''Determination of the Photometric Quantities by the Silicon Photodiode Self-Calibration Using Lasers,'' J. Light. Vis. Env. **15**(2), 126–133 (1991).
25. Sapritsky, V. I., ''Blackbody Radiometry,'' Metrologia (in press).
26. *The Measurement of Luminous Flux*, CIE Technical Report, **84** (1989).
27. Onho, Y., ''Integrating Sphere Simulation: Application to Total Flux Scale Realization,'' Appl. Opt. **33**, 2637–2647 (1994).
28. Ohno, Y., ''Realization of NIST Luminous Flux Scale using an Integrating Sphere with an External Source,'' in *Proceedings of the CIE 23rd Session, New Delhi, 1995*, CIE Publication No. 119 (1995).
29. Czibula, Gy., ''Eine präzise Methode zur Justage und Kalibrierung von Photometer und Goniophotometer,'' Licht **9**, 760–763 (1994).
30. Rea, M. S., and Jeffrey, I. G., ''A New Luminance and Image Analysis System for Lighting and Vision, I. Equipment and Calibration,'' J. Illum. Eng. Soc. **19**, 64–72 (1990).
31. Orfield, S. J., ''Photometry and Luminance Distribution: Conventional Photometry versus CapCalc,'' Lighting Design Applic. **20**, 8–11 (1990).
32. Giczi, I., and Schanda, J., ''Kontrastmessung und Herstellung eines CRF Diagrams mit Hilfe von Videokamera und Bildverarbeitungssystem,'' Licht '90 Conference, Rotterdam, 1990.
33. Rossi, G., and Soardo, P., ''A Mobile Laboratory for Lighting Measurements,'' in *Proceedings of the CIE 22nd Session, Melbourne, 1991*, Vol. 1, pp. 68–69.
34. Brusque, C., ''Development of Visual Noise Metrics for Urban Scenes,'' in *Proceedings of Lux Europa, Edinburgh, 1993*, pp. 102–109.

Appendix A

Measurement Conversion Tables

English-to-Metric Conversion Table

English unit	Multiplied by	Equals metric unit
inches (in.)	2.54	centimeters (cm)
inches (in.)	25.4	millimeters (mm)
feet (ft)	0.305	meters (m)
miles (mi)	1.61	kilometers (km)
degrees Fahrenheit (°F)	(degrees °F−32)×0.556	degrees Celsius (°C)
pounds (lb)	4.45	newtons (N)

Metric-to-English Conversion Table

Metric unit	Multiplied by	Equals English unit
centimeters (cm)	0.39	inches (in.)
millimeters (mm)	0.039	inches (in.)
meters (m)	3.28	feet (ft)
kilometers (km)	0.621	miles (m)
degrees Celsius (°C)	(degrees °C×1.8)+32	degrees Fahrenheit (°F)
newtons (N)	0.225	pounds (lb)

Absolute Temperature Conversion

kelvin (K)=degrees Celsius+273.15
degrees Celsius=kelvin−273.15

Area Conversion

1 square meter = 10.76 square feet = 1550 square inches
1 square kilometer = 0.3861 square miles

Metric Prefixes

zetta 10^{21}

tera 10^{12}

giga 10^{9}

mega 10^{6}

kilo 10^{3}

hecto 10^{2}

deca 10^{1}

deci 10^{-1}

centi 10^{-2}

milli 10^{-3}

micro 10^{-6}

nano 10^{-9}

pico 10^{-12}

femto 10^{-15}

atto 10^{-16}

zepto 10^{-21}

Non-SI Units

Because photometry is among the oldest of the physical sciences, there have been different notations used over the years to describe luminance, illuminance, etc. The following table provides a reference guide to non-SI units that have been used in photometry for conversion purposes only; the use of SI units, defined in Appendix C, is strongly encouraged.

There are several historical references available to this field. For a more detailed account of the history of photometry and the units adopted prior to the candela, refer to J. Walsh, *Photometry*, 3rd. ed. (Dover, New York, 1958) and its associated bibliography. For a more contemporary overview of the history of optics in general, including photometry concepts, see Chap. 1 of E. Hecht and A. Zajac, Optics, 4th ed. (Addison-Wesley, Reading, Mass., 1979), and its associated bibliography.

Definitions of Non-Si Units

Luminous energy:	talbot=lumen·second (lm s)
Illuminance:	meter candle (mc)=lumen/square meter (lm/m^2) (old notation for lux)
	footcandle (fc)=lumen/square foot (lm/ft^2)
	nox=10^{-3} lumens (lm)
	phot (ph)=lumen/square centimeter (lm/cm^2)
Luminous intensity:	candlepower=old notation for candela (cd)
Luminance:	nit (nt)=candela/square meter
	stilb (sb)=candela/square centimeter
	foot·lambert (ft L)=candela/π square foot
	apostilb (asb)=candela/π square meter
	skot (sk)=10^{-3} apostilbs=10^{-3} candela/π square centimeter
	lambert (L)=candela/π square centimeter

Non-SI Unit Conversion Table

Non-SI unit	Multiplied by	Equals SI unit
meter·candle	1	lux
foot·candle	10.76	lux
nox	10^{-3}	lux
phot	10 000	lux
candlepower	1	candela
stilb	10 000	candela/square meter
foot·lambert	3.426	candela/square meter
apostilb	0.3183	candela/square meter
skot	0.3183×10^{-3}	candela/square meter
lambert	3183	candela/square meter

Illumination Conversion Factors

(value in units in left-hand column)×(conversion factors)=(value in units at top of column)

	lux	phot	foot·candle
lux	1	0.0001	0.0929
phot	10 000	1	929
foot·candle	10 764	0.001 076	1

Luminance Conversion Factors

(value in units in left-hand column)×(conversion factor)=(value in units at top of column)

	cd/m^2	stilb	cd/in.^{2}	cd/ft.^{2}	mL	ft L	asb
cd/m^2		0.0001	0.000645	0.0929	0.3142	0.2919	3.1416
stilb	10 000	1	6.452	929	3141.6	2919	31 416
cd/in.^{2}	1 550	0.155	1	144	486.9	452.4	4869
cd/ft.^{2}	10 764	0.001 076	0.006 94	1	3.382	3.1416	33.82
mL	3.183	0.000 318 3	0.002 054	0.295 7	1	0.929	10
ft L	3.426	0.000 342 6	0.002 211	0.318 3	1.0764	1	10.764
asb	0.3183	0.000 031 83	0.000 205 4	0.029 57	0.1	0.0929	1

Photographic Terms

Speed of Photographic Materials

E=exposure measured in lux/second

ASA index=5/E

DIN index=(10 log E)−5

Illumination of Camera Image Plane

B=brightness of object measured in candela per square foot

I=illumination at image plane=π B/4 (f/#)^2

where illumination at object plane is π B, f/# is the f-number of the camera lens, also called *f*-stop or speed of the lens because photographic exposure time is proportional to the square of the f/#. A smaller f/# allows more light to reach the image plane. The flux density at the focal plane varies as (f/#)$^{-2}$.

$$\text{f/\#}=(\text{focal length})/(\text{diameter of lens})=1/(\text{relative aperture}).$$

Camera lens are usually specified by their focal length and largest possible apertures, such as 50 mm, f/1.5.

A camera lens diaphragm varies the amount of light that can reach the lens; each consecutive diaphragm setting increases the f/# by a multiplicative factor of $1/\sqrt{2}$, which in turn decreases the flux density by $\frac{1}{2}$. For example, the same amount of light reaches the film if the camera is set for f/1.4 at $\frac{1}{500}$ s exposure, f/1.4 at $\frac{1}{250}$ s exposure, or f/2.8 at $\frac{1}{125}$ s exposure.

Appendix B

Physical Constants

c	speed of light	$2.997\ 924\ 58 \times 10^{8}$ m/s
k	Boltzmann's constant	1.3801×10^{-23} J/K
h	Planck's constant	6.6262×10^{-34} J/s
σ	Stephan–Boltzmann constant	5.6697×10^{-8} W/m^2 K
e	charge of an electron	1.6×10^{-19} C
c_1	first radiation constant	3.7418×10^{-12} W/cm^2
c_2	second radiation constant	1.4388 cm K
K_m	maximum spectral efficacy for photopic vision	683 lm/W
K'_m	maximum spectral efficacy for scotopic vision	1700 lm/W
λ	wavelength of greatest sensitivity to the human eye (for photonic vision)	550 nm

Appendix C

Radiometric and Photometric Quantities

RADIOMETRIC AND PHOTOMETRIC QUANTITIES

Radiometric Quantities

Quantity	Symbol	Definition	Unit
Radiant energy	Q_e		J
Radiant flux, radiant power	Φ_e	$\Phi = dQ/dt$	W
Irradiance	E_e	$E = d\Phi_R/dA_R$	W/m^2
Exitance	M_e	$M = d\Phi_S/dA_S$	W/m^2
Radiant intensity	I_e	$I = d\Phi/d\Omega$	W/sr
Radiance	L_e	$L = d^2\Phi/(d\Omega\ dA \cos\theta)$	$W/(sr\ m^2)$

Photometric Quantities

Quantity	Symbol	Definition	Unit
Quantity of light, luminous energy	Q_v	$Q_v = K_m \int V(l) Q_l dl$	lm s
Luminous flux	Φ_v	$\Phi_v = dQ_v/dt$	lm (lumen)
Illuminance	E_v	$E_v = d\Phi_{v,R}/dA_R$	lx (lux)
Luminous excitance	M_v	$M_v = d\Phi_{v,S}/dA_S$	lm/m^2
Luminous intensity	I_v	$I_v = d\Phi_v/d\Omega$	cd (candela)
Luminance	L_v	$L_v = d^2\Phi_v/(d\Omega\ dA \cos\theta)$	cd/m^2

Appendix D

Electromagnetic Spectrum

Units most commonly used for the wavelength of light.

Name	Symbol	Value (meters)	Range
Micrometer	μm	10^{-6}	0.380–0.780
Nanometer	nm	10^{-9}	380–780
Ångstrom	Å	10^{-10}	3800–7800

Electromagnetic spectrum.

Region	Wavelength
Electric power	$\infty - 5\times10^6$ m
Radio or television	1×10^4 m– 1 cm
Infrared	1000–0.78 μm
Visible (light)	0.78–0.39 μm
Red light	6470–7000 Å
Orange light	5850–6740 Å
Yellow light	5750–5850 Å
Max. visibility	5560–5750 Å
Green light	4912–5560 Å
Blue light	4240–4912 Å
Violet light	4000–4240 Å
Ultraviolet	0.39–0.032 μm
X ray	0.032–0.000 01 μm
γ radiation	0.000 01–0.000 000 6 μm
Cosmic rays	0.0005 Å

Units of frequency

1 Hz=1 cycle per second

1 mHz=1 000 000 Hz=1000 kHz

Units of wavelength

1 Å=1 Ångstrom=10^{-10} m

1 μm=1 micrometer=1×10^{-6} m=10^4 Å

Regions adjacent to the visible portion of the electromagnetic spectrum.[a]

Name	Wavelength range
UV-C	100–280 nm
UV-B	280–315 nm
UV-A	315–400 nm
VIS	380–780 nm[b]
IR-A, near IR (NIR)	780–1400 nm
IR-B	1.4–3 μm
IR-C, far IR	3 μm–1 mm

[a]The wavelength boundaries of all of these regions are not consistent in all references. The boundaries used here are from the CIE.

[b]Varies with the individual. The overall limits extend from 360 to 800 nm.

Appendix E

Photometric and Radiometric Values

Spectral Response of the Normal Human Eye with Luminous-to-Radiometric Conversion Adjusted to Unity at their Peaks

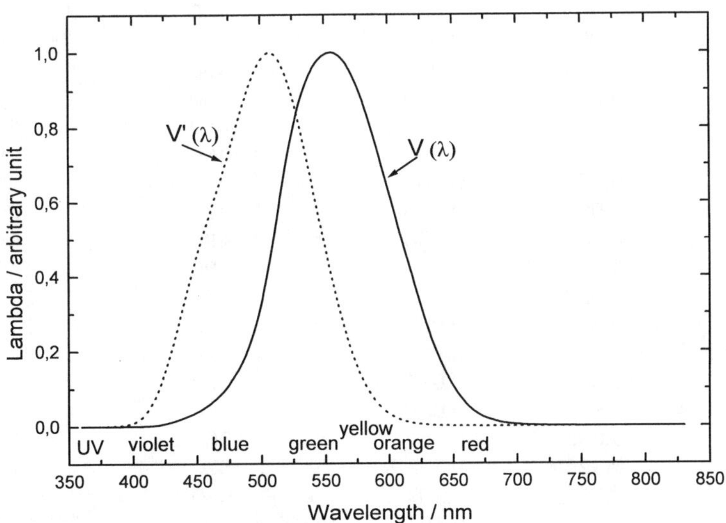

The $V(\lambda)$ and the $V'(\lambda)$ functions.

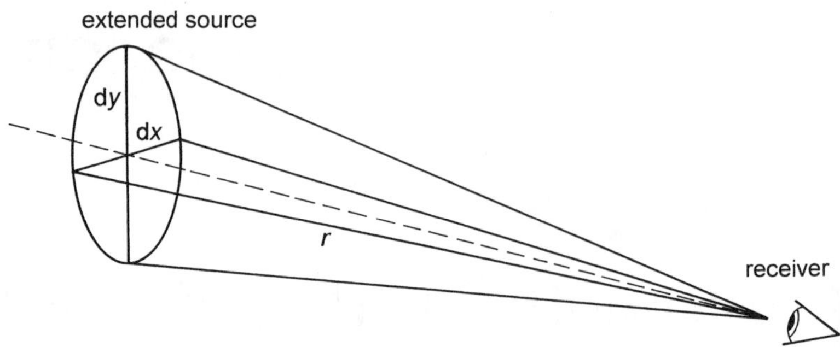

Geometry for the definition of a point source.

Wavelength (nm)	Photopic response V_λ	Scotopic response V'_λ	Wavelength (nm)	Photopic (lm/W) conversion factor
380	0.000 04	0.000 589 2	380	0.05
390	0.000 12	0.002 209	390	0.13
400	0.000 4	0.009 29	400	0.27
410	0.001 2	0.034 84	410	0.82
420	0.004 0	0.096 6	420	2.73
430	0.011 6	0.199 8	430	7.91
440	0.023	0.328 1	440	15.7
450	0.038	0.455	450	25.9
460	0.060	0.567	460	40.9
470	0.091	0.676	470	62.1
480	0.139	0.793	480	94.8
490	0.208	0.904	490	142.0
500	0.323	0.982	500	220.0
510	0.503	0.997	510	343.0
520	0.710	0.935	520	484.0
530	0.862	0.811	530	588.0
540	0.954	0.650	540	650.0
550	0.995	0.481	550	679.0
560	0.995	0.328 8	570	679.0
570	0.952	0.207 6	580	649.0
580	0.870	0.121 2	690	593.0
590	0.757	0.065 5	600	516.0
600	0.631	0.033 15	610	430.0
610	0.503	0.015 93	620	260.0
620	0.381	0.007 37	630	181.0
630	0.265	0.003 335	640	119.0
640	0.175	0.001 497	650	73.0
650	0.107	0.000 677	660	41.4
660	0.061	0.000 312 9	670	21.8

Wavelength (nm)	Photopic response V_λ	Scotopic response V'_λ	Wavelength (nm)	Photopic (lm/W) conversion factor
670	0.032	0.000 148 0	680	11.6
680	0.017	0.000 071 5	690	5.59
690	0.008 2	0.000 035 33	700	2.78
700	0.004 1	0.000 017 80	710	1.43
710	0.002 1	0.000 009 14	720	0.716
720	0.001 05	0.000 004 78	730	0.355
730	0.000 52	0.000 002 546	740	0.170
740	0.000 25	0.000 001 379	750	0.820
750	0.000 12	0.000 000 076 0	760	0.041
760	0.000 06	0.000 000 042 5		
770	0.000 03	0.000 000 241 3		
780	0.000 015	0.000 000 139 0		

CIE Standard Illuminant A and D65 relative spectral power distribution. 300–830 nm at 5-nm intervals; extract from Table 1.1, CIE Publication No. 15.2-1986.

λ	Illuminant A	Illuminant D65
300	0.930483	0.03410
305	1.128210	1.66430
310	1.357690	3.29450
315	1.622190	11.76520
320	1.925080	20.23600
325	2.269800	28.64470
330	2.659810	37.05350
335	3.098610	38.50110
340	3.589680	39.94880
345	4.136480	42.43020
350	4.742380	44.91170
355	5.410700	45.77500
360	6.144620	46.63830
365	6.947200	49.36370
370	7.821350	52.08910
375	8.769800	51.03230
380	9.795100	49.97550
385	10.899600	52.31180
390	12.085300	54.64820
395	13.354300	68.70150
400	14.708000	82.75490
405	16.148000	87.12040
410	17.675300	91.48600
415	19.290700	92.45890
420	20.995000	93.43180

λ	Illuminant A	Illuminant D65
425	22.788300	90.05700
430	24.670900	86.68230
435	26.642500	95.77360
440	28.702700	104.86500
445	30.850800	110.93600
450	33.085900	117.00800
455	35.406800	117.41000
460	37.812100	117.81200
465	40.300200	116.33600
470	42.869300	114.86100
475	45.517400	115.39200
480	48.242300	115.92300
485	51.041800	112.36700
490	53.913200	108.81100
495	56.853900	109.08200
500	59.861100	109.35400
505	62.932000	108.57800
510	66.063500	107.80200
515	69.252500	106.29600
520	72.495900	104.79000
525	75.790300	106.23900
530	79.132600	107.68900
535	82.519300	106.04700
540	85.947000	104.40500
545	89.412400	104.22500
550	92.912000	104.04600
555	96.442300	102.02300
560	100.000000	100.00000
565	103.582000	98.16710
570	107.184000	96.33420
575	110.803000	96.06110
580	114.436000	95.78800
585	118.080000	92.23680
590	121.731000	88.68560
595	125.386000	89.34590
600	129.043000	90.00620
605	132.697000	89.80260
610	136.346000	89.59910
615	139.988000	88.64890
620	143.618000	87.69870
625	147.235000	85.49360
630	150.836000	83.28860
635	154.418000	83.49390
640	157.979000	83.69920
645	161.516000	81.86300
650	165.028000	80.02680
655	168.510000	80.12070

λ	Illuminant A	Illuminant D65
660	171.963000	80.21460
665	175.383000	81.24620
670	178.769000	82.27780
675	182.118000	80.28100
680	185.429000	78.28420
685	188.701000	74.00270
690	191.931000	69.72130
695	195.118000	70.66520
700	198.261000	71.60910
705	201.359000	72.97900
710	204.409000	74.34900
715	207.411000	67.97650
720	210.365000	61.60400
725	213.268000	65.74480
730	216.120000	69.88560
735	218.920000	72.48630
740	221.667000	75.08700
745	224.361000	69.33980
750	227.000000	63.59270
755	229.585000	55.00540
760	232.115000	46.41820
765	234.589000	56.61180
770	237.008000	66.80540
775	239.370000	65.09410
780	241.675000	63.38280
785	243.924000	63.84340
790	246.116000	64.30400
795	248.251000	61.87790
800	250.329000	59.45190
805	252.350000	55.70540
810	254.314000	51.95900
815	256.221000	54.69980
820	258.071000	57.44060
825	259.865000	58.87650
830	261.602000	60.31250

CIE 1931 Standard Colorimetric Observer and CIE 1964 Supplementary Standard Colorimetric Observer color matching functions, 360–830 nm at 5-nm intervals; extract from tables taken from CIE Publication No. 15.2-1986.

λ	$\bar{x}_2(\lambda)$	$\bar{y}_2(\lambda)$	$\bar{z}_2(\lambda)$	$\bar{x}_{10}(\lambda)$	$\bar{y}_{10}(\lambda)$	$\bar{z}_{10}(\lambda)$
360	0.0001	0.0000	0.0006	0.0000	0.0000	0.0000
365	0.0002	0.0000	0.0011	0.0000	0.0000	0.0000
370	0.0004	0.0000	0.0019	0.0000	0.0000	0.0000
375	0.0007	0.0000	0.0035	0.0000	0.0000	0.0001
380	0.0014	0.0000	0.0065	0.0002	0.0000	0.0007
385	0.0022	0.0001	0.0105	0.0007	0.0001	0.0029
390	0.0042	0.0001	0.0201	0.0024	0.0003	0.0105
395	0.0077	0.0002	0.0362	0.0072	0.0008	0.0323
400	0.0143	0.0004	0.0679	0.0191	0.0020	0.0860
405	0.0232	0.0006	0.1102	0.0434	0.0045	0.1971
410	0.0435	0.0012	0.2074	0.0847	0.0088	0.3894
415	0.0776	0.0022	0.3713	0.1406	0.0145	0.6568
420	0.1344	0.0040	0.6456	0.2045	0.0214	0.9725
425	0.2148	0.0073	1.0391	0.2647	0.0295	1.2825
430	0.2839	0.0116	1.3856	0.3147	0.0387	1.5535
435	0.3285	0.0168	1.6230	0.3577	0.0496	1.7985
440	0.3483	0.0230	1.7471	0.3837	0.0621	1.9673
445	0.3481	0.0298	1.7826	0.3867	0.0747	2.0273
450	0.3362	0.0380	1.7721	0.3707	0.0895	1.9948
455	0.3187	0.0480	1.7441	0.3430	0.1063	1.9007
460	0.2908	0.0600	1.6692	0.3023	0.1282	1.7454
465	0.2511	0.0739	1.5281	0.2541	0.1528	1.5549
470	0.1954	0.0910	1.2876	0.1956	0.1852	1.3176
475	0.1421	0.1126	1.0419	0.1323	0.2199	1.0302
480	0.0956	0.1390	0.8130	0.0805	0.2536	0.7721
485	0.0580	0.1693	0.6162	0.0411	0.2977	0.5701
490	0.0320	0.2080	0.4652	0.0162	0.3391	0.4153
495	0.0147	0.2586	0.3533	0.0051	0.3954	0.3024
500	0.0049	0.3230	0.2720	0.0038	0.4608	0.2185
505	0.0024	0.4073	0.2123	0.0154	0.5314	0.1592
510	0.0093	0.5030	0.1582	0.0375	0.6067	0.1120
515	0.0291	0.6082	0.1117	0.0714	0.6857	0.0822
520	0.0633	0.7100	0.0782	0.1177	0.7618	0.0607
525	0.1096	0.7932	0.0573	0.1730	0.8233	0.0431
530	0.1655	0.8620	0.0422	0.2365	0.8752	0.0305
535	0.2257	0.9149	0.0298	0.3042	0.9238	0.0206
540	0.2904	0.9540	0.0203	0.3768	0.9620	0.0137
545	0.3597	0.9803	0.0134	0.4516	0.9822	0.0079
550	0.4334	0.9950	0.0087	0.5298	0.9918	0.0040
555	0.5121	1.0000	0.0057	0.6161	0.9991	0.0011
560	0.5945	0.9950	0.0039	0.7052	0.9973	0.0000
565	0.6784	0.9786	0.0027	0.7938	0.9824	0.0000
570	0.7621	0.9520	0.0021	0.8787	0.9556	0.0000

λ	$\bar{x}_2(\lambda)$	$\bar{y}_2(\lambda)$	$\bar{z}_2(\lambda)$	$\bar{x}_{10}(\lambda)$	$\bar{y}_{10}(\lambda)$	$\bar{z}_{10}(\lambda)$
575	0.8425	0.9154	0.0018	0.9512	0.9152	0.0000
580	0.9163	0.8700	0.0017	1.0142	0.8689	0.0000
585	0.9786	0.8163	0.0014	1.0743	0.8256	0.0000
590	1.0263	0.7570	0.0011	1.1185	0.7774	0.0000
595	1.0567	0.6949	0.0010	1.1343	0.7204	0.0000
600	1.0622	0.6310	0.0008	1.1240	0.6583	0.0000
605	1.0456	0.5668	0.0006	1.0891	0.5939	0.0000
610	1.0026	0.5030	0.0003	1.0305	0.5280	0.0000
615	0.9384	0.4412	0.0002	0.9507	0.4618	0.0000
620	0.8544	0.3810	0.0002	0.8563	0.3981	0.0000
625	0.7514	0.3210	0.0001	0.7549	0.3396	0.0000
630	0.6424	0.2650	0.0000	0.6475	0.2835	0.0000
635	0.5419	0.2170	0.0000	0.5351	0.2283	0.0000
640	0.4479	0.1750	0.0000	0.4316	0.1798	0.0000
645	0.3608	0.1382	0.0000	0.3437	0.1402	0.0000
650	0.2835	0.1070	0.0000	0.2683	0.1076	0.0000
655	0.2187	0.0816	0.0000	0.2043	0.0812	0.0000
660	0.1649	0.0610	0.0000	0.1526	0.0603	0.0000
665	0.1212	0.0446	0.0000	0.1122	0.0441	0.0000
670	0.0874	0.0320	0.0000	0.0813	0.0318	0.0000
675	0.0636	0.0232	0.0000	0.0579	0.0226	0.0000
680	0.0468	0.0170	0.0000	0.0409	0.0159	0.0000
685	0.0329	0.0119	0.0000	0.0286	0.0111	0.0000
690	0.0227	0.0082	0.0000	0.0199	0.0077	0.0000
695	0.0158	0.0057	0.0000	0.0138	0.0054	0.0000
700	0.0114	0.0041	0.0000	0.0096	0.0037	0.0000
705	0.0081	0.0029	0.0000	0.0066	0.0026	0.0000
710	0.0058	0.0021	0.0000	0.0046	0.0018	0.0000
715	0.0041	0.0015	0.0000	0.0031	0.0012	0.0000
720	0.0029	0.0010	0.0000	0.0022	0.0008	0.0000
725	0.0020	0.0007	0.0000	0.0015	0.0006	0.0000
730	0.0014	0.0005	0.0000	0.0010	0.0004	0.0000
735	0.0010	0.0004	0.0000	0.0007	0.0003	0.0000
740	0.0007	0.0002	0.0000	0.0005	0.0002	0.0000
745	0.0005	0.0002	0.0000	0.0004	0.0001	0.0000
750	0.0003	0.0001	0.0000	0.0003	0.0001	0.0000
755	0.0002	0.0001	0.0000	0.0002	0.0001	0.0000
760	0.0002	0.0001	0.0000	0.0001	0.0000	0.0000
765	0.0001	0.0000	0.0000	0.0001	0.0000	0.0000
770	0.0001	0.0000	0.0000	0.0001	0.0000	0.0000
775	0.0001	0.0000	0.0000	0.0000	0.0000	0.0000
780	0.0000	0.0000	0.0000	0.0000	0.0000	0.0000
785	0.0000	0.0000	0.0000	0.0000	0.0000	0.0000
790	0.0000	0.0000	0.0000	0.0000	0.0000	0.0000
795	0.0000	0.0000	0.0000	0.0000	0.0000	0.0000
800	0.0000	0.0000	0.0000	0.0000	0.0000	0.0000

λ	$\bar{x}_2(\lambda)$	$\bar{y}_2(\lambda)$	$\bar{z}_2(\lambda)$	$\bar{x}_{10}(\lambda)$	$\bar{y}_{10}(\lambda)$	$\bar{z}_{10}(\lambda)$
805	0.0000	0.0000	0.0000	0.0000	0.0000	0.0000
810	0.0000	0.0000	0.0000	0.0000	0.0000	0.0000
815	0.0000	0.0000	0.0000	0.0000	0.0000	0.0000
820	0.0000	0.0000	0.0000	0.0000	0.0000	0.0000
825	0.0000	0.0000	0.0000	0.0000	0.0000	0.0000
830	0.0000	0.0000	0.0000	0.0000	0.0000	0.0000

Components $S_0(\lambda)$, $S_1(\lambda)$, and $S_2(\lambda)$ of daylight used in the calculation of relative spectral power distribution of daylight illuminants of different correlated color temperatures, for wavelengths 300–830 nm at 5-nm intervals (see CIE Publication 15.2 Table 1.2 [19]).

λ (nm)	S_0	S_1	S_2
300	0.04	0.02	0.00
305	3.02	2.26	1.00
310	6.00	4.50	2.00
315	17.80	13.45	3.00
320	29.60	22.40	4.00
325	42.45	32.20	6.25
330	55.30	42.00	8.50
335	56.30	41.30	8.15
340	57.30	40.60	7.80
345	59.55	41.10	7.25
350	61.80	41.60	6.70
355	61.65	39.80	6.00
360	61.50	38.00	5.30
365	65.15	40.20	5.70
370	68.80	42.40	6.10
375	66.10	40.45	4.55
380	63.40	38.50	3.00
385	64.60	36.75	2.10
390	65.80	35.00	1.20
395	80.30	39.20	0.05
400	94.80	43.40	−1.10
405	99.80	44.85	−0.80
410	104.80	46.30	−0.50
415	105.35	45.10	−0.60
420	105.90	43.90	−0.70
425	101.35	40.50	−0.95
430	96.80	37.10	−1.20

λ (nm)	S_0	S_1	S_2
435	105.35	36.90	−1.90
440	113.90	36.70	−2.60
445	119.75	36.30	−2.75
450	125.60	35.90	−2.90
455	125.55	34.25	−2.85
460	125.50	32.60	−2.80
465	123.40	30.25	−2.70
470	121.30	27.90	−2.60
475	121.30	26.10	−2.60
480	121.30	24.30	−2.60
485	117.40	22.20	−2.20
490	113.50	20.10	−1.80
495	113.30	18.15	−1.65
500	113.10	16.20	−1.50
505	111.95	14.70	−1.40
510	110.80	13.20	−1.30
515	108.65	10.90	−1.25
520	106.50	8.60	−1.20
525	107.65	7.35	−1.10
530	108.80	6.10	−1.00
535	107.05	5.15	−0.75
540	105.30	4.20	−0.50
545	104.85	3.05	−0.40
550	104.40	1.90	−0.30
555	102.20	0.95	−0.15
560	100.00	0.00	0.00
565	98.00	−0.80	0.10
570	96.00	−1.60	0.20
575	95.55	−2.55	0.35
580	95.10	−3.50	0.50
585	92.10	−3.50	1.30
590	89.10	−3.50	2.10
595	89.80	−4.65	2.65
600	90.50	−5.80	3.20
605	90.40	−6.50	3.65
610	90.30	−7.20	4.10
615	89.35	−7.90	4.40
620	88.40	−8.60	4.70
625	86.20	−9.05	4.90
630	84.00	−9.50	5.10
635	84.55	−10.20	5.90
640	85.10	−10.90	6.70
645	83.50	−10.80	7.00
650	81.90	−10.70	7.30
655	82.25	−11.35	7.95
660	82.60	−12.00	8.60

λ (nm)	S_0	S_1	S_2
665	83.75	−13.00	9.20
670	84.90	−14.00	9.80
675	83.10	−13.80	10.00
680	81.30	−13.60	10.20
685	76.60	−12.80	9.25
690	71.90	−12.00	8.30
695	73.10	−12.65	8.95
700	74.30	−13.30	9.60
705	75.35	−13.10	9.05
710	76.40	−12.90	8.50
715	69.85	−11.75	7.75
720	63.30	−10.60	7.00
725	67.50	−11.10	7.30
730	71.70	−11.60	7.60
735	74.35	−11.90	7.80
740	77.00	−12.20	8.00
745	71.10	−11.20	7.35
750	65.20	−10.20	6.70
755	56.45	−9.00	5.95
760	47.70	−7.80	5.20
765	58.15	−9.50	6.30
770	68.60	−11.20	7.40
775	66.80	−10.80	7.10
780	65.00	−10.40	6.80
785	65.50	−10.50	6.90
790	66.00	−10.60	7.00
795	63.50	−10.15	6.70
800	61.00	−9.70	6.40
805	57.15	−9.00	5.95
810	53.30	−8.30	5.50
815	56.10	−8.80	5.80
820	58.90	−9.30	6.10
825	60.40	−9.55	6.30
830	61.90	−9.80	6.50

Appendix F

Index of Professional Organizations

AIP	American Institute of Physics
ANSI	American National Standards Institute
APS	American Physical Society
ASTM	American Society for Test and Measurement
BIPM	Bureau International des Poids et Measures (International Bureau of Weights and Measures)
CCPR	Consultatif de Photometrie et Radiometrie (International standards body that developed an intercomparison of photometric units in 1985)
CEN	Convention National European (European National Convention on Standards for Traffic Signs)
CGPM	Conference Generale des Poids et Measures (International standards body, adopted the ''international candle'' in 1948)
CIE	Commission International De L'Eclairage (International Commission on Illumination)
CIPM	Commission International des Poids et Measures (International Commission on Weights and Measures)
CORM	Council of Optical Radiation Measurement
IALD	International Association of Lighting Designers
IEC	International Electrotechnical Commission
IEEE	Institute of Electrical and Electronics Engineers
IES	Illuminating Engineering Society (also see IESNA, Illuminating Engineering Society of North America)
ISO	International Standards Organization
ITE	Institute of Traffic Engineers
NBS	National Bureau of Standards
NCSL	National Conference of Standards Laboratories
NIST	National Institute of Standards and Technology

NPL	National Physical Laboratory (United Kingdom)
NRC	National Research Council (Canada)
NREL	National Renewable Energy Laboratory
NTIS	National Technical Information Service
OSA	Optical Society of America
PTB	Physikalisch Technische Bundesanstalt (German standards organization)
SAE	Society for Automotive Engineers
SI	System International (International System of Units)
SPIE	Society of Photo-Optic Instrumentation Engineers
UL	Underwriters Laboratories

Subject Index

M

N

T

U

V

W